PLATINOTYPE

Making Photographs in Platinum and Palladium with the Contemporary Printing-out Process

Platinotype: Making Photographs in Platinum and Palladium with the Contemporary Printing-out Process describes the mechanisms and chemistry of platinum/palladium printing in safe and practical ways. Clearly presented formulae allow the printer to work with platinum, palladium, or varying combinations of both. The printed-out image appears fully during exposure, and only requires simple and safe steps for clearing to a stable, archival state.

The authors explain what makes the image, how all necessary components are prepared and used, and the kind of paper and negative needed to make prints. More than just a technical manual, the book underscores the authors' belief that printing is a creative, scientific, and philosophic way of working. The book presents an outstanding collection of prints by over 40 artists, all made with this printing-out process. The artists' notes and comments offer insights into their methods and thinking, and a large number of full-page reproductions serve as a valuable reference for the aspiring printer.

The book includes:

- A list of supplies and equipment
- A detailed chemical glossary
- Summary sheets and workflows for each step of the process
- Instructions for making traditional negatives with Pyro PMK and digital negatives
- Explanation of the chemistry and dynamics of paper, and how to use buffered papers
- Instructions for controlling hydration processes and humidity
- Instructions for preparing each chemical solution needed for the process
- Discussion about the aesthetics of the platinum/palladium print
- Explanation of the relationships between light, image, and expression
- A detailed troubleshooting list
- Recommendations from conservators about processing, handling, and conservation
- Contemporary artists using the printing-out platinum/palladium process

Learning how to make platinum/palladium prints has been cloaked in a mystique of difficulty. *Platinotype* presents the process as a set of clearly explained and defined steps. Like other books in the series, *Platinotype* is a detailed and inspiring manual, accessible to both novices and experts, and illustrative of the contemporary arts.

Pradip Malde graduated from the Glasgow School of Art (MA, 1980). He is a professor of Art at the University of the South, Sewanee, Tennessee, teaching photography and documentary studies. A Scottish Arts Council Bursary recipient (1990) and a Guggenheim Fellow (2018), he has worked extensively within small rural communities using photography as a citizen-based development tool. His works are held in numerous public collections, including the Scottish National Galleries, Princeton University Museum, and the Victoria and Albert Museum.

Dr. Mike Ware graduated in chemistry at the University of Oxford (1962). He is an Honorary Fellow in Chemistry, University of Manchester, UK, and a recipient of the Hood Medal of the Royal Photographic Society. His researches on improving historic photographic processes have had a profound impact on creative and historical aspects of photography, and are described in 70 publications. He has consulted for national museums and galleries, exhibited his personal photographic work and conducted workshops worldwide. In 2016 he was awarded the Special Recognition of the American Institute for Conservation.

Contemporary Practices in Alternative Process Photography Series

Series Editor: Christina Z. Anderson

The Contemporary Practices in Alternative Photographic Processes series focuses the lens on a variety of alternative, historical processes from the medium's 180-year history. Each book outlines a step-by-step approach to a particular medium, and features contemporary artists who use that particular process regularly in their practice. The richly illustrated books in this series serve as guidebooks for those new to alternative processes, refresher courses for professionals already familiar with each medium, and a source of inspiration for all.

Titles in the Series

Chrysotype: A Contemporary Guide to Photographic Printing in Gold, Leanne McPhee

Digital Negatives with QuadToneRIP: Demystifying QTR for Photographers and Printmakers, Ron Reeder and Christina Z. Anderson

Carbon Transfer Printing: A Step-by-Step Manual, Featuring Contemporary Carbon Printers and Their Creative Practice, Sandy King, Don Nelson, John Lockhart

Cyanotype: The Blueprint in Contemporary Practice, Christina Z. Anderson

The Handmade Silver Gelatin Emulsion Print: Creating Your Own Liquid Emulsions for Black & White Paper, Denise Ross

Polymer Photogravure: A Step-by-Step Manual Highlighting Artists and Their Creative Practice, Clay Harmon

Salted Paper Printing: A Step-by-Step Manual Highlighting Contemporary Artists, Christina Z. Anderson

Gum Printing: A Step-by-Step Manual Highlighting Artists and Their Creative Practice, Christina Z. Anderson

PLATINOTYPE

Making Photographs in Platinum and Palladium with the Contemporary Printing-out Process

Pradip Malde with Mike Ware

Routledge
Taylor & Francis Group
NEW YORK AND LONDON

First published 2021
by Routledge
52 Vanderbilt Avenue, New York, NY 10017

and by Routledge
2 Park Square, Milton Park, Abingdon, Oxon OX14 4RN

Routledge is an imprint of the Taylor & Francis Group, an informa business

Library of Congress Cataloging-in-Publication Data
Names: Malde, Pradip, 1957– author. | Ware, Mike, author.
Title: Platinotype : making photographs in platinum and palladium with the contemporary printing-out process / Pradip Malde with Mike Ware.
Description: London ; New York : Routledge, 2020. |
Series: Contemporary practices in alternative process photography |
Includes bibliographical references and index.
Identifiers: LCCN 2020029437 (print) | LCCN 2020029438 (ebook) |
ISBN 9780367415938 (hbk) | ISBN 9780367415952 (pbk) | ISBN 9780367815394 (ebk)
Subjects: LCSH: Photography–Printing processes–Platinotype. | Palladium.
Classification: LCC TR420 .M35 2020 (print) | LCC TR420 (ebook) | DDC 772/.3–dc23
LC record available at https://lccn.loc.gov/2020029437
LC ebook record available at https://lccn.loc.gov/2020029438

ISBN: 978-0-367-41593-8 (hbk)
ISBN: 978-0-367-41595-2 (pbk)
ISBN: 978-0-367-81539-4 (ebk)

Typeset in Adobe Caslon Pro and Century Gothic
by Newgen Publishing UK

Notice to readers:

Exercise caution in the handling of all photographic chemicals. Use of any such chemicals constitutes some risk, and some are poisonous. The publisher and authors accept no responsibility for injury or loss arising from the procedures or materials described in this book whether used properly or improperly. The workplace should be well ventilated. Chemicals should be mixed only in the manner described. Avoid contact between the chemicals and eyes, skin, clothing, and furniture. Do not eat or drink while using chemicals. Keep them away from pets and children. Wear protective eye wear and gloves if necessary. In short, be mindful of all safety procedures for yourself and others.

For Rachel and for Marie

Contents

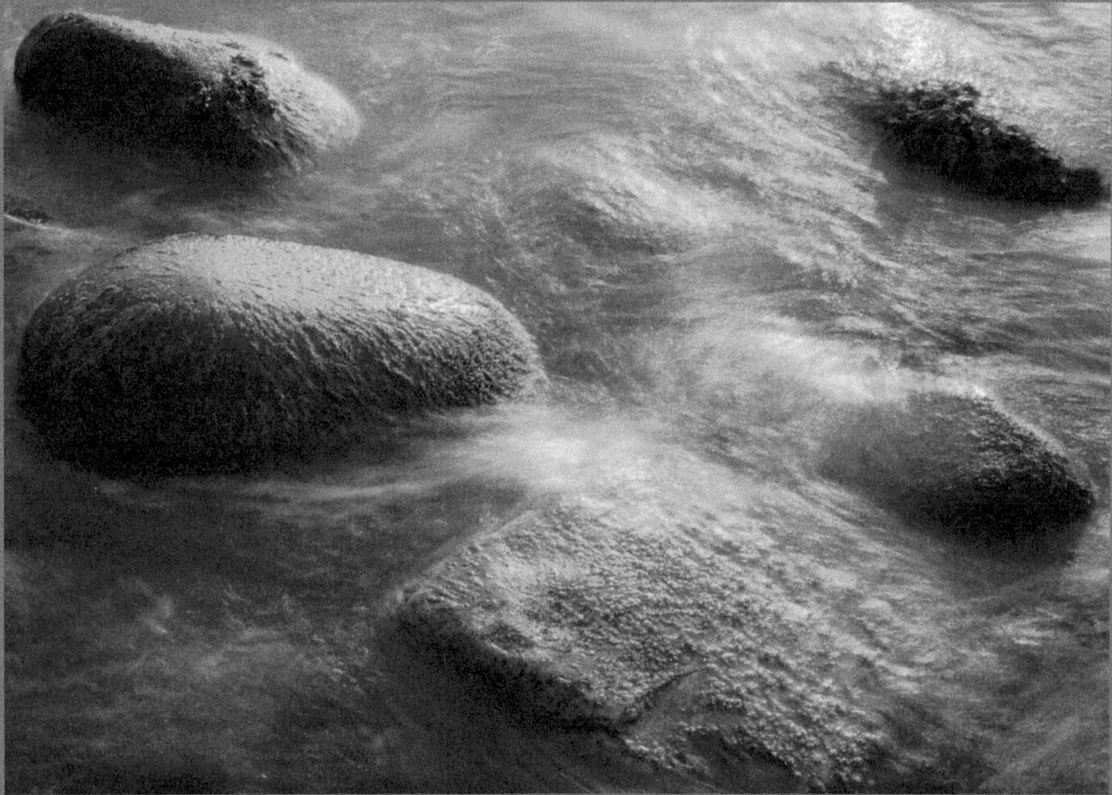

Figure F.1 Mike Ware, *Boulders in the Flow, Scapa,* 1984. Platinum-palladium print on Revere Platinum paper from digital enlarged negative of the original 6 x 9 cm rollfilm negative. 20.2 x 28.8 cm. Fe3:Pt:Pd = 2:1:1, hydration for 90 min at 76% RH / 20°C. Courtesy Mike Ware

Foreword

My first encounter with Pradip Malde, the chief author of this work, was in August 1983 at the harbour of Stromness in the Isles of Orkney—an historic archipelago lying to the north-east of the Scottish mainland, where the Atlantic Ocean meets the North Sea. As I disembarked from the ferry, after a 500-mile drive from Manchester in England, I reflected that it was only a year since we had first learned of each other's existence: as probably the only two people in the British Isles then currently practicing the Platinotype process. In the previous year we had each visited—unknowingly only a week apart—the same specialist shop in London, Falkiner Fine Papers, in search of artists' paper of sufficient quality and purity to support the handmade process. Falkiner's sales expert, Julie Norris, put the two of us in touch, thus leading to our Orkney meeting, from which sprang a collaboration intent on improving the historic process of printing in platinum and palladium. Today, half a lifetime later, the results of our joint endeavor now form the subject of this book.

Our revision of the photochemistry soon yielded a more accessible and controllable sensitizer, and processing procedure, which we published around 1986 in photographic periodicals, both amateur and professional. We also had the pleasure of teaching the process jointly in workshops, notably at Paul Hill's celebrated *Photographer's Place* from 1985 to 1995, and at California State University's Summer Arts program in 1998, courtesy of Professor Roger Vail. He is one of a number of notable exhibiting artists who have been practicing our platinum/palladium printing-out method using ammonium salts since the late 1980s, as a more viable alternative to the widely known historical method of platinum printing by development.

As Professor of Art at the University of the South, in Sewanee, Tennessee, Pradip's 26-year academic tenure now underscores his leading role in the teaching of photographic arts in the USA. He is a master of photography, using both analogue photochemical and digital electronic disciplines, and of the hybrid practice so fruitfully born as their offspring. New darkrooms and wet laboratory space have recently been set up in Pradip's teaching facility at Sewanee. Over the years he has created many enduring bodies of exhibited work in platinum and palladium, some of which may be viewed on his website. His remarkable artistic output has most recently been acknowledged by the award of a prestigious Guggenheim Fellowship, which has just culminated in the production of a book illustrated entirely by platinum-palladium prints—the first such volume in modern times.

Ever since the commercial production of platinotype paper ceased in 1937, the process has tended to be shrouded in historical obscurity, but this began to be dispelled in 2010, when the Mellon Foundation generously funded a major research project into the chemistry and conservation of historic platinum and palladium prints at the National Gallery of Art in Washington DC, which holds many notable examples in its own collections. I was then fortunate enough to be invited to join this endeavor as a scientific consultant to the NGA, which, over the ensuing six years, became the vital hub for an interdisciplinary research collaboration by curators, conservators, photographers, and scientists, distributed over two dozen major institutions across the USA. We self-styled "Platinistas" presented the first fruits of our researches at an international symposium held in Washington DC in October 2014, entitled *Platinum and Palladium Photographs: Technical and Aesthetic History, Chemistry, and Connoisseurship*. Pradip played a conspicuous role in this conference, both as an invited speaker and workshop demonstrator. Many diverse and previously little-understood aspects of the curation and conservation of historic platinotypes were brought to light during this conference, so its proceedings—greatly extended by further researches—were eventually published in 2017 under the aegis of the Photographic Materials Group of the American Institute for Conservation of Historic and Artistic Works, in a volume having 46 contributing authors, entitled *Platinum and Palladium Photographs: Technical History, Connoisseurship, and Preservation*. The chief editor of this landmark publication was the Head of Photograph Conservation at the NGA, Constance McCabe, whose energy, expertise, and dedication had provided the initiative for driving this collaboration through to a major international conference and an authoritative multi-author publication. This scholarly work of reference, beautifully illustrated, now constitutes a (weighty!) cornerstone in the literature on the process,[1] but it is not suited—nor was it ever intended—as a working instruction manual for the artist-practitioner of platinum/palladium printing.

Throughout this seven-year period of collaborative investigation, experimental testing, and peer-reviewed authorship by my co-researchers, I was assembling an online resource, explicitly chemical, to provide reference background information on the scientific underpinnings of the process. This compilation is now freely downloadable from my website as a book entitled *Platinomicon*[2] but much of its content does presume that the reader is acquainted with the language and formulae of chemistry.

The *raison d'être* for this present instructional manual, and its place in the literature of photography, should now be clear: it has been carefully written by Pradip in accessible language, and it is his voice that the reader will hear rather than mine. He avoids much scientific jargon, but devotes all due attention to explaining technical and practical details, keeping in mind a readership mainly with backgrounds in the arts and humanities rather than the sciences. I have long regarded the study of photography as one of the best meeting grounds for the "two cultures" of science and art, which are so often sadly divided. As a professional chemist, but only an amateur artist *manqué*, my adage has always been that "art deserves the best science," which should work reliably, seamlessly, and transparently beneath the surface of the artistic endeavor. While the NGA publication and my *Platinomicon* delve deeply into the underlying history and science of platinotype, it is the purpose of the present volume to provide the reader with a full and lively awareness of its contemporary artistic practice, and to impart an aesthetic sense of the "truth to materials" that is implicit in creative engagement with this beautiful photographic medium.

Mike Ware

Notes

1 https://shop.nga.gov/item/748259/platinum-palladium-photographs-technical-history-connoisseurship-and-preservation/
2 https://www.mikeware.co.uk/mikeware/downloads.html

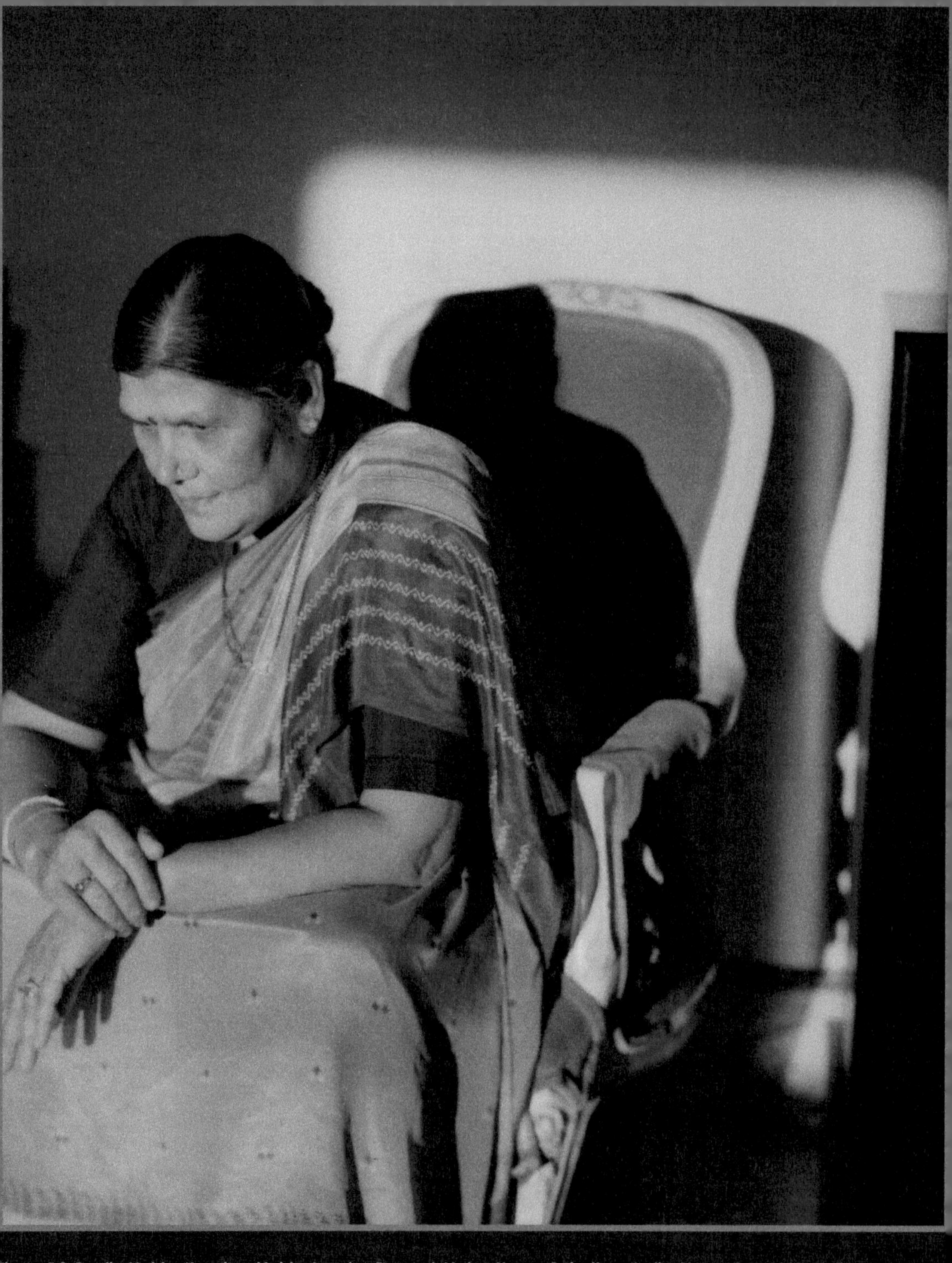

Figure F.2 Pradip Malde, *Kanchan Malde, Leaning Forward, Maloo House, Rajasthan, India*, November, 1995. Platinum-palladium print (1:1) on Wyndstone vellum, from original 8 x 10 inch negative, 9.75 x 7.75 inches. Courtesy Pradip Malde

Preface

Despite its function, and the title, a preface is hardly ever written first. It is better shaped by a backward look. Strange, that to introduce readers into new territory, or enhance what may be familiar, the writer ends by framing what has already been shaped. It is a bit like inserting a door after building an entire house. This form of reflecting on a traverse, in order to establish paths for others, is not confined to writing. We, as image-makers, do the same thing. The print shapes a photograph, and gazing at it, finished in a mat, hand-bound book, or frame, is akin to writing a preface. We look back in order to move forward. Much has been said about memory and photography, but this, to look closely for summaries of new-found patterns, is different. This book is about shaping a print. It is about making objects that hold attention, because they promise a rich preface. To paraphrase the seminal photographer, Minor White (1908–1976), this book is not just about making a platinum/palladium print but for what else a platinum/palladium print becomes.

Other books have inspired my approach here. Among these, especially important is Robert Bringhurst's *The Elements of Typographic Style* (1992). This gem is a bedside book because it is about much more than typography, and has added another level of experience to the *act* of reading a printed (or even on-screen) page. John Blakemore's *Black and White Photography Workshop* (2005) and Robin Gillanders's *The Photographic Portrait* (2004) set high standards for me too. This book strives to describe the technical details of a particular version of platinum/palladium printing, where the image is physically and visibly formed during exposure. I refer to it as the "printing-out" platinotype or platinum/palladium process in order to distinguish the method from other recipes which require development. Printing-out platinum/palladium prints are simply cleared after exposure. There is some elegance in this. The following pages describe a precise approach so that you can engage with the matter of giving voice to your photographs. As my long-term collaborator and most valued friend, Mike Ware often reminds me, there is much to be gained from putting good science to the service of artistic experience. So it is that the book also strives to connect technicalities with expression and meaning. I write about the experience of Making as well as Looking in the hope that this will hone your creative endeavors.

I first looked at platinum prints in an exhibition around 1978, and was so moved by them that I walked out of the London gallery and headed straight for a well-known photo supply store. Naively, I asked the shop assistant for a box of platinum paper. It took some five years or so before I held a print in my hand that came partway to satisfying me; and this, only because of having connected with Mike Ware in 1982. You may not want to wait that long, or even a day. Stores certainly no longer have shelves stocked with Platinotype paper. Yet, it is possible to start printing within a few days from now, and peruse this book during the ensuing effort. If, as I was in London those many years ago, you are itching to make prints right away, then do so! Pre-mixed sensitizer, which is a one-bottle solution (forgive the pun — this is the only one in the book), may be purchased from a vendor along with the three clearing-bath chemicals (see the appendix). Find or build a contact-printing frame (but a sheet of

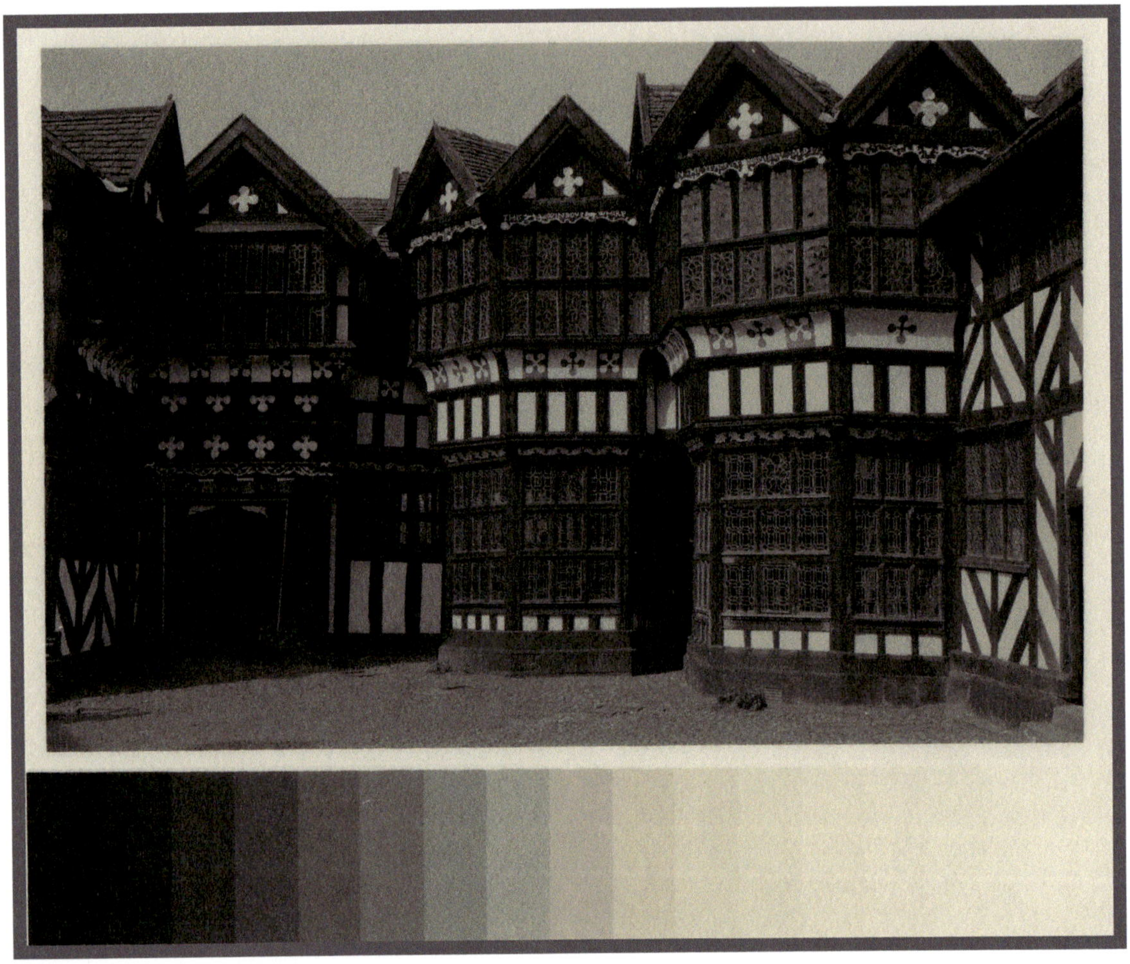

FIGURE F.3 Mike Ware, *Gawsworth Hall, Cheshire*, 1982. Palladium Print on Crane's AS 8111, from original 6 x 9 cm rollfilm

glass and some clips can work too), and acquire Arches Platine, Revere Platinum or any one of several papers listed in Chapter 6. Have a suitable negative on hand (Chapter 7). Coat the paper and hang it up to dry in the dark, at 40 to 80% RH and around 68°F / 20°C, for about 90 minutes. Place the negative and dried, coated paper in the contact-printing frame and expose to ultraviolet (UV) light—the sun is just fine. Every two minutes or so, inspect the print by opening one half of the frame in UV-protected lighting until the image looks right, then process it in clearing solutions as described in Chapter 8.

The final manuscript for this book came about as the COVID-19 pandemic was breaking out, and while I write the preface, there is a further curdling of a sense of catastrophe as the inequities of power erupt into civic action around the globe. Our climate and environmental crises form a persistent background to all of this. To be thinking about the practices that bring us together in this context—platinum/palladium printing and photography—may seem trivial, even a decadent luxury. I would argue against that, though. Even in some small, unintentional way, a fine print brings with it an experience of the beautiful, and with that, hope. It seems self-evident that we need hope in any dark time, but I don't want to leave it there. Instead, I cast back to John Berger's wonderful meditation, Bento's Sketchbook,[1] about which he said:

> Hope is a contraband passed from hand to hand, and story to story. … I'd rather reject the terms optimistic and pessimistic. They suggest a calculation of how things are going to evolve, and if it's going to evolve in the way you want, you're optimistic. That has very little to do with despair and hope. Hope is not a form of guarantee, it's a form of energy, and very frequently that energy is strongest in circumstances that are very dark.

As printers, as practitioners of a craft, we are defined by optimism. We make this process work, and make it evolve in the way we want. And yet, the result is exactly what Berger ends with: it is a form of energy. Our work is needed now, at this dark moment, as transactions in the currency of hope.

Pradip Malde

Note

1 Berger, John. *Bento's Sketchbook: How Does the Impulse to Draw Something Begin?* New York: Pantheon Books, 2011.

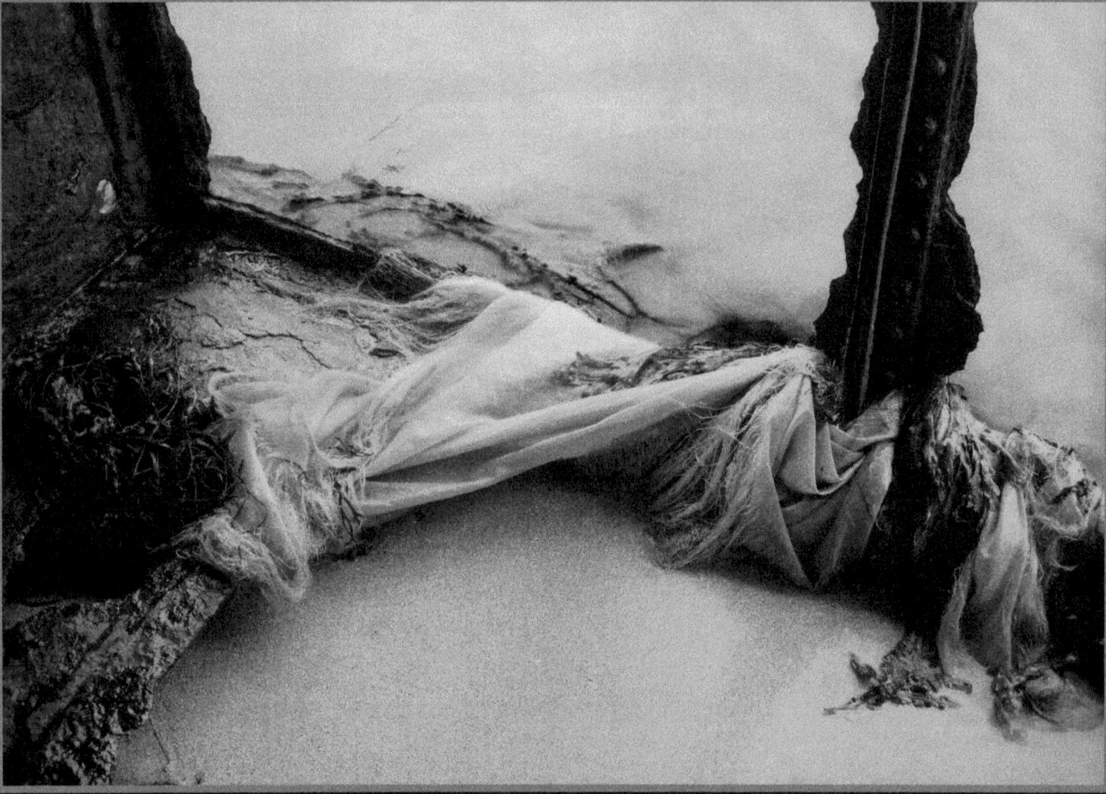

Figure F.4 Mike Ware, *Shrouded Blockship, Orkney*, 1983. Platinum-palladium print from enlarged copy negative (Kodak SO 015) of original 6 x 9 cm rollfilm negative, 28 x 41 cm. Courtesy Mike Ware

Acknowledgements

Historically, our united thanks should go, first and foremost, to Paul Hill and his late wife, Angela, for having the confidence to invite us both to conduct workshops in our process, and for all their wonderful hospitality at their celebrated *Photographer's Place* in Derbyshire from 1985 on, as mentioned in the Foreword.

In the ensuing years, workshops hosted at Sue and the late Peter Goldfield's Duckspool Workshops helped us fine-tune how to communicate the links between practice and expression.

We are grateful to Roger Vail, Professor Emeritus of California State University (Sacramento), for inviting us to contribute a workshop to the 1998 CSU Summer Arts program at Long Beach, and to all the participants in that event for critically, and so enthusiastically 'road-testing' our process.

Our work was supported at the earliest stages by grants from Kodak, The Scottish Arts Council, and the Norwegian Arts Council. The latter funded an initial collaboration with life-long friend and photographer Alf Magne Andreassen, whose work is represented in Chapter 11. More recently, we thank the Andrew W. Mellon Foundation for funding a major research project into the preparation and conservation of platinum and palladium prints at the National Gallery of Art, Washington DC, over the years 2010–2015, as described in the Foreword. We are indebted especially to Constance McCabe, Head of Photograph Conservation at the Gallery, and to her colleagues, Matthew Clarke, Adrienne Lundgren, Christopher Maines, and Sarah Wagner, for all their very generous help and professional expertise. We are grateful to Christina Z. Anderson for not only tapping us to write *Platinotype*, but then, along with the Focal / Routledge team, headed up by Sarah Pickles, for efficiently guiding us through this project; we could not wish for a better group of people to work with. An important part of this volume consists of images, generously provided by courtesy of Stephen Alvarez, Alf Magne Andreassen, Angela Berry, Stuart Clook, Cathy Cone, Jon Cone, Ivey Dahlstrom, Regina DeLuise, David Dennard, Virgil DiBiase, Danielle Edwards, Natasha Erdman and Patrick Ravines, Jill Enfield, Emily Gomez, Pamela Heemskerk, Allie Horick, Brooke Irvine, Jane Izard, Skip Klein, Pablo Kolodny, Gilles Lorin, Paula Luttringer, Rachel Malde, Caroline Minchew, Elizabeth Motlow, Shaun O'Dell, Greg Petropoulos, Stavros Pippos, Robert Poole, Allan Razo, Whitt Sellers, Mark Severson, Sara Silks, Larry Smukler, Chandler Sowden, Mark Stracke, Jamie Steele, Hunter Swenson, Mary Swisher, Basil Tahan, Roger Vail, William Watson, Elisabeth Weissensteiner, Kirk Williams, Lucy Wimmer, and Ellie Young; and by Dr. Matthew Clarke and The American Institute for Conservation, Teri and Bill Block, John Falconer and the British Library, the Getty's Open Content Program, the Imogen Cunningham Trust, the National Gallery of Art, Washington, DC, the National Museum of American History, the Rondal Partridge Archive, and the US Library of Congress. We would be anchorless without the valuable visual references they have all provided for reproduction here.

Mike gratefully acknowledges the perceptive guidance and encouragement he received, during the early years of the 1980s, from two notable practitioners and scholars of fine-art photography: Jane Routh of Lancaster University, and John Malcolm of Manchester Polytechnic.

Many have, knowingly or otherwise, supported our platinistic efforts since 1982, and brought us to this volume. Among them, Pradip would like to thank, in no particular order save for the tenuous threads of memory, Zelda Cheatle, Helena Srakocic and the late Sue Davis, then of The Photographers' Gallery in London for their early encouragement and support; Mr. Ian Shearer of Clydesdale Bank's Kirkwall Branch, for extending an overdraft in 1982 for an initial purchase of platinum salts, with only a portfolio of photographs as collateral; Morag Tweedie and Mary Jane Reyner for their patience and sharing studio spaces; Tam McPhail's Stromness Books and Prints, from which he prescribed a formative reading list; Gunnie Moberg who inspired and was a fellow beach comber; Colin Hamilton and the late Kulgin Duval who commissioned and bound two books of platinum-palladium prints; Thomas Joshua Cooper who introduced me to Ron and Elizabeth Partridge of the Imogen Cunningham Trust, and also, along with Tessa Jackson, Harry Kerr and Martha McCulloch, and a cluster of other Glasgow friends, who supported and helped produce my first exhibition of platinum-palladium prints at the Collins Gallery, Glasgow in 1984; Kay and Kit Richie for bringing me to Scotland; Robin Gillanders, Dave Williams, Marjory Gillanders Wilson, Murray and Kate Johnston for their passion and commitment to our art; Elizabeth Motlow for her beautiful work; the Partridges and Craig Withrow for enfolding me into their projects at the Imogen Cunningham Trust; the sweet memory of Richard Lorenz and his special insight into Imogen Cunningham and Margrethe Mather; Sara Stevenson for unveiling photo-historical riches, time after time; Celina Lunsford and the Frankfurt Fotografie Forum for decades of support and encouragement; Peter Bunnell for his clarity and adherence to the poetic vision; Beth Batton for a fostering voice; Chris Bucklow for unfolding William Blake; Roger Vail for the Goldberg Variations (most of this book was written while listening to various renderings of Bach's piece) and for making some of the finest platinum-palladium prints I have ever seen; Jon Cone, Cathy Cone, and Walker Blackwell for transforming my work methods through their extraordinary digital negative processes and workshop program; Jeff Wheeler of Quality Camera Company in Atlanta, Georgia for years of technical support and hard-to-find equipment; Pam Roberts and Hope Kingsley, then at the Royal Photographic Society, for an extraordinary platinum-palladium printing commission; and Wendy Ewald for being the compass rose. The University of the South has, over some 30 years, given me a family that extends far beyond the walls of my home. I am especially grateful for the support of my dean, Terry Papillon, and departmental colleagues Jeff Thompson, Jessica Wohl, Greg Pond, and Greg Petropoulos, for having directly and indirectly supported the preparation of this book. I will never be able to sufficiently thank hundreds of Sewanee students who have given me the greatest of gifts—to let me be a student too, to tolerate me as I learn to teach; among them, Caroline Minchew, Lucy Wimmer, Hunter Swenson, Matthew Hambree and Isabel Smith have made valuable contributions to this volume. Angela Berry and Amy Johnston, as students and as alumni, have been perpetual guides.

There are some without whom I strongly doubt that this book would ever have come about. It seems both awkward and unorthodox for one co-author to thank the other: Mike's peculiar mix of being a scientist and artist have shaped me, and defined my career, for these past 40 or so years. He has, with generosity and patience, brought me to not only understand the chemical intricacies of platinum/palladium printing, but also taught me to experiment methodically, write clearly, and work poetically. I cherish him. Most certainly, I would not be writing this book had we not met. My parents, Moti and Kanchan Malde, and brothers Bharat and Mayur, gifted me with much more than support during the early 1980s, when much of the experimental work on platinum/palladium printing was carried out—they had faith in me, and passed on a sense of persistence.

Their presence is imbued in this volume and also throughout my photographic work. Many years ago, the late Ron Partridge told me about how he would help his photographer-mother, Imogen Cunningham, in her small darkroom. He processed prints as a six-year-old while standing on a footstool. Imogen was fortunate in seeing her son grow into an accomplished and influential photographer. Similarly, I am grateful to my sons Luca and Kiran, who, at the ages of 15 and 11 respectively, helped in the processing of many of the prints in this book. They are accomplished cyanotypists and platinistas in their own right.

Finally, we gratefully acknowledge the forbearance of our respective wives, Rachel Malde and Marie Ware, for tolerating the installation of chemical laboratories in their homes, and for continuing nevertheless to provide lifegiving support over the four decades of our friendship.

Pradip Malde and Mike Ware

PART 1

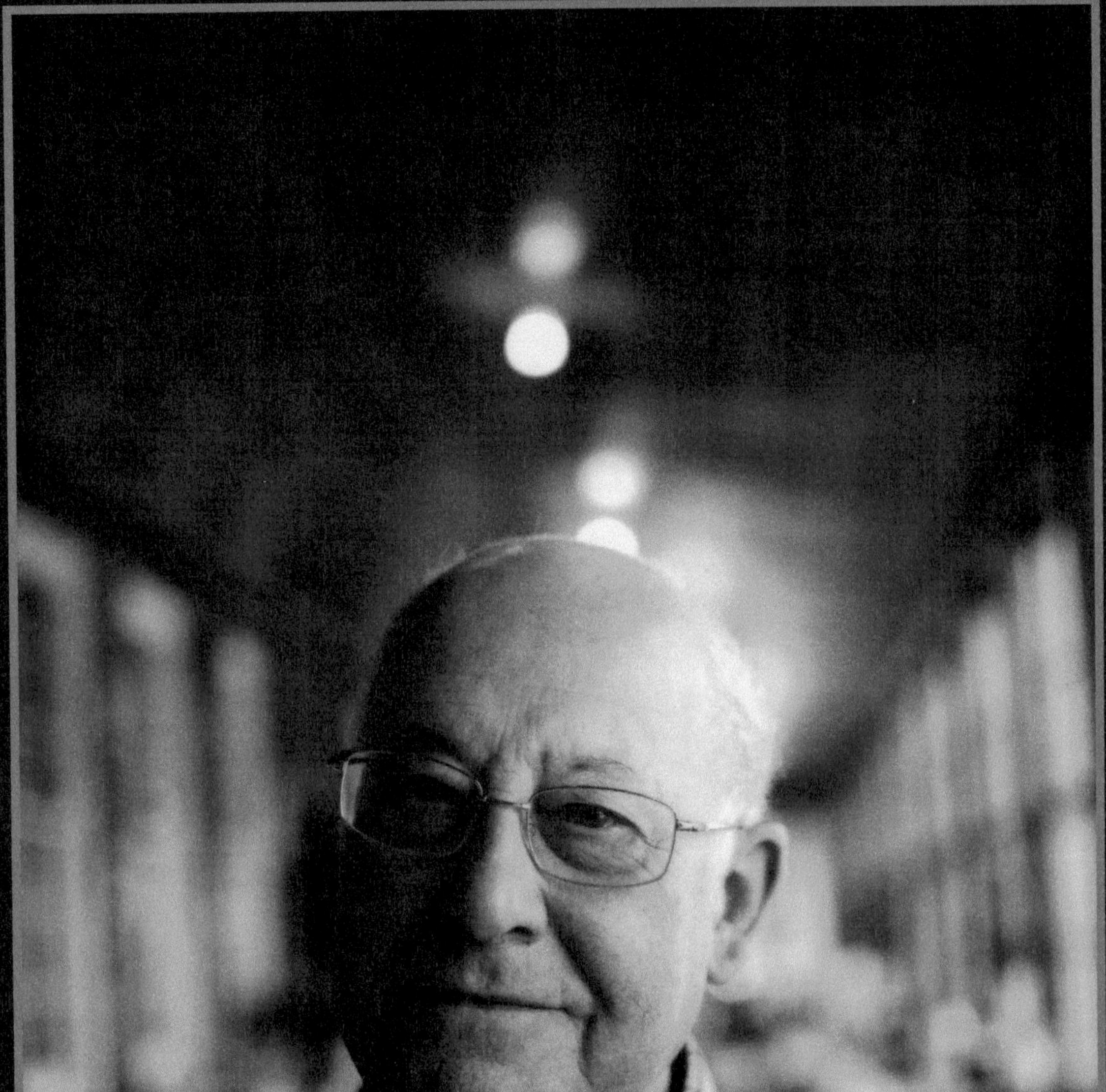

Figure 1.0 Pradip Malde, *Mike Ware, Madison County, VA*. 2014. Platinum-palladium print on Reich CT 24#, from enlarged PiezoDN digital negative of full-frame digital original (Leica M9 / Summilux 35mm), 25 x 35 cm. Courtesy Pradip Malde. Fe1:Pt:Pd = 2:1:1, 20 min hydration in 100% RH / 20°C. A photographic portrait echoes the experience of looking at a fine print: it is like a contract between the beholder and the beheld, and the print is the language by which the contract is defined.

Chapter 1

Contexts and Expectations

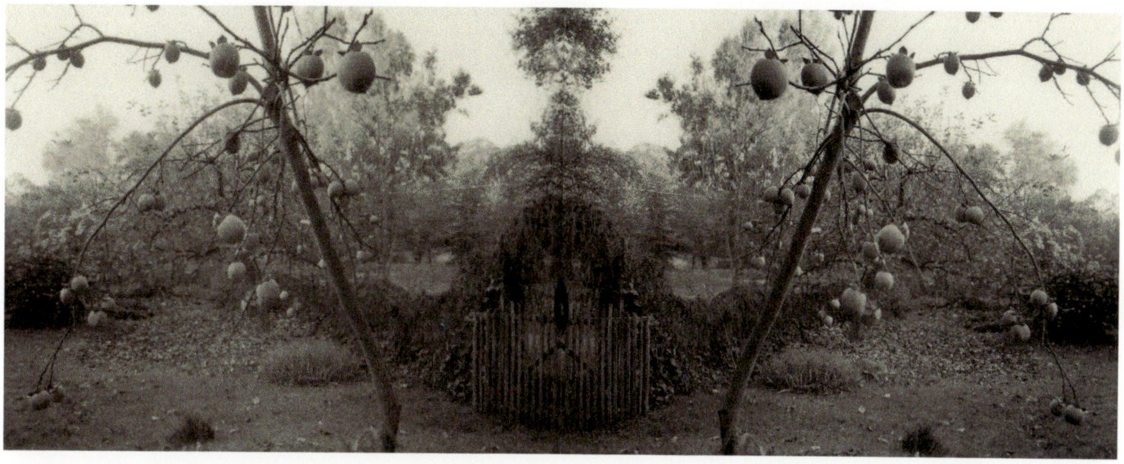

Figure 1.1 Mary Swisher, *Fantasy Gardens*, 1996. Platinum-palladium print on Crane's paper, from recto and verso of two original 4 x 5 inch negatives. Courtesy Mary Swisher. Fe1:Pt:Pd = 2:1:1, 20 min hydration in 100% RH / 20°C.

Every journey begins with an itch, with a sense of discomfort about the way things are, a feeling that something has shifted and everything around it needs to be readjusted. The first time I saw platinum prints, by Imogen Cunningham (1883–1976), Gertrude Käsebier (1852–1934), Paul Strand (1890–1976), and Edward Weston (1886–1958), at an exhibition in London in late 1970, I knew this was a moment of change, that I was about to embark on a long journey. The way I had been making prints on silver-gelatin papers suddenly felt insufficient. Now, I wanted to make photographs that were like the platinotypes before me: rich in dimensional and tonal information; fusing image and mark, place and moment, surface and paper, and all together behaving like an alloy, complete and inseparable from its components. Like an alchemical philosopher's stone, I wanted to make photographs that would be transformative in the same way that Gertrude Käsebier's platinum print, *Blessed Art Thou Among Women, 1899*, shook me as I looked at it. Some 40 years later, this print has continued to direct my work. It is my hope that you have picked up this book out of a similar desire, feeling motivated to make and know more about prints from platinum and palladium.

Some Recent Historical Context

At this point, you most probably just want to flip forward through the book and start printing (Chapter 8). I certainly urge you to do so: go ahead and get on with the process, assemble equipment and order supplies, prepare a workspace, make negatives, and print. There is, finally, the moment, so sweetly described by William Crawford

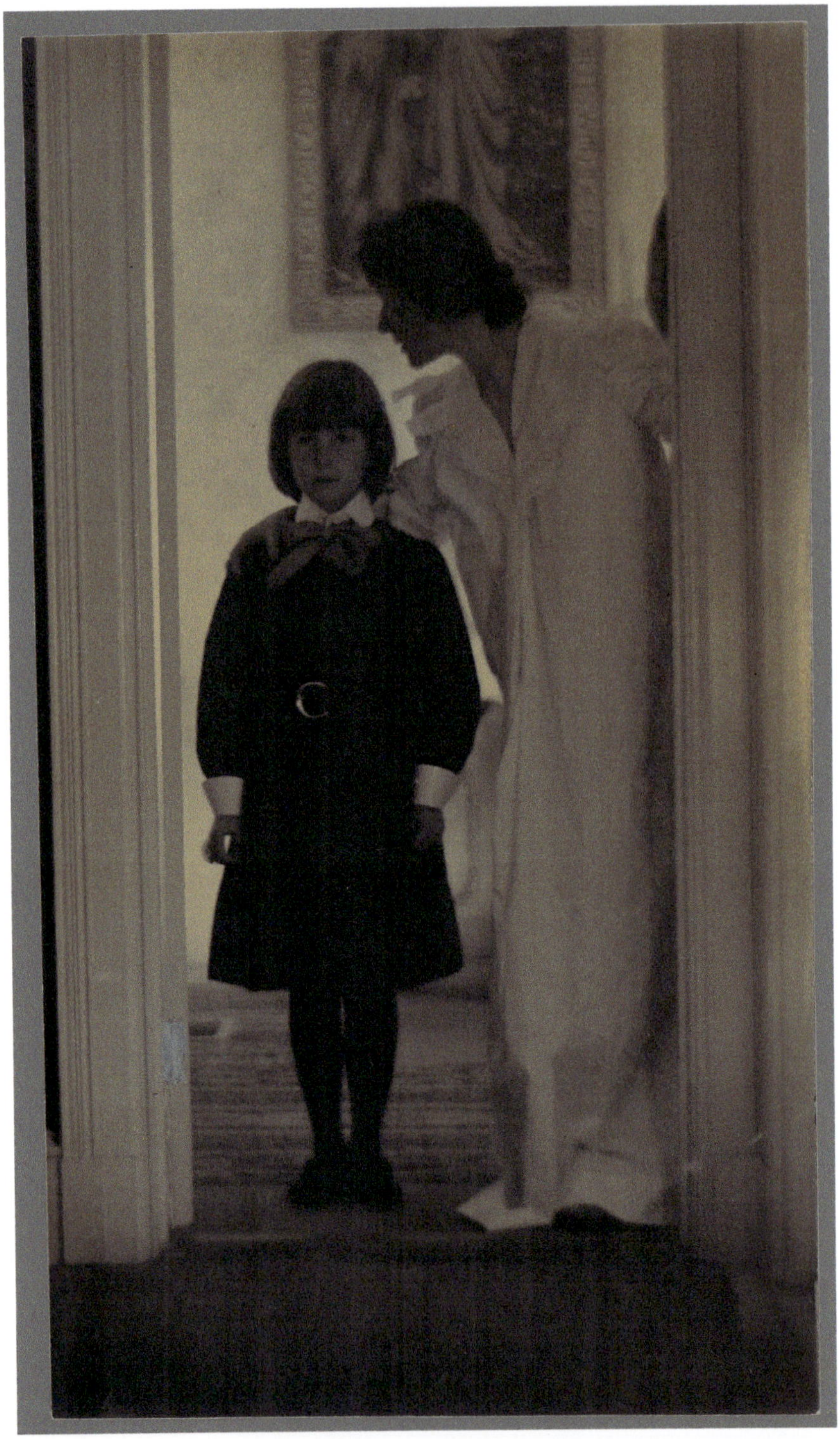

Figure 1.2 Gertrude Käsebier, *Blessed Art Thou Among Women*, 1899. Platinum print, sheet 25 x 14 cm. Retrieved from the Library of Congress, https://www.loc.gov/item/2015647712/ call number PR 13 CN 2015:052, no. 542 [P&P]. One of the first platinotypes I saw first-hand, this photograph shook my emotions by fusing image and mark, place and moment, surface and paper, and all together behaving like an alloy, complete and inseparable from its components.

(b. 1948) when looking at a platinum print, when you "take in a deep breath, and then sometimes you hear yourself let out a sigh."[1]

Platinotype: Making Photographs in Platinum and Palladium with the Contemporary Printing-out Process is designed to serve as an instructional guide and a reference. It will help, however, to read through these introductory pages before jumping elsewhere. This chapter may seem superfluous, and a diversion from the real work of making prints, but it offers a summarized context, both technical and cultural, for our endeavors. This section also, I hope, serves as an impetus, a jumping-off point, a springboard. There is a lovely proverb, I believe it is Chinese, which calls us to train hard and learn how to jump high, so that one day, the winds of inspiration may catch and transport us to places of wonder. The desire for wonder and magic have brought you and me to this point, to this book. The decisions that a photographer makes about printing, or a collector or curator makes about gathering and presenting work, are fundamentally subjective and passionate. A fine photographic print is shaped by aesthetic motivations, and because beauty reminds us of higher ground, of greater callings, a good print stands to make a difference in our lives. *Platinotype* is as much a technical guide as it is a collection of inspiring work, as much a chemical journey as an alchemical reveille. Knowing what brings us to a waypoint on a journey matters, because history shapes the roads we walk on.

Digital technology has had a profound impact on photography. Yet, until the beginning of the twenty-first century and the advent of digital imaging, the bulk of photographic practice, both amateur and professional, was based on surfaces coated with light-sensitive chemicals. This was, essentially, a technology that evolved directly from the earliest photographic processes: paper and celluloid films, coated with gelatin emulsions containing silver compounds. In 2001, the first digital single-lens reflex cameras entered the popular market. Six years later, smartphones with built-in cameras became widely available. By 2010, Kodak was phasing out film production. Roughly within the first decade of the twenty-first century, what was considered to be normal photographic practice shifted to being called "traditional," "wet lab" or, most accurately, "analogue" photography, as a result of the rapid spread of digital photographic capture and output technology. Digital photography became the affordable norm, while the analogue approaches *seemed* more expensive, time-consuming, technically challenging, and environmentally damaging. Yet, there remains a persistent murmur about the qualitative difference between digital and analogue; not about one being better or worse than the other, but that one form does not carry all of itself over to the other. The analogue just does not convert to the digital, or vice versa—it is more like a translation across different languages. This difference also implies, as so often happens with closely-related phenomena, that the sum of the two can be far greater than the separate parts.

Over the past ten years, theorists and educators have recognized the cognitive values of experience-based learning. Simply put, thinking and learning with all of our senses, as opposed to passively absorbing information through the spoken and written word, is a good thing. With this has come a greater acknowledgment of the value in tactile learning, and of art processes that are taught in "studio" classes. This up-swell has included a call for a return to analogue photography. More broadly, the internet, and publishing ventures such as this book series, have contributed to a sustained and growing interest in analogue photographic practices. Out of this has formed not only an affirmation of the benefits of combining digital with analogue, but also a certain refutation: that analogue approaches do not necessarily suffer from the cost, time, technical, and environmental disadvantages heretofore perceived in relation to digital photography. It is worth noting that analogue photography itself went through a major transformation during the years leading up to, and into the first decades of the twentieth century. There was a growing use of smaller cameras, roll-based acetate film, and silver gelatin printing papers. By the 1950s, any form of photographic printing other than silver gelatin on paper was highly unusual. Yet here again, just as the qualitative murmur of past processes inserted itself into the digital era, so did, thankfully, the practices and techniques of nineteenth-century processes persist among a few *cognoscenti*. It is because of them that we now find ourselves at an unusual and exciting moment. Analogue photography, or some hybrid with digital photography, is certainly still unusual, but not unheard of. The most commonly and widely used analogue form is still that of film-based capture with silver gelatin printing, but there has been a resurgent interest in combining digital with the so-called historic processes which date back to the nineteenth century. The reasons for this growing interest in making prints such as the salt print, argyrotype, and cyanotype, to name but a few, are largely out of a desire to engage with the physical and qualitative aspects of photography. The experience of working with these processes is quite different from a strictly digital approach, which usually stops at an on-screen viewing of the image.

Why Platinum and Palladium?

Satisfying and extremely fine results may be produced by all historic photographic printing processes. Each has a distinguishing set of characteristics. Most obviously, for instance, cyanotypes are known for their remarkable blue tones. The salt print is formed out of silver compounds rendering a range of hues from warm brown to blue-black, depending on how it is developed and toned. Argyrotypes are similar to salt prints, with a more nuanced tonal scale. Other than the most obvious color differences, however, it is ultimately quite difficult to clearly delineate the optical *advantages* of one process over another, because much depends on the negative from which a print is made. Ultimately, it becomes a matter of preference and feeling as to which process is best suited to the photographer's expressive endeavor. And it is at this point that the tactile and long tonal scale of the platinum and palladium printing process strongly inserts itself. The process has the capacity to evoke an almost synesthetic experience. In William Crawford's words: "The physical relationship between the individual particles of metallic platinum gives the print their characteristic 'tactility' ... This look can be quite lovely, but it is like certain minor keys in music: You hear them and right away they put you in a definite mood."[2] I subscribe enthusiastically to the musical analogy, but prefer to think of the platinum/palladium print as being like music played on the clavichord or harpsichord versus, say, the silver gelatin print as being akin to the same music played on an upright or grand piano.

Almost all traditional, or analogue photographic prints, such as the processes referred to here, are actually composed of clusters of finely divided metal or pigment particles, distributed in varying amounts across a supporting structure like gelatin or paper. As these clusters increase in size, they absorb more light. Small and sparsely distributed particles appear as light tonal values, and large and denser clusters of metal appear darker, even black. Thus, we end up seeing an image that approximately echoes our experience of the visual world. The vast majority of photographic processes use silver as the primary image-forming metal. Most metals are unstable: iron reacts with oxygen in the air, for instance to form oxide compounds better known as rust. Sometimes this instability takes years to manifest, such as rust forming on iron in dry conditions. In other cases, the reaction is quite dramatic, such as potassium igniting in oxygen. The least reactive metals, such as silver, gold, platinum, and palladium, are also the ones that have been most widely used in photography, and thankfully do not react directly with oxygen. Of these, platinum, and palladium[3] are the most stable, and, photographically speaking, most practical. In other words, images made from platinum and/or palladium are, potentially, much more permanent than images made from silver. The fading of silver-based photographs has been of concern from the earliest stages of photographic invention through to current times. Throughout the

1840s, and certainly by the late 1850s, techniques were being devised to stabilize photographs by toning them with gold, and then by 1856, with platinum.[4] It was perhaps this quest for a permanent, stable, photograph that brought the English innovator, William Willis, junior (1841–1923), to begin experiments with platinum in 1872. By 1878, Willis had formulated and patented a method for making a pure platinum photograph, "altogether dispensed with"[5] silver compounds, and even skipped a dependency on any colloidal binding, or emulsion layer that could undermine the structural integrity of the image. He called it the Platinotype. Along with the promise of permanence came something else: because of the absence of an emulsion layer, the image was formed within the paper fibers, and not on a colloidal layer above the paper. The optical effect of this is significantly more nuanced than the dramatic ways in which emulsion-based prints look, with their smooth and often highly reflective surfaces. Remember my harpsichord vs. piano analogy? It was this quality, along with their neutral black tones, that made platinum prints stand apart from other processes then, and what attracts many to the process now.

Early History of Platinotype

With Willis's patent of the Platinotype in 1878 came the establishment of The Platinotype Company, which sold pre-sensitized paper and processing solutions to the public. The Willis platinotype was a development process, with the image forming fully only after immersion in a chemical bath. Along with Willis's own modifications to his product line and processing methods, others were looking into alternative recipes. In 1882, two Austrian army captains in the photographic department of the military technical administration, Giuseppe Pizzighelli (1849–1912) and Baron Arthur von Hübl (1853–1932), were the first to publish detailed recipes and instructions, in German, for their version of platinum printing, which required the practitioner to mix the chemicals from raw ingredients and hand-coat the paper.[6] In 1887 came a key innovation, especially for the authors of this book: Pizzighelli, by then working alone in Banjaluka, Bosnia, made the significant discovery of an alternative iron sensitizer that yielded a platinum image directly during the exposure. The ready-formed picture did not need a developer, and the print only required bathing in dilute hydrochloric acid to clear the residual chemicals.[7] This "printing-out" method further amplified qualities that were already influencing the expressive endeavors of photographers and the musings of aestheticians. At about the same time, Peter Henry Emerson (1856–1936), a physician, photographer, and vociferous commentator on photographic aesthetics, recommended the process to others for being "as suitable and as beautiful for soft gray-day effects as for brilliant sunshiny effects."[8] The printing-out platinotype potentially rendered, for reasons that will be explained later in the book, tones that were even smoother and more nuanced than the development method for making platinum/palladium prints (which I will refer to as the "development Pt/Pd" process).

The first few years of the twentieth century were perplexing ones for aficionados of the platinotype. On the one hand, there was a growing fascination with, and use of, the process by photographic artists, from Henry Peach Robinson (1830–1901) for his fantastic multi-negative tableaux, through Frederick H. Evans (1853–1943) for his architectural interiors, Fred Holland Day (1864–1933) with his deeply personal and expressive portraits, and on to Alfred Stieglitz (1864–1946) and members of the Photo-secession Movement which he founded in 1902, including Edward Steichen (1879–1973), Frank Eugene (1865–1936), Alvin Langdon Coburn (1882–1966), Clarence H. White (1871–1925), and Gertrude Käsebier. Records indicate that more platinum prints than any other printing process were being shown in photography salon exhibitions in the UK around the turn of the century.[9] Willis's Platinotype Company listed an almost bewildering range of formulations, surfaces, and paper weights, and was exporting paper to the USA, where it had established an affiliate company. Platinum was

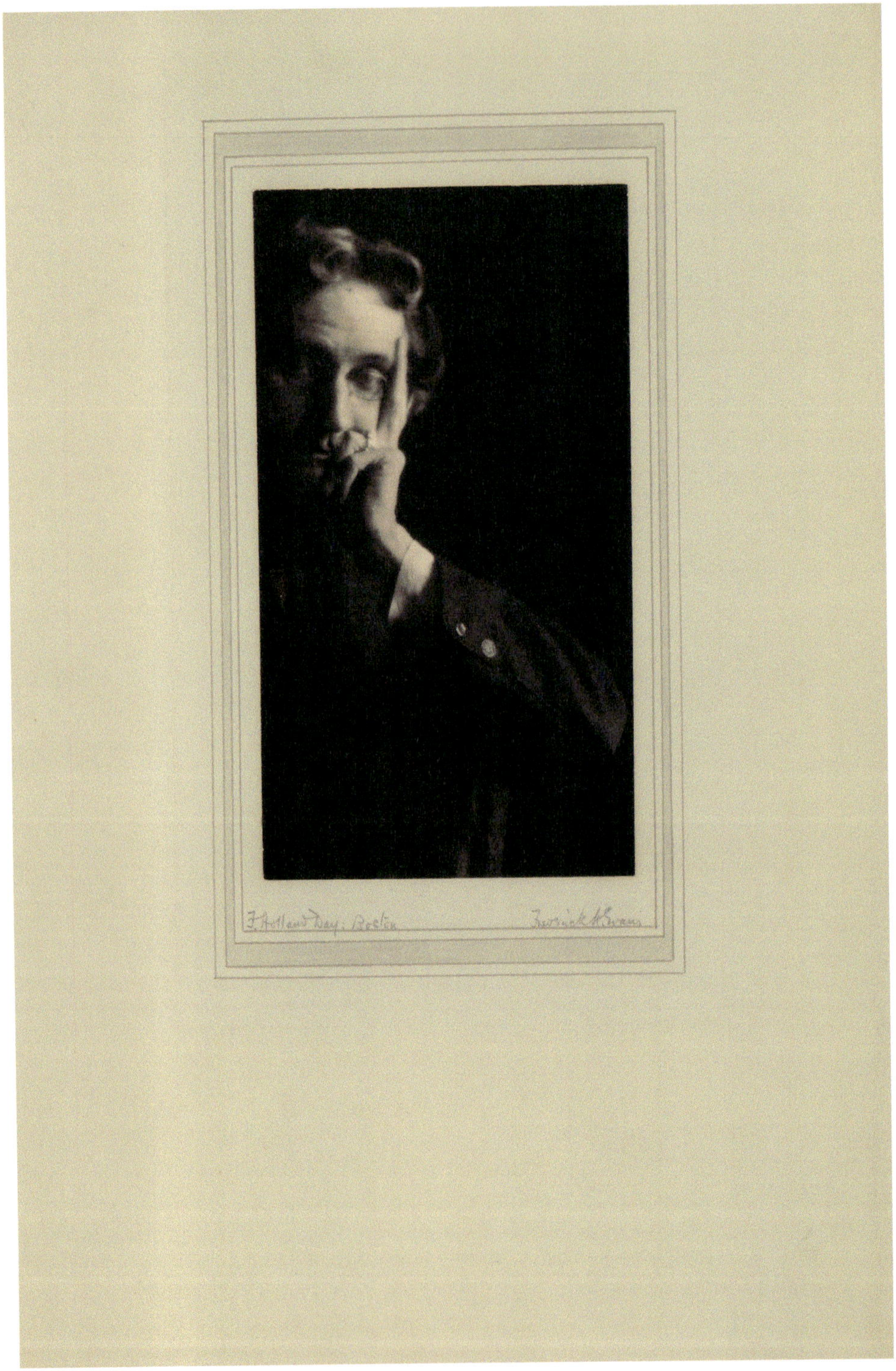

Figure 1.3 Frederick H. Evans, *F. Holland Day*, 1900. Platinum print on French-ruled mount, 20.3 x 11.4 cm. National Museum of American History, Smithsonian Institution, 67.165. The tactile and long tonal scale of the platinum printing process has the capacity to evoke an almost synesthetic experience.

getting a lot of attention for its photographic worth, and more importantly, ultimately, for its chemical properties and behavior. On the other hand, this rise to prominence was, ironically, the beginning of the undoing of the platinotype. There is almost a pun here, and it relates to a particular and rather special property of this precious metal.

Platinum its Own Undoing

Platinum is a catalyst. It has the ability, in other words, to accelerate a chemical reaction between certain substances, while remaining unchanged itself. Some would say that catalysis is akin to a good dating service, which brings two people together while helping them avoid glacial musings about whether or not one should ask the other for a romantic evening out. Of course, some money is exchanged, and the dating service profits from that, but I hope you get the idea. In 1902, the Latvian/German chemist and philosopher, Friedrich Wilhelm Ostwald (1853–1932), discovered platinum's ability to catalyze the oxidation of ammonia to nitric acid.[10] He later received a Nobel Prize for his work on "catalysis, chemical equilibria and reaction velocities."[11] In itself, Ostwald's seemed a fairly innocent discovery: it made it easier to manufacture nitric acid, and nitric acid is essential, in turn, for the manufacture of the fertilizer ammonium nitrate. However, more ominously, ammonium nitrate is also an explosive. Even more significantly, there are a number of other explosives such as nitroglycerine, picric acid (trinitrophenol), and TNT (trinitrotoluene), which require nitric acid for their manufacture. Along with catalysis, other strategic applications for platinum were discovered, such as its use in light bulb filaments. In short, platinum—a metal previously only of scientific interest for constructing chemically resistant apparatus—was swiftly elevated in the first decade of the twentieth century to the status of a highly protected, and expensive, strategic material. "Catalysis" was coined after the Greek word, καταλυειν, which means "to unbind" or, more loosely, to undo a situation. And so it was that platinum's commercial and strategic utility severely compromised its use as a photographic material; its value to others was the cause of its own undoing as a photographic material.

All commercial production of papers sensitized with platinum and/or palladium had stopped by the end of the 1930s. At this point, very few photographers could afford to work with the paper anyway. The advent of small-format cameras and the need for enlargement silver-gelatin papers, along with the economic and cultural shockwaves of the Great Depression, the Second World War, and post-war recovery, must have further contributed to the almost complete demise of the platinum/palladium printing processes. Despite all indications of a complete discontinuation, the period from the 1930s to 1960s was one of dormancy,[12] at the end of which an up-swell of interest began to revitalize the process.

Revival

It was a slow awakening, and the lack of well documented methods made it difficult to revive the process. Today, thankfully, there are numerous online resources about how to make photographic prints with a variety of traditional, or "wet," techniques. There are also, as may be seen in the bibliography of this volume, a plenitude of texts and manuals about platinum/palladium printing. Suppliers of materials and equipment abound. Yet this has not always been the case. Few texts in the 1970s, when I saw platinum/palladium prints for the first time, provided detailed practical instructions. Just as scarce were suppliers of chemicals and equipment, and that most crucial of ingredients: papers that were suitable for coating with sensitizer. Despite this, a few collectors, curators, historians, and practitioners during the 1960s and 1970s recognized the particular aesthetic pull of platinum and palladium prints, and began to bring their fascination for this medium into the public realm. Almost all of the notable early to mid-twentieth-century photographers using the technique had ceased making platinum/palladium prints by the time industrial production of the paper stopped in 1938. Frederick H. Evans, who many would say made some of the finest platinum prints in existence, gave up photographing when the paper was no longer commercially available.[13] However, a very small number of photographers persisted by hand-coating papers using platinum and palladium printing recipes. Laura Gilpin (1891–1979) is said to have brushed sensitizer onto sheets of paper laid out on an ironing board in her living room. She had learned about the process through Gertrude Käsebier and while studying with Clarence White at his school in New York. Irving Penn embarked on a remarkable project in 1963, and after having gleaned recipes from extensive research[14] produced a body of large platinum prints some ten years later. George Tice (b. 1938) published instructions in issue 11 of *Album Magazine*,[15] and also provided an attractive description of how to make a platinum print for the Time-Life series of books about photography.[16] It was only in 1977 that Nancy Rexroth (b. 1946), during a postgraduate internship at the Smithsonian Institution, published what may be the first comprehensive set of updated recipes. These detailed guidelines were in a small volume called "The Platinotype 1977" and formed the basis of our earliest practical work with the process. It is Rexroth's gentle caveat that I hope this volume stands to counter: "The whole process is of course trial and error. Don't be surprised if you sometimes get unexplainable gremlins in your prints" and she finishes with what became a portent for the contents of this book, "Always avoid the contamination that results from your chemicals coming into contact with metal. *This is also true of moisture*" (my italics). As will become apparent, moisture, or more accurately, hydration, performs a crucial function in platinum/palladium printing, and was only presented as a hazard by Rexroth and others because its role was not fully understood. Despite this, it is thanks to her efforts, and the endeavors of many others, that this volume exists.

What Makes this Book Different

At the very least, this book aspires to add to the historic and contemporary literature pertaining to platinum/palladium printing. The vast majority of existing writing concentrates on development methods of making platinum/palladium prints, using sensitizer and developer recipes that are largely derivatives of those formulated by William Willis. More specifically, most recipes utilize potassium salts of platinum and palladium, and ferric oxalate as the light-sensitive compound. While some contemporary texts refer to the printing-out method as a curiosity and a variant worth investigating, this book is the first practical manual devoted entirely to providing a detailed account of the printing-out method using ammonium salts. It results from a thorough re-examination of the platinotype and palladiotype chemistry. Mike Ware and I began to collaborate in 1982. Our research led us to re-formulate a modernized printing-out process resembling Pizzighelli's "direct printing platinotype" of 1887.[17] We also employed some up-to-date chemistry, and the result provided greater accessibility, economy, chemical reliability, and health and environmental safety. Our printing-out method, initially referred to as the "Ammonium System" and later referred to as the Malde-Ware method, was published in 1986,[18] and taught in workshops at Paul and Angela Hill's Photographer's Place in Derbyshire, Peter and Sue Goldfield's Workshops at Duckspool in Somerset, and elsewhere. It is important to know how this volume differs from other texts. This book outlines methods for making printing-out platinum/palladium prints using an integrated recipe of *just* ammonium salts of iron, platinum, and palladium, which have the most desirable solubility properties. Additionally, the processing solutions, which are essentially clearing baths, have been formulated to maximize the efficient removal of compounds which might otherwise cause stains, leaving behind a fully archival print. *Platinotype: Making Photographs in Platinum and Palladium with the Contemporary Printing-out*

Process serves as a manual, a resource, an aesthetic guide and is designed to fill something of a lacuna in the current literature around platinum/palladium printing. This contemporary printing-out process offers the following distinct advantages over other methods:

- Convenience: a single, large volume of sensitizer solution may be prepared at one time, then used, and stored for years at a time—traditional, development methods for making platinum/palladium prints always require sensitizer solutions to be combined just before coating the paper
- Efficiency: an integrated mix of highly soluble, ammonium-based iron, platinum, and palladium salts provides a cost-effective sensitizer—there is no need to double-coat
- Printing-out: the image appears, and may be inspected multiple times, during exposure, allowing for an accurate assessment of the proper final density
- Contrast: because of printing-out, as exposure progresses, the densest areas of the positive image become self-masking, while highlight values continue to be exposed at a relatively normal rate—so contrast control is more relaxed
- Image quality: printing-out platinum/palladium images are especially nuanced in the extreme densities and highlights, while rendering particularly smooth mid-tones—there is no sensitizer-induced granularity
- Permanence: a thoroughly tested and scientifically verified processing sequence ensures the production of permanent, archivally sound prints
- Environmental: all the processing solutions are environmentally safe, and used at high dilutions
- Availability: ammonium ferric oxalate, the light-sensitive ingredient, is globally sold by chemical houses and is also much purer, and easier to get into solution, than ferric oxalate[19]

The following chapter takes us deeper into the fascinating chemical "landscape" of platinum/palladium printing, and through this, prepares us for making beautiful prints. Beauty, as Elaine Scarry has noted, is a compact or contract between the beholder and the thing beheld,[20] and this contract aspires, as all contracts should, to improve the human condition. Understanding how to make a beautiful thing then, lies at the heart of being human, of being creative; an understanding that is needed if one is to spring into the winds of inspiration. Otherwise, as Nancy Rexroth cautions, we will remain at a baser level, dancing with gremlins.

Notes

1 William Crawford, *The Keepers of Light: A History and Working Guide to Early Photographic Processes*, Dobbs Ferry: Morgan & Morgan, 1979, p. 77.

2 Ibid.

3 Platinum and palladium occupy vertically adjacent locations in the same group of the Periodic Table.

4 Ernest de Carranza, "Procédé de tirage des positives par le chlorure de platine," *Bulletin de la Société Française de Photographie*, 15, April 1856, 82–84; *La Lumière*, 6 (8): February 23, 1856, 29; translated in *Journal of the Photographic Society*, 3 (40): March 21, 1856, 14; see also: H.L.T. Haakman, "Toning paper positives with platinum," *The Photographic News*, 1, January 28, 1859, 251; 2, 1859, 204, 263; and Mike Ware, *Platinomicon: A Technical History of Photographic Printing in Platinum and Palladium*, 2016, p. 18.

5 W. Willis Jun., "A New Process of Photo-Chemical Printing in Metallic Platinum," *The British Journal of Photography*, 25, August 23, 1878, 400. Also https://books.google.com/books?id=WIE7AAAAMAAJ&pg=PA400&lpg=PA400&dq="A+New+Process+of+Photo-Chemical+Printing+in+Metallic+Platinum"&source=bl&ots=lfPBE4ye2C&sig=ACfU3U0gW26WtCML_PTqM7fUhBG1NOL6lQ&hl=en&sa=X#v=onepage&q&f=false

6 Giuseppe Pizzighelli and Arthur Freiherr von Hübl, *Die Platinotypie*, Vienna and Leipzig: 1882. Also Ware, *Platinomicon*, p. 66.

7 Capt. Giuseppe Pizzighelli, "The Platinum Print," *Anthony's Photographic Bulletin*, 1888, 410 – 421. See also: Capt. Giuseppe Pizzighelli, *Photographische Correspondenz*, 24, October 1887; 25, January 1888; *The Photographer's Book*, eds. Holmes and Griswold, 1888. Also, Ware, *Platinomicon*, p. 66.

8 Peter Henry Emerson, *Naturalistic photography for students of the art*, London: Dawbarn & Ward, 1899, p. 137.

9 Ware, *Platinomicon*, p. 42.

10 Ware, *Platinomicon*, p. 62.

11 https://www.nobelprize.org/prizes/chemistry/1909/ostwald/biographical/

12 Tatiana Cole, "The Platinum Renaissance: Oral Histories of Platinum-Palladium Printers and Artists," in *Platinum*

and Palladium Photographs: Technical History, Connoisseurship, and Preservation, edited by Constance McCabe. Washington DC: Photographic Materials Group of the American Institute for Conservation of Historic and Artistic Works, 2017, p. 433.

13 Anne Kelsey Hammond, "Frederick H. Evans: The Interior Vision," in *Creative Camera*, March 1985, 243, 12; see also Anne Kelsey Hammond, "Aesthetic Aspects of the Photomechanical Print," in *British Photography in the Nineteenth Century*, edited by Mike Weaver, pp. 176, 256, Cambridge: Cambridge University Press, 1989.

14 Vasilios Zatse and Constance McCabe, "Irving Penn's Platinum-Palladium Prints," in *Platinum and Palladium Photographs*, pp. 405–423.

15 George Tice, "The Lost Art of Platinum," *Album 11*, December 1970, 12 – 13, 23. Also https://billjay.com/album-11/?fbclid=IwAR2gHnY77MECxATI-mGmVid9NHKLdxBZlDPjePbF1YQZ27vXVnSoQ6-BzYk

16 George Tice, "Caring for Photographs," in *Life Library of Photography*, 1972, pp. 86–93.

17 Michael J. Ware, "An Investigation of Platinum and Palladium Printing," *Journal of Photographic Science*, 34, 1986, 165 – 177. http://www.scribd.com/doc/51088563/An-Investigation-of-Platinum-and-Palladium-Printing. Also Ware, *Platinomicon*, pp. 66–67.

18 Pradip Malde and Mike Ware, *The Ammonium System: a contemporary method for making platinum and palladium prints, A Workshop Manual*, 1985; Mike Ware, "Platinum Reprinted," *The British Journal of Photography*, 133 (41): 10 October 1986, 1165 – 1167; ibid., 133 (42): 17 October 1986, 1190 – 1194; Pradip Malde, "New Solutions for Platinum Printers," *View Camera*, Sept/Oct 1994, 36-41.

19 Mike Ware, "Photography in Platinum and Palladium: Platinum & Palladium Printing, Second Edition, an essay book review," in *Platinum Metals Review*, 2005, 49, 190–195. https://technology.matthey.com/article/49/4/190-195

20 Elaine Scarry, *On Beauty and Being Just*, Princeton: Princeton University Press, 1999, p. 90.

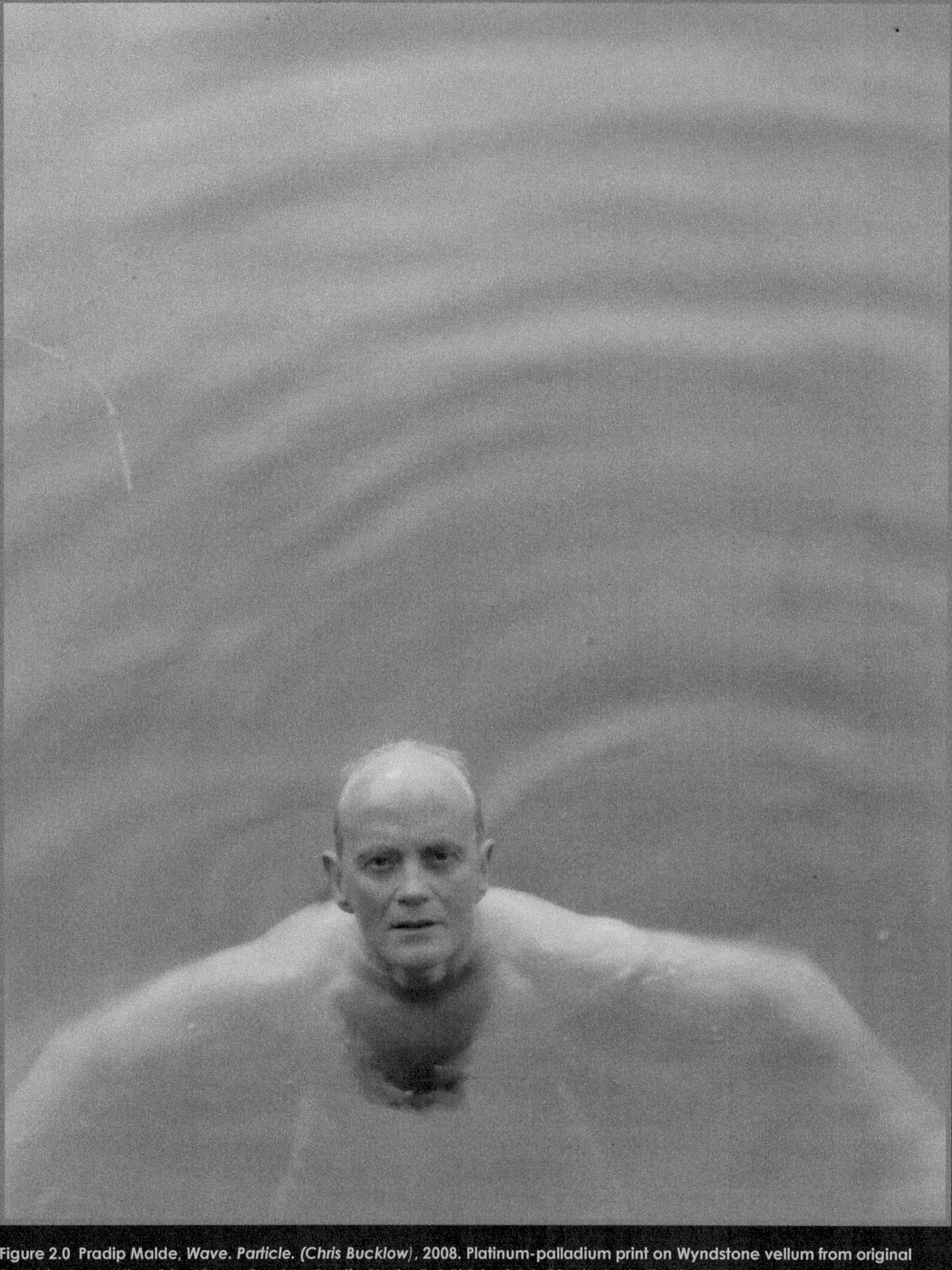

Figure 2.0 Pradip Malde, *Wave. Particle. (Chris Bucklow)*, 2008. Platinum-palladium print on Wyndstone vellum from original 8 x 10 inch negative, 24 x 19 cm. Courtesy Pradip Made. Fe1:Pt:Pd = 2:1:1, 20 min. hydration in 100% RH / 20°C.
It helps to understand, at the very minimum, how the platinum-palladium "mark" is made on paper, and how such marks eventually shape the photograph, letting water become body, head, thought waves, and the place from where light itself emanates—all of these are expressed by the *form* of the platinum-palladium print.

Chapter 2
Chemistry of the Platinum/Palladium Print

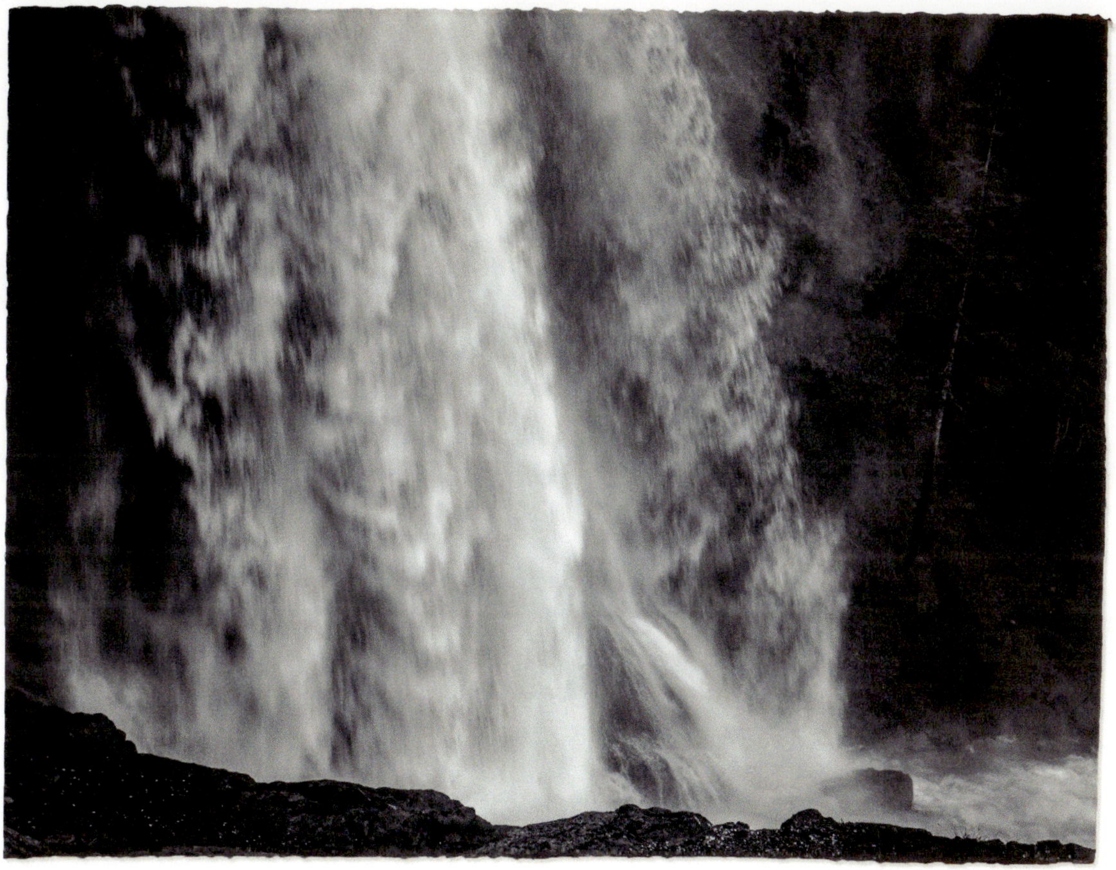

Figure 2.1 Gilles Lorin, *Waterfall, study no. 1*, 2018. Palladium print on Arches Platine, 56 x 76 cm. Fe1:Pt:Pd = 1:0:1, hydrated to 70%.

The connection between form and content is intertwined. Thinking of an image in these terms brings us to consider the interplays between what a photograph is *of* and *about*. Another way to think of this dynamic is as an interplay between the print (marks on paper) and the photograph (in all its complexities: window, focus, moment, etc.). Altogether, this amounts to meaning and revelation. Our aim is to make a print so fine that it will leave the viewer taking in that proverbial deep breath, and be caught in a visual enchantment: shuttling back and forth between the print and the photograph. The messenger, in other words, is transparent, and exists to deliver the message. And yet, the message is a consequence of the messenger. This problem, the relationship between medium and message, is not too different from the typographer's or translator's dilemma. A translator should, from the utterance of one language, create an echo of itself in the distant forest of another, a "reverberation of the work in the alien one."[1] A good translator, or typographer, becomes transparent, as should the technical, chemical details of the process.

Knowing how the printed image is formed will help you access the language of the print, to move towards a poetry without words, a photography beyond the representational. This chapter intends to clarify the underlying chemical processes so that we, as printers, may discern between medium and message, and be capable of creating the proper tension between process and expression. The following paragraphs take us into the mechanisms of platinum/palladium printing so that we may work with clarity, intention, and vision. The decisions we make in order to arrive at a fine print are certainly subjective, but the print is determined also by its material. The material, in turn, affects and conditions the experience of looking at a print. There is a recursive relationship between the physical and the mental, between form and content. We need to engage with the chemical aspects of the process with some understanding and in a methodical way. We need to understand at the very minimum, how the platinum/palladium "mark" is made on paper, and how such marks eventually shape the photograph. It seems logical then, that we should begin at the beginning: the atomic level.

Naming Conventions

It is important that we first take a side-step and clarify some terminology. Prints may be made with either of the two precious metals platinum or palladium, or a mixture of both. Throughout this book, the term "platinum/palladium" will denote all of these recipes, resulting in a single-metal print or a combination of both metals. The hyphenated form, "platinum-palladium" is understood to describe a print made from a combined recipe of platinum and palladium salts, where the greater portion of metal in the finished print is platinum, or platinum and palladium are present in almost equal amounts. By this logic, a "palladium-platinum" print is one of combined metals where palladium is the predominant ingredient. This approach allows for clearer instructions; it presents a more accurate description of the process to collectors, curators, and conservators; and helps practitioners engage in more efficient conversations. There is also an ethical issue here: many contemporary prints made with a combined recipe are rendered from a predominance of palladium and should, strictly speaking, be described as suggested here. I also occasionally refer to prints made with platinum or palladium as "Platinotype" and "Palladiotype" or when combined, as platino-palladiotypes. To summarize:

- Platinum print or platinotype = a print made only with platinum sensitizer
- Palladium print or palladiotype = a print made only with palladium sensitizer
- Platinum-palladium print or platino-palladiotype = a print made with equal amounts of platinum and palladium sensitizer or a greater amount of platinum sensitizer than palladium sensitizer
- Palladium-platinum print or palladio-platinotype = a print made with a greater amount of palladium sensitizer than platinum sensitizer
- Platinum/palladium print or platino/palladiotype = a general way of referring to all of these recipes

You will notice an inconsistency, for which I apologize: in accord with the above guidelines, the title of this book *should* be "Platino/Palladiotype," but this seems ahistorical, an awkward mouthful, and presents a typographic moraine. I hope that the subtitle redresses the matter and provides a

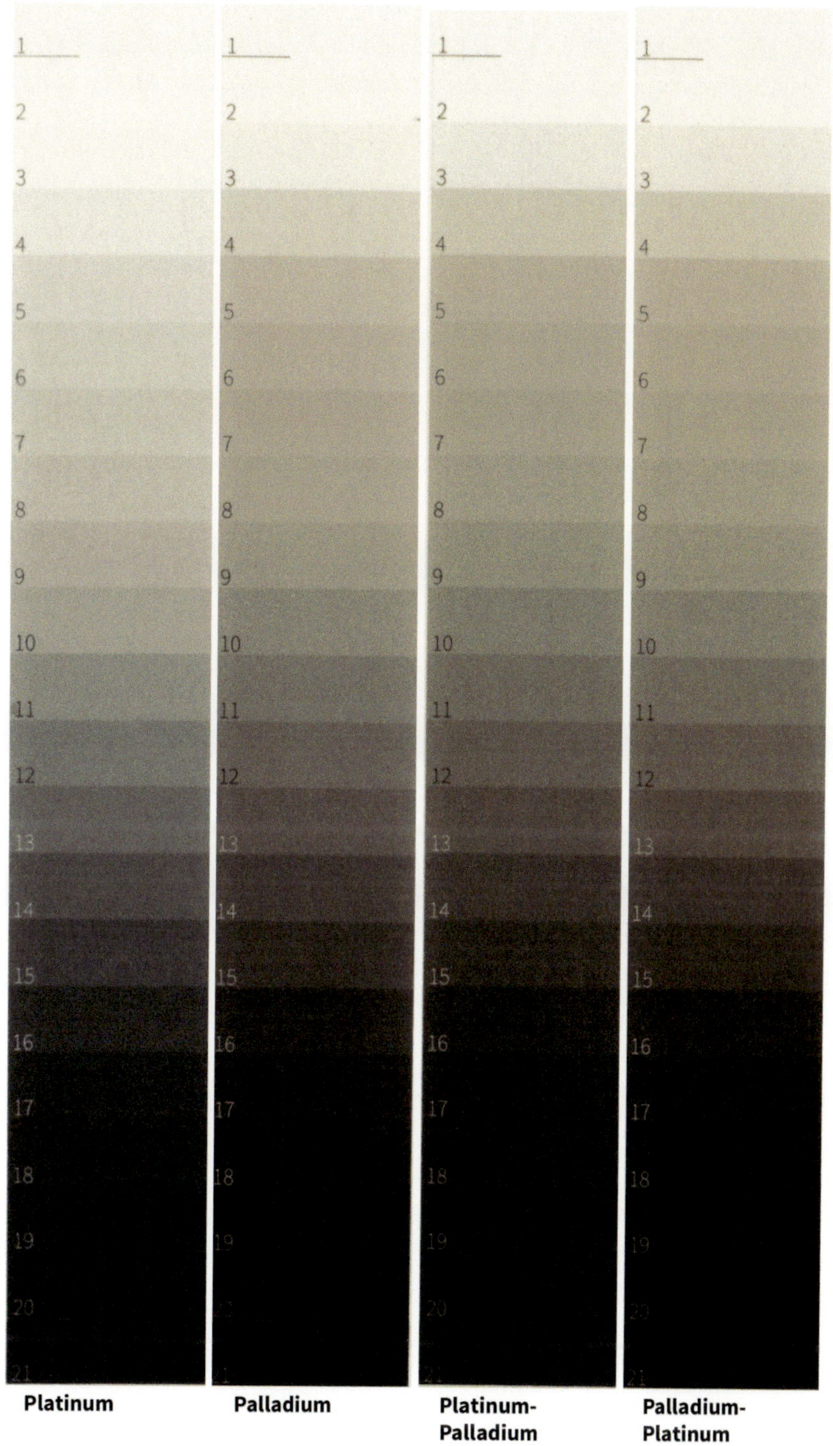

Figure 2.2 Examples of each variation of metal in the printing-out process, demonstrated with four step-tests from left to right, respectively: platinum, palladium, platinum-palladium, and palladium-platinum. Throughout the book, the ratios of ammonium ferric oxalate (Fe), platinum (Pt) and palladium (Pd) salts are denoted in that order. With these prints, Fe Method 1 was used (Fe1). Thus, from left to right, the ratios for Fe1:Pt:Pd = 1:1:0; 1:0:1; 4:3:1; 4:1:3.

sufficiently clear indication of the path we are about to walk on.

The title also makes clear what this book is about: making prints from platinum and palladium such that they appear during exposure. I refer to this as the "printing-out" platinum/palladium process. While there are other printing-out recipes, this book focuses only on the printing-out process using ammonium salts of iron, platinum, and palladium. Sometimes referred to as the "Ammonium System," or more recently as the "Malde-Ware process," it is capable of rendering prints by any combination of platinum and palladium. Other methods for making platinum/palladium prints are quite different: they require development in order to fully form the image. For the sake of clarity, these platinum/palladium printing processes by development will be referred to as "development Pt/Pd."

While here, there is one more matter to address about nomenclature: we should refer to the light-sensitive water solution of iron and platinum/palladium salts as a "sensitizer," not as an "emulsion"—an emulsion is strictly a fine dispersion of a liquid, suspended in another (immiscible) liquid. In other words, an *emulsion* is understood to be a combination of two or more substances that are finely dispersed but not dissolved, such as egg yolk, vinegar and oil combined to make mayonnaise. Even the familiar photographic combination of silver halide with gelatin is not truly an emulsion—it is a *suspension* of solids. Here, with platinum/palladium, we are using a liquid that will sensitize paper to ultraviolet light *and* produce a visible result. The coated paper does not have an "emulsion layer" or an "emulsion side" but rather a sensitized surface or side. *A sensitizer is understood as being the combination of solutions of iron and platinum/palladium salts, and therefore capable of rendering a visible image. I refer to all other liquids as "solutions"; i.e. the iron solution, platinum solution, palladium solution, etc.*

Finally, the chemical symbols for iron, palladium, and platinum are Fe, Pd and Pt respectively, and in certain clearly established contexts such as when labeling sensitizer bottles (see Chapter 5, *Preparation of Sensitizer and Processing Solutions*), these symbols will be used as a shorthand for the full names of ammonium ferric oxalate, ammonium tetrachloropalladate(II) and ammonium tetrachloroplatinate(II).

The Light that Marks

The word "photography" was coined by Sir John Herschel (1792–1871) in 1839.[2] It was a name given to any process in which an image is created by the action of light on photochemically sensitized surfaces. Light, although life-defining, seems immaterial, while chemicals, also life-defining, are certainly anything but immaterial. However, we know now, more than ever, that light and matter are profoundly interlaced. Even more perplexingly, light in the form of energetic particles (photons) is also light as the conveyor of this energy (electromagnetic waves). Together, this particle-wave duality is a part of the structure of matter while simultaneously determining the behavior of chemicals.[3] At an elemental level, matter is made of atoms, and all atoms have negatively charged, very light particles called electrons, "smeared out" in orbitals surrounding the positively charged, much heavier and smaller nuclei. As atoms interact with each other, these electronic states are redistributed. Chemistry is governed, in other words, by the redistribution of electrons between atoms. William Henry Fox Talbot (1800–1877) famously described photography as the "pencil of nature"[4] implying that nature, in effect, was drawing itself. Perhaps inadvertently, his description anticipated the most confounding discoveries about light and quantum physics by some 60 years. Even now, what remains fascinating is that photography arises from a dance in the deepest, most mysterious folds of phenomena: energy and matter.

Siderotype Processes

The underlying mechanisms for platinum (and a bit later, palladium) printing were well established by the 1880s. Sir John Herschel, in that early paper of 1839, even considered an image made "of platina and lime-water, by solar light, forming an insoluble compound, which might afterwards be blackened by a variety of agents."[5] He was attracted to platinum because its chemical nature suggested a much more permanent photographic image than silver. There is no record that Herschel went on to make prints with platinum, but while investigating color photography he found that an iron salt, ammonium ferric citrate, was highly sensitive to light. His discovery led to a broader investigation of similar iron salts, and generated a whole new class of photographic processes[6] such as cyanotype, chrysotype, argentotype, and kelainotype, which Herschel dubbed collectively as "siderotype," from the Greek word for iron, σιδηρος (sideros).[7] In 1872, the siderotype family came to include prints made with platinum, and by 1917, with palladium.

All siderotypes are made when light-sensitive iron salts, such as ferric oxalate, are exposed to ultraviolet (UV) light. When iron is chemically combined with other elements, it usually ends up in one or the other of two possible states of oxidation: either "ferrous" or "ferric" iron. The difference between each may be thought of in terms of positive charges. Ferrous iron has two of these, and ferric iron has three positive charges. Thus, ferrous iron, also called iron(II) [spoken as "iron-two"], is symbolized by Fe^{2+}. Similarly, ferric iron, also called iron(III) ["iron-three"] is symbolized by Fe^{3+}. Ferric oxalate is a compound made of iron(III) *cations* and oxalate *anions* ($C_2O_4^{2-}$). Hence the formula, $Fe_2(C_2O_4)_3$, in which the positive and negative charges are balanced for neutrality. Iron(III) and oxalate are chemically bound to each other via the attraction of appropriate positive and negative charges. When the dry ferric oxalate compound is exposed to UV light, it undergoes a fascinating "dovetailed" reduction-oxidation reaction. The energy received from photons in ultraviolet light transfers electrons (negative charges) from the oxalate to the iron, while releasing carbon dioxide gas. The reaction[8] may be described in its simplest terms like this:

$$2Fe^{3+} + C_2O_4^{2-} \xrightarrow{UV} 2Fe^{2+} + 2CO_2\uparrow$$

ferric cations + oxalate anion → ferrous cations + carbon dioxide gas

which in English, would read something like "Under UV light, ferric cations react with oxalate anion to give ferrous cations and carbon dioxide gas."[9]

This light-induced change occurs in solid ferr**ic** oxalate, to give solid ferr**ous** oxalate. However, while a *photochemical* change has occurred, it is only slight, barely visible (from pale yellow-green to pale yellow-brown), and not permanent. Oxygen in the air will eventually re-oxidize[10] the iron(II) back to the iron(III). In contrast to iron, metals such as platinum, palladium, and gold are very resistant to oxidation, don't rust or even tarnish, and are designated as "noble." To make a satisfactory *photographic* image, the iron(II) oxalate which is formed by the UV light must be reacted promptly with something else. Now, iron(II) is a reducing agent, because it wants to readily give up an electron and revert to its more stable, more positive, iron(III) state. Keep this in mind as we next consider how the situation can be used to form an image.

Development Pt/Pd

(Platinum/Palladium Printing Processes by Development)

The traditional platinum and palladium printing process, as described in literature since the nineteenth century, consists of two stages, or more accurately, reactions. In these recipes, the platinotype sensitizer consists of water-based (aqueous) solutions of ferric oxalate and potassium tetrachloroplatinate(II),[11] its formula being written as K_2PtCl_4. After coating and drying, the sensitized paper is exposed to UV light. As explained above, the first reaction occurs when iron(III) gains an electron and is reduced to iron(II). However, the second reaction cannot take place in the dry sensitizer, because the charged atoms and molecules

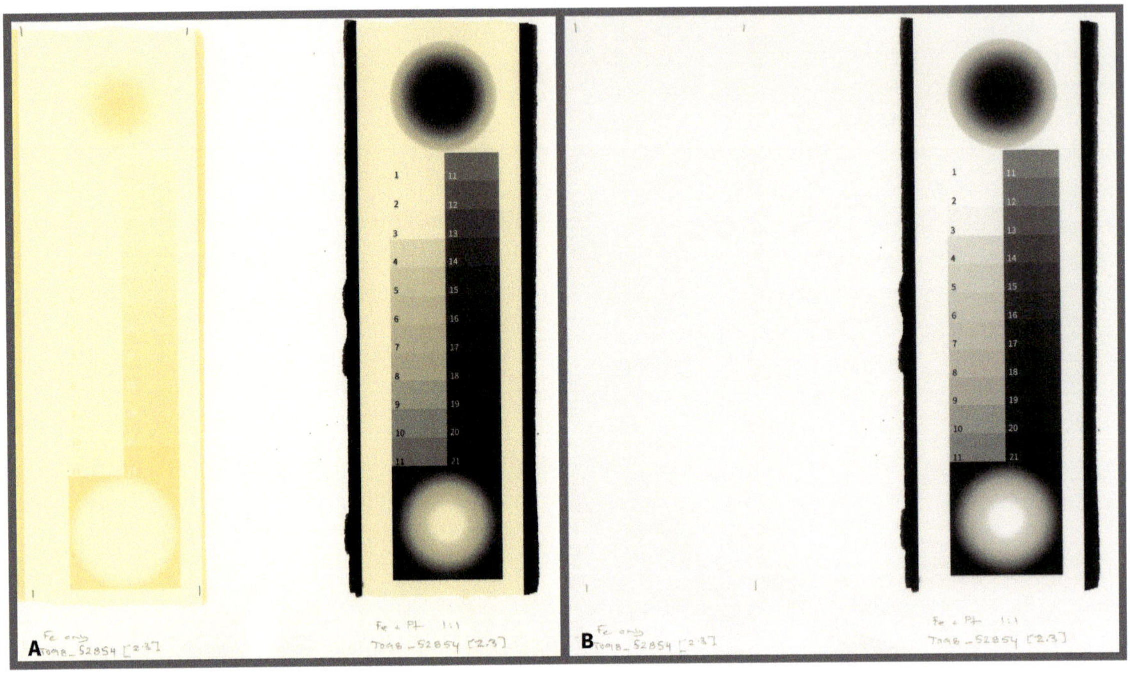

Figure 2.3 Image formation and processing. On the left of each frame coated with ammonium ferric oxalate only, and on the right with ammonium ferric oxalate and ammonium tetrachloroplatinate(II) 1:1.
A. image formed after exposure.
B. the processed and cleared print.

are, literally, immobile in the solid and cannot encounter one another. Only when dissolved in water are they freely mobilized to react. Here too, there is another problem: iron(II) oxalate produced by the exposure to UV light is insoluble in plain water. Thus, we end up needing a 'developer': an aqueous solution of the salt, potassium oxalate, which supplies oxalate anions that bind to, and dissolve the ferrous oxalate. The result, still as an iron(II) complex, can then react with the water-soluble tetrachloroplatinate(II) to precipitate tiny particles (more properly called "nanoparticles") of platinum metal.[12] These appear black, and are trapped in the paper fibers to manifest as a photographic image—the number of metal particles formed locally is in direct relation to the intensity of the UV light at that point.

Finally, all the excess unreacted sensitizer chemicals and soluble reaction products are removed from the paper by baths of dilute hydrochloric acid, thus "fixing" or, more accurately, "clearing" the platinum image. The closely related element, palladium (Pd), reacts in a similar way. Contemporary versions of this traditional recipe, generally referred to as the development (but also known as "developing-out" or "development-up") method of platinum/palladium printing, also utilize ammonium citrate as a developer, and citric acid plus various combinations of tetrasodium EDTA salt with sodium sulfite for clearing the print.[13] See Chapter 10 for more about this.

Printing-out Method

An alternative to the development Pt/Pd method was formulated by Giuseppe Pizzighelli in 1887.[14] He tested sodium and ammonium ferric oxalate with potassium tetrachloroplatinate(II), but preferred the sodium salt because of its greater

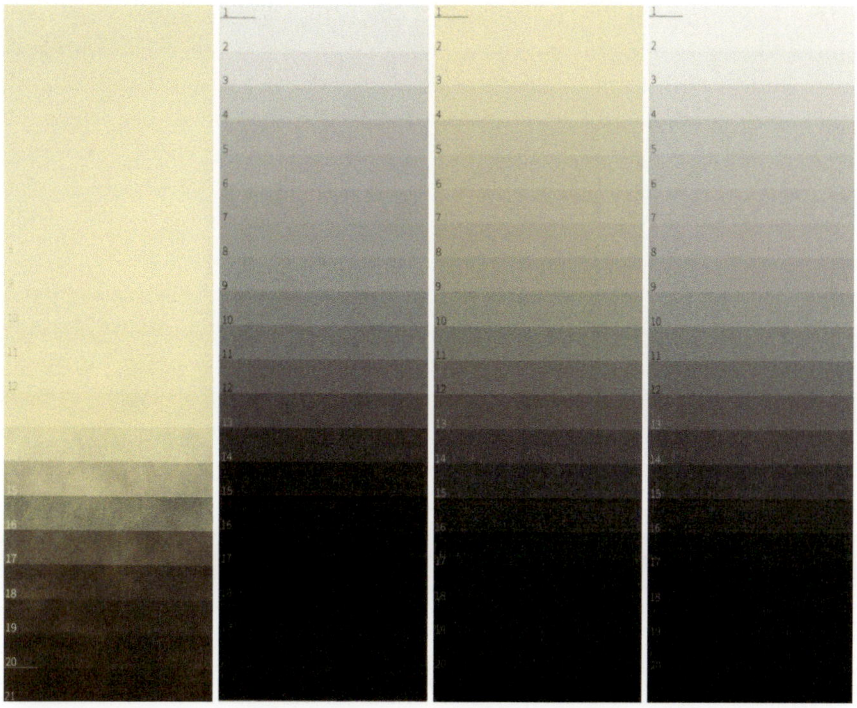

Development Platinum Print After Exposure | Development Platinum Print After Development and Clearing | Printing-out Platinum Print After Exposure | Printing-out Platinum Print After Clearing

Figure 2.4 The difference between Development Pt/Pd and Printing-out Pt/Pd, illustrated with four step tests, left to right: Development Pt/Pd after exposure; and after development and clearing; Printing-out Pt/Pd after exposure; and after clearing.

contrast. It gave a printing-out platinum image directly during exposure, needing no developer, and only washing in dilute hydrochloric acid to clear the image. Pizzighelli's "direct printing" or "water developed" platinotype formulation was similar in its concentrations of iron and platinum salts to the development Pt/Pd process.[15] Pizzighelli's recipe was taken up commercially by a number of manufacturers but, after an initial flush of enthusiasm, these papers did not remain on the market for very long. It is possible that they had limited storage life, and printers may have been frustrated by the need to control humidity in their use. This early recipe has been updated and redefined to give us a contemporary printing-out process. As presented in this book, the sensitizer is now composed of solutions of ammonium iron(III) oxalate, formula $(NH_4)_3Fe(C_2O_4)_3$, and ammonium tetrachloroplatinate(II), formula $(NH_4)_2PtCl_4$, or the analogous palladium compound. Note that potassium salts are not used because of their lower solubility and greater likelihood of crystallizing out of the solution.

The sensitized paper is not fully dried but allowed to acquire a controlled degree of humidity prior to exposure. At normal relative humidity, around 50%, cellulose paper already has about 6% of its weight made up by water. Additionally, these ammonium salts are hygroscopic, readily forming bonds with water molecules from the air. Under these conditions the platinum or palladium image is formed during the exposure, because ammonium iron(II) oxalate resulting from exposure to UV light *is* highly water soluble. So, if the paper is slightly humid, there is sufficient mobility for an immediate reaction with the platinum(II) salt. Platinum and/or palladium metal is precipitated and an image is formed as exposure progresses. There is little or no development of the image after exposure. The traditional clearing agent of dilute hydrochloric acid tends to dissolve palladium and weaken the cellulose structure of the paper (Chapter 10). A better clearing agent is disodium EDTA, short for ethylenediaminetetraacetate, which is effective in binding iron(III) strongly under mildly acid conditions. Subsequent baths in sodium sulfite and an alkaline tetrasodium EDTA, finished with a water wash, complete the processing, leaving an archivally permanent print. In sum, we have an elegant sequence of events with fortunate results. The ferric and platinum/palladium salts, combined in one solution, are fairly stable, and continue to be so after being coated and in a drier state on paper. They cohabit well. During exposure however, energy, in the form of UV light, shifts the stable iron complex to an unstable state, which in turn destabilizes the noble metal salts and causes the platinum/palladium to precipitate and be trapped by the cellulose fibers in the paper. Voilà. A photograph.

Paper

A platinum/palladium print is made of pure metal particles embedded in paper fibers. There is no binder layer. This is unlike a contemporary silver gelatin print where silver particles are suspended in a gelatin layer above a sheet of supporting paper. The preceding paragraphs describe the mechanics of how the printing-out process creates an image in a way that is different from the development Pt/Pd method. It is important next to delve into the optical effects of the printing-out approach. Keep in mind that the development Pt/Pd processes share most of the following qualities because essentially, they too do not have a binder layer.

Paper is largely taken for granted, even though it is ubiquitous and impacts our lives deeply. Paper has facilitated the dissemination of knowledge (where would Gutenberg's press be without paper?), the spread of civilizations, and served as the vehicle for much that is good as well as bad. It can even be argued that paper, like most far-reaching technologies, is not just a means to an end, but "a way of revealing" experiences and phenomena that would otherwise have remained unknown and unknowable.[16] Both as a technology and for the cultural role it has played, the chemistry and two-thousand-year history of paper is a fascinating one.[17]

The production of paper, whether crafted by hand in artisan molds, or mechanically in huge industrial mills, relies on hundreds of years of technical and scientific knowledge. It is a primary component of the platinum/palladium printing process because, along with providing a pleasing support for images, paper's chemical environment allows the metals to precipitate within its fibrous, organic structure. The exact makeup of papers selected for sensitizing has a profound effect on the way these noble metals actually precipitate and form an image. In other words, paper, for our purposes, is a highly reactive part of the process. We will examine this matter in greater detail later, but for now, it will help to consider how the platinum/palladium print's special and particular qualities are conditioned by paper.

Structurally, paper is mostly made of cellulose. The majority of cellulose is derived from softwoods, but archivally sound papers, such as the ones used in photographic processes, tend to be sourced from cotton, linen, hemp, bamboo, and other non-woody fibers. Cellulose fibers, once pulped, processed, drawn through a grid-like mesh and dried, form a network of bonded, interlocking fibers. Each strand of cellulose is translucent, and may be thought of as resembling tubular pasta, such as macaroni; but when crushed and overlaid into a network, these same fibers receive, scatter, and reflect light. Extremely fine platinum and palladium metal particles embedded within such a network absorb light, and are perceived by the naked eye

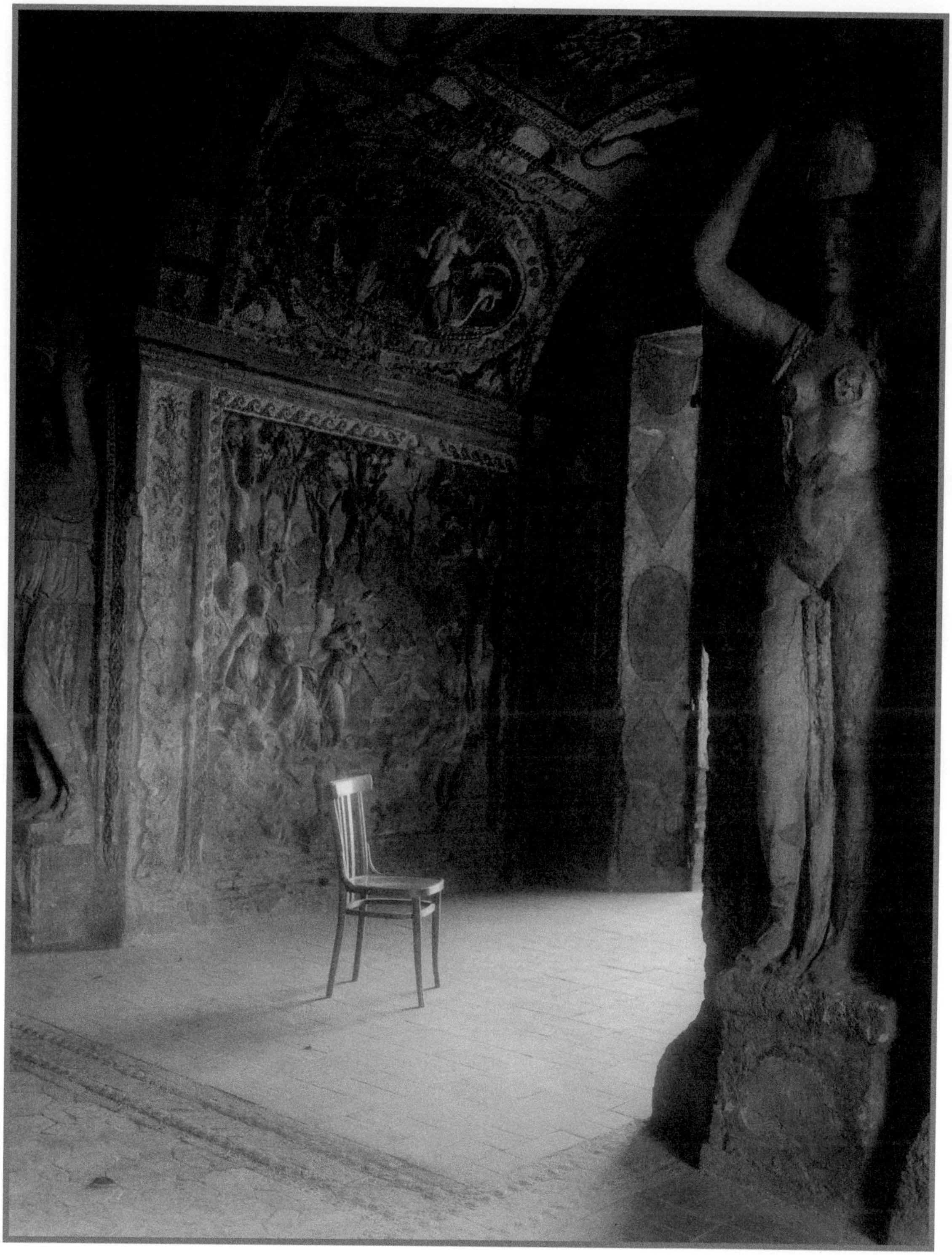

Figure 2.5 Regina DeLuise, *Diana's Grotto, Villa d'Este, Tivoli, Italy,* 1994. Development Pt/Pd print on Bienfang Layout Bond #360, from the original 8 x 10 inch negative, ca. 24 x 19.5 cm. Courtesy Regina DeLuise.

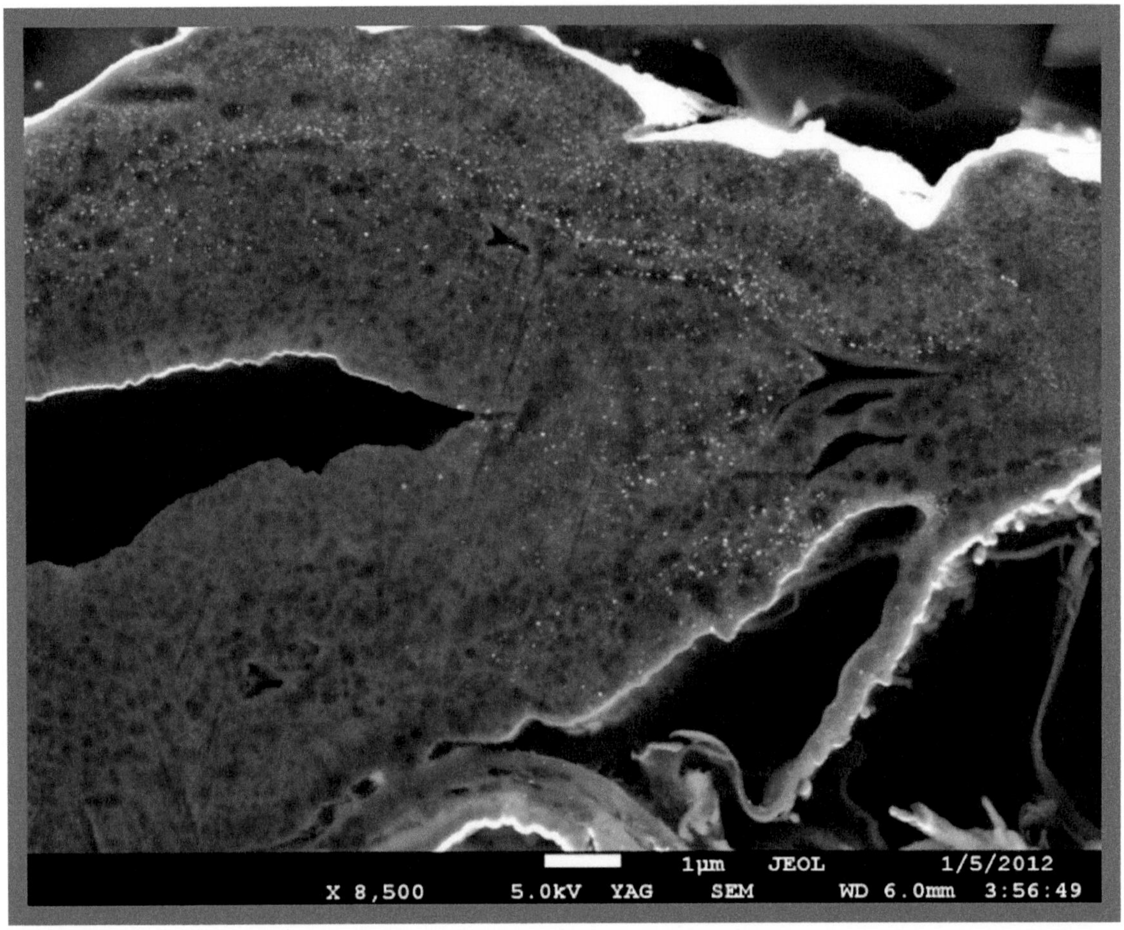

Figure 2.6 Cross-section or micro photo of a Pt/Pd print. Note that in this transmission EM image showing the structure, the platinum and palladium nanoparticles appear as white specks, not black, because they are electron-dense. Courtesy Natasha Erdman and Patrick Ravines.

as areas of less-than-white; in other words, tones ranging from very light to deep black. The particles, depending on their size and nature (platinum or palladium), also reflect particular wavelengths of light, i.e. color. The defining dynamic here is one of light entering the superficial layers of cellulose, scattering, being absorbed in places and scattered in others, and then bouncing back out. The net result of this behavior is a soft, slightly diffused array of tones that have been generated by nanoparticles of platinum and/or palladium. To the eye, this appears as a well-defined and dimensional image.

Looking Back, Walking Forward

The platinum/palladium print is noted for its archival robustness. Prints made over 100 years ago can look freshly made. Many of the most enduring and moving platinotypes from the Pictorialist movement, such as F. Holland Day's (1864–1933) self-portraits, are soft-edged and devoid of temporal indicators. The process of making the print itself, as we shall soon find, is a gentle one. The range of expressions, however, the variety of "voices"[18] that the process is capable of rendering, is vast. The optical softness of

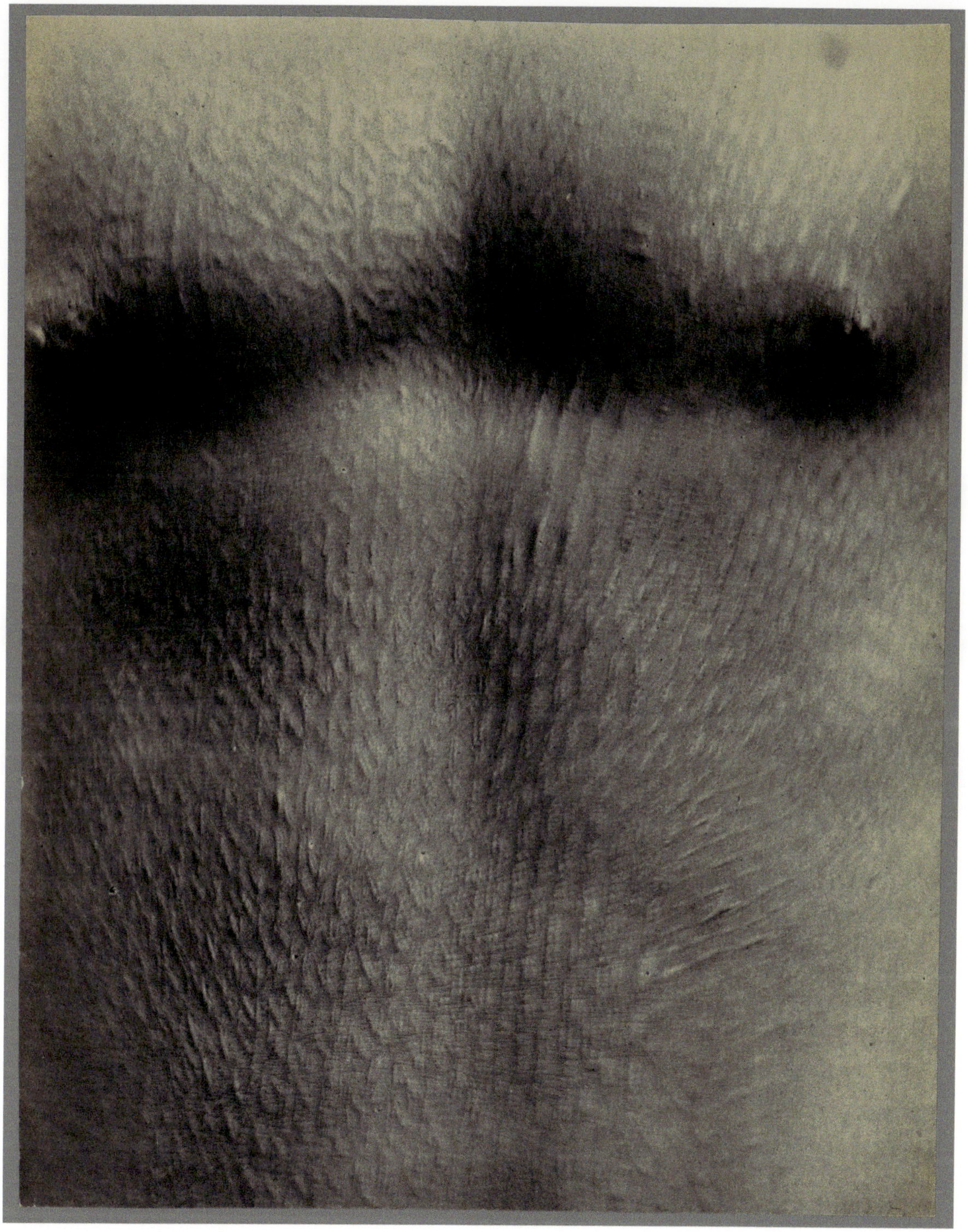

Figure 2.7 Maurice Vidal Portman, *The Chest and Stomach of a Man of the South Andaman Group of Tribes, Showing the Pattern of Tattooing, or Scarifying Common to This Group*, c. 1890–93. Platinum print, probably development Pt/Pd, 18.8 x 15 cm. Courtesy British Library, Photo 125/2 (44).

platinum/palladium has given voice to prints as diverse as Alfred Stieglitz's luminous-but-murky portraits of Georgia O'Keeffe (1887–1986) and Regina De Luise's (b1959) radiant Tuscan interiors; from Maurice Vidal Portman's (1860–1935) gut-wrenching, *The Chest and Stomach of a Man of the South Andaman Group of Tribes, Showing the Pattern of Tattooing, or Scarifying Common to This Group,*[19] to Irving Penn's (1917–2009) vastly magnified and equally disorienting portraits of cigarette butts. To think of this process just in terms of its chemical history would be limiting. To be attracted to making platinum/palladium prints only as a photographic craft or curiosity seems a disservice, a denial of revelation. And yet, neither history nor craft can be ignored here. Thus, perhaps it behooves us to look back while pushing into the realms of expression and beauty (in itself a matter of justice)[20] – walking backward into the future.

Notes

1 Walter Benjamin, ed. Marcus Bullock and Michael W. Jennings. "The Task of the Translator" in *Selected Writings*, Cambridge, Mass.: Harvard University Press, 1996, pp. 258–259

2 This paper was published only as an abstract: Sir John F.W. Herschel, "Note on the Art of Photography, or the Application of the Chemical Rays of Light to the Purposes of Pictorial Representation" [Abstract] in *Proceedings of the Royal Society*, 4 (37): 1839, 131–133. It was also briefly reported in *The Athenaeum*, 595, 23 March 1839, 223. For the complete text of the full paper which was submitted to the *Philosophical Transactions*, but withdrawn before publication, see Larry Schaaf. "Sir John Herschel's 1839 Royal Society Paper on Photography," in *History of Photography*, 3 (1): 1979, 47–60.

3 If you want to dig deeper into this, a good starting point may be Richard P. Feynman, *QED The Strange Theory of Light and Matter*, Princeton: Princeton University Press, 2014.

4 Larry Schaaf, *Out of the Shadows: Herschel, Talbot and the Invention of Photography*, New Haven, Conn.: Yale University Press, 1992

5 Schaaf, *Shadows*, p. 1.

6 See https://mikeware.co.uk/mikeware/main.html for details and instructions on many of these processes.

7 Herschel first proposed this word in a modest footnote to a report in *The Philosophical Magazine* of the meeting of the Royal Society: "Proceedings of Learned Societies: Royal Society, June 16 1842," in *The London, Edinburgh and Dublin Philosophical Magazine and Journal of Science*, series 3 (21): 1842, 227.

8 The actual equation for the photochemical reaction in the solid is usually written like this:

$$Fe_2(C_2O_4)_3 \xrightarrow{UV} 2FeC_2O_4 + 2CO_2\uparrow$$

ferric oxalate → ferrous oxalate + carbon dioxide gas

9 In chemical equations, the large + signs simply mean "...added to..." or "...mixed with..." (they have nothing to do with electric charges) and the horizontal arrow → means "...react chemically to give the products...". Written above → is the source of energy promoting this reaction: UV light. The upwards arrow ↑ on carbon dioxide means "...is evolved as a gas."

10 **Oxidation** is making an atom or molecule more positive (or less negative) in the electrical sense. When iron forms compounds, it is oxidized. **Reduction** is the converse: making an atom or molecule less positive (or more negative). e.g. oxygen gas (O_2) is reduced to form oxides, which contain the oxide anion O^{2-}. Both processes involve the transfer of electrons, which are the fundamental particles of negative electricity and make up the outer spheres of all atoms. Positively-charged atoms or molecules are called **cations**, Negatively-charged atoms or molecules are called **anions.** Oxidation is the removal of electrons—formation of cations. Reduction is the addition of electrons—formation of anions.

11 Nineteenth-century literature referred to "potassium chloroplatinite," which we now prefer more accurately to call potassium tetrachloroplatinate(II).

12 The sensitizer contains platinum in the oxidation state +2. The iron(II) formed as described above by the exposure to UV light is capable of reducing the platinum(II) to its solid metallic state, which is oxidation state zero (the 0 is not usually written):

$$2Fe^{2+} + Pt^{2+} \rightarrow 2Fe^{3+} + Pt\downarrow$$

iron(II) cations + platinum(II) → iron(III) cations + platinum metal

The downwards arrow ↓ means "precipitated as a solid." Note that it takes two iron(II) cations to make one atom of platinum, because two electrons are needed to balance the charges on each side of the equation. These ratios are important in formulating the sensitizer solution correctly. This is a simplified version of the chemistry–the actual molecules involved are complex. See Mike Ware, *Platinomicon*, 2016, Appendix I.

13 Dick Arentz, *Platinum & Palladium Printing*, Focal, 2000; Jill Enfield, *Jill Enfield's Guide to Photographic Alternative Processes*, Routledge, 2nd edition (February 3, 2020); Christopher James, *The Book of Alternative Photographic Processes*, Cengage, 3rd edition 2016; Ian Leake, *The Platinum Printing Workshop*, 3rd edition, 2017.

14 Capt. Giuseppe Pizzighelli, "The Platinum Print," in *Anthony's Photographic Bulletin*, 1888, pp. 410–421. See also: Capt. Giuseppe Pizzighelli, *Photographische Correspondenz*, 24, October 1887; 25, January 1888 and *The Photographer's Book*, eds. Holmes and Griswold, 1888.

15 Ware, *Platinomicon*, 2016, pp. 119–120.

16 Martin Heidegger, *The Question Concerning Technology and Other Essays*, translated by William Lovitt, Garland, 1977.

17 J.C. Roberts, "The Chemistry of Paper," in *The Royal Society of Chemistry*, 1996, and Mark Kurlansky, *Paper: Paging Through History*, Norton, 2016.

18 Paul Caponigro, *The Voice of the Print*, Hadley: Allethaire Press, 1990.

19 John Falconer, "Maurice Vidal Portman and the Platinotype in India," in *Platinum and Palladium Photographs: Technical History, Connoisseurship and Preservation*, Washington DC: American Institute for Conservation, 2005, pp. 318–331.

20 Elaine Scarry, *On Beauty and Being Just*, Princeton, N.J.: Princeton University Press, 1999.

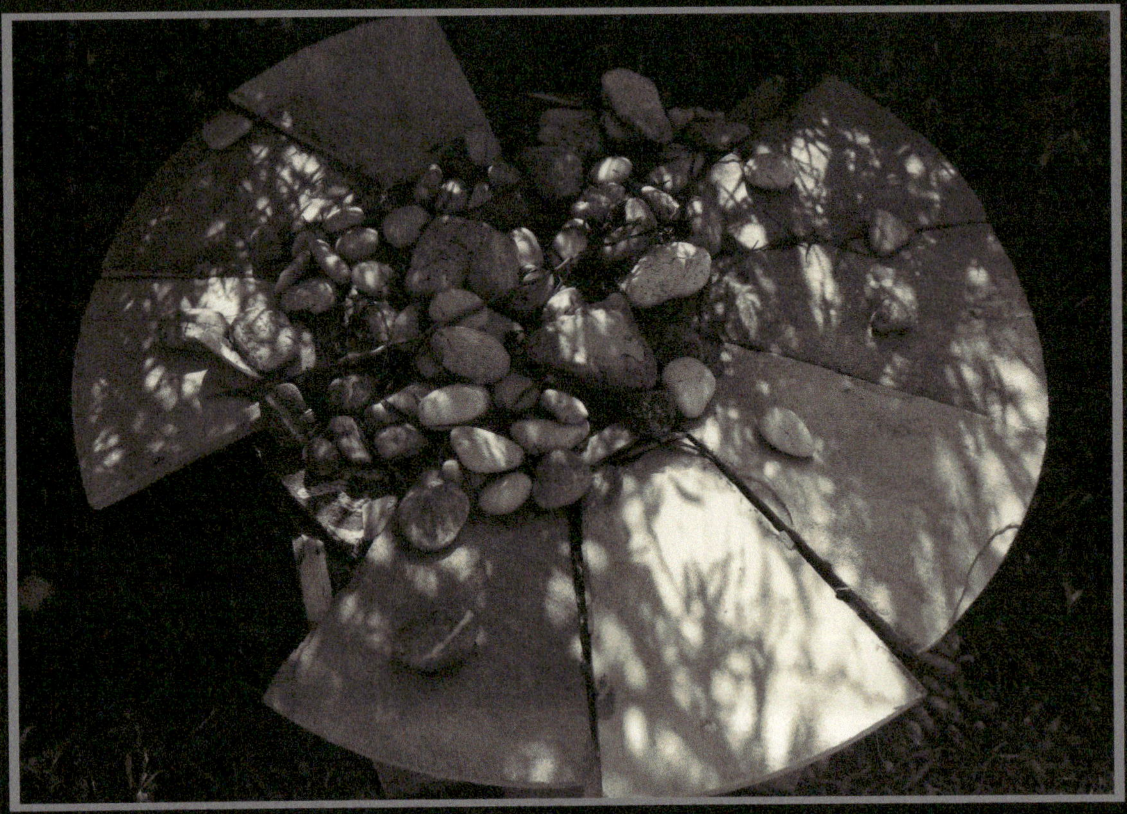

Figure 3.0 Mike Ware, *Stone Table, Tilos*, 2011. Platinum-palladium print on Buxton from enlarged digital negative. ca. 30 x 40 cm. Courtesy Pradip Malde. Fe:Pt:Pd = 2:1:1, 20 min. hydration in 100% RH / 19°C.

Chapter 3

Safety, Workspaces, and Procedures

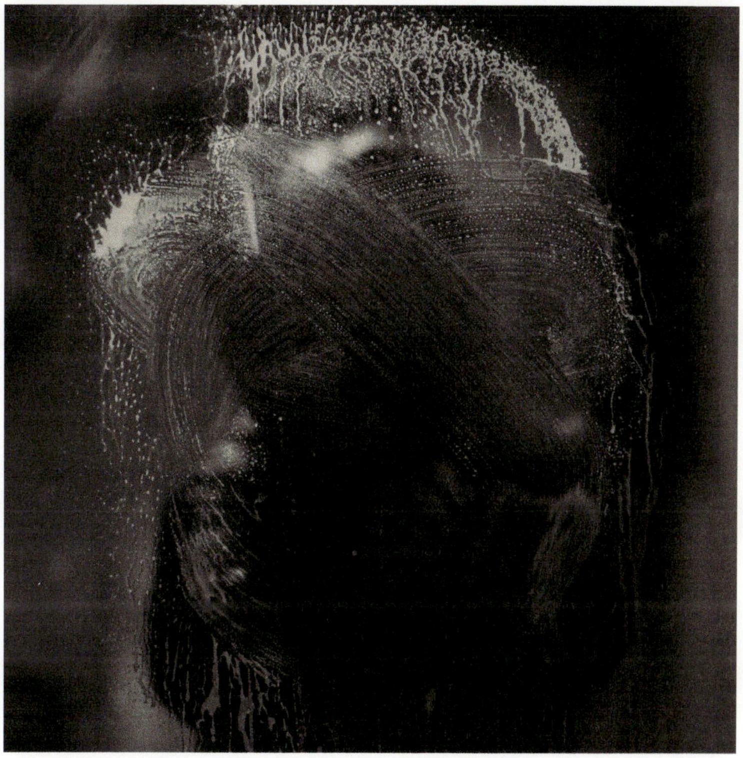

Figure 3.1 Cathy Cone, *Rewinding Forward, "Soap Smear"*. 2019. Platinum-palladium print on Revere Platinum paper, from PiezoDN negative of full frame digital original, 23 x 19 cm. Courtesy Cathy Cone.
Fe3:Pt:Pd = 2:1:1, pre-coating humidity 50% for 2 hours at 68°F, rod coated, hydration 50% RH at 68°F, 60 minutes. Exposure using NuArc FT32. Post-exposure hydration 10 minutes 50% RH at 68°F. Cleared using the M-W clearing method.

Most discussions and thinking about photography attend to the photograph as an image. We talk predominantly about the what, where, and when of a photographic image, the subject/object relationship, photography's documentary and expressive function, the camera, its lens and sensor, and so on. There is good reason for this. A photograph and its means of production and viewing are deeply embedded in our lives. Photography, and photographers, have a social function and, by extension, one could argue that anyone wishing to live a photographic-visual life[1] is tasked with a social imperative. Along with these important approaches to thinking about the medium, we are always confronted, I would argue, with a more nuanced and less apparent stimulus. The photograph is always perceived and experienced as a physical object. Even when it seems virtual and on the digital screen of a smartphone, it is a glass-bound object. As we consider this, photography goes beyond the telling of how things are, and also becomes about the telling of how else things can be. And here lies its special and wonderful relationship with science, with physics and chemistry. Photography is the result of a chemical response to energy in the form of light. The functions and meanings of photography, in other words, are first determined by light and chemicals. The physical, chemical existence of this "pencil (drawing) from nature"[2] affects our emotional perception of image, and consequently, our aesthetic responses to it. On looking at a photograph, we are responding to light bouncing off chemicals which have also responded to light—what an exquisite cycle! Amidst this swirl, and the wonder of seeing a window into another time and place, it is easy to take for granted the chemical aspects of the medium, to have them seem innocuous and benign. This chapter hauls us up, and brings attention to photography's physicality, to the photographic process, and for now at least, all to the service of healthy and safe working practices.

Health and Safety

As we embark on a journey that is about crafting an image, almost from raw materials, we would do well to establish safe and standardized methods. The chemicals and the workspaces for making platinum/palladium prints can be hazardous. If you approach all photographic procedures with this cautionary note in mind, then you have taken the most basic and important step in safe and healthy practices. All chemicals can be harmful if used improperly, even the most benevolent and life-giving: water. Badly designed and ventilated work areas can also present health hazards. It is important, then, that we spend a bit of time becoming aware of potential hazards and proper procedures for handling chemicals and the design of workspaces. While this chapter provides guidance, it is the reader's responsibility, or that of anyone who chooses to work, or who causes others to work with hazardous materials, to be informed in advance about the dangers, and to ensure that precautions are taken to safeguard themselves, other people, and the environment. The authors and publishers of this book hereby deny liability for any consequent injury, sickness, damage, or loss resulting from the use of the information contained herein.

This chapter considers those aspects of the workplace, equipment, and chemicals that present potential hazards. Best practices for the measurement and preparation of chemical solutions are suggested, and we conclude with some ideas about proper note-taking and record keeping. While the following is presented within the context of platinum/palladium printing, much of it may be applied to any similar analogue or "wet" photographic process.

Materials Safety Data Sheets

Regulations originating in Canada and the USA in the 1980s, and now adopted as a global practice, require chemical manufacturers to provide technical information with each product. The data sheet lists the product's name, formula, physical data, hazardous ingredients and decomposition products, effects of exposure, protective measures, emergency and first aid procedures, and steps to take if the material is released or spilled, among other things. These sheets are called Materials Safety Data Sheets (MSDS or SDS). You should obtain and keep an MSDS for each chemical in your workspace, ideally in a binder that also contains an up-to-date inventory of all substances that you work with. This is a required procedure in all teaching spaces. Place the clearly labeled binder in an accessible location. Full information regarding hazards may be obtained from suppliers or by consulting any reference manual on chemical safety. They may also be consulted on-line: lists of websites carrying the MSDS may be found at:

- http://www.ilpi.com/msds/
- http://msds.com
- http://www.msdsonline.com

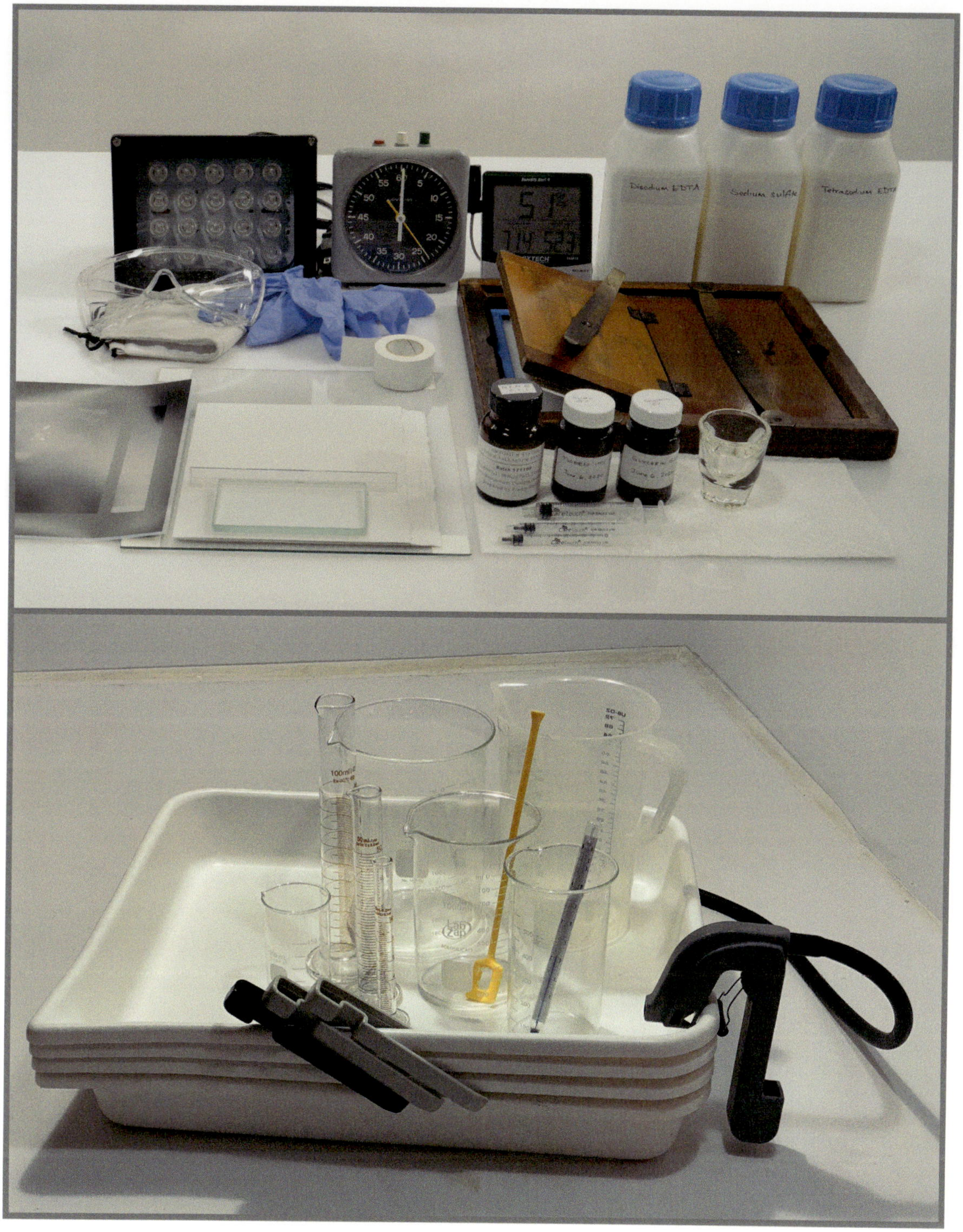

Figure 3.2 Proper materials and equipment needed for this process help establish safe and standardized methods.

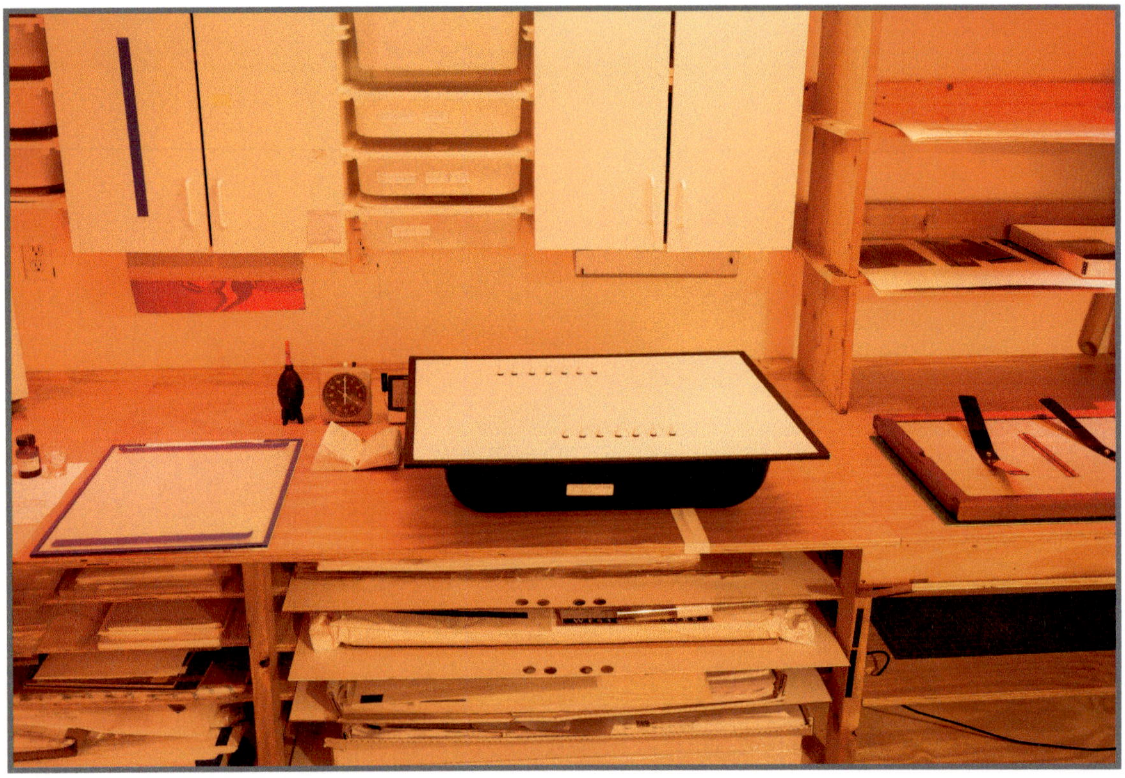

Figure 3.3 A well-organized and arranged work space, including coating area, hydration chamber, contact printing frame, below which are shelves for paper storage and on the far right, a UV exposure unit. The entire area is well-lit with a mixture of 'white' and RubyLith®-covered light sources.

Workspace

The design of work areas should include a careful consideration of airflow, illumination, and cleanliness, along with storage space, work surfaces, water supply, disposal, and of course, space for an ultraviolet exposure source. Workplaces for platinum/palladium printing should be clean, and as with most traditional darkroom spaces, generally separated into "dry" and "wet" areas. The dry area is used for electrical equipment such as an ultraviolet exposing unit, negative and paper storage, notebooks, coating the paper, and inspecting prints during exposure if working with a contact-printing frame. The wet area, normally a darkroom sink with high side walls and a rear splash panel, is used for the preparation of chemical solutions and print processing. There are a couple of hybrid zones, and while they occupy the dry area, they are best kept segregated from the aforementioned functions. The first is a surface for a print hydration chamber, the use of which is optional, and described in much greater detail in Chapters 4 and 5. The second is essential: a cupboard or high shelf for the safe storage of dry chemicals.

Air Quality

Airflow and ventilation, while being two of the most important factors for a healthy work area, are usually the most difficult to customize to our needs. Spaces are rarely installed with infrastructure designed specifically for photographic darkrooms. This is just as true for domestic darkrooms in basements, bathrooms, and kitchens as it is for professional and institutional facilities in commercial buildings and refurbished teaching spaces. While printing-out platinum/palladium presents far fewer air-quality hazards than most other versions of the process, and is certainly safer than silver-gelatin printing, there must be a mechanism for removing and replacing room air in such a way that vapors and dust from any exposed chemicals pass away from the face and out towards an exhaust duct. Proper ventilation, for instance, would move air in from the floor and up and out across a sink towards an exhaust vent placed just above the sink height. This exhaust should also be close to the chemical storage shelf or area. If the location of input and exhaust vents cannot be modified, consider using deflectors and small fans to direct air movement. Along with airflow, we also need to think about temperature, and if possible, humidity. Fortunately, the human comfort zone also presents an optimal environment for platinum/palladium printing. An ideal workspace will be at around 70°F / 21°C, with a relative humidity (RH) of between 40 and 70%. Finally, you will find that a safe working space and predictable results are best obtained by a consistent environment: avoid sudden and large fluctuations of airflow, temperature, and relative humidity.

Lighting Conditions

Strictly speaking, we will not be working in a darkroom, since the process does not require total blackout, or even red/orange safelight, conditions that characterize traditional photographic practices. Platinum/palladium sensitizers, like all siderotype formulae, require rather large doses of ultraviolet exposure to render an image. But small doses, such as stray ultraviolet coming in from a window, or even that emanating from household halogen or fluorescent tube lights, can fog coated papers. It is best, then, to work in a "dim room," illuminated with 40- to 60-watt tungsten or 5-watt LED bulbs, no closer than 4´ / 1.3 m to the work surface, but bright enough to let you work with ease. Light switches with dimmers are ideal. More specifically, the overall illumination should be 50–100 lux (EV of about 4.5 to 5 incident reading, or reflected off a grey card at the work surface, with the light meter set to 100 ISO).

Fogging Test

An ultraviolet meter is useful for ensuring the absence of stray radiation, but if such a device is not available, you can carry out a fogging test with a small strip of sensitized paper, as follows.

1. Working in dimmer-than-normal conditions (the dimmer the better), coat a sheet of workable paper (e.g. Revere Platinum or Arches Platine), and prepare it for exposure as described in Chapter 8.
2. Lightly expose this sheet to your UV source, giving it ca. 1% of base exposure to render a perceptible tone, but do not process it yet. Partially exposed paper is preferable to totally unexposed paper for this test because it is more sensitive to change, the fogging exposure having surmounted the "toe" of the characteristic curve.
3. Cut this sheet in several pieces, placing one portion into a drawer or light-tight box, and the others under your normal ambient light. Arrange an inert (not metal), opaque disc in the center of each piece.
4. Wait for 10 minutes, and then process one piece of the test, along with the first piece that was placed in total darkness as normal. Do the same with succeeding sheets at 10-minute intervals up to at least 30 minutes, but ideally an hour.
5. Compare the tests. If the lighting conditions are optimal, there should be no indication of the disc's imprint, and all sheets of paper should look identically clear. If there is an image, that portion of the test will give a measure of the degree of "unsafety" of the lighting within a known time.

Water

Most of the water needed for platinum/palladium printing may be supplied from the household faucet (tap). Water should be clean, as close to neutral pH 7 as possible, and not heavily chlorinated. If using well water, or if filtration is necessary, about

30 micron cartridges will be adequate. Rinse and wash water temperatures are best kept around 68°F / 20°C, ± 5%. Bottled distilled or purified water will be needed for the preparation of some solutions. Procedures for water conservation are described in Chapters 8 and 9. All of the processing solutions, along with rinse and wash water, may be disposed of via normal wastewater drainage systems. However, it is the reader's responsibility to check and comply with any local regulations.

Dry Storage

Notebooks, negatives, and paper should be stored in conditions that fall within our normal comfort zone: around 70°F / 21°C and around 50% relative humidity (RH). Negatives are best organized in archival containers and sleeves. Papers for sensitizing and printing should be well wrapped in archival or acid-neutral paper and stacked flat. It is best not to store these materials in high airflow areas, where temperatures and humidity may be much more varied than in static pressure zones. Thus, it is best to use drawers or enclosed shelves here. See the following section about airflow and ventilation. Be sure to label all of these systematically.

Chemical Storage

The storage of chemicals presents a significant hazard and requires careful planning and attention. Label all containers with the full chemical name of its contents, hazard warnings, and other information such as a supplier batch number, date prepared, and concentration if the chemical(s) is in solution. Dry chemicals should be kept at least slightly separate from solutions. Both should be stored in a secure cupboard or high shelf to prevent easy access by children, animals, and those who are compelled to ignore safety warnings. Ideally, locate the chemical storage area close to the exhaust or air outflow, and farthest from the ventilation inflow.

Work Surfaces

Coating paper and inspecting prints during exposure require a large, flat, and clean surface that has an even and safe illumination (see section on Light). The surface's height should be such that it is comfortable to work at either standing or sitting. The area should be large enough to coat the maximum size of paper you plan to work with. This area can also double up as a place where contact printing frames, if used, are loaded with negatives and sensitized paper, and where exposures that are in progress may be inspected. If there is enough space however, it is far more convenient to have some extra work space next to the coating area for loading, inspecting and unloading activities. The ideal dimensions for this work surface are around 30″ x 72″ / 75 cm x 180 cm, at a height between 30–36″ / 75–90 cm.

Print hydration chambers are sealed trays containing a salt slurry or water, and should be arranged on a clean, dry surface at an easily accessible height. One or more of these may be installed. Each should be at least 4″ / 10 cm larger than the dimensions of the paper being used. If constructed according to designs provided in this book, they are best located on a separate surface from the coating and inspection area, but at a similar working height. A typical hydration unit may have a footprint of about 14″ x 16″ / 35 cm x 40 cm (for 10″ x 12″ / 25 cm x 30 cm sheets of paper with an image of 8″ x 10″ / 20 cm x 25 cm). Units may be placed in sink areas if there is enough room, but with the caveat that coated paper can be easily damaged by stray drops of liquid.

The wet area is for preparing chemical solutions, processing and washing prints, and is usually adjacent to a print-drying area. A large sink is ideal, but a table covered in waterproof material can also suffice. As with the print hydration chamber, processing trays should be slightly larger than the dimensions of the paper being used. This is especially important with vellum, or transparentized, papers since they expand significantly when immersed in solutions. Thus, a wet area for processing 10″ x 12″ / 25 cm x 30 cm sheets of paper will need to accommodate at least three 11″ x 14″ / 28 cm x 36 cm trays for processing, one tray for rinsing, and one for final washing. In situations where space is very tight, it is possible to use even fewer trays if you are willing to pour solutions in and out between storage containers and the trays. A standard darkroom sink of 30″ x 72″ / 0.75 m x 2 m works well.

Housekeeping

An organized work habit manifests as a clean space, which in turn makes for a healthy and safe environment. Housekeeping includes the management of all your materials, notes, negatives, and prints. It also means that surfaces and floors are kept clean, that vents and air filters are well-maintained, inkjet-printers and ultraviolet units are covered and dust-free when not in use, and so on. Water and power infrastructure should be regularly inspected. Spent chemicals and garbage must be regularly disposed of in compliance with local regulations. Combined, these habits foster a process and workflow consistent enough to permit predictable results, and such that when you have problems, or choose to experiment, solutions may be arrived at with efficiency. The salient goal here is to create an efficient space in which science and craft coexist to support your creative and expressive endeavors.

Procedures

It must be apparent by now that there is a continuum between considerations about workspaces and the way we work, between safe practices and well-crafted prints. We now need to think about practical procedures.

Ultraviolet Radiation

The platinum/palladium process requires exposure to ultraviolet light. Extensive and unscrupulous exposure to this invisible part of the electromagnetic spectrum can damage human skin and vision, whether its source is the sun, a tanning bed, or a professional-grade metal-halide exposure unit. Some types of ultraviolet are much more harmful than others. Sources like table-top tanning devices, quartz-halogen tungsten lamps, and the sun can present hazards and also emit significant infrared radiation, which has the undesirable effect of heating the sensitized paper as the exposure proceeds. Changes in temperature lead to changes in humidity, which consequently affects image color. Artificial ultraviolet sources are preferable and come in a variety of designs and configurations—see Chapter 4. All units emit heat to some extent, and the workspace's air flow must be designed to facilitate the efficient cooling of exposure units. The most active, or "actinic," part of the spectrum for our purposes is what is called the near ultraviolet. It is divided into three regions, referred to as UVA, UVB, and UVC. There is no advantage, and much additional risk, in working with the more dangerous 'short-wave' ultraviolet lamps emitting UVC and UVB. Both of these, with wavelengths ranging from 100 to 315 nanometers (nm), rapidly damage living tissue. It is best to work with long-wave UVA lamps, which chiefly emit a wavelength range from 315 to 400 nm, which includes the 370 to 350 nm wavelengths to which iron-based siderotype processes are most responsive. Despite being the safest portion of the near ultraviolet region, we still have to be cautious. UVA causes our skin to tan but can also lead to melanoma. Thus, all ultraviolet units should be shielded so as to minimize stray exposure. Protect skin from stray exposure by wearing appropriate clothing. Wear UV-blocking eyeglasses or goggles if exposed to ultraviolet for lengthy periods. Do not look directly into a light source and be just as cautious of this when light reflects off glass surfaces such as from a contact printing frame, or reflective work surfaces. Wear UV-blocking eyeglasses or goggles if you are working in unshielded conditions.

Chemical Exposure

Some of the solids and liquids that we work with can present hazards, against which there is one simple general guiding principal: protect yourself and others. Just as I urge you to minimize exposing your body to ultraviolet radiation, I present a general guideline regarding chemicals: do not handle them without a barrier in the form of a lab coat or apron, gloves, face mask, or safety goggles as the situation demands. All the chemicals listed here are non-volatile, so respirators are not needed. At most, a simple face mask protects against the inhalation of dust. Your work area should have a supply of all of these protective items, along with access to fresh, clean water in case of an emergency. Be aware (using MSDS) of the proper cleanup of accidental spills and maintain an easily accessible spill kit. Just as with all work surfaces and sinks, aprons, clothing, and towels should be regularly cleaned. Trash cans must be emptied and their contents disposed of in compliance with municipal code. Together, these practices will help render a consistently safe environment.

Glassware, or labware, is used to prepare or store chemical compounds and solutions. Well-marked and accurate measuring containers, and storage bottles with handles and tight seals, contribute to safe working methods and environments. Use trays designed for photographic processing, each with a pouring lip and a convenient depth for handling prints and emptying solutions. All items should be promptly and thoroughly washed after use and stored so that they drain and dry safely. Strong detergent, bottle brushes, and hot water should be used to remove films of grease from glassware and storage bottles. Particularly resistant deposits of solid may need a scouring powder. The criterion for a clean glass surface is that water can be seen to wet it and drain off in a uniform film without beading into droplets. Most chemicals used in this work are highly water-soluble, so apparatus should only need thorough rinsing to clean it—rinsing three or four times should suffice to dilute any residual chemical to an insignificant amount.

Measurement and Precision

Installing and using equipment is only part of an overall safe practice. Along with sustaining a healthy work environment, you need to be aware of standard procedures for the handling, measurement, and preparation of chemicals. As indicated above, I assume you will always use appropriate protective clothing and equipment, that your work surfaces and labware are clean, that you are in an adequately lit and well-ventilated space, and that you are aware of any hazards as stated in the MSDS. From this baseline of good practice, we now consider how to measure and prepare chemicals precisely and consistently. Remember, always treat chemicals with care and mindfulness. Use, label, and store them under the assumption that others, such as children or pets, could gain access. Protect yourself and others from direct contact with chemicals and wash your hands thoroughly after handling them. Chapter 5 provides detailed formulae for all the necessary chemical solutions needed to make platinum/palladium prints. Each formula includes specific precautions and handling guidelines. General guidelines for the preparation of chemical solutions include attention to the measurement of volumes and weights, mixing solutions, preparing proper dilutions or concentrations, filtering, and storage. The printing-out platinum/palladium process described in this book is much safer, and more forgiving of minor variations and inaccuracies in preparation and measurement, than traditional approaches. However, if you aspire to achieve the finest results, with a high degree of reproducibility and the maximum economy, then you should be prepared to make careful measurements of the relevant quantities, and maintain a record of your printing procedures for future use. The following instructions are written on that assumption.

Success comes from the use of our instruments with accuracy and precision. While precision is the consequence of similar successive results, accuracy is about rendering exactly what we expect. Here, we hope to achieve both. The hallmark of intelligent chemical practice is to work to a precision appropriate for the task—neither too low, nor too high. Don't use a sledgehammer or a scalpel to crack a walnut, but rather, a nutcracker. To assist you in deciding what is appropriately *precise*, the following notes attempt to explain some of the conventions of quantitative chemistry,[3] or that branch of chemistry devoted to calculating the amounts of chemicals in any given sample.

- The abbreviation "ca." will be frequently encountered in this text. Sometimes just written c., it is short for circa, the Latin for "about." It does not have a generally accepted definition in scientific use, but may be taken here to imply that a precision in the order of ± 10% is appropriate. E.g., a weight specified as ca. 30 g means any amount between 27 and 33 g should do; a temperature specified as ca. 50°C implies that any temperature between 45°C and 55°C would be acceptable.
- In everyday scientific practice, the manner in which a numerical quantity is written also indicates the precision to be observed in measuring it. In other words, the precision should match the last significant figure, which has usually been rounded up or down. E.g.: a weight specified as 30.0 g implies weighing to a precision of 0.1 g, i.e. the weight should lie between 29.95 g and 30.05 g. A weight written as 30 g implies a precision of 1 g, i.e. any weight between 29.5 and 30.5 g is acceptable.
- When scaling quantities of substances up or down, keep in mind that the precision of measurement is also scaled. In other words, it is determined by the number of significant figures, not the number of decimal places. For example, suppose a volume of 1.25 ml (measured with a precision of 0.01 ml) is to be scaled up by a factor of 10: this results in a volume of 12.5 ml (measured with a precision of 0.1 ml). In both cases, the precision of measurement is 1 part in 125 parts, or 0.8% (100/125%).
- Volumes of liquids are expressed throughout this book in metric units, as liter[4] and milliliter.[5] Larger volumes are sometimes specified in liters, abbreviated to "l," and smaller volumes in milliliters abbreviated to "ml."

Labels and Notes

Labels, notes and records should be maintained in a systematic way. This habit will help maintain your equipment and materials, journal your activities while preparing solutions and printing, and build a catalogue of your negatives and prints. It also allows repeatable and consistent results. Detailed notes are best prepared in advance of mixing chemicals. They should include a description of every step taken for preparing a formula—this is especially important when making a sensitizer or one of its component solutions for the first time. These "studio notes" as I prefer to call them, should be dated, and organized in a folder, along with the MSDS. Records describing the preparation of negatives and platinum/palladium prints are essential for editioning as well as problem-solving purposes. These records should correspond to unique identifiers marking every print. See Chapters 5 and 9. All chemicals and solutions must be securely contained, and clearly labeled with the chemical name, chemical formula, supplier, batch number or date prepared, and concentration and amount. Similarly, it is advisable to label packets of paper, negative sleeves and folders, as well as any notebooks.

This seems pedantic, and obvious, but in our haste to make prints, these basic steps are often left for another time, and then forgotten.

Platinum Hazard

For legal reasons, and because information changes over time and across regions, you must refer to MSDS and local regulations with regard to specific hazard warnings and disposal guidelines. That said, all of these chemicals are harmful if swallowed, inhaled, or if they come into direct contact with skin or eyes. Platinum deserves a special mention here. Contact with ammonium tetrachloroplatinate(II), and other chloro-complexes of platinum, is known to cause symptoms of asthma, urticaria, and dermatitis; some allergic individuals may become particularly sensitized to these biologically active chemicals. The symptoms disappear on removing the cause: if you develop this allergy, then platinum printing is not for you—but you could still use palladium. Appropriately, "platinum allergy" was first observed in 1911 as an occupational disease of photographic factory workers handling platinotype paper. Never touch the surface of platinum-sensitized paper or immerse ungloved fingers in the processing solutions. It is better not to store large amounts of dried sensitized paper. Platinum metal itself is not implicated in this—so take comfort that there is no risk in handling fully-processed platinotypes!

Notes

1 Dorothea Lange, audio of her voice, in *Dorothea Lange: A Visual Life*, by Meg Partridge, Pacific Pictures, 1995

2 William Henry Fox Talbot

3 See Mike Ware, *The Chrysotype Manual: the science and practice of photographic printing in gold*, Brighton: ffotofilm publishing, 2006, pp. 67 passim

4 The International Bureau of Weights and Measures spells the unit as "litre," and it is used as such by almost all English-speaking countries. The spelling "liter" is predominantly used in American English

5 A milliliter is almost the same as a cubic centimeter, for which the approved symbol is cm^3, but often misleadingly written as cc.

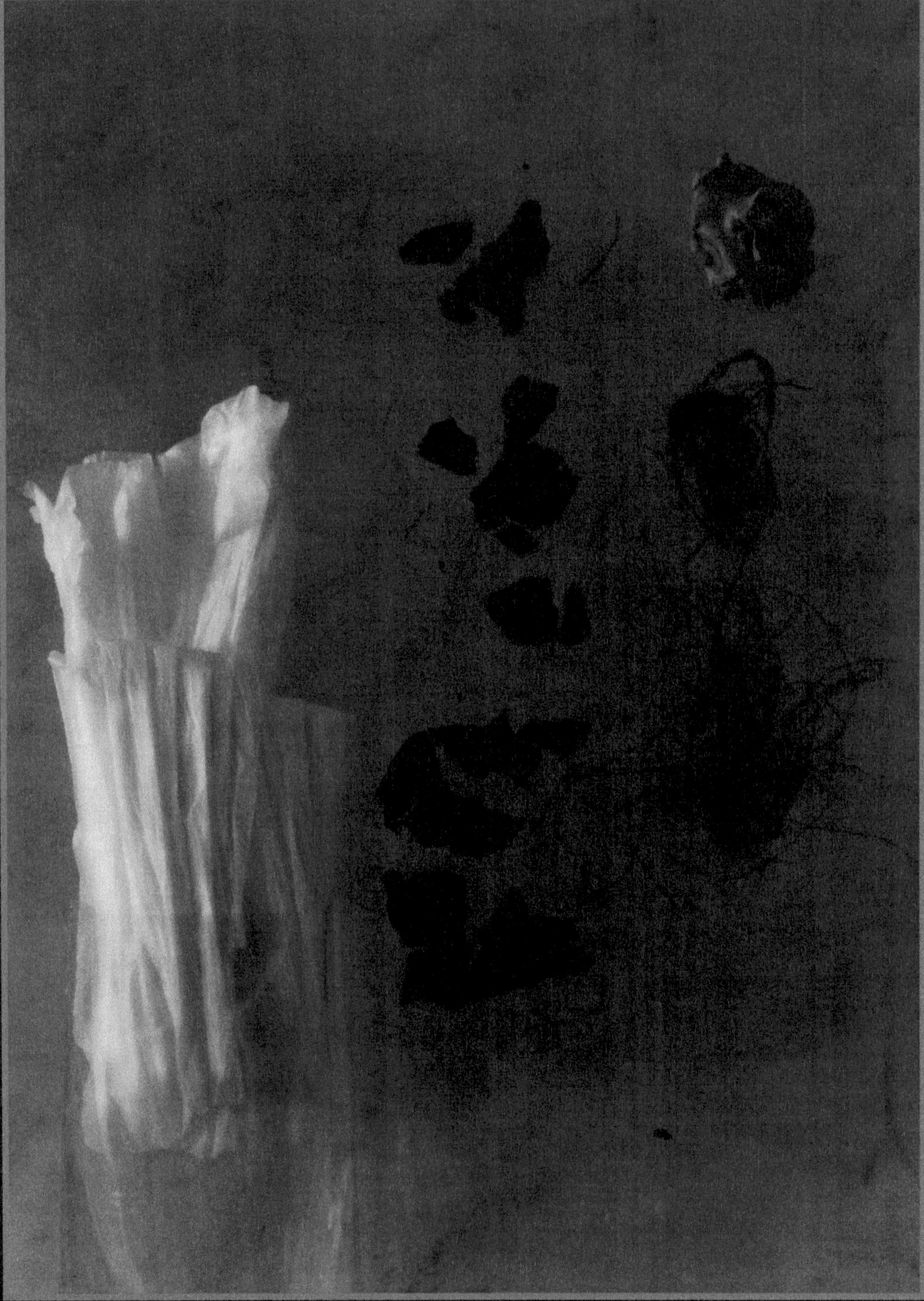

Figure 4.0 Pradip Malde, *Imitation Krakatoa: for the armchair button-pusher, Orkney*, 1982, Platinum-palladium print on Van Gelder Simili Japon paper from original 8 x 10 inch negative. 24.75 x 19.5 cm. Courtesy Pradip Malde. Fe1:Pt:Pd = 2:1:1, 20 min hydration in 100% RH / 20°C.

Chapter 4

Equipment, Materials, and Chemicals

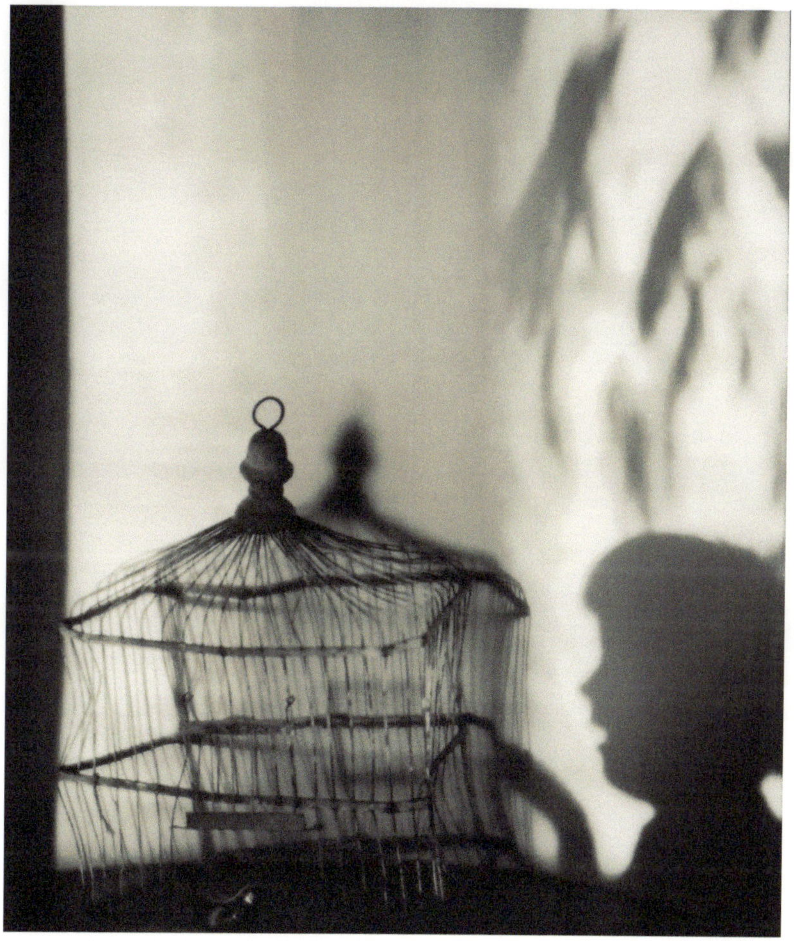

Figure 4.1 Imogen Cunningham, *Birdcage and Shadows,*1921. Platinum-palladium print on Revere Platinum paper, from PiezoDN version of the original 8 x 10 inch negative, 22.5 x 19.5 cm. Printed by Pradip Malde. Courtesy Imogen Cunningham Trust. Fe3:Pt:Pd = 2:1:1, hydration for 90 minutes at 76% RH / 20°C.

It is likely that you may already have some portion of the setup needed to make platinum/palladium prints, but it does no harm to provide a comprehensive list of all essential items. If you are new to wet photographic processes, or to making platinum/palladium prints, it may be most helpful to skim through this chapter, jump to the introductory section of Chapter 8: *Making the Print*, and then return to this point for a closer reading.

This chapter is organized in two alphabetical segments. The first portion offers descriptions, and where appropriate, the proper assembly, arrangement and use of equipment and materials. The second segment presents a complete list of all the chemicals that are referred to in this book's formulae (Chapter 5: *Preparation of Sensitizer and Processing Solutions*). Up-to-date versions of both portions are maintained on the book's web resource site, pradipmalde.com/platinotype.

Equipment and Materials

Acetate Film

(See *Archival Polyester Film.*) Acetate film is used to make quick printing masks by "tracing" areas of the negative with permanent markers. Place one or more masks, in register, above the negative, on the outside of the contact-printing frame, and remove as necessary during exposure. Since the printing mask's shadow should have diffuse edges, this method is best suited to a light array rather than a point source exposure unit. Thickness and optical clarity are not critical; materials cheaper than archival polyester film may be used here. See Chapter 7, Masks.

Adhesive Tape

(See *Coating brush* and *Glass film cleaner.*) Use a low-tack, pressure sensitive tape such as Scotch Removeable 3M811 or Scotch Delicate Surface Painter's Tape 2080 for holding paper to a coating surface.

Amber Bottles

Store light-sensitive chemicals in dark brown, technically referred to as "amber," glass bottles; this minimizes their long-term exposure to ultraviolet light. Additionally, storing containers in a dark cupboard or box should increase shelf-life to years. Being glass, it is also almost completely inert. Cheaper plastic containers mimicking some of these properties are now available, but they may not be as chemically inert as glass. It is best to use wide-mouth necks for storing and dispensing sensitizer solutions, with plastic caps and liners. The design is sometimes referred to as a "packer." Avoid using "child-safe" lids.

Apron

A cotton chef's apron, fabric-lined vinyl, or a lab coat protects clothing.

Archival Polyester Film

(See *Acetate film.*) This is not the same as material generically known as "acetate film" or "PET film." Chemically, this is polyethylene terephthalate, an inert plastic, free of plasticizers, residual solvents, and active sulfur compounds, and sold under the trade names Mylar and Melinex. Type "D" archival polyester film is optically more clear than other acetate films, archival and pH neutral, does not yellow, and has greater dimensional stability than the cheaper versions of film. For printing, it is advisable to use as an interleaf, to protect negatives from damage if placed in contact with crystallized or overly-humid sensitized paper. Some printers prefer to always use a sheet of very thin film between negative and coated surface. The film's thickness will have some effect on image resolution. With an increase of distance between the negative's emulsion and the printing surface there is a corresponding decrease of resolution, or sharpness. This effect will be less with a point source of ultraviolet, such as the sun or a single-bulb metal halide lamp, and greater with a diffuse source such as a bank of fluorescent tubes. Film thickness is commonly measured in "mil" (1/1000″) or microns (1/1000 mm, symbol: μm), but sometimes also described on packaging in "gauges," millimeters, and inches. Of the most readily available thicknesses, I recommend anything close to 1 mil (25 μm/ 100 gauge / 0.0254 mm / 0.0010″), but not thicker than 2 mil (50 μm, etc.).

Blotting Paper

Blotting paper is for drying prints, and is especially important when using vellum and tracing papers. Small strips of this are also useful for blotting up excess sensitizer when coating paper. It should be 100% high alpha cellulose and have a neutral pH ca.7.

Blower

Blowers or bulb blowers are useful for removing surface dust and fibers when preparing to coat paper or print digital negatives. Avoid using compressed air canisters because they can sometimes cause condensation on the surface due to a gas phase change, and even deposit propellant if the can is shaken. The larger bulb blower designs are environmentally sound and provide better control.

Clips

A "bull-dog" style, or "chip-clip" such as those wide-mouthed grippers used for closing food bags, serves as a general-purpose item (not a paper-clip). It can be useful for holding paper to a glass surface while sensitizing, keeping vellum papers flat, or hanging prints up to dry. Avoid metal clips.

Coating Brush

(See *Glass coating rod.*) Printers sometimes prefer a brush to a glass coating rod. It is better to use a brush for very large areas, and essential for papers with textured surfaces or Japanese gampi. I do not recommend foam-rubber applicators as these are wasteful by holding back a substantial amount of sensitizer which never reaches the paper. Avoid brushes with metal ferrules because the sensitizer chemicals will react with the oxidized metal and render problematic results. If using brushes with metal ferrules, cover these parts with a relatively inert plastic tape or similar material. Japanese hake brushes, made from synthetic fibers, wood and binding thread, are best for this reason. Their design, being broad and flat, makes it easy to distribute the sensitizer solution in even, broad strokes. Japanese hake-style brushes such as the Richeson 9000™ or Kobayashi Soft Synthetic™ series, designed for applying sensitizer washes, are ideal because they are made with synthetic fibers held together by relatively inert ferrules. Widths of 1.5–3″/ 3–8 cm are good for prints up to 8″ x 10″ / 20 cm x 25 cm, and 4″ / 10 cm brushes will serve to coat most areas larger than that.

The sensitizer is poured along the paper, just outside one edge of the picture area, and spread by means of a brush, previously dampened with distilled water, making long even strokes in a consistent direction to cover the area, then repeating the brushing at right angles. For coating areas larger than 150 sq. inches / 1 sq. meter, pouring and distributing the sensitizer on two or more edges may be necessary. If sufficient sensitizer remains, the brushing may be continued until it is all absorbed, but avoid abrading the paper surface. Sensitizer volumes will vary, depending on the area to be coated, sensitizer, brush's absorptivity, and paper. Brush-coating always requires more sensitizer than coating with a glass rod. See Chapter 8, *Printing Workflow*, step 9.

Color Calibration Target

Useful when creating high-resolution digital scans or photographs of prints. These targets are, to put it simply, small swatches of color and tone. They are precisely manufactured to match an international color standard maintained by organizations such as the International Color Consortium (ICC) and Committee for Graphic Arts Technology Standards (CGATS). A calibration target, when included in a scan or photograph of a platinum/palladium print, provides a standardized reference for viewing and reproduction purposes. These targets can also help with calibrating computer monitors, scanners, and digital printers. They can be expensive, but for our purposes, a target such as the DGK Color Tools DKK 5″ x 7″ will suffice, from which smaller targets may be made by cutting and reassembling.

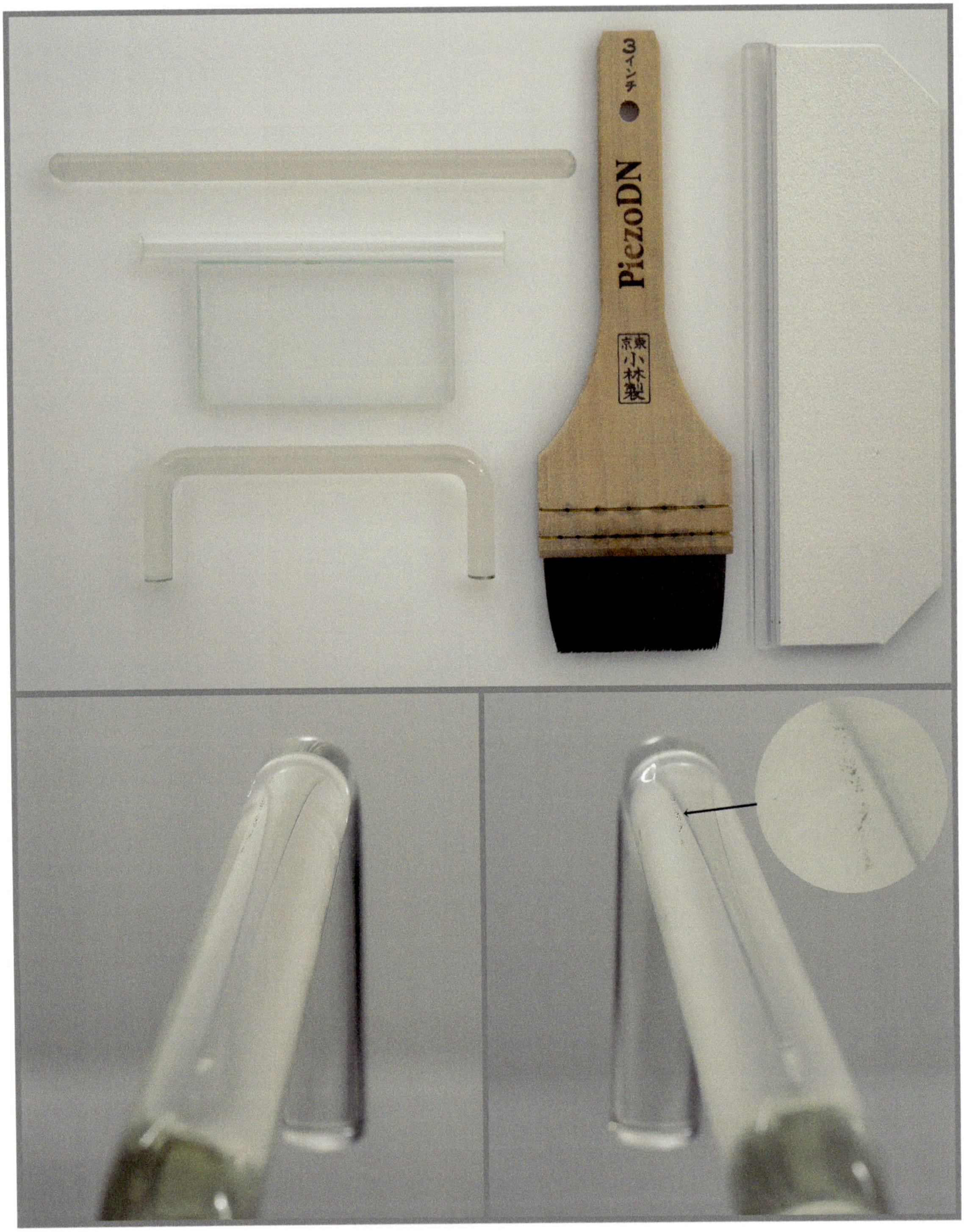

Figure 4.2 An assortment of coating tools. Note the absence of a metal ferrule on the brush. Coating rods must be smooth. Glancing a rod against light and moving it around can reveal minute scratches, as shown in the lower images of the same rod. These small aberrations are enough to create blemishes in the print.

Contact-Printing Frame

The negative and the sensitized paper must be in close contact with each other during exposure. Under certain situations, it helps to be able to inspect the print as it is exposed, without compromising registration between negative and print. Printing-out is the primary characteristic of the ammonium-based platinum/palladium process—the image appears almost completely by the end of exposure. Thus, a hinged-back printing frame permits one part of the print to be inspected while the rest remains under pressure. Batons or steel springs ensure close contact. A sheet of glass provides optical clarity and transparency to UV–although most "picture glass" will block between 10% and 30% of UV. For practical reasons, use a sheet of plate glass 4–6 mm thick, but avoid the 2 mm thick "picture" glass, which may bend or fracture under the pressure. Be sure to avoid the special UV-absorbing glass (it has been done!). Plexiglass that is completely transparent to UV is cheaper to purchase than similarly transparent "quartz glass," but bows and scratches much more easily.

It is important to place an impervious sheet of rubber or plastic closest to the hinged back. This acts as a vapor barrier and evens out any variances of pressure, and should be the same size as the back and glass. It must not be stuck to the hinged back and divided into two, otherwise diffusion of air or moisture through the hinged joint may cause a mark down the print, or modify the color in undesirable ways. Foamed neoprene, or Plastazote™, a black plastic foam sheet, of archival quality, and ca. 1/8″ / 3 mm thick is a good material.

Also, a sheet of 100pt (0.10" / 723gsm) cotton blotting paper, or 1/8″ / 3 mm thick papermakers felt, is required to maintain full pressure during exposure. Carbon dioxide is a product of the siderotype photochemical reaction (see Chapter 2, *Siderotype Processes*). The print is actually "evolving" small amounts of gas during exposure, and these usually coalesce into bubbles or pockets of separation between the negative and print surface. The result is a loss of sharpness, and its visibility varies depending on the paper, area coated, contact pressure, and light source (see *Ultraviolet light source*). This effect can be made less conspicuous or entirely negated if the carbon dioxide has somewhere to go. Some of it is absorbed into the small voids within the sensitized paper itself. The rest may be adequately absorbed by placing a sheet of dry blotting paper or felt between the printing paper and impervious barrier.

Frames can be bought from specialist suppliers, and modern versions are often of high craftsmanship and cost. The Arista Ultimate Contact Frame II or Photographers' Formulary Select Contact Print Frame work well. If it is absolutely necessary, construct a serviceable version from a strong commercial picture frame kit of extruded, mitered aluminum alloy with a deep groove to allow fixing of hinges. In its simplest form, a negative and print may be sandwiched between a stiff board and glass sheet, all of which is held together by strong clips.

Cheap alternatives that do not apply pressure from the central portions of the back are generally to be avoided. Also see *Vacuum frame*, below.

Cotton Gloves

Use cotton gloves when handling negatives, paper that has been trimmed to the sensitized edge (see Chapter 3, *Platinum Hazard*), cleaning glass, scanning, or even as a liner within a pair of waterproof gloves used for extensive periods. Thin, white gloves, available in multi-packs, are usually available from suppliers of archival materials.

Face Mask

A simple, disposable-type dust mask covering the nose and mouth will suffice. All the chemicals listed here are non-volatile, so respirators are not needed.

Filtration Funnel

Any one of several conical designs will work for filtering small volumes of solutions: a long, short or stemless type; glass is recommended but plastic will work too. Used with filter paper, see *Filter papers* below.

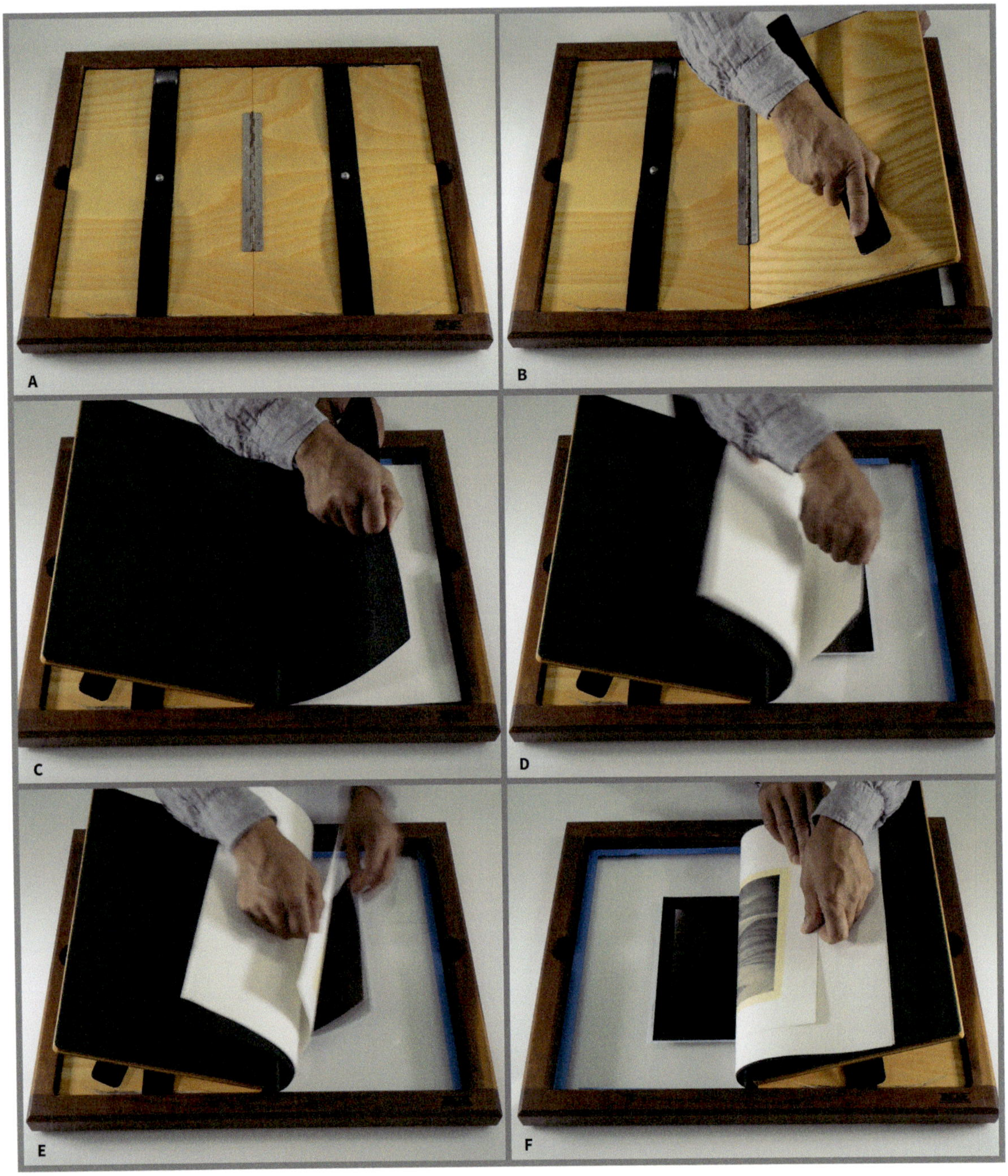

Figure 4.3 Component parts of a contact-printing frame.
A. fully closed
B. one spring clip releases hinged back
C. opaque black backing layer
D. permeable felt or blotting paper backing layer
E. sensitized paper contacting negative
F. inspect exposed and printed-out image

Filter Papers

Whatman Grade 1, Qualitative Filter Paper Standard Grade, circle, 125 mm, used with a suitable filtration funnel, see *Filtration funnel* above.

Glass Beakers

Use heat-resistant Pyrex™ or Corning™ glass beakers for the preparation of sensitizer and small volumes. Recommended quantities and sizes are one each of 50 ml, 100 ml, and 250 ml. These are available in "low" or "tall" form. Either design will work, but the low form may be more convenient.

Glass and Film Cleaner

At least 98% Isopropyl alcohol, or a product like Aztek RC 2001 Film cleaner, used with lint free wipes to clean negatives and glass surfaces on scanners, contact-print frames and exposing units. Do not use fluids containing ammonia, such as Windex™.

Glass Coating Rod

(See *Coating brush*) In place of the traditional brush, a more economic coating is achieved by distributing sensitizer with a glass rod. Mike Ware and I introduced this method in 1986, and it has gained much traction since then. However, it is said "there is nothing new under the sun," and we have since discovered three historical references, from 1853, 1856, and 1861 to the use of glass rods for coating, suggesting that it was standard practice at the time![1] An appropriately sized rod will be needed for each format to be printed: coating is more easily done with a rod length that is equal to the shorter dimension of the desired print, and this is particularly recommended in the case of panoramic formats. Various commercial versions of this device are marketed, usually with a central flat handle cemented onto an open-ended glass tube, such as the Puddle Pusher from The Photographer's Formulary. Avoid devices made out of plastic, as these materials tend to have a lower surface-wetting effect than glass, and consequently inhibit the even spreading of sensitizer as coating progresses.

Coating tools may be made from thick-walled glass capillary tube or "gage glass"; these are usually manufactured to higher standards of straightness than the solid rod. An external diameter between 0.25 and 0.5″ / 6 and 12 mm is suitable, with a straight central section equal in length to the width of the coating area. A handle is necessary, and may be formed by having extended portions of the rod bent at an angle, using a powerful gas torch; a bicycle handlebar shape is simple and ergonomically effective in use, although more exotic shapes may be tried if you are a capable glass worker. The center portion of the rod acts as the spreader, and the distance between the bends determines the width of the coated area; the limbs serve as handles. Alternatively, a rectangular plate of thick glass may be attached to the rod with a suitable glue. The rod's coating surface must be flat and smooth so as to provide an even distribution of sensitizer and prevent the coating process from abrading the paper. It must be kept scrupulously clean and free of grease film. Store coating rods flat, and on a soft surface. To avoid scratching the rod, do not place it directly onto another glass or ceramic surface. See Chapter 8, *Printing Workflow*.

Glass Sheet

A heavy (0.25″ / 6 mm) glass plate forms an ideal platen for coating paper with sensitizer. It should be at least 10% larger than the maximum dimension of paper being coated. Arrange the sheet on a work surface so that it is completely level, especially left-to-right, so that sensitizer will be evenly distributed as one spreads it with a glass coating rod from top to bottom.

Glass Stirring Rods

Simple glass rods, or glass cocktail swizzle sticks for solution preparation.

Gloves, Disposable

Always wear gloves when handling dry chemicals and preparing solutions from concentrates. Wear gloves, or use print tongs, when processing prints through the first and second clearing baths, disodium EDTA and sodium sulfite respectively, where most of the residual chemicals accumulate. Gloves should also be used when processing wet-lab negatives.

Hotplate / Magnetic Stirrer

A hotplate allows for preparing small volumes of heated solutions. Some hotplates include a magnetic stirrer, which, when activated, rotates a small plastic-coated metal rod placed inside a beaker. Alternatively, and more cheaply, a hot-water bath that comes up to the same height as solution contained inside a beaker will function just as well. *Never use an open flame for heating.*

Hydration Chambers

The amount of water present in the fibers of sensitized paper at the time of printing is a crucial factor in determining the characteristics of the final print (see Chapters 5, 6 and 8). It affects the extent of printing-out, the color, resolution, sensitivity, and the contrast. Being hygroscopic, the water content of sensitized paper may be controlled by defining the relative humidity of the air around it. If the air is less humid than freshly coated paper, the sheet will begin to dry out until it reaches a point of stasis. On the other hand, if the air is more humid than already dried paper, the sheet will begin to absorb moisture over time.

If you want reproducible results in your prints, or your working environment has an unsuitable or somewhat variable RH, as many workspaces do, you will have to employ methods to regulate the humidity of sensitized paper. There is no alternative to careful control of this parameter. Fortunately, this is not as challenging as trying to control the ambience of an entire room, and can be accomplished with a simple piece of inexpensive equipment that amounts to a tray with a lid.

The following instructions are for constructing a hydration chamber big enough for making 8″ × 10″ prints on 11″ × 14″ paper, and can easily be scaled up for larger sizes. Note that the suggested polycarbonate sheet (of the kind used for framing, box construction, etc.) is light and convenient, but the material may be dimensionally unstable on extended exposure to humid conditions and prone to slight bowing. Plexiglass (Perspex) or acrylic sheet can be used instead of glass, but this material also tends to be dimensionally unstable in humid conditions and therefore warps—unless the sheet is very thick, in which case it may be too heavy to handle. Glass may be used instead, but with a metal sheet sandwiched in between, creates a heavy lid, and at larger sizes this may become unwieldy. Another alternative is to simply use one thin sheet of glass, made opaque on one side with a thin plastic sheet or some other material, while the other side has strips of self-adhesive magnetic tape, to which is attached a sheet of sensitized paper.

- Opaque (preferably black or red) plastic tray, at least 2″ / 5 cm larger than the paper size. With depths of at least 5″ / 12 cm, cat litter trays, oil-drip pans or small utility mixing trays are ideal for this purpose; photographic dishes are usually too shallow and have a lip which makes sealing difficult
- A plastic grid, such as a suspended light ceiling panel, cut to fit inside the tray and about 2″ / 5 cm below its rim
- Closed-cell sponge draught sealing tape, enough to place all the way around the rim of the plastic tray
- Two sheets of twin-walled polycarbonate sheet, at least 2″ / 5 cm larger than the length and width of the plastic tray.
- Sheet of zinc-plated metal, 26 gauge, same size as the glass sheets
- Masking tape or gaffer tape, 2″ / 5 cm wide
- Water, or a salt slurry that has the consistency of a thick pancake mix
- 6 to 8 strong magnetic "pushpins" or "buttons"

To make a hydration chamber:

A. Clean the plastic tray.

B. Cut the plastic grid so that it will rest inside the tray, and about 2″ / 5 cm below the rim.

C. Adhere the closed-cell draught seal all the way around the rim of the tray, and carefully splice it to form a gapless, airtight seal at the point where it meets itself.

D. Wear gloves and be cautious of sharp edges. Thoroughly clean the polycarbonate sheets. Clean the zinc-plated metal sheet with a household scouring powder such as Bon Ami and dry it thoroughly.

E. Sandwich the metal between the plastic sheets. Align all edges and tape them up to create a bonded and safe lid.

F. Remove the plastic grid, and pour room-temperature water, or already prepared salt slurry (see below), into the tray. The liquid or slurry should easily cover the entire bottom of the tray to a depth of about 0.5″ / 1.25 cm. Replace the grid.

G. Use the magnets to hold a coated sheet of paper, face up, to the lid. Flip the lid in order to suspend the sensitized surface over the grid. The grid serves as a safety net, just in case the magnets unclasp and the coated sheet drops.

H. Place this hydration chamber on a flat surface, in a draught free area, and arranged so it can be accessed easily.

There are three methods for regulating the humidity in a sheet of coated paper, which are referred to, for good scientific reasons, as the *Equilibrium method*, the *Timed method*, and the *Room hydration method*. The first two methods rely on a closed container, the hydration chamber, that has a known concentration of water vapor in its atmosphere. Each has advantages and disadvantages. The equilibrium method, although accurate and rendering the most consistent results, requires a long hydration time. To fully and evenly hydrate a print, a sheet of sensitized paper should be left in the chamber for one to two hours. This means that one either works at an extremely gentle pace using a single hydration chamber, or works through some other strategy, such as having two or more chambers, or having a chamber large enough to hold two or more sheets of paper (never in contact with each other). Production is then staggered to allow a higher frequency of throughput. The timed method, while providing much shorter hydration times, requires that sensitized paper is dried, ideally at 100°F / 40°C for 10 minutes, then suspended in a chamber with an atmosphere of high RH. This approach is harder to control. The third method of ambient, or room hydration is the simplest but offers the least by way of controls, unless, of course, one has the luxury of a consistent and favorable workspace. Each of these are described in more detail below.

Equilibrium Method

An aqueous, or sludge-like, saturated salt solution can provide an atmosphere of known relative humidity, and at a consistent level within a comfortable temperature range on either side of 68°F / 20°C. Keep in mind, however, that absolute humidity, measured as the weight of water per unit volume of air, does increase with temperature. Thus, it is best to work in a consistent ambient temperature.

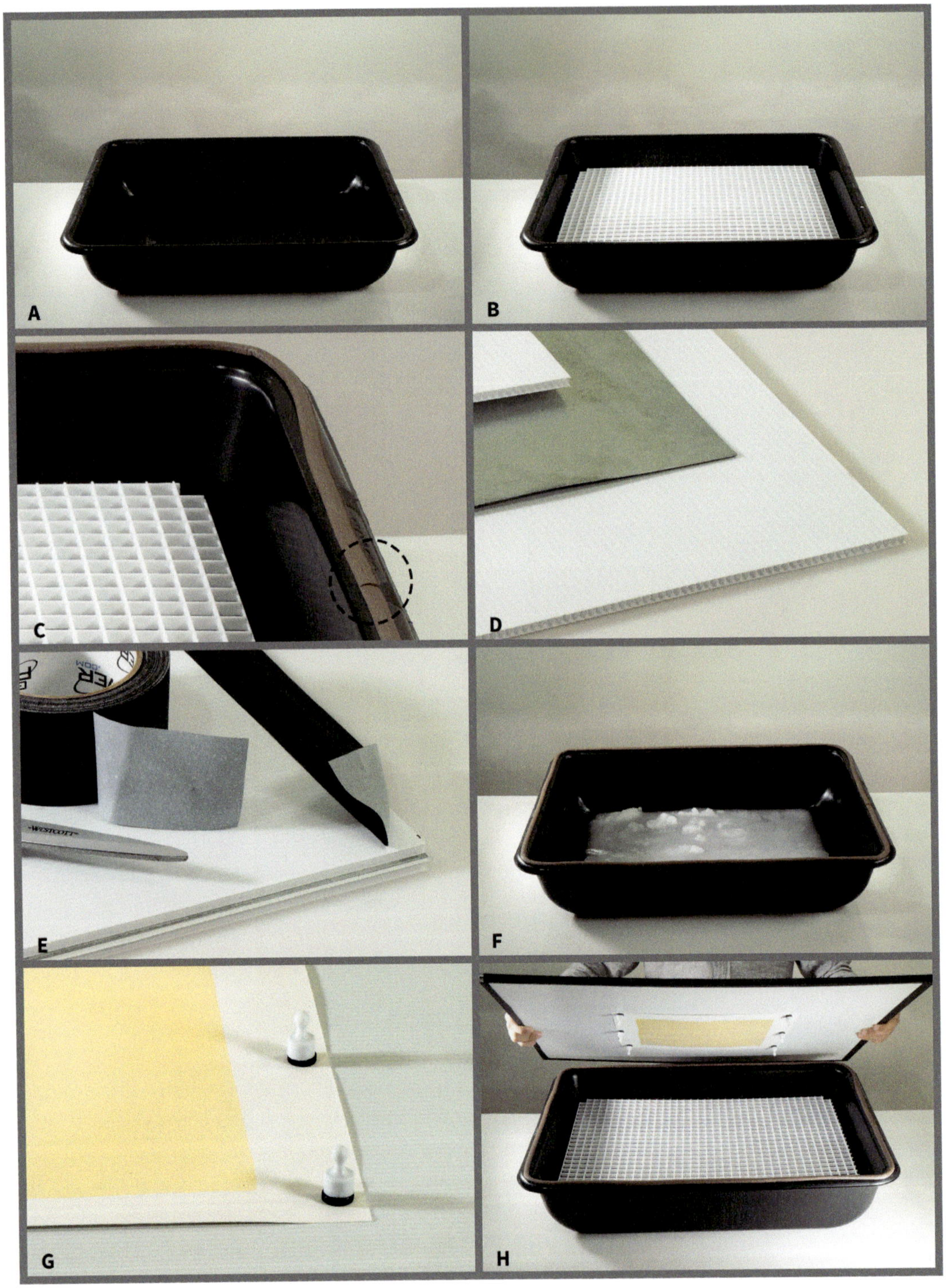

Figure 4.4 Instructions for making a hydration chamber, step-by-step. Note that the slurry (F) looks like small puddles of liquid among an excess of salt clumps.

It is important that there is always a significant excess of salt in these solutions. Distribute enough salt to cover the bottom of a hydration chamber evenly to a depth of about 0.5″ / 1.25 cm, then add small amounts of slightly warm water across the area while swirling the mixture and allowing time for the salt to dissolve to saturation before adding more water. The solution should look as if there are small puddles of liquid among an excess of salt clumps.

The RH values of many such solutions have been determined, but the salts listed in Table 4.1 are the most readily available and practical. Many more salts, and mixtures, have been used to obtain intermediate RH values, see the literature on aqueous salt solutions for extensive lists.

Hydrate for one to two hours.

Timed Method

This is also sometimes referred to as the *Water Hydration Method.* Instead of aqueous salt, fill the chamber to a depth of ca. 0.5″ / 1.25 cm of water that is at ca. 1°F / °C higher than ambient, room temperature. This creates an atmosphere of RH 100% so long as the ambient air temperature matches that of the water inside the chamber. This approach offers a simpler and cheaper alternative method of regulating humidity and the degree to which paper is hydrated—but it is less reliable. It first requires drying the coated paper to a known level, then promptly placing the paper in the hydration chamber.

Unlike the equilibrium salt solution chambers, where the paper can be left for up to two hours, this exposure to 100% RH must be timed for reproducibility, and it should not be excessively lengthy, up to a maximum of 30–40 minutes, or the paper will imbibe far too much water, with consequent degradation of the image quality. If left in such a 100% RH bath, cellulose goes on absorbing water into its capillaries until it is perceptibly "soggy" and becomes useless for image-making, while also increasing the chance of damaging negatives. The effects of low RH are achieved by exposing the sheet for only a minute or two, but it can introduce problems of non-uniformity. Note that it is important to keep the humidifying chamber at a uniform and constant temperature, about 18–20°C, because the vapor pressure of water increases steeply with rising temperature.

Room Hydration Method

Sometimes also called the *Ambient Hydration Method*, here hydration can be accomplished by just hanging sensitized paper in a (darkened) room, provided that you are willing to accept the results the ambient RH will give. It helps to have a hygrometer to check the prevailing relative humidity.

Hygrometer

An instrument that will measure relative humidity levels, and essential for controlling the extent to which sensitized paper is hydrated. Digital models are widely available and the easiest to use. They usually combine a thermometer, and are often referred to as hygro-thermometers. Look for a design with

Table 4.1 Substances for making hydration chambers using an aqueous salt slurry (except with the 9% calcium chloride chamber, where the chemical is used as a solid) to create an atmosphere of a known, and stable, relative humidity.

Substance	*Formula*	*Relative Humidity at ca. 68–72°F / 20–22°C*
Calcium chloride, anhydrous	$CaCl_2$	ca. 9% NB: use solid
Calcium chloride	$CaCl_2$	32%
Calcium nitrate	$Ca(NO_3)_2$	56%
Sodium chloride	$NaCl$	76%
Ammonium chloride	NH_4Cl	80%

a remote probe, with a humidity sensitivity range from RH 10% to 99% and accurate to ±4%. It is best to opt for the slightly more expensive models that can be calibrated, as sensors tend to give erroneous readings over time.

Labels

Self-adhesive, should be easy to write on, and smudge-proof.

Light Table

Convenient for examining negatives and aligning masks. LED versions come in many sizes, are convenient to work with, and emit little heat.

Masking Film

For creating paper-base white borders around a printed image. Rubylith® film is best for this, but may be hard to find.

Measuring Cylinders

A range of sizes will be needed for measuring out liquid volumes, the most convenient being two each of 1 liter, 500 ml, 100 ml, and 50 ml. It is likely that a couple of 2 liter sizes are needed if working with paper larger than 11″ x 14″ / 30 cm x 35 cm, but this is only for convenience. The plastic type of measuring cylinder is adequate, but glass is preferred for better precision of measurement and greater ease of cleaning. Measuring cylinder sizes of 100 ml capacity and lower should be graduated so that it can be read to within 1 ml—a precision approaching ± 1%. Use accurate and well-marked containers—it is worth spending a little bit more for professional-grade equipment. These come in low or tall form, and I recommend the tall shape for the 100 and 50 ml size.

Mixing Glass

A small liqueur, or shot glass makes an ideal vessel for combining sensitizers and additives just prior to coating the paper. To minimize wastage, use a glass that has a deeply concave internal base, allowing you to syringe up almost all of the liquids.

Mortar and Pestle

A ceramic mortar and pestle designed for chemistry laboratories is useful for grinding and crushing solids prior to making sensitizer solutions. A 300 ml capacity should suffice.

Negative-Making

Making traditional film-based or digital negatives for platinum/palladium printing requires specific instruments and equipment. The following list identifies the most salient general features, recognizing that much depends on space, budget, and of course, individual needs and objectives. See Chapter 7.

Camera Equipment and Film

Platinum/palladium printing is a contact process, as far as current technologies permit. The print size is limited to the negative size. If working with original, "wet-lab" negatives, note that the camera equipment, film, and subsequent processing decisions will determine the size and quality of the resulting platinum/palladium prints. Satisfactory prints can be made with negatives as diminutive as $2^1/_4$″ x $2^1/_4$″ / 6 cm x 6 cm. This approach will require, of course, a traditional darkroom for film processing. See Figure 10.9.

Computer

It is likely that image files, whether originated digitally or from scans of traditional negatives, will be large—in excess of 25MB or even ten times that size. For a digital negative that has been output to a size of 8″ x 10″/ 20 cm x 25 cm, for instance, an image file may well approach 500MB. Images are prepared with software such as Photoshop, Lightroom, or Capture One, which themselves make heavy demands on processors. Add to this the high likelihood of a burgeoning collection of large image files; you will need to spend a significant sum on some combination of computer, monitor, external hard drive and, crucially, an external backup system.

Digital Scanner

For digitizing traditional film negatives and positives, a unit with a head for transparency scanning will be needed, such as the Epson V600 or higher. For the highest resolution scans, use a commercial drum-scanning service. Digital cameras with copying rigs may also be used.

Digital Printer

(See *Transparency Film for Inkjet*) Inkjet printers can be loaded with either the manufacturer's recommended/original inks (materials supplied by the "original equipment manufacturer" or OEM), or with a customized inkset option. Both approaches will work—OEM inks are usually more affordable, but certain (not all!) custom inksets offer a more tunable approach to making digital negatives. All of Epson's SureColor and Large Format printers will render suitable negatives with their OEM inks when printed on appropriate material. For optimal quality, consider using customized combinations of print-driver software and inksets with Epson printers, such as the Piezography™ and PiezoDN™ system made by Cone Editions.

Transparency Film for Inkjet

Several manufacturers make transparent acetate films specially coated to receive inkjet pigments. They vary in cost, which is often directly related to quality. Pictorico's Ultra Premium Overhead Projection (UP OHP) film seems to have the greatest capacity to absorb pigments, and therefore render a wider tonal range, than all other products. Do not confuse Pictorico Premium OHP with UP OHP—they are not the same.

Non-woven Spunbound Polyester

Non-woven, spunbound polyester is a fabric with high tensile strength and tear resistance, and very useful for supporting very thin, low wet-strength papers during processing. It also has a porous structure, good dimensional stability, and a very smooth surface, which allows translucent ("vellum") papers such as Reich CT and Wyndstone Vellum to dry flatter when they are placed between a sandwich of this material and blotting paper. Commercially available as Hollytex™, it is used in conservation and restoration work because of its exceptional combination of physical and chemical properties. It is pH-neutral. See Chapter 8, *Printing Workflow*, step 25.

Notebooks

Whatever size(s) notebook you use, establish a cataloging and indexing system that corresponds to your images, negatives, prints, and chemicals. In other words, at the very least, date everything, and make notes as you go.

Optical Wipes

Lint and streak free disposable optical wipes, used with 98% isopropyl alcohol, may be used for cleaning glass surfaces, film, and other optical and delicate equipment.

Paper for Sensitizing

After having tested a few sheets of paper from a known production batch, buy as much paper as you can afford; a 11″ x 14″ / 30 cm x 35 cm sheet is an ideal size for printing 8″ x 10″ / 20 cm x 25 cm images, with 30 such sheets using about 45 ml of total sensitizer solution. See Chapter 6.

Paper Towels

Used for cleaning, wiping, and when cut into thin strips, for blotting up excess sensitizer solution at the end of a coating action. Use a robust towel such as Bounty™.

Pencils

2B or not 2B, it is your choice, but dispose of shavings with care, else the native hue of resolution be sicklied o'er with the pale cast of graphite, and enterprises of great print and image with this regard their currents turn awry.[2]

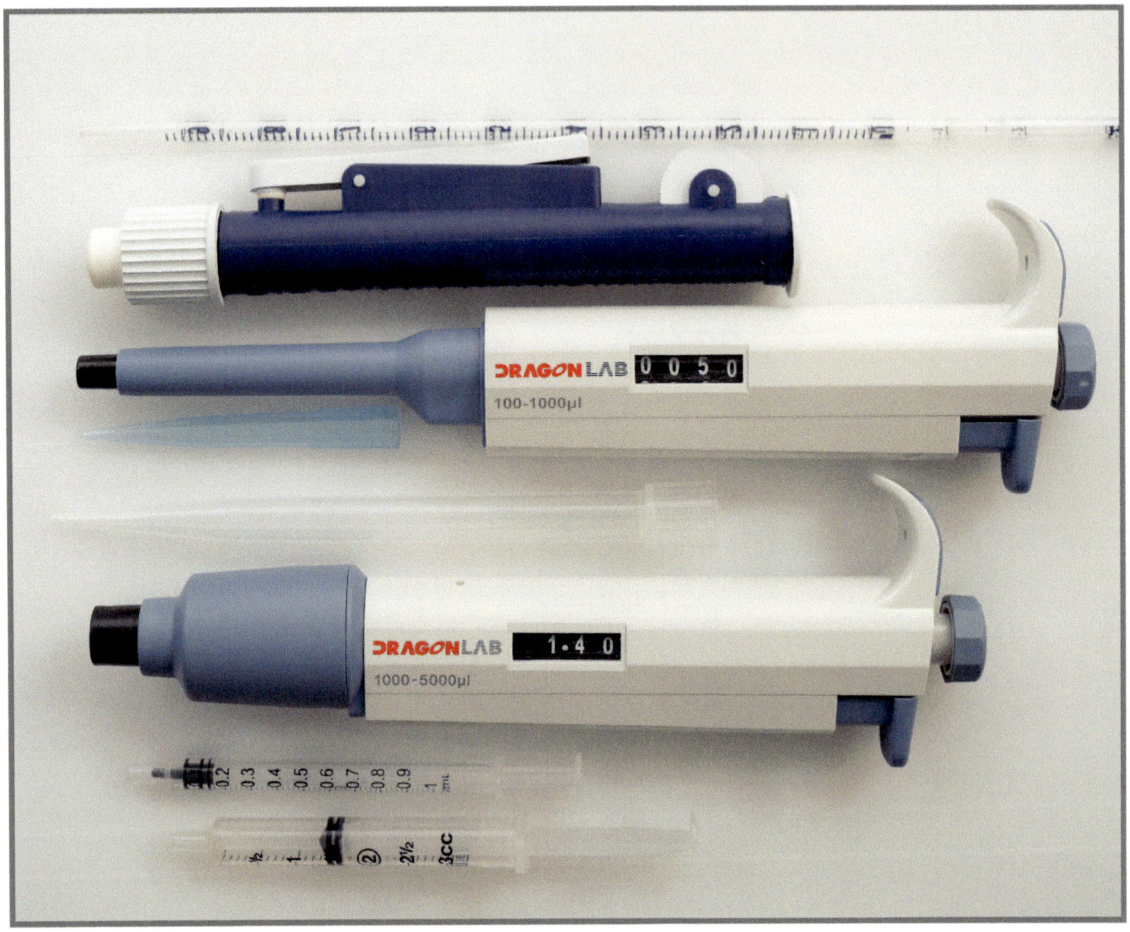

Figure 4.5 Pipette with filler, adjustable micropipettes, and syringes

Permanent Markers

(See *Acetate film*) Permanent markers come in a variety of tips sizes, and are used for labels and manually drawing printing masks. Fine tipped markers may be most useful.

Pipettes

There are numerous sizes and designs of pipettes; it is helpful to have one each of 100–1000 µl and 1000 –5000 µl sizes. They can be used with or instead of syringes (see *Syringes*). Adjustable micropipettes, manufactured in a variety of sizes, provide extremely accurate and repeatable measurements of small volumes. Micropipettes are graduated in microliters, abbreviation "µl." One milliliter is 1000 µl. The 100–1000 µl size (0.1 to 1.0 ml) is good for extremely small volumes such as those needed for adding Tween and/or glycerin to the sensitizer, or coating 4″ x 5″ / 10 cm x 12 cm areas, and the 1000–5000 µl sizes (1 ml to 5 ml) may be most appropriate for volumes for areas up to 11 × 14 inches. Slip tips will be needed for each size, but only a few—they last a long time. Yet another, but fragile, alternative for precise volume measurement is provided by graduated glass pipettes, especially the kind used in hematology, called serological pipettes, usually of 1 or 2 ml capacity. These must be used with a pipette filler (a rubber bulb or piston type), or simply a syringe of similar or larger volume with a short connection of rubber tube 1 cm long. *Operation of pipettes by mouth is now strictly forbidden in all chemical practice.*

Print Drying Screen/Rack

An inert vinyl mesh in an aluminum frame, much like insect screen window inserts, will suffice. These screens can be layered with clean sheets of blotting paper for drying very thin or translucent papers.

Print Spotting Kit

Use high-quality gouache pigments in tube or tablet form for spotting prints. Lamp Black, Prussian Blue, and Burnt Umber in various blends will match almost all paper/print tones. Do not use Spotone™ or any other dye-based retouching material, as this will seep sideways and down into the paper, creating a stained halo.

Print Washer

The washer should have an exchange flow rate of ca. 2 gallons / 8 liters per minute. Use a tray at least 30% larger than the paper size, so that it will allow the print to lay flat within a gentle flow of water. Kodak Automatic Tray Siphons are ideal, although they are no longer manufactured. Avoid washers where prints are arranged vertically; gravity could damage papers, especially those with low wet-strength. For water conservation, about four prints at a time may be washed in three changes of fresh, room-temperature water for ten minutes per bath, with gentle agitation.

Processing Tongs

Avoid skin contact with the first and second clearing baths, disodium EDTA and sodium sulfite respectively, where most of the residual chemicals accumulate. In itself, these baths are not toxic, but as unreduced platinum salts come off the paper and go into solution, safety and non-toxicity become less certain. When in doubt, it is best to minimize skin exposure by using print processing tongs. There is no harm in continuing to use tongs, or gloves, for the entire clearing sequence, even though the solutions are entirely benign by the final tetrasodium EDTA bath. Plastic tongs are more favorable than steel. The wooden varieties should be avoided entirely because they tend to be harder to keep clean.

Processing Trays

Four trays will be needed, with one for each of the three clearing baths plus another for the intermediate rinses. If space is at a premium, just one tray may be used, with solutions being poured in and out of the tray, while leaving the print stationary. This latter method is also the least likely to damage papers that are very thin or have low wet-strength.

Rulers

Scissors

Spatula

For handling small amounts of chemicals when weighing and measuring.

Spirit Level

A bubble level, used to arrange a level coating surface. If using a smartphone's built-in level, be sure that the device itself is sitting flat on the surface.

Step Wedge / Target

Also known as a step tablet or transmission step wedge. A transparent film, with at least 12 zones of density, or steps, progressing in regular stages from film-base transparency to almost completely opaque. This can be purchased calibrated or not, and used to compare platinum/palladium prints made under varying conditions. Standard step wedges are made by Stouffer, and may also be printed digitally with an inkjet printer setup using calibration targets, such as those available from Cone's Piezography™ system. See Appendix IV Resources for a link to download a target file.

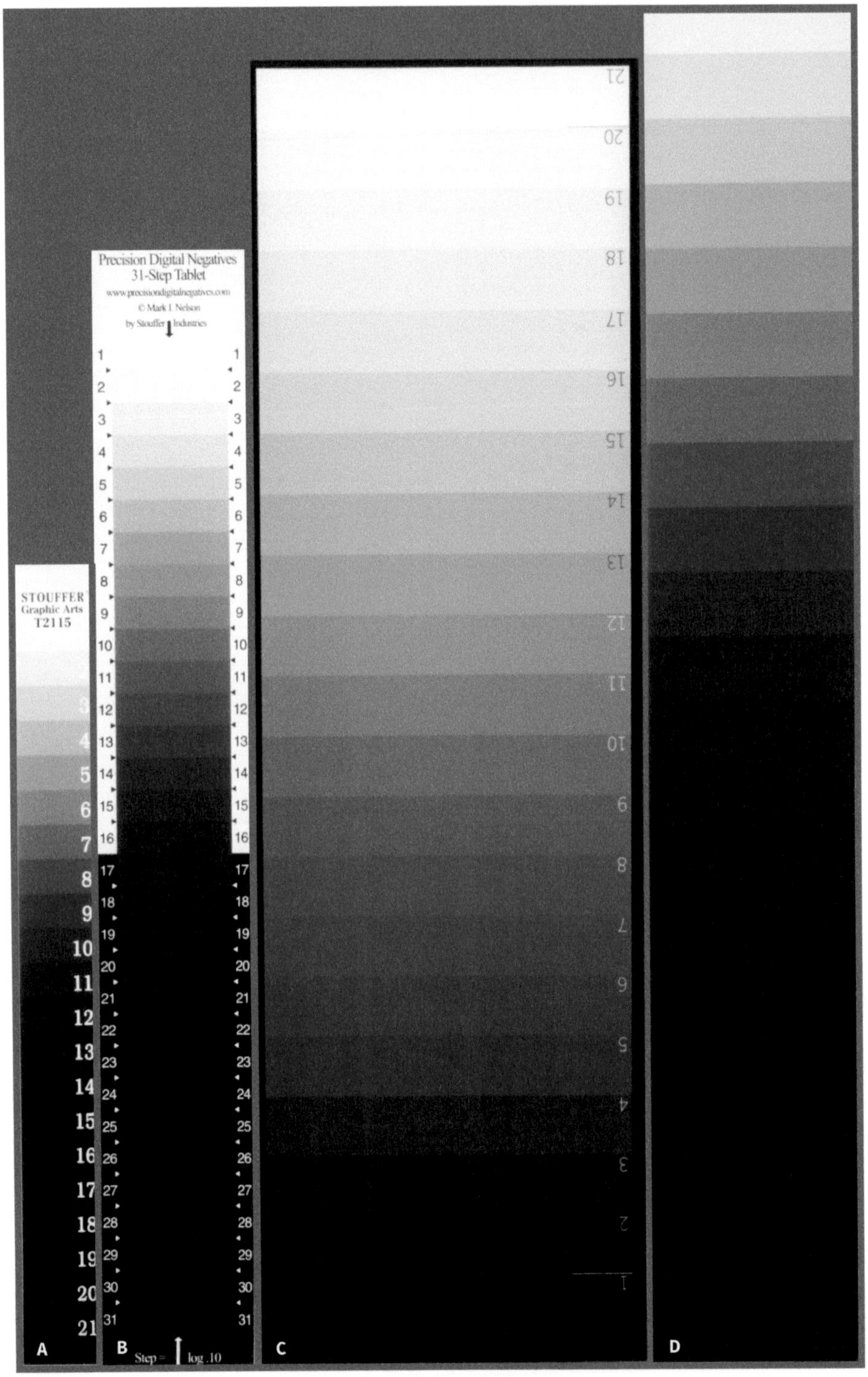

Figure 4.6 Varieties of step wedges: A: Stouffer; B: Precision Digital Negatives; C: digitally printed (downloadable from pradipmalde.com/step-wedge); and D: Kodak No. 2.

Storage Bottles

A wide-mouthed bottle for each of the two EDTA clearing baths, with sufficient capacity to hold two to four liters of liquid each. They can be simple, clear plastic containers.

Syringes

Disposable, *Luer* slip tip plastic syringes for measuring and dispensing small volumes of sensitizer. Even when using an automatic pipette for measuring (see Chapter 5), dispensing the solution onto paper may be better accomplished with a syringe. Have on hand several of each of the various capacities, of which the 1, 3, 5, and 10 ml are most useful. If kept clean, they may be reused dozens of times. Do not use the "lock" tip, and most emphatically, don't fit syringes with the hypodermic needles for which they were originally designed.

Thermometer

For measuring air and liquid temperatures; designed for photographic use with a waterproof tip or probe, and a range covering room temperatures to at least 125°F / 50°C.

Timers

Darkroom wall or desk timers, stopwatches, or smartphone apps are essential for timing various stages of the printing process. It is likely that the workflow will be at several different stages simultaneously, especially as the process becomes familiar, and will require multiple timers. A convenient, and ultimately cheaper, tool is the MultiTimer™ app by Serghei Astakhov for iOS—similar apps are available for other platforms.

Ultraviolet Light Source

Also See Chapter 3, Ultraviolet Radiation

Any light source with a substantial ultraviolet content will serve for exposing platinum/palladium prints, or any other siderotype process. However, sources that also emit significant infrared radiation, such as "sun-guns," should be avoided because of the undesirable effect of heating the sensitized paper as the exposure proceeds, so affecting the humidity and consequently the color of the image (Chapter 5 describes and explains the correlative effects of heat, humidity, and color). The preferred sources are therefore: an LED array; a "cold" bank of fluorescent tubes containing mercury vapor; or a single-point metal halide lamp. For reasons already pointed out in Chapter 3, regardless of the source, use long-wave UVA lamps emitting chiefly a wavelength range of 315–400 nm, with a maximum output centered on a wavelength around 350–370 nm.

LED Arrays

A fairly new, but rapidly evolving technology presents some exciting possibilities for platinum/palladium printing. LEDs have low heat emission, can be designed to peak at very specific wavelengths, and emit far more intense radiation than the more traditional technologies, and all at lower, manageable power. Commercial manufactured units are available, but at a high cost. We can only assume that as usage increases, costs will come down. For now, I refer you to Appendix IV: Resources.

Fluorescent Light Sources

In Europe, one of the major manufacturers of UV fluorescent tubes is Philips PLC, whose designations for suitable lamps are: actinic /05 (peaking at 370 nm) or actinic /09 (peaking at 355 nm). These are available in various lengths and power ratings, but are usually in the order of 10−20 watts. In the USA the tubes made by Sylvania, called 350BL, have the desired emission, as do the BL tubes of General Electric. Users in the USA should note that the manufacturers' designation "BL" indicates a suitable tube; this stands for "black light"—even though these sources do also emit much visible light. The US designation "BLB" stands for "black light blue," as in the General Electric BLB tube, and signifies tubes that

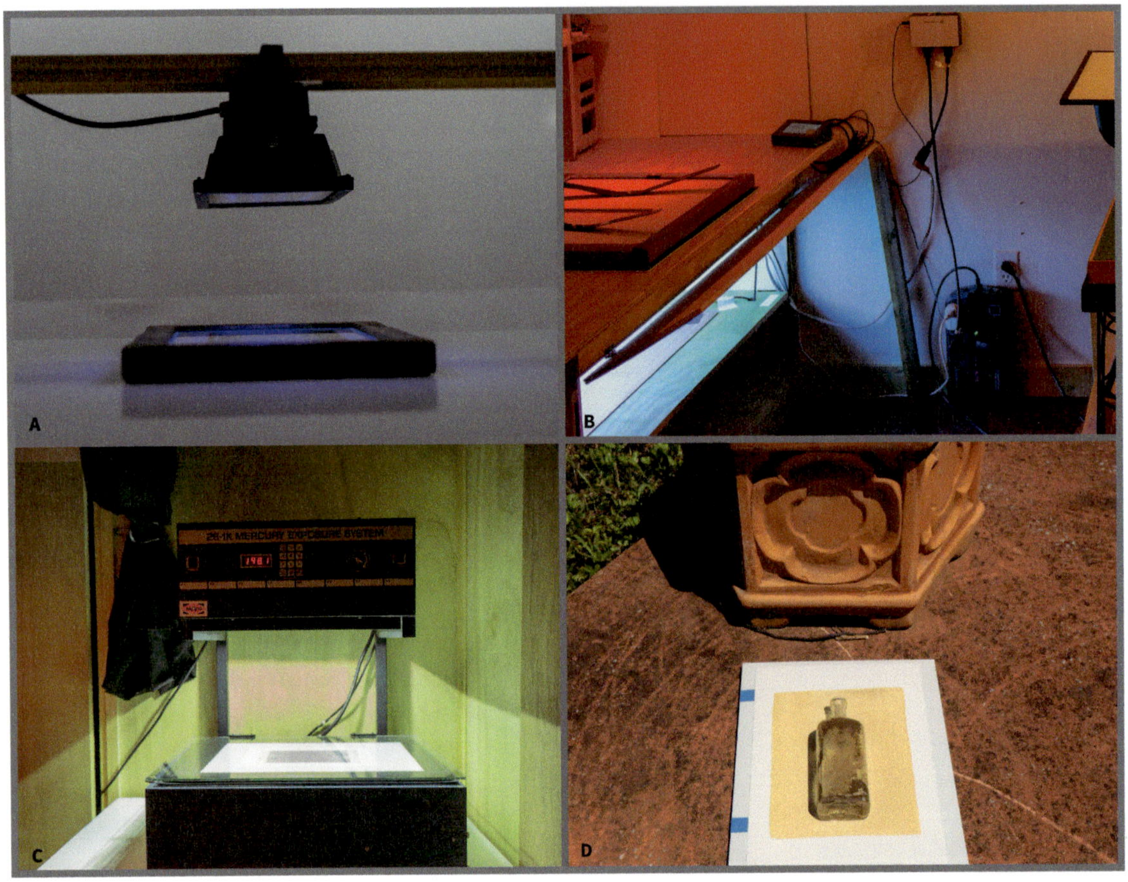

Figure 4.7 Four types of UV sources:
A. LED types are becoming more affordable and widely available.
B. fluorescent tube sources may be readily purchased, and easy to use and install.
C. more expensive metal halide units may also require professional installation and extra safety precautions.
D. the most available source, the sun, is best used along with an ultraviolet light meter.

are fitted with a deep blue glass envelope that filters out almost all the visible light, transmitting only a band of UV around 350 nm. While these BLB tubes may create a desirable ambiance in a club, and they do perform quite acceptably for print-making, their use is not advantageous, either in efficiency of light output or their greater cost.

Also available is a type of fluorescent called "Super-Actinic" (SA) or "Aqua" (these are intended for aquarium use), with a peak output in the blue, around 420 nm (this corresponds with Philips lamps designated as actinic /03). Such lamps are not, however, very efficient for platinum/palladium printing, and even cyanotype. Claims to the contrary have largely been based on work with related processes such as kallitype and chrysotype, where the internal filter effect[3] is not as strong as with other siderotype processes.[4] The standard SA tube does have the advantage of being nearly four times as powerful (75 W instead of 20 W) which compensates somewhat for its lower actinic efficiency.

Fluorescent tubes are usually connected to "instant start" ballasts that are designed to reduce start-up time to just a few seconds. Each on/off cycle releases some emissive material[5] and there is a corresponding reduction in the effective life of the lamp. Printing requires frequent on and off switching, and with these standard ballasts, lamps can wear out rapidly. Two other kinds of ballasts are available. Ideally, units should be designed with either the "rapid start," or the preferable "programmed start" ballasts. Both extend lamp life by almost 50%, with the latter design running a short, cathode-heating cycle before full lamp voltage is applied.

Consider the use of an integrator—see *Ultraviolet integrator* .

Metal Halide Lamps

Metal halide lamps are usually single-point light sources and physically much smaller than the fluorescent arrays, and must be used at a greater distance to achieve uniformity of illumination. They therefore run at much greater power (typically 1000 watts or more) to deliver a practical flux of ultraviolet at the print. Their only advantage, and a significant one for some, is that as point sources, the acutance of the print is higher. This is especially important if there is a plastic film intervening between negative and paper. Equipment using these lamps should also have an integrator to compensate for the significant "warm up period" required by metal halide lamps, during which the output is not constant. See the note on *UV light meter (integrator).*

Ultraviolet Exposure Units

The following ultraviolet sources are listed in approximately ascending order of cost. Inspection of the printed-out image, formed during exposure, provides an effective way of determining density. However, for fine-tuned decisions, and for those cases where similar versions of the same print are desired, I strongly recommend the use of a UV light meter, even when a unit is equipped with an integrator. See *Ultraviolet integrator (or ultraviolet light meter)* below.

Sun

This is of course free, and pleasant to employ, but it is an uncertain and variable source. It is effectively a point source, subtending an angle of half a degree of arc (0.01 radians), consequently giving high acutance.[6] While difficult to compare to artificial light sources because of its variability, the sun can be about four times (2 stops) faster than a typical fluorescent tube printer. There is a significant heating effect, however, which may have a detrimental effect on both the negative and the print. The use of an ultraviolet light meter will help produce more consistent results.

Domestic Sunlamp

Such as a Pifco 300-watt UV lamp No. 1012 or similar. These small, single-bulb domestic sunlamps can still sometimes be purchased cheaply from yard sales and thrift shops, especially since there is a greater awareness of the dangers of melanoma and people are switching to non-UV emitting light therapy lamps. They should be used at a distance of ca. 12″ to 20″ / 30 cm to 50 cm from the printing frame to achieve uniformity of illumination over the print, and consequently will be rather slow in printing.

Mercury Discharge Reprographic Lamps, such as the Phillips HPR 125

These should be used at a distance of ca. 12″ / 30 cm from the frame, or at a greater distance if an array is employed to cover a large area. They are high pressure mercury arcs, contained in an outer glass envelope fitted with a standard ES connection. They require the usual mercury discharge control gear for the ballast, and are commonly employed in commercial "plate burners" and "screen-printers" for the graphic arts.

LED and Coiled Fluorescent UV Lamps

There are numerous lamps for screen printing, or illuminating reptile houses, aquariums, and plants, and even for other purposes such as hardening adhesives and fingernails. Lamps come either already housed with a plug, or ready to be fitted into a standard lamp socket. Multiple units can easily be conjoined to cover a desired printing area. Be sure that specifications indicate a peak that includes the 370 to 350 nm wavelengths, to which platinum/palladium printing is most responsive.

A Commercial Luminaire, Fitted with UVA Tubes

Lighting equipment suppliers sell such ready-made housings for arrays of fluorescent tubes, equipped with all the sockets and control gear to take four or more 24″ / 60 cm or larger, fluorescent tubes mounted in a metal case, which may be provided with reflectors. The plastic diffusers also commonly supplied should not be used.

Graphic Arts and Screen-Printing Exposure Units

Designed for hobbyists, look for units that have at least four light arrays and adequate coverage for at least 20% more than the contact print frame or full paper size of the print.

NuArc System

This is a commercial UV exposure system, with a small high intensity mercury or metal halide lamp, and a vacuum easel. A variety of sizes and designs are all equipped with a light integrator. Manufactured for the graphic arts and screen-printing industry, they are steadily being replaced by LED versions, and can often be bought used. However, even these can be expensive, and the cost of transport, installation, and space needed to house the equipment should be considered.

UV Light Source Made for Photographic Processes

Several manufacturers are now making exposure units. It is usually an array of fluorescent tubes, with or without a vacuum frame and integrated timers.

Commercial Graphic Arts Units

State-of-the-art commercial UV exposure systems now come with LED arrays, vacuum frames, timers and even UV light meters, and can be recommended for platinum/palladium printing when high cost is no obstacle.

Ultraviolet Equipment Design Guidelines

It is not in the scope of this book to provide instructions for fabricating an exposure unit, but the following details will help specify a purchase, or design a suitable piece of equipment. Generally, a source should provide intense and even coverage, not generate excessive heat, allow close contact between negative and paper, and avoid unnecessary spillage of ultraviolet; all of this in a controlled and repeatable manner.

Coverage

You will be using one of two light sources: either a "point" source such as a single metal halide lamp, or a "diffuse" array made from multiple lamps. In either case, we need to establish a bright and even distribution of light. Placing the printing frame close to the light source increases non-uniformity of illumination and the degree of heating. Increasing the distance between source and print surface can project a more even pool of light, with a cooler temperature. However, this adjustment leads to light loss and longer exposures.

Point Sources

It is well-known that the intensity of light from a *point* source falls off predictably as the inverse square of the distance—for example, when the distance between a lamp and the printing frame is doubled, the intensity of ultraviolet reaching the printing frame's surface is reduced to one quarter of its origin. So, the ideal arrangement is to have the shortest possible distance while maintaining a homogenous distribution of light at the printing surface. For a point source, expect significant changes of exposure as distances between it and the printing surface are adjusted to arrive at an even and cool distribution of light.

Diffuse Sources

Things get a little bit more complicated mathematically, however, when designing diffuse, or light bed sources. The separation between the fluorescent tubes, or LEDs, in a light bed array has some bearing on the uniformity of illumination at the printing surface. When the tube separation, for instance, is too large compared with the distance to the print surface, "barring" or "banding" of the print will occur, due to parallel regions where the light intensity is greater. From a line source (such as a single tube light—but keep in mind there is no single "point" of light emanating from it), it may be proved that the intensity falls off inversely to the distance, just as with the point source dynamic we discussed above. However, we are dealing with a source made of multiple tubes, a "light bed," an array of non-finite[7] points to which the inverse square law only applies neatly after a considerable distance. For small amounts, sufficient to even out illumination, the intensity does *not* fall off at all with increasing distance! For a diffuse source, exposure changes are almost insignificant as adjustments of up to ca. 6″ / 15 cm are made between it and the printing surface.

Heat

The undesirable effect of heat on print quality is discussed in Chapter 8. Lamp(s), housing, and the printing surface need to be maintained at as close to room temperature as possible. Arrange ventilation systems within and around the unit to ensure a well-circulated air supply.

Contact Pressure

Maintain contact across the entire surface of the negative and print assembly. This is especially important with diffuse sources. The use of vacuum frames may be necessary for large prints. Units and lamp-to-printing frame coverage and distances should be designed to accommodate the height of the printing frame.

Protection

Exposure assemblies should include optical shielding so that stray ultraviolet emissions are kept to a minimum. Ballasts, integrators, fans, and vacuums can create significant noise, and if possible, steps should be taken to reduce these levels.

Ultraviolet Integrator (or Ultraviolet Light Meter)

As has been noted above in *Ultraviolet light source*, bulbs require some warm-up time before emitting a steady amount of ultraviolet. This irregularity can be frustrating, even with a printing-out process, if one wishes to make prints in a predictable and controlled manner. Normally, exposure is measured with time, on the assumption that the light source's intensity is consistent. If the light source fluctuates in intensity, however, then exposures are going to vary from print to print despite of all of them having been placed in UV for the same duration. For finer controls, it is helpful to determine exposures based on the amount of ultraviolet being emitted, rather than for how much time a unit has been switched on. An integrator measures ultraviolet continuously, and allows us to numerically determine exactly how much ultraviolet is being emitted or arriving at the printing surface. Some integrators, which are essentially small sensors, are coupled with switches that control the power supply. On completion of a desired amount, or dose, of ultraviolet—previously dialled into the unit's display—power is automatically switched off. This kind of setup is usually found in the NuArc Exposure System units. External integrators can be coupled with one or more sensors and a power supply, and incorporated into custom-built or premade light array units. This latter type is available from companies like Lightmeasure.com. I have found Ian Leake's[8] LC series of integrators to be particularly useful.

UV-Blocking Eye Glasses

Frames with clear, optical quality UV-blocking lenses are ideal. These glasses are important to wear if working in areas where there is any stray ultraviolet light.

Vacuum Frame

(See *Contact-printing frame*) The common, spring-backed frame does not provide adequate pressure for very large contact prints. 11″ x 14″ negatives or larger, for instance, are better printed with more sophisticated technology such as a vacuum easel, but this precludes the possibility of inspecting the progress of the print during exposure, unless the negative is taped (at two points) to the underlying sheet, or the easel is fitted with registration pins. Care must be taken that the application of the vacuum does not change the humidity of the sensitized paper, by "pumping off" the water during the exposure; once again a vapor barrier sheet, placed just below the sensitized paper, is needed.

Wash-Bottle

Called a goose-necked plastic wash bottle, this is useful for dispensing purified water for cleanup of the coating equipment.

Weighing Scale

Single-pan digital scales functioning to a precision of ± 0.1 gram are affordable and practical. For making up large quantities of processing solutions, a precision of ± 1 g, obtainable with some kitchen scales, will usually suffice. Electronic balances or kitchen scales have a "tare" facility which enables the scale reading to be reset to zero with the empty vessel in-situ. The scale reading then gives the weight of the added material alone. You should always protect the balance pans from direct contact with chemicals, which may corrode them, by containing the substance in a glass or plastic beaker, or some kind of "boat."

Chemicals List

The following section lists alphabetically all the chemicals that are referred to in this book. Each entry is organized by name, chemical formula, alternative names, and usage. Chapter 5 provides detailed information about the preparation and actual usage of each of the following chemicals.

Ammonium Chloride

NH_4Cl

Also known as sal ammoniac, the salt of ammonia and hydrogen chloride. This is combined with palladium(II) chloride as an alternative method for making an ammonium tetrachloropalladate(II) solution. Ammonium chloride may also be used to generate a constant RH 80% chamber for hydrating sensitized paper.

Ammonium Iron(III) Oxalate Trihydrate

$(NH_4)_3[Fe(C_2O_4)_3].3H_2O$

Referred to in this text as ammonium ferric oxalate, but also commonly known as ferric ammonium oxalate, ammonium ferrioxalate, or ammonium trisoxalatoferrate(III) trihydrate. You may also find it being called ammonium iron(III) oxalate. This is the light-sensitive ingredient in the sensitizer.

Ammonium Ferric Oxalate

See above.

Ammonium Oxalate Monohydrate

$(NH_4)_2C_2O_4.H_2O$

Used with ferric nitrate to make ammonium ferric oxalate. Formulated this way, the sensitizer will render cooler print tones than when made with just pure ammonium iron(III) oxalate trihydrate.

Ammonium Nitrate

NH_4NO_3

Used as an optional additive to ammonium ferric oxalate solution to produce cooler print tones.

Ammonium Tetrachloroplatinate(II)

$(NH_4)_2[PtCl_4]$

Also known (inaccurately) as ammonium chloroplatinite or ammonium tetrachloroplatinite,

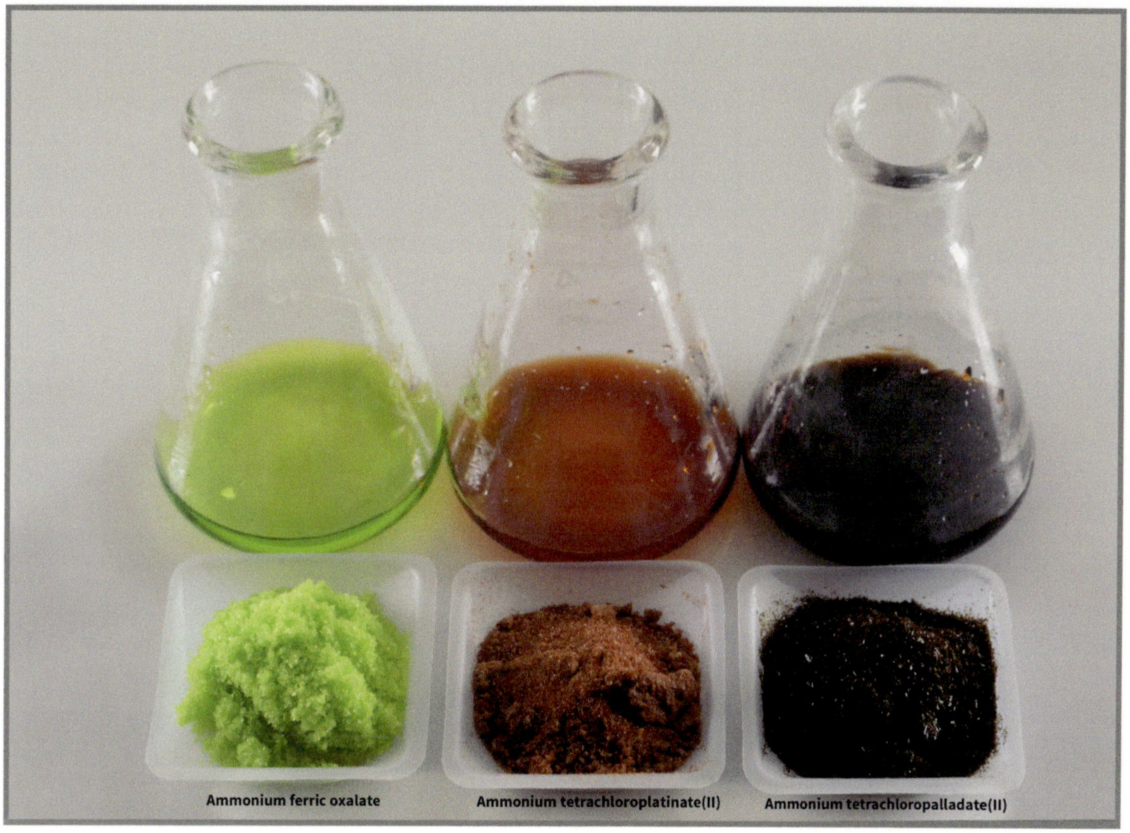

Figure 4.8 Left to right, ammonium ferric oxalate, ammonium tetrachloroplatinate(II), and ammonium tetrachloropalladate(II), in crystalline and solution forms.

and not to be confused with ammonium hexachloroplatinate(IV). Ammonium tetrachloroplatinate(II) is the metal salt that ultimately renders the platinum image. A solution of this salt is combined with an ammonium ferric oxalate solution to make a working sensitizer for making platinum prints. To produce platinum-palladium prints, ammonium tetrachloroplatinate(II) is combined with ammonium tetrachloropalladate(II) and ammonium ferric oxalate solutions, to make a working sensitizer solution.

Ammonium Tetrachloropalladate(II)

$(NH_4)_2[PdCl_4]$

Also known as ammonium tetrachloropalladite or ammonium chloropalladite. This is the metal salt that ultimately renders the palladium image. A solution of this salt is combined with an ammonium ferric oxalate solution to make a working sensitizer for making palladium prints. To produce palladium-platinum or platinum-palladium prints, ammonium tetrachloropalladate(II) is combined with ammonium tetrachloroplatinate(II) and ammonium ferric oxalate solutions, to make a working sensitizer solution.

Calcium Chloride

$CaCl_2$

Also known as calcium dichloride or calcium(II) chloride. Calcium chloride is used to generate a constant RH 32% chamber for hydrating, or a RH ~9% chamber for desiccating sensitized paper.

Calcium Nitrate

$Ca(NO_3)_2$

Not to be supplanted with fertilizer-grade calcium nitrate, which usually also contains ammonium nitrate. Calcium nitrate tetrahydrate is used to generate a constant RH 56% chamber for hydrating sensitized paper.

EDTA

$C_{10}H_{16}N_2O_8$ or $C_2H_4(N(CH_2COOH)_2)_2$

Known as ethylenediaminetetraacetic acid[9] or H_4EDTA or free tetrabasic acid H_4EDTA or edetic acid. If disodium or tetrasodium EDTA is not available (see the next entries), H_4EDTA is mixed, as described below, with one of several sodium alkalies to prepare initial and final clearing baths.

Dissolve **one** of the following amounts of alkali per 1000 cc of water, then add ~35 g of H_4EDTA in small portions with stirring, allowing (in the case of the carbonates) the effervescence of CO_2 to subside:

Alkali	**to make: Na_2EDTA / Na_4EDTA**
Sodium hydroxide	NaOH 9.6 / 19.2 grams
Sodium carbonate	Na_2CO_3 (anhyd.) 12.7 / 25.4
Sodium carbonate	$Na_2CO_3.10H_2O$ 34.3 / 68.6
Sodium bicarbonate	$NaHCO_3$ 20.2 / 40.3

EDTA, Disodium

$C_{10}H_{14}N_2O_8Na_2.2H_2O$ or $C_2H_4(N(CH_2COOH)_2)(N(CH_2COONa)_2)$

Most commonly known as disodium EDTA, or as the disodium salt of ethylenediaminetetraacetic acid. Disodium EDTA is used for the first clearing bath.

EDTA, Tetrasodium

$C_{10}H_{12}N_2O_8Na_4.2H_2O$ or $C_2H_4(N(CH_2COONa)_2)_2$

Most commonly known as tetrasodium EDTA. Tetrasodium EDTA is used for the third and final clearing bath.

Ferric Ammonium Oxalate

$(NH_4)_3[Fe(C_2O_4)_3].3H_2O$

See Ammonium iron(III) oxalate trihydrate above.

Ferric Nitrate Nonahydrate

$Fe(NO_3)_3.9H_2O$

Also known as iron(III) nitrate nonahydrate. Used with ammonium oxalate to make ammonium ferric oxalate with excess ammonium nitrate. Formulated this way, the sensitizer will render cooler print tones than when made with just pure ammonium ferric oxalate.

Glycerin

$C_3H_8O_3$

Also known as glycerine or (more correctly) glycerol. Glycerin is used as an additive to the sensitizer to produce smoother print tones with certain papers, as well as for certain special "development" techniques.[10]

Hydrobromic Acid

HBr

Also known as Hydronium bromide. Hydrobromic acid may be used to reduce ("bleach") out black spots of palladium in the print, as a retouching strategy.

Palladium(II) Chloride

$PdCl_2$

Also known as palladium dichloride. This is combined with ammonium chloride as an alternative method for making an ammonium tetrachloropalladate(II) solution.

Magnesium Chloride

$MgCl_2$

Magnesium chloride is used to generate a constant RH 33% chamber for hydrating sensitized paper.

Sodium Bicarbonate

$NaHCO_3$

Also known as baking soda. If disodium or tetrasodium EDTA is not available, sodium bicarbonate is mixed with H_4EDTA to prepare the initial and final clearing baths.

Sodium Carbonate Anhydrous

Na_2CO_3

Also known as anhydrous sodium carbonate, and not to be confused with sodium bicarbonate. If disodium or tetrasodium EDTA is not available, anhydrous sodium carbonate is mixed with H_4EDTA to prepare the initial and final clearing baths.

Sodium Carbonate Decahydrate

$Na_2CO_3.10H_2O$

Not to be confused with sodium bicarbonate. If disodium or tetrasodium EDTA is not available, sodium carbonate decahydrate is mixed with H_4EDTA to prepare the initial and final clearing baths.

Sodium Chloride

NaCl

Common, table, or Kosher salt. Sodium chloride is used to generate a constant RH 75% chamber for hydrating sensitized paper.

Sodium Sulfite

Na_2SO_3

Also known as anhydrous sodium sulfite or disodium sulfite. Sodium sulfite is used in the second clearing bath.

Sulfamic Acid

H_3NSO_3 or H_2NSO_3H

Also known as amidosulfonic acid, amidosulfuric acid, aminosulfonic acid, and sulfamidic acid. It is a colorless, water-soluble compound used to decalcify buffered papers.

Tween 20

$C_{58}O_{26}H_{114}$

Also known as polyoxyethylene(20)-sorbitan-monolaurate or polysorbate 20. Tween is a surfactant (or wetting agent) used as an additive to the sensitizer to produce smoother print tones with certain papers.

Notes

1 John A. Spencer, "Instrument for Spreading Solutions," *Journal of the Photographic Society*, 7 (July 21,1853) 88-9; Joseph Sidebotham, "The Printing Process in Photography," *Recreative Science* 2, London: Groombridge and Sons 1861, p. 120; Marcus Sparling, ed. James Martin, "The Theory and Practice of the Photographic Art," *The Circle of the Sciences: 7 Practical Chemistry*, London: Griffin, Bohn and Co., 1856, p. 142.

2 For more on this, see William Shakespeare, *Hamlet*, Act 3, scene 1.

3 The internal filter effect is the absorption of actinic light by a component of the sensitizer that is not photoactive. These long wavelength lamps may be fairly effective for chrysotype, because at the 420nm wavelength there is little or no "internal filter" effect due to the lack of absorption by the colourless gold(I) complex.

4 See Mike Ware, *The Chrysotype Manual: The Science and Practice of Photographic Printing in Gold*. Brighton: ffotofilm publishing, 2006, p. 156.

5 Universal Lighting Technologies. *Programmed Start: the New Way to Start a Fluorescent Lamp*. Unvlt.com, https://unvlt.com/resources/literature-finder/articles-and-tech-papers/programmed-start-the-new-way-to-start-a-fluorescent-lamp.html, accessed: 10.26.2019.

6 The apparent relative angular motion of the sun of 15°/hour (i.e. it moves through its own apparent diameter in 2 minutes) is not a serious consideration for printing, unless the object (diaphane or negative) is not in contact with the paper, such as when making photograms; in which case exposures longer than one or two minutes can result in blurred edges to the shadows.

7 In practice, of course, no real source is infinite in extent, and that is why I use the word "non-finite."

8 Also see his blog article about light integrators: Ian Leake, *Why I Built My Light Integrator – And Why Calibration Is So Important*, https://ianleake.com/blog/why-i-built-my-light-integrator-and-why calibration-is-so-important/, accessed 11.01.2019.

9 'diaminoethanetetraacetic' is pronounced diy-uhmeenoh-eethayn-tehtruh-uhseetik, based on https://en.wikipedia.org/wiki/Wikipedia:Pronunciation_(simple_guide_to_markup,_American).

10 Adrienne Lundgren, "The History and Use of Glycerine in Platinum Printing," in *Platinum and Palladium Photographs: Technical History, Connoisseurship, and Preservation*, edited by Constance McCabe, pp. 204–217. Washington DC: Photographic Materials Group of the American Institute for Conservation of Historic and Artistic Works, 2017.

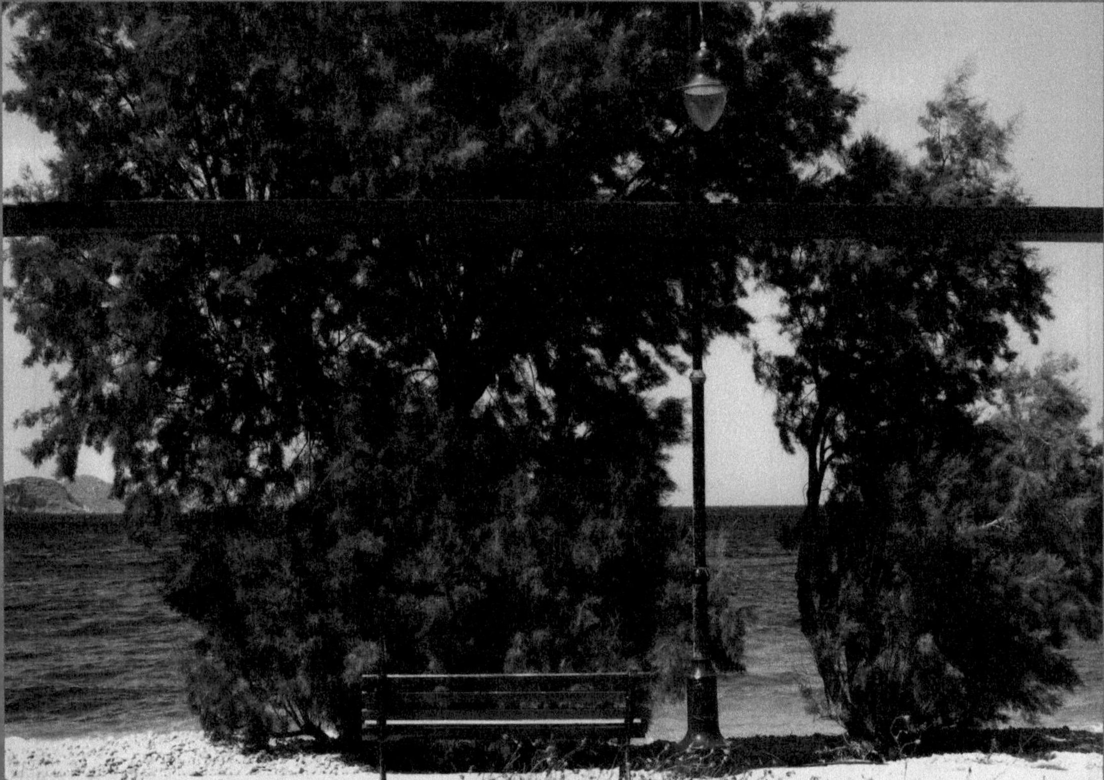

Figure 5.0 Mike Ware, *Sea View, Tilos*, 2011. Platinum-palladium print on Buxton from enlarged digital negative. ca. 30 x 40 cm.

Chapter 5

Preparation of Sensitizer and Processing Solutions

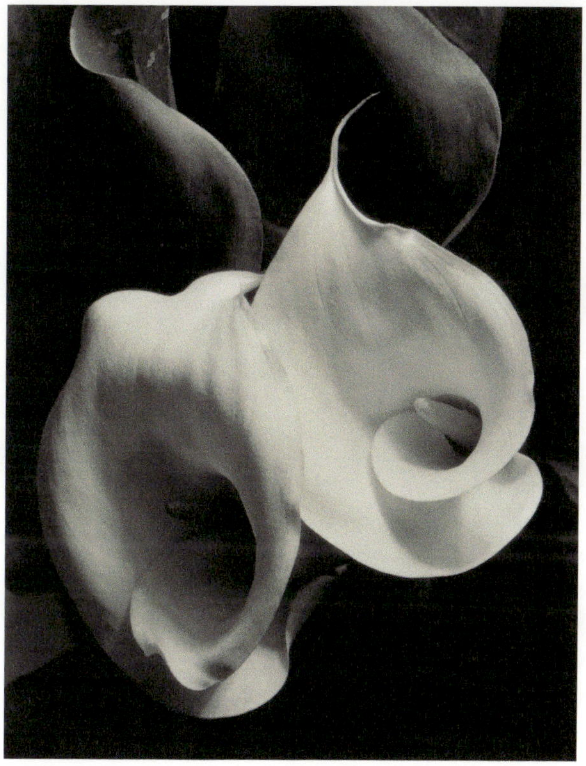

Figure 5.1 Imogen Cunningham, *Two Callas*, ca. 1929. Platinum-palladium print on Revere Platinum paper, from PiezoDN version of the original 4 x 5 inch negative, 22.5 x 19.5 cm. Printed by Pradip Malde. Courtesy Imogen Cunningham Trust. Fe3:Pt:Pd = 2:1:1, hydration for 90 minutes at 76% RH / 20°C.

A couple of points should be noted before going much further. The first is a reiteration: the general procedural guidelines in the following paragraphs about handling chemicals may be applied to any traditional photographic process. The other is that the printing-out process described in this book is much safer, and more forgiving of minor variations and inaccuracies in preparation and measurement, than traditional platinum/palladium methods. However, if you aspire to achieve the finest results, with a high degree of reproducibility and the maximum economy, then you should be prepared to make careful measurements of the relevant quantities, while maintaining a record of your procedures. The following pages provide exact formulae for preparing all the solutions you may need to make platinum/palladium prints. Some of these, especially the sensitizer solutions, may be purchased in pre-mixed volumes from suppliers such as Cone Editions in Vermont, USA, and Gold Street Studios in East Victoria, Australia. All solutions must be handled safely and with care. Installing and using equipment, as we have examined in the previous two chapters, is only part of an overall safe practice. Along with sustaining a healthy work environment, you need to be aware of standard procedures for the handling, measurement and preparation of chemicals. *Please read through this first part of the chapter before preparing any solutions.*

As indicated above, we assume you will always use appropriate protective clothing and equipment, that your work surfaces and labware are clean, that you are in an adequately lit and well-ventilated space, and that you are aware of any hazards as stated in the MSDS. Remember, always treat chemicals with care and mindfulness. Use, label, and store them under the assumption that others, such as children or pets, could gain access. Protect yourself and others from direct contact with chemicals by wearing gloves and goggles, and wash your hands thoroughly after handling them. From this base-line of good practice, we now consider how to measure and prepare chemicals precisely and consistently.

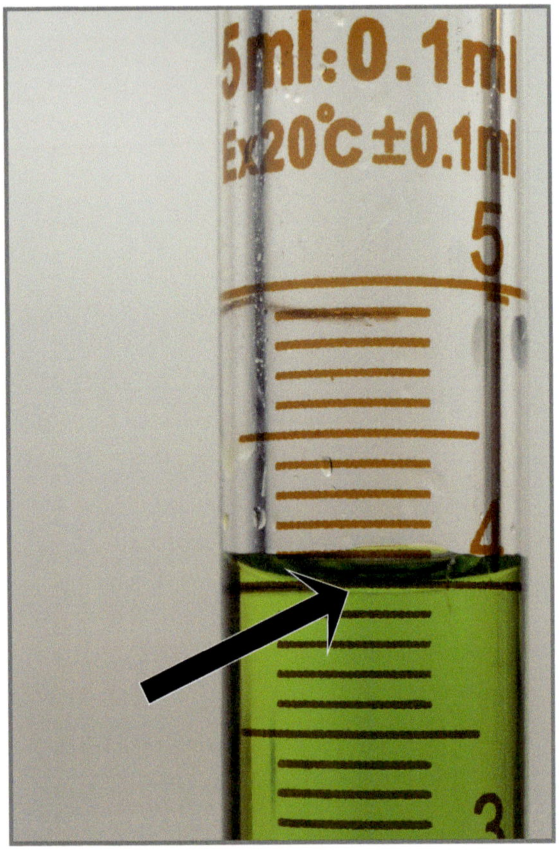

Figure 5.2 The arrow indicates the point from which a measurement is taken on a meniscus.

Measuring Liquids

Large volumes of liquid, greater than ca. 10 ml are measured out with sufficient precision for our purposes by means of graduated measuring cylinders. The plastic type of measuring cylinder is adequate, but glass is preferred for better precision of measurement and because it can be cleaned more thoroughly. A good quality measuring cylinder of 100 ml capacity is graduated so that it can be read to within 1 ml—a precision approaching ± 1%. Remember that the calibration marks on all volumetric chemical glassware are based on the rule of use that it is the bottom of the curved liquid meniscus that should just touch the calibration line.

Use accurate and well-marked containers—it is worth spending a little bit more for professional-grade equipment, especially for measuring volumes between 10 to 100 ml.

Very small volumes of less than 10 ml are best measured by means of graduated syringes or pipettes. They come in a variety of tips, but it is recommended that you use the *Luer* slip tip, not the lock tip, and most emphatically, these are not to be fitted with the hypodermic needles for which they were originally designed. Cheap, disposable plastic syringes can be used many times before they need to be discarded and are available in various capacities, of which the 1, 3, 5, and 10 ml sizes will be found most useful. Their precision of measurement is quite adequate, and the wastage of precious solutions is very small. Always use a syringe that is minimally larger than the volume you wish to measure, so as to maintain the maximum precision. So, to measure 0.75 ml, use a 1 ml syringe, not a 5 ml or 10 ml size, which have coarser graduations.

A slightly more expensive but extremely convenient alternative for precise volume measurement is provided by adjustable micropipettes, also manufactured in a variety of sizes. They provide extremely accurate and repeatable measurements of small volumes. Micropipettes are graduated in microliters, abbreviation "µl." One milliliter is 1000 µl. The 100–1000 µl size (0.1 to 1.0 ml) is good for extremely small volumes such as those needed for adding Tween and/or glycerin to the sensitizer, or coating 4″ x 5″ / 10 cm x 12 cm areas, and the 1000–5000 µl sizes (1 ml to 5 ml) may be most appropriate for volumes for areas up to 11 × 14 inches. Slip tips will be needed for each size, but only a few—they last a long time. Yet another, but fragile, alternative for precise volume measurement is provided by graduated glass pipettes, especially the kind used in hematology, called serological pipettes, usually of 1 or 2 ml capacity. These must be used with a pipette filler (a rubber bulb or piston type), or simply a syringe of similar or larger volume with a short connection of rubber tube 1 cm long. *Operation of pipettes by mouth is now strictly forbidden in all chemical practice.* Each sensitizer stock solution should have its own dedicated syringe(s) or pipette, which should be labeled on the barrel and plunger to avoid cross-contamination (use small self-adhesive colored dots or similar tagging system). Regardless of the exact type of instrument used for measuring small volumes, attention needs to be placed on its method of use.

1. Ensure that the plunger of the syringe is fully depressed (or for micropipette, follow specific usage instructions—they are usually plunged in two steps, with the first stage being the measurement preset, and the second stage for ejecting).
2. Place the nozzle well below the liquid level in the bottle, and draw the liquid into the syringe by slowly withdrawing the plunger to a point above the desired volume. Do not accidentally draw air into the syringe by surfacing the nozzle tip.
3. With the syringe held vertically, nozzle down, over the bottle, gently expel the liquid until you reach the desired volume. Take the volume reading from the bottom of the plunger, not from the meniscus of the liquid.
4. Ignore the small air bubble that is present in the syringe between plunger and liquid. Do not attempt to expel the air bubble. The "missing" volume of liquid corresponding to the air gap is exactly compensated by the extra liquid contained within the uncalibrated tip of the syringe, a volume which is included when practically all the liquid is expelled by the plunger, thus there is little or no waste by this method.
5. Wash out syringes thoroughly after use by pumping them in pure water, and pull the plunger out to dry separately from the barrel.

Since the earliest days of alternative processes, in the nineteenth century, when calibrated equipment was less available, it has become a common practice among workers to measure small volumes of solutions by counting the number of drops from a dropping pipette or "medicine-dropper" bottle. This is a relative measurement. While the method may be fairly consistent for the practice of a given individual, who always uses the same droppers in the same way, you should bear in mind that the volume of one person's drop may differ considerably from another person's drop—much depends on variables such as temperature, location on the planet (gravitational force varies), and the steadiness of one's hand. It is worth taking a quick deep-dive: according to the physics of

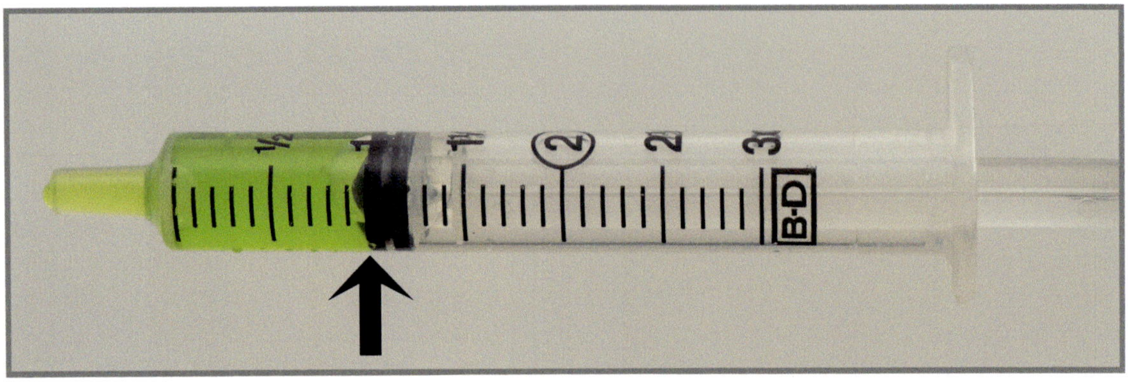

Figure 5.3 Arrow indicates the bottom of the syringe plunger, from which point volumes are read against the graduated marks.

stalagmometry, the volume of a drop is proportional to the radius of the dropper outlet, and also depends on the density and surface tension of the liquid being dispensed. A water drop can vary between 0.05–0.06 ml (a "minim"). A drop of equal parts of platinum-palladium sensitizer will have a different volume not just from water, but also from a pure platinum sensitizer. Add to this the tedium of carefully counting out, for example, 43 drops of expensive sensitizer solutions and it quickly makes sense to employ a proper scientific method of measuring volume in the interests of consistency, economy, and the accurate sharing of information. Bulb droppers or dropping pipettes do have their uses, however, for transferring very small volumes of solution, or as a convenient means of making additions where precision is low, e.g. for Tween.

Weighing Solids

All weights in this book are expressed in grams, universally abbreviated as "g."[1]

Electronic scales with a single pan, functioning to a precision of ± 0.1 g are now inexpensive, and are ideal for our purposes. Traditional beam balances may also be used, provided they have sufficient precision for the sensitizer formulae. A coarser precision of ± 1g, measured with some types of kitchen scales, is adequate for preparing larger quantities such as those needed for clearing solutions. You should always protect the balance pans from direct contact with chemicals, which may corrode them, by containing the substance in a glass or plastic beaker. Alternatively, disposable containers may be used: a creased paper sheet will serve, but paper cupcake containers, plastic tubs, or cleaned yoghurt pots are less likely to lead to spillages. Small amounts of solid should be transferred using a spatula or plastic spoon. Remember to subtract the weight of the empty vessel or paper from the total via the scale's tare function. Always ensure that the tare on such balances is reset to zero at the start of each weighing procedure.

Preparing Solutions

The preparation of chemical solutions is carried out according to their formulae, or recipes. A properly written formula, as with good culinary recipes, sometimes starts with a list of needed equipment, but always begins with a list of ingredients and their required amounts and temperatures (where applicable), arranged in the order of use. This is followed by a narrative set of instructions. *Hazards are emphasized in an italicized font.* Read a formula thoroughly before carrying out any instructions. Take notes as you prepare, and then assemble equipment and ingredients. If you are using hazardous materials or equipment, following procedures for the first time, or carrying out hazardous procedures, let someone know ahead of time, and ask them to check in with you at a suitable later time. Work with care and think ahead.

The sequence of instructions in preparing a solution is important. For instance, if directed to add sodium sulfite to water, do not add water to sodium sulfite. Always dissolve chemicals before adding another and follow temperature guidelines. Generally, while it is unlikely that great physical harm can come from not following formulae exactly, it is likely that you will encounter unexpected or unpredictable results if the solutions are not prepared correctly.

Solution Concentrations

The concentration of a solution can be specified in a number of ways, with perhaps the most common approach being given in "parts," or ratios. This was widely used in earlier photographic literature, and the practice persists to this day. It is not uncommon to have fixer or developer labels provide ratios for dilution, such as "1 + 4."

Concentrations are more accurately defined in terms of percentages: the weight of a solid (in grams) that has been dissolved and made-up to a final volume of 100 ml of the solution is denoted as "% w/v." Similarly, the volume of a neat (i.e. pure) liquid (in ml) in a final volume of 100 ml of the solution is denoted as "% v/v." For instance, when a solution is said to have a concentration of 5% w/v, it means that 5 g of solid is contained in 100 ml of the solution. Or, when a solution of 10% v/v is prepared, it means that 10 ml of one liquid is contained in 100 ml of the final solution. The advantage of this approach is that it provides a scalable formula. All of the formulae in this book are described in these terms, while also providing actual measurements for specific results.

Stirring and Filtration

In practice, always add the solid or solute to a volume of solvent that is appropriately less than the final desired volume. When a solid has dissolved in water, or other solvent, density variations can occur due to incomplete mixing. Stir the mixture until all solids are completely, or almost completely dissolved, then add more solvent to make the final volume. For all sensitizer solutions it is very important to exclude any suspended particles of solid matter that could cause unwanted marks on the final print, so filtration is a standard procedure in the final stages of sensitizer solution preparations. Only the simplest apparatus is required: a conical filter funnel, made of glass or inert plastic. The size should be appropriate to the total volume of solution to be filtered—about a quarter to a tenth of that volume, because more solution is added little by little to the paper in the funnel as the filtration proceeds. The diameter of the filter paper chosen should be no more than twice the diameter of the funnel; use of too large a sheet of filter paper will cause a proportionally greater loss of solution due to absorption. The standard technique for folding filter papers is to fold a sheet in half, then half again, and open just one "leaf" of the triangular form to create a conical cup. Place this in the funnel. Filter papers are manufactured in various grades for different purposes; generally, grade #1 is adequate for the solutions to be prepared here.

Labels and Laboratory Notebooks

After preparing any solution, it is important to store it promptly in a suitable container, and to label this fully. Keep a record of how and when each solution was made up in a laboratory notebook (although this step is not necessary for the day-to-day clearing bath solutions). The security of your chemical stocks is an obvious responsibility, especially if there are children on the premises. Thus, every label should clearly indicate, at the very least, the contents and date of preparation—and these two bits of data should correspond to a more detailed entry in your notebook. Additionally, a label may need to have one or more batch numbers, concentration information, and hazard warnings. If working with multiple printing methods, a label may even indicate a specific process. MSDS should be kept in or near laboratory notebooks, and all of this paperwork should be clearly accessible and well organized. See Chapter 9.

Formulae for Making Platinum/Palladium Prints

The following instructions are organized by their sequence of usage. The three primary components of iron, palladium, and platinum salts are prepared first. Then, a stock, or sensitizer solution may be combined from these into a single, larger volume; this can be stored without any detrimental effect for several months at a time. This sensitizer solution is used to make platinum, or palladium, or platinum-palladium prints. Alternatively, you may prefer to combine the components on a coat-by-coat basis, for which a method is outlined here, but more details are made evident in Chapter 8: *Making the Print*. Some papers require Tween, glycerin or a combination of the two, to be added to the sensitizer just prior to coating (see Chapter 6). Instructions for preparing these and the three clearing bath solutions conclude this chapter.

Iron Solution

Note: Ammonium iron(III) oxalate trihydrate is also referred to as ammonium ferric oxalate, and we use this latter nomenclature here because it is less of a mouthful than the former.

This is the only light sensitive solution. Prepare it in a "dim room," illuminated with 40 to 60-watt tungsten or 5-watt LED bulbs, no closer than 4′ / 1.3 m to the work surface, but bright enough to let you work with ease. Specifically, the overall

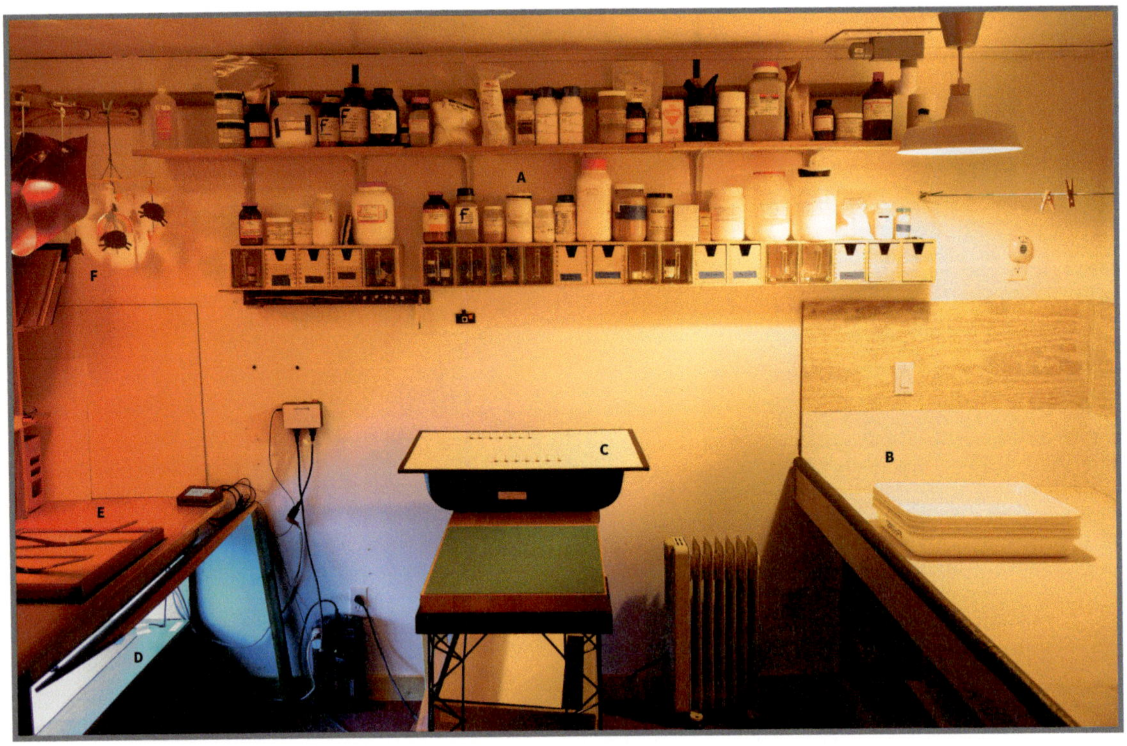

Figure 5.4 Small spaces can be efficiently organized and structured.

A. a well-labeled and secure stock of chemicals on high shelves. The containers are organized alphabetically.

B. a sink made out of wood framing and ABS plastic sheet.

C. a hydration chamber on a wheeled table.

D. a fluorescent UV source built into the underside of a work surface — note that its door has been opened for this photograph!

E. a print inspection area shielded from UV.

F. Light sources are made UV-safe with red Rubylith® acetate film.

illumination should be ca. 50–100 lux (EV of ca. 4.5 to 5 incident reading, or a reflected reading off a gray card at the work surface, at 100 ISO), with negligible to no stray ultraviolet. Do not work in unshielded fluorescent or daylight, since these can have a significant amount of UV radiation.

Three methods for preparation are described below, with each one rendering a slightly different final print tone, from warm to cool black, the degree of which will also depend on hydration levels and paper type.

The sensitizer solution obtained by Method 1 makes the warmest tone. A cooler tone can be achieved by the addition of ammonium nitrate as described in Methods 2 and 3, which produce almost identical results. Ammonium nitrate is a humectant—it holds on to water. The presence of water molecules in the sensitizer, during exposure, permit the image to form. The greater the amount of water molecules, the larger the size of platinum, and particularly palladium, particles. And as these particle sizes increase, they tend to shift towards a more neutral to cool color. This effect only happens with the printing-out process. So it is that prints made with a sensitizer using pure ammonium ferric oxalate (Method 1) have warmer tones than prints made with ammonium ferric oxalate with an excess of ammonium nitrate (Methods 2 and 3). There are some subtle differences in contrast and sensitivity, and these can be easily compensated by eye, or by customized digital negative printing profiles. It should be noted that Method 1 is the simplest to prepare, and presents the clearest practical first step in making a sensitizer. However, if you want to be in full control of your sensitizer, then Method 3 may be more attractive, as it ensures that it is fresh and prepared from raw ingredients; it may also be easier to obtain its ingredients.

Ammonium Ferric Oxalate: Method 1

To make a final volume of 50 ml, 60% w/v ammonium ferric oxalate

Equipment & Ingredients

Prepare in a dim room
Gloves, disposable
Pyrex™ or Corning™ glass beakers, 50 ml and 100 ml
Measuring cylinder, 50 or 100 ml
Glass stirring rod or Hotplate / Magnetic stirrer
Warming bath or Hotplate / Magnetic stirrer
Thermometer
Weighing scale
Label
Amber glass bottle, 50 ml capacity
Ammonium ferric oxalate, $(NH_4)_3 [Fe(C_2O_4)_3] . 3H_2O$, 30 g
Distilled or purified water at 122°F / 50°C, 50 ml

Instructions

1. Weigh out 30.0 g of the ammonium ferric oxalate and set it aside.
2. With a cylinder, measure 30 ml of warm distilled water (104–122°F / 40–50°C) and pour this into a beaker.
3. Add the ammonium ferric oxalate to this, while stirring with a clean glass rod until all the crystals have dissolved. The result is a clear bright green liquid. This should take no more than about 10 minutes.
4. Pour back into the measuring cylinder. Make up to a final volume of 50 ml with distilled water.
5. Filter the solution into its brown storage bottle.
6. Label the bottle.
7. If, after about 24 hours, a few white needle-like crystals (probably of ammonium oxalate) have appeared, re-filter the solution to remove them.
8. In your notes, also include the source and batch number of the ammonium ferric oxalate.

This solution should be made up at least 24 hours before use, and stored in the dark at room temperature. The solution will keep under these conditions for at least a year. If it cools too much below 68°F / 20°C, green crystals may form. Re-dissolve them before use by gently warming the solution in a water bath.

Label

Ammonium ferric oxalate 60% w/v
Method 1 or Fe1
Date

Fe3 1:0:1 56% RH Fe3 2:1:1 56% RH Fe3 1:1:0 56% RH

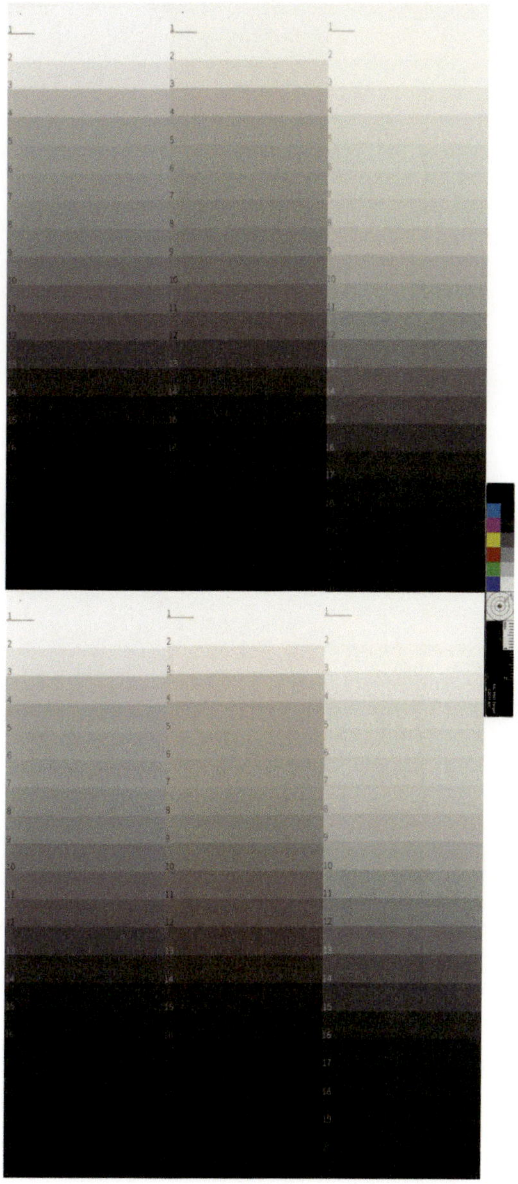

Fe1 1:0:1 56% RH Fe1 2:1:1 56% RH Fe1 1:1:0 56% RH

Figure 5.5 Differences between each of the two methods for preparing ammonium ferric oxalate: prints made with a sensitizer using pure ammonium ferric oxalate (Method 1) have warmer tones than prints made with ammonium ferric oxalate with an excess of ammonium nitrate (Method 3). All prints are hydrated for 90 minutes, 20°C / 56% RH. The prints are identical to Figure 5.7 but organized by different criteria. Note: the ratios are for Fe:Pt:Pd

Ammonium Ferric Oxalate + Ammonium Nitrate: Method 2

To make a final volume of 50 ml, 60% w/v ammonium ferric oxalate + 34% w/v ammonium nitrate

Equipment & Ingredients

Prepare in a dim room
Gloves, disposable
Pyrex™ or Corning™ glass beakers, 50 ml and 100 ml
Measuring cylinder, 50 or 100 ml
Glass stirring rod or Hotplate / Magnetic stirrer
Warming bath or Hotplate / Magnetic stirrer
Thermometer
Weighing scale
Label
Amber glass bottle, 50 ml capacity
Ammonium ferric oxalate, $(NH_4)_3\,[Fe(C_2O_4)_3].3H_2O$, 30 g
Ammonium nitrate, NH_4NO_3, 17 g
Distilled or purified water at 122°F / 50°C, 50 ml

Instructions

1. Weigh out 30.0 g of the ammonium ferric oxalate and set it aside.
2. Weigh out 17.0 g of ammonium nitrate and set it aside.
3. With a cylinder, measure 30 ml of warm distilled water (104–122°F / 40–50°C) and pour this into a beaker.
4. Add the ammonium nitrate to the distilled water and stir with a clean glass rod until all the crystals have dissolved.
5. Add the ammonium ferric oxalate to this. Stir until all the crystals have dissolved. The result is a clear bright green liquid. This should take no more than about 10 minutes.
6. Pour back into the measuring cylinder. The presence of a few crystals can be ignored. Make up to a volume of 50 ml with distilled water, stirring well, and set aside in a very dark place, to cool to room temperature, for about 1–2 hours.
7. Filter the solution into its brown storage bottle.
8. Label the bottle.
9. In your notes, also include the source and batch number of the ammonium ferric oxalate and ammonium nitrate.

This solution should be made up at least 24 hours before use, and stored in the dark at room temperature. The solution will keep under these conditions for at least a year. If it cools too much below 68°F / 20°C, green crystals may form. Re-dissolve them before use by gently warming the solution in a water bath.

Label

Ammonium ferric oxalate 60% w/v +
Ammonium nitrate 34% w/v
Method 2 or Fe2
Date

Ammonium Ferric Oxalate with Excess Ammonium Nitrate: Method 3

To make a final volume of 50 ml, 60% w/v ammonium ferric oxalate + 34% ammonium nitrate

Equipment & Ingredients

Prepare in a dim room
Must have robust exhaust air movement away from the face
Face mask
Gloves, disposable
Pestle and mortar
Pyrex™ or Corning™ glass beakers, 50 ml and 100 ml
Measuring cylinder, 50 or 100 ml
Glass stirring rod or Hotplate / Magnetic stirrer
Freshly boiled water bath or Hotplate / Magnetic stirrer
Thermometer
Weighing scale
Filtration funnel and filter paper
Label
Amber glass bottle, 50 ml capacity
Ammonium oxalate monohydrate, $(NH_4)_2C_2O_4.H_2O$, 30.0 g
Ferric nitrate nonahydrate, $Fe(NO_3)_3.9H_2O$, 28.3 g
Distilled or purified water 20 ml

Instructions

1. *Poisonous! Take care not to inhale dust.* Weigh out 30.0 g of the ammonium oxalate monohydrate and powder it in the mortar with a pestle. Set aside.
2. *Corrosive! Avoid inhaling the fumes.* Weigh out 28.3 g of the ferric nitrate nonahydrate, and place in a 100 ml beaker that is either standing in a bath of very hot water, or on a hotplate (this is why I recommend that you use a Corning ware flask). Stir the crystals gently with a clean glass rod, as they appear to melt. Actually, the ferric nitrate is dissolving in its own water of crystallization. Keep the liquid warm in the water bath or on the hot plate and stir until a clear red-brown liquid is obtained. You may have to replenish the water bath to maintain the high temperature. This stage should take about 5–10 minutes.
3. Now add the powdered ammonium oxalate to the ferric nitrate by stirring it in bit by bit. Keep the solution hot, at a temperature of about 50°C (122°F). When all the ammonium oxalate has been added and dissolved (5-10 minutes), the liquid should be a clear, bright green syrup.
4. Pour this into the measuring cylinder. The presence of a few crystals can be ignored. Make up to a volume of 50 ml with distilled water, stirring well, and set aside in a very dark place, to cool to room temperature, for about 1–2 hours.
5. Filter the solution into its brown storage bottle.
6. Label the bottle.
7. In your notes, also include the source and batch number of the ammonium oxalate and ferric nitrate.

This solution should be made up at least 24 hours before use, and stored in the dark at room temperature. The solution will keep under these conditions for at least a year. If it cools too much below 68°F / 20°C, green crystals may form. Re-dissolve them before use by gently warming the solution in a water bath.

Label

Ammonium ferric oxalate 60% w/v +
Ammonium nitrate 34% w/v
Method 3 or Fe3
Date

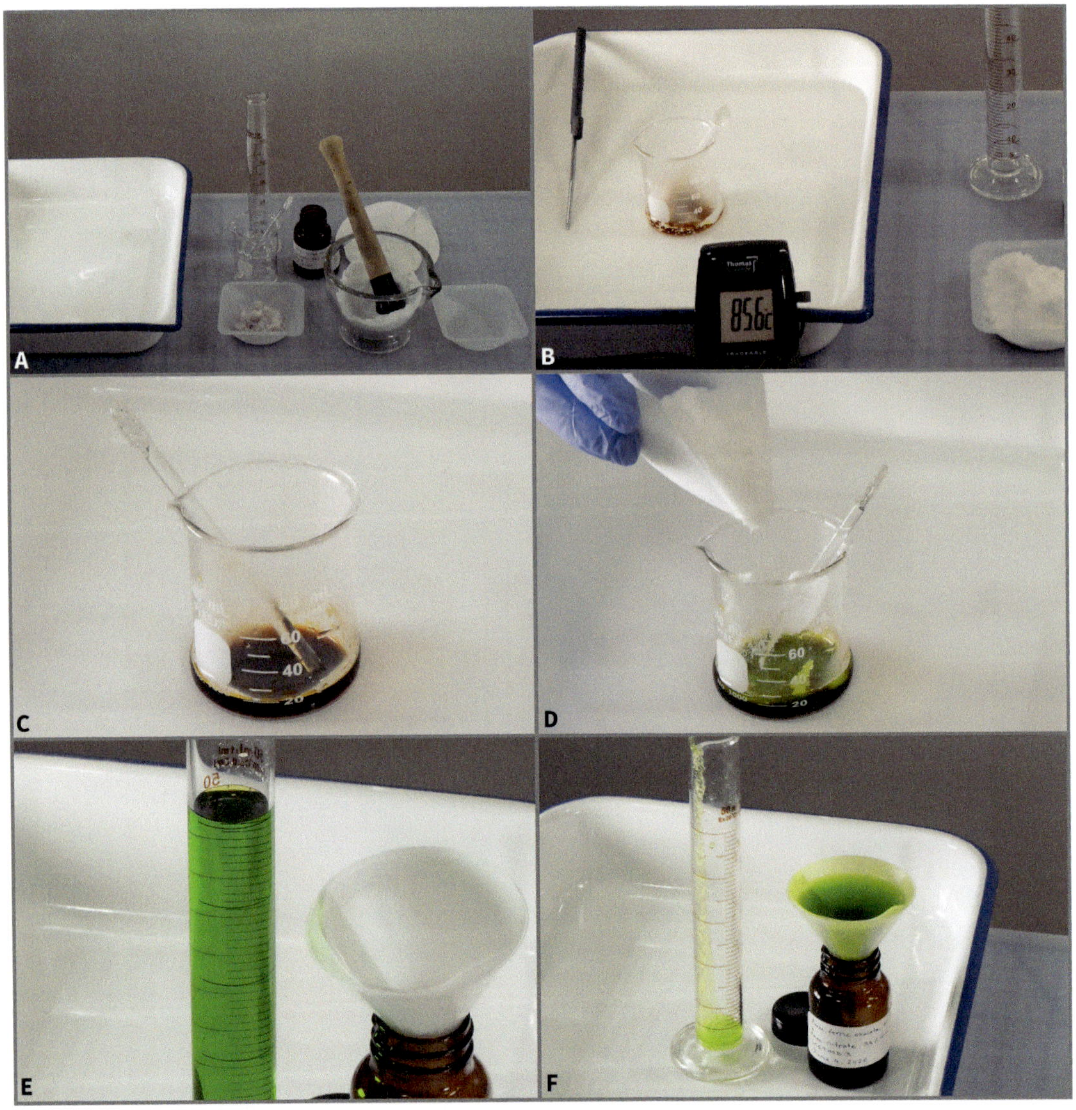

Figure 5.6 Step-by-step instructions for preparing Fe Method 3.

A. all necessary equipment
B. water-bath for heating ferric nitrate
C. ferric nitrate in solution
D. addition of ammonium oxalate
E. making up volume to 50 ml
F. filtration of solution

Palladium Solution

Two methods for preparing a palladium solution are described here, with no difference in the final result. While the first method is simpler to execute, it may be that materials for the second approach are easier to obtain and slightly more economical. I generally refer to this component of the sensitizer as the palladium solution, but use the full chemical name below for the sake of clarity.

Ammonium Tetrachloropalladate(II): Method 1

To make a final volume of 26 ml, 19% w/v

Equipment & Ingredients

Gloves, disposable
Measuring cylinder, two, 50 or 100 ml
Glass stirring rod
Thermometer
Weighing scale
Filtration funnel and filter paper
Label
Amber glass bottle, 25–50 ml capacity
Ammonium tetrachloropalladate (II), $(NH_4)_2[PdCl_4]$, 5 g
Distilled or purified water at 158–176°F / 70–80°C, 26 ml

Instructions

1. Weigh out 5.0 g of ammonium tetrachloropalladate(II), and transfer into a 50 ml measuring cylinder.
2. Add ca. 15 ml of hot distilled or pure water to fully dissolve the solid while stirring at room temperature. This will take about five minutes.
3. Make up the solution with pure water to a volume of exactly 26 ml in the measuring cylinder.
4. Filter the solution into its brown storage bottle.
5. Label the bottle.
6. In your notes, also include the source and batch number of the ammonium tetrachloropalladate(II).

This solution should be made up at least 24 hours before use, and stored in the dark at room temperature. The solution will keep under these conditions indefinitely. If it cools too much below 68°F / 20°C, some crystals may form. Re-dissolve them before use by gently warming the solution in a water bath.

Label

Ammonium tetrachloropalladate(II), 19% w/v
Date

Ammonium Tetrachloropalladate(II): Method 2

To make a final volume of 25 ml, 19% w/v

Equipment & Ingredients

Face mask
Gloves, disposable
Pestle and mortar
Pyrex™ or Corning™ glass beakers, 50 ml and 100 ml
Measuring cylinder, two, 50 or 100 ml
Glass stirring rod or Hotplate / Magnetic stirrer
Freshly boiled water bath or Hotplate / Magnetic stirrer
Thermometer
Weighing scale
Filtration funnel and filter paper
Label
Amber glass bottle, 25–50 ml capacity
Ammonium chloride, NH_4Cl, 1.8 g
Palladium(II) chloride, $PdCl_2$, 3.0 g
Distilled or purified water, 25 ml

Instructions

1. Weigh out 3.0 g of palladium(II) chloride. Transfer to a mortar, and *wearing a dust mask*, grind down to a fine powder. Set aside.
2. Accurately weigh out 1.8 g of ammonium chloride, and transfer into a 100 ml Pyrex™ or Corning™ glass beaker.
3. With a cylinder, measure 20 ml of distilled or pure water and pour into the beaker. All the ammonium chloride solid should dissolve easily at room temperature with stirring.
4. Heat the solution to ca. 158°F / 70°C.
5. *Wearing a dust mask,* add the palladium(II) chloride, a little at a time, with stirring.
6. Keep hot and stir until all the brown solid has dissolved to give a very dark red solution—which may take up to an hour. Carefully view the solution from below to see if any solid remains.
7. Allow the solution to cool and transfer it to a 50 ml measuring cylinder, and make up to a volume of exactly 25 ml with pure water (use some of this water to wash out any solution in the beaker from the previous step).
8. Filter the solution into its brown storage bottle.
9. Label the bottle.
10. In your notes, also include the source and batch number of the ammonium chloride and ammonium tetrachloropalladate(II).

This solution should be made up at least 24 hours before use, and stored in the dark at room temperature. The solution will keep under these conditions indefinitely. If it cools too much below 68°F / 20°C, some crystals may form. Re-dissolve them before use by gently warming the solution in a water bath.

Label

Ammonium tetrachloropalladate(II), 19% w/v
Date

Platinum Solution

Ammonium Tetrachloroplatinate (II)

To make a final volume of 20 ml, 25% w/v

Equipment & Ingredients

Face mask
Gloves, disposable
Pyrex™ or Corning™ glass beaker, 50 ml
Measuring cylinder, 50 ml
Glass stirring rod
Weighing scale
Filtration funnel and filter paper
Label
Amber glass bottle, 25–50 ml capacity
Ammonium tetrachloroplatinate(II), $(NH_4)_2\ PtCl_4$, 5 g
Distilled or purified water at room temperature, 20 ml

Instructions

1. *Wearing a dust mask and gloves*, weigh out 5.0 g of ammonium tetrachloroplatinate(II), and transfer into a 50 ml beaker.
2. With a cylinder, measure 18 ml of distilled or pure water, pour into the beaker and dissolve the solid by stirring at room temperature. This takes between 5–10 minutes. The final volume should be exactly 20 ml, if not, make it up to this amount with pure water.
3. Filter the solution into its brown storage bottle.
4. Label the bottle.
5. In your notes, also include the source and batch number of the ammonium tetrachloroplatinate(II).

This solution should be made up at least 24 hours before use, and stored in the dark at room temperature. The solution will keep under these conditions indefinitely. If it cools too much below 68°F / 20°C, some crystals may form. Re-dissolve them before use by gently warming the solution in a water bath.

Label

Ammonium tetrachloroplatinate(II), 25% w/v
Date

Sensitizer Solution

Note: *a sensitizer is understood as being a combination of solutions of iron and platinum/palladium salts, and therefore capable of rendering a visible image. I refer to all other liquids as "solutions"; i.e. the iron solution, platinum solution, palladium solution, etc.*

To make a platinum/palladium print, paper needs to be sensitized to light. To do this, we find ourselves at yet another one of the many wonder-inducing steps of this process. From the preceding instructions, you should have at least two, but probably three bottles of solution: the light-sensitive ammonium ferric oxalate; and one or both of the noble metal salts, ammonium tetrachloroplatinate(II) and ammonium tetrachloropalladate(II). It is sometimes convenient to refer to each of these as the iron (or Fe), palladium (or Pd), and platinum (or Pt) salt, respectively—but always keep in mind that these are abbreviated forms of reference. These salts are combined into a single liquid called the *sensitizer*, which is then coated onto paper in order to ultimately create a photographic print. The sensitizer may be prepared from each of its component solutions just before coating. For instance, to make an 8″x 10″ platinum-palladium print, you would begin by mixing 0.7 ml of the iron salt with 0.35 ml of the platinum salt and 0.35 ml of the palladium salt. Then, after waiting for about an hour (in itself not mandatory, but this pause renders a slightly darker maximum black), the paper is coated, and on its way to becoming a platinum-palladium print. By staggering one's work-flow, it is feasible to make five to ten prints in one day with this approach, even though measuring out such small amounts, print by print, can become tedious.

Traditional, development Pt/Pd processes always require sensitizer solutions to be combined just before coating the paper. A distinguishing, and advantageous, property of the ammonium-based printing-out process is that a large volume of sensitizer may be prepared at one time, and used over several weeks or even years, without any detrimental effect. It does not need to be mixed on a print-by-print basis. This single-bottle solution makes for a convenient approach, and once set aside to "mature" for an hour, does not require any lengthy pauses in the work-flow. One can, quite simply, measure the appropriate amount of the sensitizer from its bottle, immediately coat the paper and proceed from there.

Each of the three component solutions have been formulated to work efficiently in a specific ratio. The platinum and palladium salts may be mixed in any proportion, *so long as their combined volume is equal to that of the iron salt.* We see this in the preceding example, where the volume of ammonium ferric oxalate (the iron salt) solution matches the sum of platinum and palladium solutions. Additionally, the iron salt may be any combination of ammonium ferric oxalate: Method 1 with Method 2 or 3, so long as their combined volume equals the combined volume of platinum and palladium salts. From all of these variations one can render a wide range of print qualities in terms of color, contrast and more. We will consider these in greater detail in Chapter 8. For now, here are examples of five combinations of sensitizer solutions which may be prepared with volumes derived from the preceding formulae.

The prints are identical to Figure 5.5 but organized by different criteria.

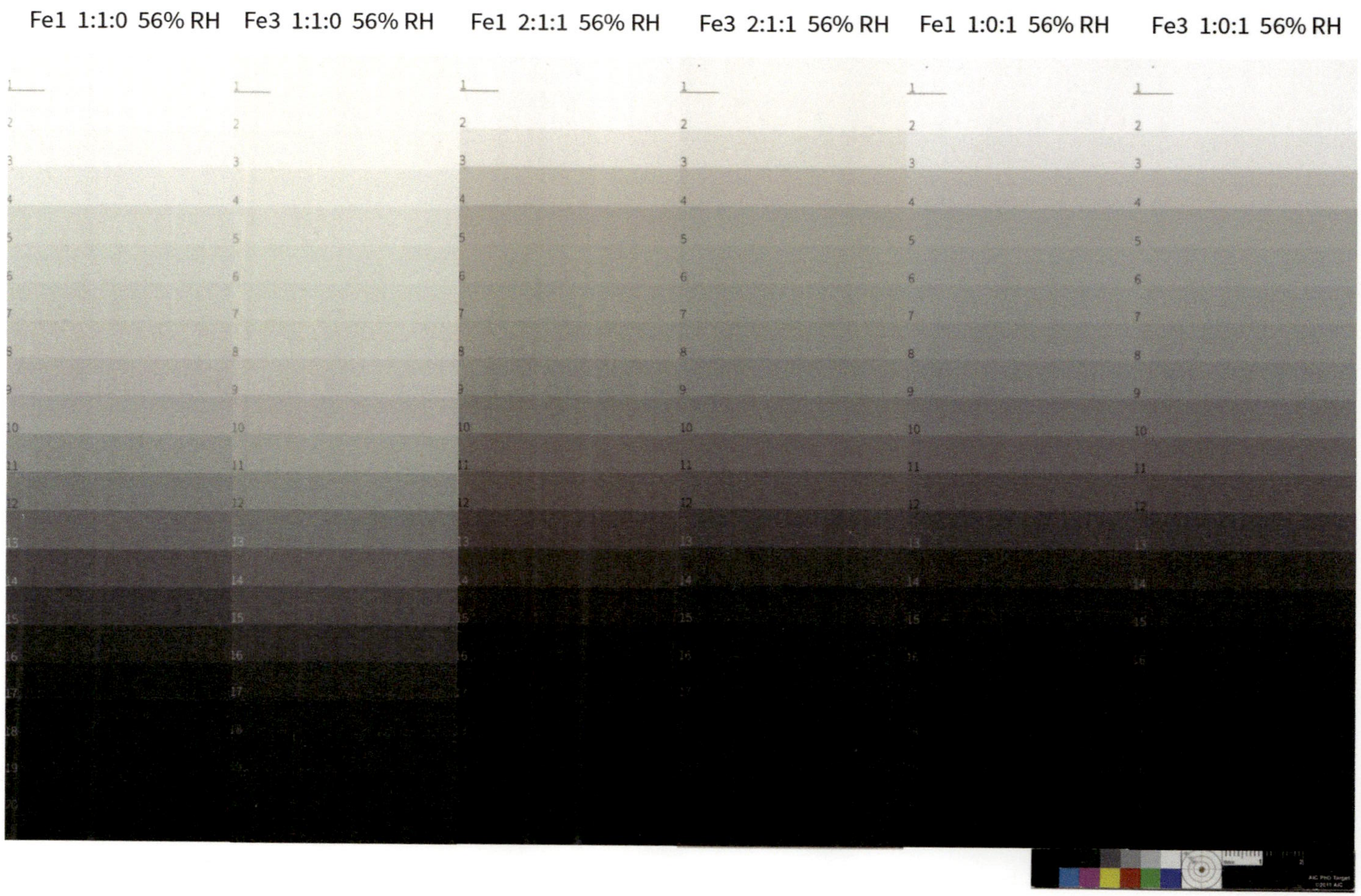

Figure 5.7 Prints illustrating the differing effects of Pt and Pd combinations. All prints are hydrated for 90 minutes, 20° C / 56% RH. Note: The ratios are for Fe:Pt:Pd

Palladium Sensitizer

To make a final volume of 50 ml

Equipment & Ingredients

Measuring cylinder, 50 ml
Label
Amber glass bottle, 50 ml
Ammonium ferric oxalate 60% w/v, 25 ml
Ammonium tetrachloropalladate(II) 19% w/v, 25 ml

Instructions

1. Combine the two solutions in a 50 ml measuring cylinder.
2. Pour into an amber glass bottle.
3. Label the bottle.
4. In your notes, write down which ammonium ferric oxalate method was used, or combinations of these. Note that the label information abbreviates the ammonium ferric oxalate solution method # as "Fe Method #." Also include a reference to the laboratory notes you made for each component solution, with their preparation date and batch information.

The solution may be used immediately. Store tightly capped, in the dark at room temperature. The solution will keep under these conditions for at least a year. If it cools too much below 68°F / 20°C, crystals may form; if so, re-dissolve them before use by gently warming the solution in a water bath.

Label

Palladium sensitizer
Fe Method #
Fe:Pd = 1:1
Date (of mixing the sensitizer)
50 ml

Platinum Sensitizer

To make a final volume of 40 ml

Equipment & Ingredients

Measuring cylinder, 50 ml
Label
Amber glass bottle, 50 ml
Ammonium ferric oxalate 60% w/v, 20 ml
Ammonium tetrachloroplatinate (II) 25% w/v, 20 ml

Instructions

1. Combine the two solutions in a 50 ml measuring cylinder.
2. Pour into an amber glass bottle.
3. Label the bottle.
4. In your notes, write down which ammonium ferric oxalate method, or combinations of methods, were used. Note that the label information abbreviates the ammonium ferric oxalate solution method # as "Fe Method #." Also include a reference to the laboratory notes you made for each component solution, with their preparation date and batch information.

The solution may be used after an hour. Store tightly capped, in the dark at room temperature. The solution will keep under these conditions for at least a year. If it cools too much below 68°F / 20°C, crystals may form; if so, re-dissolve them before use by gently warming the solution in a water bath.

Label

Platinum sensitizer
Fe Method #
Fe:Pt = 1:1
Date (of mixing the sensitizer)
40 ml

Platinum-Palladium Sensitizer

To make a final volume of 80 ml

Equipment & Ingredients

Measuring cylinder, 100 ml
Label
Amber glass bottles, two of 50 ml
Ammonium ferric oxalate 60% w/v, 40 ml
Ammonium tetrachloroplatinate (II) 25% w/v, 20 ml
Ammonium tetrachloropalladate (II) 19% w/v, 20 ml

Instructions

1. Combine the three solutions in a 100 ml measuring cylinder.
2. Pour into amber glass bottles.
3. Label the bottle.
4. In your notes, write down which ammonium ferric oxalate, or combinations of methods, were used. Note that the label information abbreviates the ammonium ferric oxalate solution method # as "Fe Method #." Also include a reference to the laboratory notes you made for each component solution, with their preparation date and batch information.

The solution may be used after an hour. Store tightly capped, in the dark at room temperature. The solution will keep under these conditions for at least a year. If it cools too much below 68°F / 20°C, crystals may form; if so, re-dissolve them before use by gently warming the solution in a water bath.

Label

Platinum-palladium sensitizer
Fe Method #
Fe:Pt:Pd = 2:1:1
Date (of mixing the sensitizer)
80 ml

Figure 5.8 Fe1:Pt:Pd = 2:1:1 sensitizers of different ages. Tests were made on Revere Platinum, hydrated for 90 mins at 56% RH / 20°C. A sensitizer may be prepared in larger volumes than needed for just one sheet. The shelf life of mixed sensitizer can extend for at least a year with no image degradation. The results of a solution mixed on January 20, 2014 indicate a slight loss of Dmax, but this can be compensated by extending exposure.

Tween Solution

To make a final volume of 20 ml, 15% v/v

Equipment and Ingredients

Syringe, 1 ml with 0.1 ml graduations
Glass bottle, wide mouth, any size between 10–50 ml
Measuring cylinder, 50 ml
Label
Tween 20
Distilled or purified water, ca. 20 ml

Instructions

1. Measure 17 ml of distilled or purified water into a clean glass bottle.
2. Syringe up a total of 3.0 ml of Tween 20, pushing out the syrupy liquid gently into the bottle each time.
3. Syringe this mixture in and out several times to encourage mixing.
4. Tightly cap the bottle and swirl solution to ensure it is thoroughly mixed, but do so gently to minimize frothing. Tween 20 is very thick and syrupy, and should be thoroughly homogenized in water before being used.
5. Label the bottle.

Solution may be used immediately. Store at room temperatures. The solution may grow some gelatinous mold after a month. For this reason, it is best to use sensitizer with Tween within a few days. Discard and replace with a fresh solution if any such substances can be seen in the bottle.

Label

Tween 15% v/v
Date

Glycerin

To make a final volume of 20 ml, 15% v/v

Equipment & Ingredients

Syringe, 1 ml with 0.1 ml graduations
Glass bottle, wide mouth, any size between 10–50 ml
Measuring cylinder, 50 ml
Label
Glycerin
Distilled or purified water, ca. 20 ml

Instructions

1. Measure 17 ml of distilled or purified water into a clean glass bottle.
2. Syringe up a total of 3.0 ml of glycerin, pushing out the syrupy liquid gently into the bottle each time.
3. Syringe this mixture in and out several times to encourage mixing.
4. Tightly cap the bottle and swirl solution to ensure it is thoroughly mixed. Glycerin is very thick and syrupy, and should be thoroughly homogenized in water before being used.
5. Label the bottle.
6. Solution may be used immediately. Store at room temperatures.

Label

Glycerin 15% v/v
Date

EDTA Disodium

To make 1 liter, 5% w/v

Equipment and Ingredients

Weigh scale or 1 tablespoon measure
Measuring cylinder, 1 liter
Glass or plastic stirring rod
Adequately sized storage container
Label
EDTA disodium, $C_{10}H_{14}N_2O_8Na_2.2H_2O$, ca. 50 g or 2 level tablespoons
Distilled or purified water, ca. 80°F / 25°C

Instructions

1. Weigh or measure out ca. 50 g of EDTA tetrasodium.
2. With stirring, add it to 1 liter of water.

This is sufficient to process ca. thirty 8″x 10″ platinum/palladium prints, or until the volume diminishes to a level where more solution is needed to adequately immerse the paper sheets. It may be stored for several months in a container with a tight lid, at room temperature. The solution is susceptible to mold, in which case, it may be filtered and then continue to be used. Cool to room temperature before using, but only because warm water may overly soften the paper and make it more susceptible to damage during the wet processing sequence. There should be at least 0.5″ / 1.25 cm of liquid in the processing tray—scale the volume appropriately.

If only EDTA tetrasodium is on hand, it may be effectively converted to EDTA disodium by the addition of 15 g citric acid, ($C_6H_8O_7$), to 50 g of EDTA tetrasodium in 1 liter of distilled water.

Label

EDTA disodium, 5% w/v

Sodium Sulfite

To make 1 liter, 2.5% w/v

Equipment & Ingredients

Weigh scale or 1 tablespoon measure
Measuring cylinder, 1 liter
Glass or plastic stirring rod
Label
Sodium sulfite, Na_2SO_3, ca. 25 g or ca. 1 level tablespoon
(Kodak Hypoclearing Agent may be substituted for sodium sulfite)
Tap or faucet water, room temperature

Instructions

1. Weigh or measure out ca. 25 g of sodium sulfite.
2. With stirring, add it to 1 liter of water.

This is sufficient for two 8″x 10″ platinum/palladium prints, after which it should be discarded. There should be at least 0.5″ / 1.25 cm of liquid in the processing tray—scale the volume appropriately.

Label

(not necessary since you will not be storing this)

EDTA Tetrasodium

To make 1 liter, 5% w/ v

Equipment & Ingredients

Weigh scale or 1 tablespoon measure
Measuring cylinder, 1 liter
Glass or plastic stirring rod
Adequate storage container
Label
EDTA tetrasodium, $C_{10}H_{12}N_2O_8Na_4.2H_2O$, ca. 50 g or 2 level tablespoons
Tap or faucet water, room temperature

Instructions

1. Weigh or measure out ca. 50 g of EDTA disodium.
2. With stirring, add it to 1 liter of distilled water.

This is sufficient to process ca. sixty 8″x10″ platinum/palladium prints. It may be stored for several months in a container with a tight lid, at room temperature. It should always be clear, so discard if it begins to discolor. There should be at least 0.5″ / 1.25 cm of liquid in the processing tray—scale the volume appropriately.

Label

EDTA tetrasodium, 5% w/v

Sulfamic Acid Decalcifying Bath

To acidify/decalcify approximately twenty sheets of 20″x 30″ / 50 cm x 76 cm paper

Equipment & Ingredients

Gloves, disposable
Measuring cylinder, 1 liter
Glass or plastic stirring rod
Adequate storage container
Weighing scale
Label
Sulfamic acid, H_2NSO_3H, 100 g
Tap or faucet water, room temperature

Instructions

1. Weigh out 100 g of sulfamic acid and set aside.
2. Measure out 1 liter of water in the cylinder.
3. Slowly add the sulfamic acid with stirring until all particles are dissolved.
4. Store and label the container.

There should be at least 1″ / 2.5 cm of liquid in the processing tray—scale the volume appropriately.

Label

Sulfamic acid, 10% w/v
Date
<score off number of sheets processed>

Note

1 Never write this as "gr," because it will cause confusion: in the old literature the abbreviation "gr" signified the unit of "grains," now long obsolete, and a much smaller measure (15.43 gr = 1 g).

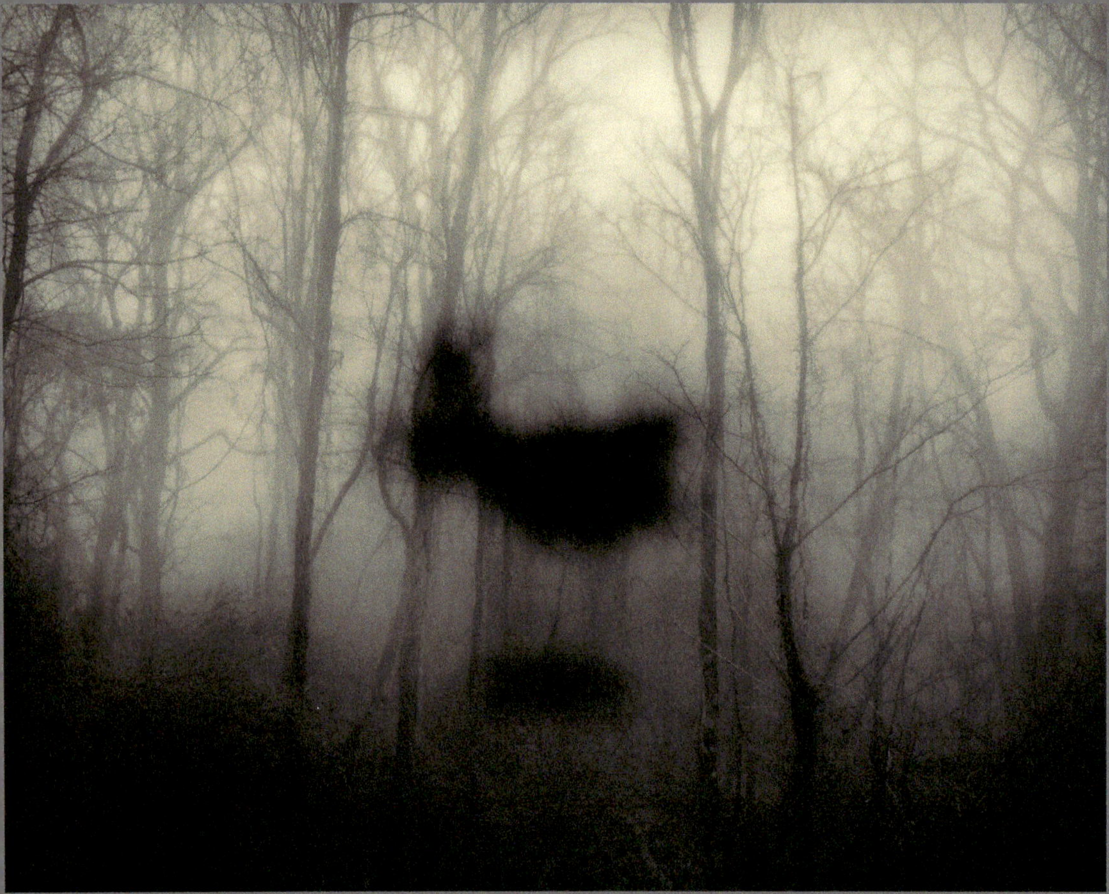

Figure 6.0 Pradip Malde, *Trees, Shadow, Fog, Sewanee, Tennessee*, 2007. Platinum-palladium print on Wyndstone Vellum, from original 11 x 14 inch negative, 27 x 34 cm. Courtesy Pradip Malde. Fe1:Pt:Pd = 2:1:1, 20 min hydration in 100% RH / 20° C

Chapter 6

Paper—Its Properties and Choice

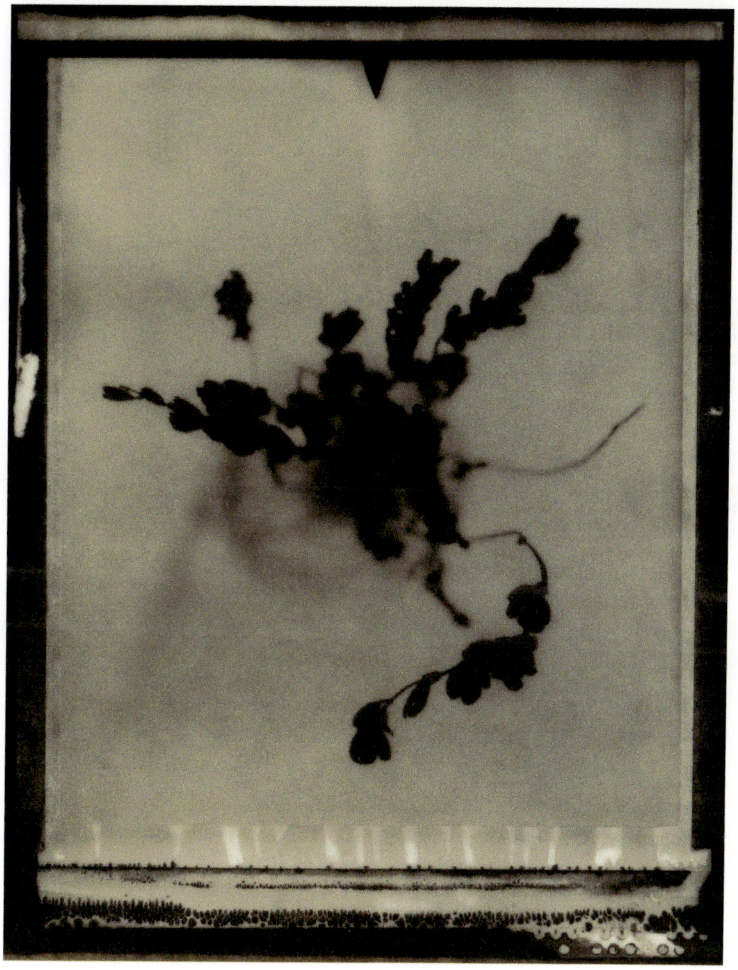

Figure 6.1 Jane Izard, *Untitled*, 2004 (printed 2006). Platinum-palladium print on Wyndstone Vellum, from original 4x5 Type 55 Polaroid neg/pos film, 13.5 x 11 cm. Courtesy Pradip Malde. Fe3:Pt:Pd = 2:1:1, hydrated for 20 min in 100% RH / 20°C (water chamber).

Many hand-coated photographic processes can render an image by the formation of metal particles, such as silver, within the flattened, macaroni-like, cellulosic structure of paper. Consider the kallitype, argyrotype, or any other siderotype process.

A platinum and palladium print, however, has a different signature from all other processes. These precious metals are capable of engaging with cellulose in the finest manner, as nanoparticles, to the very "edge" of what we call paper-base white. As a result, print highlights take on nuanced values. At the other end of the tonal scale, dark values are separated better than most other processes, and remain expressive, whispering in the shadows. These dark tones certainly do not have the deep, black values of emulsion-based silver-gelatin printing processes. Yet, blacks are rarely an unresolved mass of monotone—akin to the way we actually see. It is for these reasons that the overall experience of looking at a well-printed platinum/palladium image is often described as being subtle and intimate. The role of paper in conveying these qualities is impossible to ignore. Most obviously, there is the visual experience of paper: its color and surface texture. William Willis, Jr., the inventor of the platinotype, recognized paper's crucial optical—and even more importantly—chemical function from the earliest stages of his research. By the time his company commenced with the full production of platinotype materials, Willis had settled on a specific paper from Steinbach & Co., of Malmedy, in Rhenish Prussia, which was famed for its photographic paper base, the so-named "papier de Saxe."[1] The photographic chemist George Dawson (1827–1897), in 1880 pointedly linked the physical and chemical importance of paper to the process: "Saxe ... prepared by the makers specially to give pure whites, as nearly all other papers contain substances which combine with platinum salts, thus producing impure highlights. With other papers too, it is impossible to produce good halftones."[2] The effect that paper had on photographic processes began to be increasingly recognized from the mid-nineteenth century on, and it remains as a paramount ingredient. It behooves us to understand the chemical and physical nature of paper, and some of its most relevant characteristics in relation to platinum/palladium printing.

Paper

Paper is a material composed of cellulose fibers, matted into a compact, interlaced web. We know, use, and experience this ubiquitous material in many forms: as a robust packaging board or, at the other extreme, as a fine, gossamer-like tissue; as the utilitarian object here in your hands or as the manifest of a higher order, to be handled with respect and awe, such as a sacred text or an annotated copy of the *Origin of Species* (1859).[3] Paper can be fabricated to meet an astonishingly wide variety of needs, from which one can present a strong argument that it is among the greatest technological achievements of humankind. It is a seminal technology because it amplified and even allowed for the development of other technologies. It has a complex cultural history, parallel to its technical and chemical development, that goes back at least to 105 CE, when the Chinese technocrat Ts'ai Lun reports researches on paper to his emperor, Yüan Hsing.[4] Many pages have been devoted to paper's fascinating narrative (another sweet parallel in itself), but for our purposes, we need to consider some specific aspects. What follows applies chiefly to "western-style" papers, even though some "eastern" papers, as such, are now made by hybridized methods that are described below.

Manufacture: Machine-made, Moldmade, and Handmade

Contemporary papers are made either by machine in large quantities, or handmade, as the great craftsman and paper historian, Dard Hunter (1883–1966) put it, by the "ancient traditional methods."[5] Regardless of which particular method is used in forming the paper sheet, its fabrication begins with deriving fibers from plant materials such as cotton and wood, and culminates in a finished sheet of dry paper.

Machine-made papers are mostly made by the Fourdrinier process, which is designed for the mass-production of voluminous rolls. While being very affordable and consistent, the nature of their production often requires chemical additives

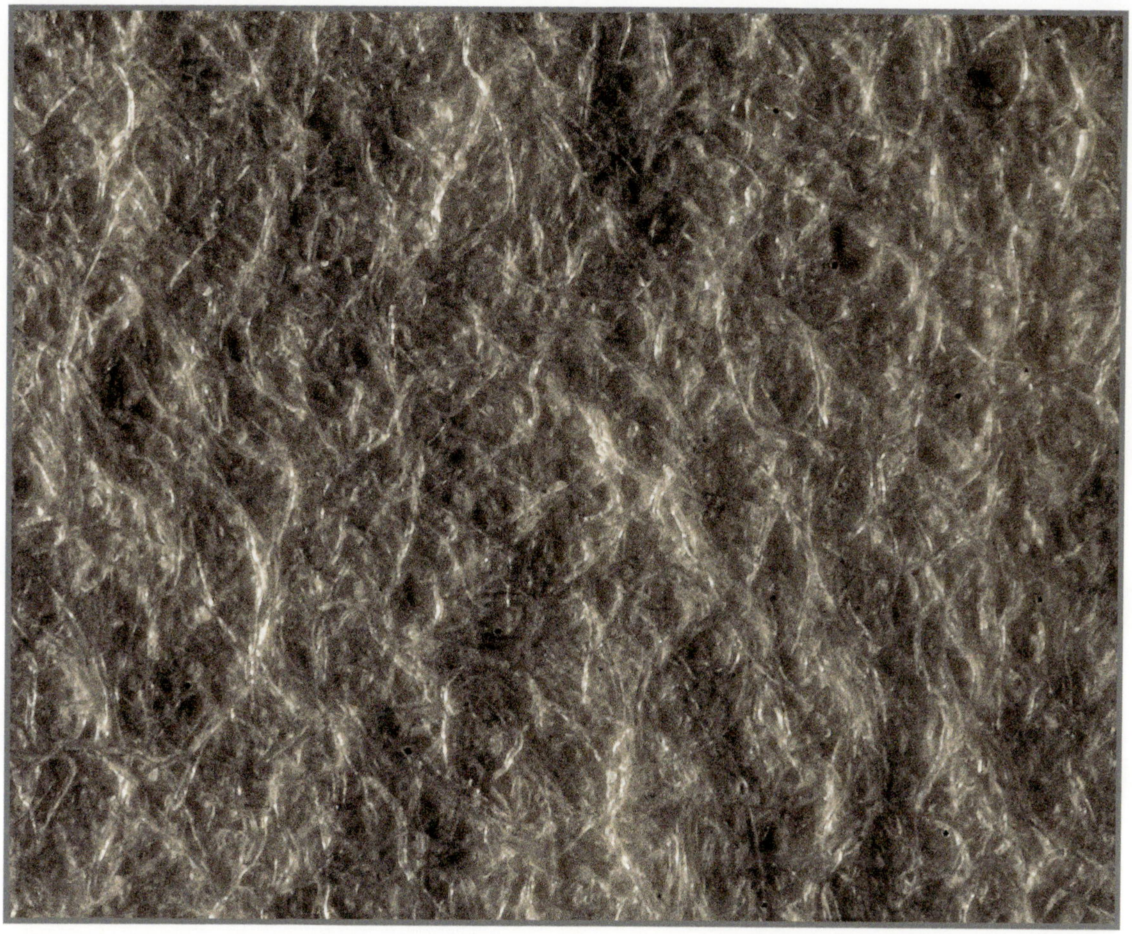

Figure 6.2 Micro-photograph of cellulose fibers matted into a compact web to form paper.

which amount to an unsuitable material for our purposes. A more refined machine process emulates traditional techniques through the use of cylindrical molds, and thereby can use pulp of a similar quality to that used by handmade methods—these types of industrially produced papers are called "moldmade." Handmade papers, with each sheet laboriously crafted with much skill and time, tend to require fewer chemical additives than moldmade productions, and consequently offer a more benevolent reactive structure for our photographic materials. On the other side however, is the physical matter of structural consistency and cost—a moldmade production, running to thousands of sheets from one recipe, is appealing as we strive for predictable and economic printing sessions. Handmade papers that are consistent tend to be expensive, and since they are produced in relatively small batches, there is greater risk of variance from sheet to sheet and batch to batch. This conundrum is somewhat resolved by offerings of artist-grade, moldmade papers from a handful of manufacturers such as Arches, Awagami, Hahnemühle, and Revere.

Regardless of the production method, the cleaned plant fibers are first suspended in water and refined or beaten to a "pulp" by machines known as Hollander Beaters—a process called

Figure 6.3 Paper surfaces: examples of Hot, Not and Rough papers. Note that the three Fabriano Artistico papers are for display purposes only, to illustrate papers surfaces. They do not work for Pt/Pd processes, even after de-calcifying.

fibrillation, which greatly increases the surface area by shredding or teasing out the fibers, and which enables better interlocking of the fiber network when the sheet is formed. Other ingredients may be added to the "vat" of pulp at this stage, as will be discussed shortly. To form a sheet of paper, an appropriate volume of the pulp is drained down (assisted by vacuum in machine processes) through a fine metal wire screen stretched over a rectangular frame—the paper "mold." This mesh acts as a sieve or strainer to retain the cellulose as the water is drawn off, and the fibers are shaken sideways to help them interlink. In hand-making paper, scooping pulp up in a mold and shaking it is the traditional skill of the "vatman"; the mold is then passed to the "coucher" who turns the sheet out of the mold onto a stack, with alternated "felts" to dry it under pressure in a pile called a "post." All of these are modified, of course, into one long flow of steps in machine manufacturing.

Characteristics of Paper for Pt/Pd

The platinum/palladium sensitizer solution is imbibed into the surface fibers of the paper sheet, where it comes into intimate chemical contact with the cellulose—and with every other substance that happens to be present in the paper. Depending on the manufacturing process, some of these substances may be hostile to the photochemistry. That paper is not an inert component of the process was recognized even from the earliest decades of photography's invention. Frustratingly, unlike all other aspects of printing-out platinum/palladium, the precise makeup of this essential material is usually beyond our control and even knowledge (papermakers are notoriously secretive about their recipes). Always be on the lookout for a suitable paper, know how to test it, and then, on finding a good stock, be prepared to buy as much of it as you can afford. With this in mind, it will help to know the ideal characteristics of paper for this process, and the terms used to describe them.

Ideal Paper Characteristics

Ultimately, an ideal paper will display the physical characteristic of resisting absorption of sensitizer solution long enough for it to be easily and uniformly distributed across the surface, and then absorbing it evenly to within about 0.1 mm, just below the paper's surface. This paper will also present a benign chemical environment that permits the formation of full tonal values (i.e. a maximum black, which usually has an absolute optical reflectance density value of around 1.4) and a thorough clearing of unexposed areas to paper base white. The following characteristics are most significant in choosing a paper that helps us meet these goals.

- Furnish: a high cellulose furnish, derived from any single source such as cotton, linen, hemp, gampi, or delignified softwood
- Texture: a wove mold rather than laid; any watermark excludable from the picture area
- Surface: smooth, hot-pressed (HP) or cold-pressed (CP) surface
- Sizing: internal sizing with Aquapel™ or similar alkyl ketene dimer (AKD)
- Buffer: no alkaline buffering agents such as calcium carbonate
- Weight: a weight of ca. 160 g/m^2 to 360 g/m^2 (heavier for large printed sheets)
- Additives: minimal, and preferably no other additives
- Absorption: the capacity to absorb between 20 and 30 milliliters of sensitizer solution per 1 square meter of surface area (20 to 30 ml/m^2)

While the following paragraphs are ordered alphabetically, it will become apparent that some characteristics are more important to platinum/palladium printing than others.

Deckle

The "deckle" is the rough, slightly wavy edge of tapering thickness formed when pulp seeps between the two parts of a mold's frame. It used to be considered an undesirable mark of papermaking, and was usually trimmed off. Since the advent of machine-made papers, the deckle is a valued indication of a handmade paper, of which a full, uncut sheet has four true deckle edges. Cylinder moldmade papers have only two deckle edges running the length of the roll, but may be cut into sheets in such a way as to produce "false deckles," such as with the Revere Platinum.

Furnish

The "furnish," or "furnishings," is the term given to the basic ingredients of paper: the fibers. Since reproducibility (as in getting predictable results from sheet to sheet) and the capacity to render homogeneous tones in a paper are essential to this and all iron-based processes, so the papers of greatest interest to us are those having a high (greater than 98%) cellulose content, and sourced from one type of material such as cotton, linen, or gampi (*Wikstroemia canescens*). High cellulose content papers are still sometimes called "rag" papers, although recycled rags, once a major industry, are no longer used in their manufacture. For strength, it is desirable to select cellulose which has long fibers, with a greater tendency to form interconnected meshes of similar material, which are very stable. Papers made from a mixed furnish of cellulose fibers from different sources can absorb sensitizer unevenly, causing a fibrous granularity or "blotchiness" in the print, and should be avoided. Pure cellulose papers have the best expectations of archival permanence but only if they are carefully processed. Specimens of "rag" paper made 500 years ago are still known in excellent condition. In contrast, mechanically ground wood-pulp papers of only a few decades ago, e.g. newsprint, are already in serious states of decomposition.

pH Buffers

The paper must be free from alkaline buffering agents such as the calcium carbonate (chalk) which is often added to "archival quality" papers to protect them from the damaging long-term effects of acid. Calcium carbonate reacts with the sensitizer chemicals in an undesirable way, precipitating insoluble calcium oxalate (the same substance that causes some types of gall stone) and causing a yellow stain from uncleared iron compounds. Ultimately, these residual iron complexes will destroy the paper. It is possible to remove buffers by pre-treating paper—see *Paper Modifications* below. The implication here is that platinum/palladium printing does well with papers that are neutral to slightly acidic, with a pH of just under 7. This runs contrary to the usual thinking on paper conservation, which favors the incorporation of buffers to prevent acidic embrittlement. However, it does not matter in our context, because in the course of wet processing, paper will be passed through baths that effectively de-acidify it and bring it to a favorable pH of 9 or so, before the print is finally dried. From the conservation viewpoint, we end up with an extremely stable print.

Sizing

Paper's primary ingredient, cellulose, is water-loving or hydrophilic. "Sizing" agents are substances included in paper that control its rate and degree of penetration by liquids, particularly water. A paper sheet that is completely unsized is called "waterleaf," and is obviously highly absorbent—examples are blotting paper and filter paper. Thin waterleaf papers may be used for platinum/palladium printing, but lack wet-strength and may disintegrate during processing. Thicker, unsized paper results in sensitizer being absorbed to depths greater than around 0.1 mm, and results in a waste of expensive materials, and inferior images.[6] A sheet that is highly resistant to water, hydrophobic, and overly sized, will similarly present problems: while the sensitizer may dry and remain on the surface during exposure, negatives can get damaged during exposure and the resulting image may wash away during processing. Ideally, sizing agents for platinum/palladium printing need to strike a balance between absorptivity, wet strength, and chemical compatibility.

There are two ways of incorporating sizing agents into paper. The first method is variously called "tub" or "surface" or "external" sizing, in which a sheet of paper is briefly immersed into a vat of a solution of the agent, and excess coating material is scraped off before drying, and possibly hardening. External sizing may also be applied by brush. The usual agents are either some form of starch, or an animal gelatin, in which case a hardener, such as alum, or an aldehyde, such as formaldehyde, glyoxal, or glutaraldehyde, is incorporated in the papermakers' sizing bath. Externally sized papers, unless prepared with great skill and care, can result in uneven image tones due to varying absorptivity. It should also be noted that gelatin sizing inhibits the formation of platinum nanoparticles, and should be avoided when working with sensitizers that are composed of more than a tenth of the 25% w/v platinum solution.

The second method is called "engine" or "body" or "internal" sizing, and uses a hydrophobic sizing agent added to the pulp or fiber suspension, before the sheet is formed. The traditional method is "alum-rosin" sizing: rosin is derived from the sap of coniferous trees and is water-repellent. When treated with sodium hydroxide this forms a water-soluble sodium salt which may be added to the pulp. This must then be treated with acid to re-form the rosin size within the pores of the paper—the substance providing this acidity is alum (potassium aluminum sulfate). Such alum-rosin sized papers seem to work well with the siderotype processes, and were employed by William Willis for his Platinotype papers, but their archival stability has been questioned owing to the potential for acidity associated with the alum. Modern methods of internal sizing use reactive chemical agents called alkyl ketene dimers, such as Aquapel™ and Hercon 40™, as additives to the pulp, which are commonly described as "neutral sizing agents" by papermakers. The fascinating molecule of this substance has a long hydrocarbon "tail" which repels water, and a reactive "head" that is attracted to the cellulose molecule. Thus, water penetration is controlled by making the fibers somewhat hydrophobic. The advantage of internal sizing is that it results in a homogeneously absorbent paper, both internally and on the surface.

Surface Texture

As a consequence of their treatment in manufacture, papers can have broadly three types of surface. "Rough" papers are not pressed in any way other than between felts after the sheets have formed, and consequently have highly textured surfaces that may prove difficult to coat with sensitizer and may cause some loss of resolution in the contact printing. "Not," meaning "not hot pressed," or "cold pressed" is the result of passing paper sheets through cold rollers, a process that is called "calendering." "Hot-pressed sheets" generally have the smoothest surface, as they are calendered between heated rollers. A fairly smooth, hot-pressed (HP) or cold-pressed (CP) surface is usually desirable for printing photographic images. However, it should be noted that some heavily calendered, hot-pressed commercial papers do not retain their smoothness on wet processing, which can cause them to swell and roughen perceptibly.

Watermark

A "watermark" is caused by a local thinning of the sheet where a wire emblem has been woven into the mold, or into an additional roller in the cylinder machine. The reading of an unsymmetrical watermark, in different papers, cannot be relied upon to indicate which side is wire and which felt (see below). Some manufacturers impress a watermark that reads correctly from the wire side, others do the reverse. The watermark itself should be excluded from the coated area, or it will become apparent in the finished print and interfere with the image. Some printers prefer the watermark to be included in the margin as a token of authenticity. This is not always economically feasible.

Weight

The thickness of paper is specified by its "weight," and is usually described in grams per square meter, abbreviated as gsm or, preferably, g/m^2. Some sources, especially in the USA, quote the weight of paper in the older measure of Imperial pounds (lbs) per ream; however, this is meaningless unless the size of the sheet is also specified, and even the number of sheets per ream can vary in its definition.[7] Papers thinner than 60 g/m^2, to even thinner tissue papers of 20-30 g/m^2, can yield exquisite translucent prints. However, skill is required to coat and manipulate this fragile material. For platinum/palladium (or any other siderotype) printing, a weight of ca. 100 to 240 g/m^2 is generally sufficient to ensure adequate wet strength for sheets smaller than 11″ x 14″ / 28 cm x 35 cm; larger formats, of 20″ x 16″ / 50 cm x 40 cm or more, may call for a heavier weight, ca. 300-360 g/m^2. Very heavy papers

Table 6.1 The equivalent descriptions of paper surfaces used by French, German, Italian, and Spanish papermakers

English	*Hot Pressed*	*Not (cold pressed)*	*Rough*
French	Satine	Fin	Torchon
German	Glatt		Rauh
Italian	Satinata or Liscia	Fina	Grossa
Spanish	Liso	Fino	Gruesso

(ca. 500 g/m^2) may be slower to completely clear of residual chemicals, and should be washed for at least 45 minutes.

Wet Strength

When paper fibers dry, they form a strongly bonded, interconnected mesh. A paper's durability is chiefly the result of cellulosic fiber strength, various additives such as sizing agents, and inter-fiber bonds that form as the paper dries. Importantly for us, these bonds are weakened as the paper is rewetted, and the ability of thicker papers to withstand immersion and handling in liquids is described as high "wet strength." Conversely, very thin papers are easily damaged and can even disintegrate during the processing sequence, and therefore are considered to have low wet strength.

Wire or Felt

The surface texture of a sheet results from a combination of several factors: the furnish; the type and degree of Hollander beating; the type of mold, or some other way of impressing a pattern either during or after its formation; and ultimately, the downward-facing surface that was formed against the mold, called the "wire" side, and its opposite, upper surface called the "felt" side. These two sides of a sheet of fine paper usually differ discernably in a manner that reflects the method of manufacture: with a cylinder moldmade paper, the wire side often shows a very fine geometrical diamond-mesh pattern, impressed by the wire screen on which the web was formed; the felt side has a completely random fibrous texture, because it has only been in contact with the felt blanket of the machine. In the USA the wire and felt sides may be referred to as the "screen" and "nap" sides, respectively. Paper may be coated with sensitizer on either side, according to your taste for the image surface, and you can expect that the behavior may differ both with regard to ease and volume of coating, and the appearance of the final print. Moreover, some additives that are heavier than the pulp may move towards the lower wire side during formation, thus creating a potentially different distribution of the sensitizer from the felt side. The nature of the side may be detected by inspection under glancing illumination, or on close examination under bright transmitted light, what the papermaker calls the "look-through."

Wove

The terms "wove" or "laid" describe the pattern of the wire mesh of a mold, which imparts a corresponding textural quality to the paper surface. A wove mold, with a structure like woven cloth, generally provides a more uniform paper texture which is preferable for pictorial purposes. A laid mold of close parallel wires, bound by more widely spaced "chain" lines, leaves ribbed lines visible as a watermark in the sheet, and is more suited for writing papers—unless a strong linear texture is desired in the image.

Other Additives

Papermakers may choose to include a variety of other chemical additives in their product—doubtless for the best of reasons, in view of the use for which their paper is intended. In this regard, it should be remembered that they very seldom have the alternative photographic process printer in mind. Not all of the following additives are necessarily deleterious, but if problems arise, these are the first things to avoid.

Optical Brightening Agents

Also known as "OBAs," these are fluorescent substances that make the paper appear extra-white, sometimes unnaturally so. Their archival stability has been questioned, and they should be avoided when selecting papers for platinum/palladium printing.

Fillers

Fillers such as clay and titanium dioxide are sometimes used to give the paper an extremely smooth or brighter look. They occupy interstitial spaces between the cellulosic fibers, and can inhibit the precipitation of platinum and palladium particles, which ultimately form the image. Additionally, fillers may also present chemical incompatibilities with the sensitizer. While it is difficult to find papers without any fillers, their inclusion should be kept to a minimum.

Coloring Agents

Dyes or pigments are added to papers for a variety of reasons, but do not serve our purposes well. Pigments, like fillers, are fine particles embedded in the interstitial spaces of paper fibers. They are generally inert, but can behave unpredictably. Dyes are more persistent, but can introduce unpredictable chemical reactions with the sensitizer. Paper base color is best if determined by the source fibers rather than via an additive.

Working with Paper

Machine Direction

Most handmade papers are intentionally made with a "roughshake" so that fibers are distributed randomly. In contrast, an important characteristic of a moldmade paper is the "machine direction," which is the direction of travel as the roll of paper is formed. The cellulose fibers naturally tend to align in favor of the machine direction. This directionality of the "grain" has a consequent effect on the degree of expansion and dimensional changes when the paper is wetted and dried, and is most noticeable with translucent or "vellum" papers. The relative change in length will be less along the fiber direction (i.e. the machine direction) than perpendicular to it (the cross direction). The paper is also more rigid along the fiber direction, so any natural curl of a sheet tends to occur in the cross direction. The paper also bends more readily in this direction, which provides a convenient test for determining which direction is which: cut two identically sized strips of paper from the sheet, at right angles to one another, marking their original orientation. Hold these together at one end horizontally; the paper strip cut perpendicular to the machine direction will bend more under gravity than the one cut parallel. As a useful, but not infallible, guide, the machine direction often coincides with the longer dimension of a manufactured sheet.

Dimensional Change

Paper dimensions change when wet, due to the expansion of cellulose fibers when moistened, and contraction once dry again. Machine-made papers are dried under some tension, so their freshly milled dry length will differ from their "stress-free dry length" to which the paper reverts after having gone through a cycle of wetting and stress-free drying. These dimensional changes, "hydroexpansivity" (as a consequence of immersion in water) and "hygroexpansivity" (due to changes in the ambient RH) have some practical implications, especially for multiple printing.

Partly oriented cellulose fibers, as are found in machine-made papers, when wetted, tend to expand more in their width than in length. When coating paper, it should be clipped or taped down across the machine direction, if possible, and coated in the machine direction, to allow freedom in the direction of greater expansivity, which is the cross direction. This is only important when working with translucent papers like Reich CT and Wyndstone Vellum.

Cutting Paper

It is worth planning the cutting of larger sheets of paper to give maximum economy in the formats, but the machine direction may also be a factor to consider, as noted above. Wyndstone Vellum and Reich CT, both being heavily fibrillated translucent papers, undergo a dramatic dimensional shift when wet. To complicate matters further, they dry down to somewhere between the wet and original package size, depending on the way a print is dried. At the very least, if sensitizing with a glass rod, these papers should be cut and oriented so that the direction of travel of the rod is the same as the machine direction; otherwise the paper begins to buckle and presents a surf-coating situation!

Many fine papers can be torn uniformly to give a sympathetic feathered edge or "false deckle," especially if they are folded first, then sheared through the fold with a blunt instrument like a paperknife. A heavy, bevel-edged steel ruler is useful to provide an edge against which large paper sheets may be torn. Some conservators consider it sacrilege to cut fine paper with a knife or guillotine, and other workers claim that these instruments can leave behind tiny particles of metal that will mark the image. This latter concern is negligible if working with clean instruments, surfaces, and spaces. Always take care to avoid touching the paper surface in the intended picture area: blotches in finished prints due to fingermarks can be very vexing. Wear cotton gloves when cutting papers.

Testing Papers for Pt/Pd

You will always, while working from the preceding desiderata, be searching for a good paper. I cannot underscore enough the effect on the reactive sensitizer chemicals of the papermakers' additives, which they are usually reluctant to disclose or discuss. Each commercial paper has its own idiosyncrasies, which may even vary from batch to batch or be changed by the manufacturer without notice. There is no substitute for personal trial and experiment. Once you have discovered a satisfactory paper, it is wise to invest in a substantial amount of the same batch, or "making." It may now be apparent, out of all the conditions and components needed for platinum/palladium printing, that paper is the one over which we have least control. Yet, paper is a crucial ingredient for the process, especially that of 100% platinum, which is more particular than any other siderotype formulation. Paper, as we have often said, is the great imponderable. Or, to put it in the parlance of politics, it is, for us printers, the most uncontrollable of controllables. Thankfully, there are a few papers that are known to work well with the process, and there is a sharing community of printers who are always seeking, testing and bringing suitable papers to the attention of enthusiasts. A list of paper suppliers is presented in Appendix IV: Resources, and this chapter concludes with currently available recommended papers.

It is likely that you will encounter a paper that has not been tested for the platinum/palladium process. This section suggests a procedure for testing papers. It relies on a set of important assumptions:

- that you have successfully made printing-out platinum/palladium prints with a standardized step-test negative (also known as a test "target"—see Chapter 7), ideally on one of the papers recommended below; and that you can refer to these prints, and use the step-test negative for subsequent tests
- that you understand how to accurately control the measurements of liquids, paper hydration, coating technique, exposure, and processing for platinum/palladium printing
- that you maintain detailed notes

What follows is presented as a guideline for the printing-out platinum/palladium process, and it may need to be modified either for your particular workflows or for other siderotype processes.

Paper Test Flow Chart

1. *Acquire paper from a known stock/supply*

Start by selecting a paper. Do you have a production, batch or "make" number from which the sample sheet came, and does the supplier have sufficient stock from this batch? If not, look for another supplier, or better, seek out the distributor/wholesaler for a list of retailers. Enough paper is needed to coat at least three sheets, with the sheet sizes depending on that of the step test. Ideally, leave a 1″/2.5 cm border all around the coated area.

2. *Acclimate paper*

Once you receive them, let the paper samples acclimate to standard room conditions (ca. 20°C, RH

40–60%) for at least one hour before coating. This is simply done by either hanging the sheets on a line in the room, or placing them on a clean work surface, or putting them in a hydration box with a RH of 40–60%.

3. *Select felt or wire side*

Repeat the entire test sequence for the other side if needed.

4. *Sensitizer completely applied in 4 to 5 passes, using 0.018 ml per 1 sq. inch / 0.0028 ml per 1 sq. cm of sensitizer as a starting point*

If Yes, go to 5

4.1. If the sensitizer runs out early, increase its amount by 10% and coat another sheet, repeating the entire sequence thus far. If sensitizer runs out again after this, the paper is too absorbent. Reject the paper.

4.2. If excess sensitizer remains after five passes, decrease the sensitizer volume by about 10% or add Tween, and coat another sheet, repeating the entire sequence thus far. Use 15% Tween, adding an amount equal to ca. $^{1}/_{30}$th of the sensitizer volume, e.g. combine 1.5 ml of sensitizer with one drop of 15% Tween. (1 drop is approximately 0.05 ml.)

If there is still an excess of sensitizer after this, the paper may be too heavily sized. Note that some textured surfaces are difficult to coat with a rod. Try the other side, consider using a brush, or reject the paper.

Coated surface is uniform and free of crystals and/or blotches after drying and humidifying

If Yes, go to 6

5.

5.1. If only very fine crystals are visible (sparkling surface in glancing light), the paper may be too dry. Give it a pre-coat hydration for 10–30 minutes in an atmosphere of RH 60–80% at ca. 68°F / 20°C (be consistent with both of these variables).

If crystals appear after coating a well-hydrated sheet of paper, then add Tween to the sensitizer and coat another sheet, repeating the entire sequence thus far. Use 15% Tween, adding an amount

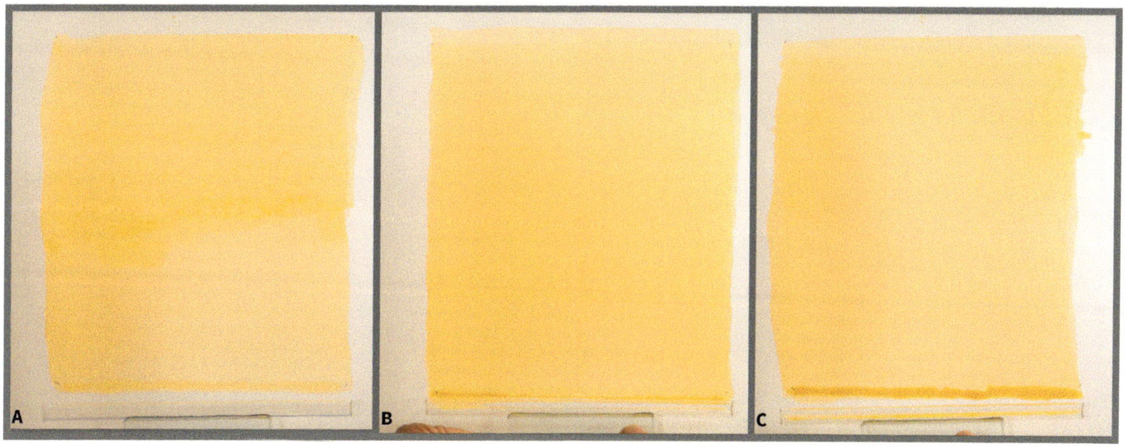

Figure 6.4 Three coatings illustrating:
A. sensitizer being absorbed too rapidly and running out before four passes are completed (see step 4.1).
B. sensitizer being evenly distributed in four to six passes with just the minimum amount left over.
C. too much sensitizer remaining after 6 passes (see step 4.2).

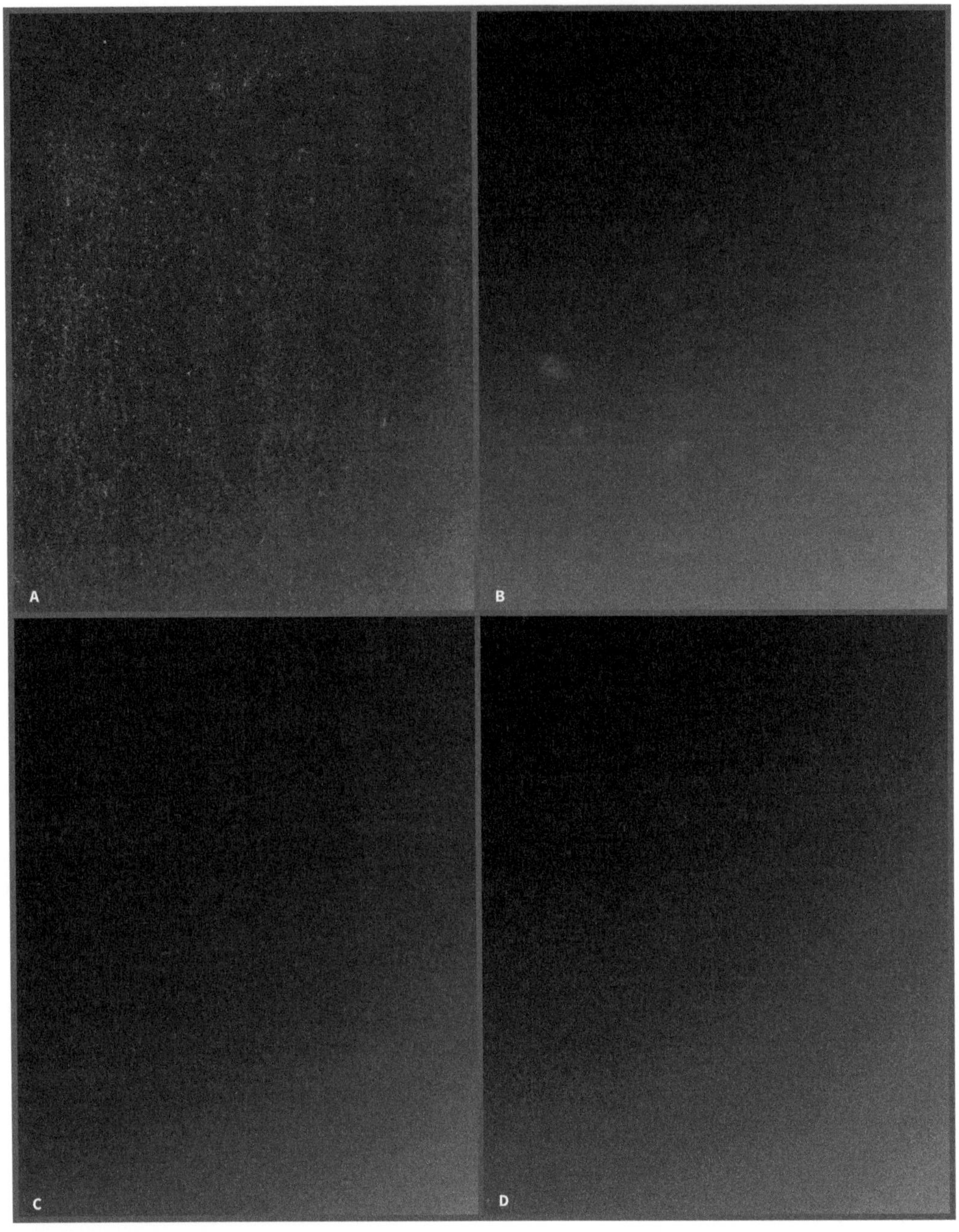

Figure 6.5 A and B show details from a print illustrating faulty coatings:
A. Revere Platinum with crystals and blotches (see step 5.1).
B. Arches Platine with blotches (see step 5.3).
These are resolved by:
C. Revere Platinum with Tween and glycerin
D. Arches Platine with glycerin.

equal to ca. 3% of the sensitizer volume as indicated in 4.2 above.

5.2. If clearly visible crystal clusters, abandon the paper.

5.3. If only blotches are visible, then add glycerin to the sensitizer and do another coat, repeating the entire process thus far. Use 15% glycerin, adding an amount equal to ca. $^{1}/_{30}$th of the sensitizer volume, e.g. combine 1.5 ml of sensitizer with one drop of 15% glycerin.

5.4. If fine crystals *and* blotches are visible, add a combination of Tween and glycerin and do another coat, repeating the test sequence with the additive(s). Use 15% glycerin and 15% Tween, adding an amount equal to ca. $^{1}/_{30}$th of the sensitizer volume for each, e.g. combine 1.5 ml of sensitizer with one drop of 15% glycerin and one drop of 15% Tween.

If blotches or crystals show at this stage despite these additives, then this is an unsuitable surface. Try the other side or reject the paper.

6. *Expose. There is a 'good' black / Dmax*

If Yes, go to 7

Dmax is that area of maximum black where any further exposure does not render a visibly darker tone. Refer to your reference step test negative to locate these areas. See Chapter 7.

6.1. Increase UV exposure to reach a maximum black. Repeat the entire test sequence with the new exposure time. To do an exposure test for maximum black, coat a small section of paper, then prepare and expose it in your normal conditions to a stepped series of exposures, running to at least 4x your normal exposure. After processing and drying, select the exposure just after the step beyond which you cannot discern any further difference in black tones. This compensates somewhat for the film's base density, and is your adjusted base exposure for this paper.

If there is no acceptable Dmax after this, then reject the paper. The paper may have buffering agents in it that are incompatible with the process. Iron compounds are not being cleared, rendering a non-archival print. Or consider repeating the test after acidifying the paper (see below).

7. *Process and dry—the white, unexposed parts of the coated area are fully cleared*

If Yes, go to 8

7.1. If whites are discolored, residual iron compounds are preventing thorough clearing. This is due to buffering agents or other incompatible compounds included in the manufacture of the paper, or faulty wet-processing baths. Check your solutions. Then consider repeating the entire test after acidifying the paper (see *Acidifying* below), or reject the paper.

8. *Inspect print: the mid- and highlight tones are smooth*

If Yes, go to 9

8.1. If fibrous, add Tween to sensitizer. If blotchy or uneven areas of tone, add glycerin. Repeat the entire test sequence with the additive(s). If the result is still blotchy or uneven, reject the paper.

9. *Felt or wire sides*

Note down differences, if any

Is the overall result acceptable? Ultimately, it comes down to how the print *feels*. If you don't like the visual feel of the print, reject the paper.

10. *Acquire*

As much as you can afford from the same batch of paper!

Keep a copy of the invoice. Upon receipt of paper, check that you have the same batch of paper as the test sheets. Double-check your testing process and, as soon as possible, replicate the test sequence on a sheet from the middle of the stack. The result should match your previous test. Rejoice. If results are mis-matched, you were sent a different batch. Contact the supplier.

Paper Modifications

It is rare to find a paper that works perfectly with the printing process. One could even argue that there is no such thing; "perfect" can vary depending on the photographer/printer, negative, and image. After all, the paper is not only a messenger but also part of the message. Thus, we may, more often than not, have to fine-tune a paper by a number of methods, which range from simple to tedious and time-consuming.

Acidifying / Decalcifying

As has been noted above, buffered papers containing calcium carbonate will degrade the image. If buffered papers are unavoidable,[8] they should be pre-treated in an acid bath, washed, and dried. It may also be desirable to calender after washing, to return the surface to its original smoothness.

Numerous approaches have been taken to remove calcium carbonate from the paper, including the use of dilute hydrochloric acid, which works satisfactorily. There have been some recommendations to decalcify paper with oxalic acid, but this should be avoided: the byproduct of calcium carbonate and oxalic acid is calcium oxalate, which is just as insoluble as the buffer we want to remove.

Recently, a bath of sulfamic acid has been recommended as a convenient substance to decalcify papers.[9] It is more readily available and less corrosive than hydrochloric acid. However, if preferred, a 5% v/v of hydrochloric acid may be used as a substitute.

Prepare enough 10% w/v sulfamic acid solution to generously cover several sheets of paper at a time, placed in a suitably sized tray. Immerse the sheets for 20 minutes, all the while with occasional, gentle agitation, moving sheets through the stacking order and flipping them over. This ensures even and thorough acidification, after which, wash the paper in five complete changes of water over 20 minutes, in fresh water at room temperature. During the wash, move sheets through the stacking order regularly, making sure that each sheet gets to be at the top of the stack once. Once the wash is complete, the paper may be calendered (see below), dried by blotting, and hanging or laying out on screens (face up). Wait at least 24 hours before sensitizing the paper.

Calendering

Some papers may provide all the necessary conditions for rendering a good print except for their texture. There is diminished contact between the negative and paper when the surface texture becomes coarse, and this begins to interfere with the image. To improve image sharpness, a textured paper may be given a slightly finer surface by pressing it against a smooth surface. This same principle is applied on a larger scale during the manufacture of paper, when sheets are passed between rollers under pressure, and is referred to as the "calender."

Here, we can apply a process of "pseudo-calendering" with the minimum of equipment. Fully immerse paper sheets in water (be sure it is clean, and at neutral pH—if in doubt, use purified water) at ca. 68°F / 20°C for ca. 1 minute. The time should be adjusted depending on the paper and desired surface. Heavily sized papers may need to be immersed longer. Sheets that have absorbed more water have increased pliancy and consequently surface textures will smooth out to a greater degree when pressed. After immersion, place one sheet at a time face down on a very clean sheet of plexiglass and run it through an etching press, having adjusted the roller pressure as desired. If an etching press is not available, place the paper on clean plexiglass or plate glass, on a flat work surface, cover with three or four layers of blotting paper, and pass a roller that is greater than the width of paper, over the entire sheet with considerable pressure. Uncover and peel off gently. Allow to air dry for at least 24 hours before coating with sensitizer.

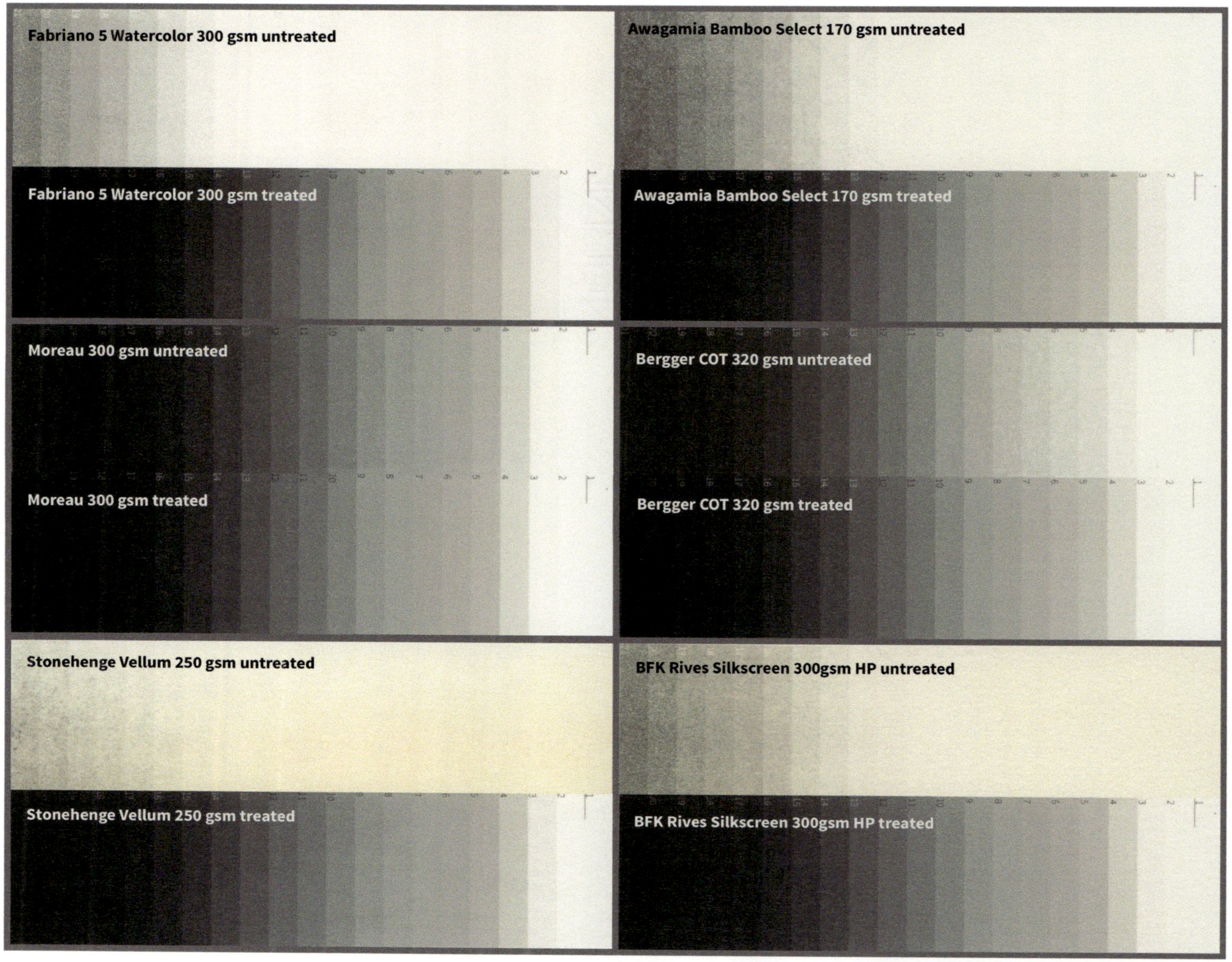

Figure 6.6 Decalcifying paper. Six buffered papers before and after de-calcifying in a 10% solution of sulfamic acid. All tests were cleared and washed by the normal methods of processing prin-ing-out Pt/Pd. All prints made with Fe1:Pt:Pd at 2:1:1, hydrated for 90 min. at 52% RH / 20°C.

Humectants and Surfactants

Paper is about 50% empty space—a void which can be partly filled by imbibing liquids such as sensitizer. There are essentially two interconnected systems of pores within the structure of cellulose paper:

1. The macroporous structure, which consists of the capillary channels between the cellulose fibers, easily visible with a 10x magnifying loupe. This provides a route for the imbibed fluid into the interior, but this space is not useful for retaining nanoparticle image substances, such as platinum and palladium, because their particles will easily be washed out of these regions.
2. The microporous structure within the cellulose fibers, which is provided by gaps between the fibrils that make them up. This region, known as the interfibrillar space, can effectively trap the platinum/palladium nanoparticles as the image is formed.

Penetration of the interfibrillar space of the cellulose structure by the sensitizer solution is therefore of paramount importance, if the image is not to wash away in the wet processing. That noted, it is also undesirable that a sensitizer should soak rapidly into the interior of a paper sheet, where it cannot contribute to the image and may resist washing out—besides being very wasteful of the sensitizer chemicals. Ideally, the sensitizer and image should be established in the interfibrillar space within ca. 0.1 mm of the coated surface. For this reason, the paper's sizing agent and other additives should make it slightly water repellent at first, allowing time for an aqueous sensitizer to be uniformly spread over the surface of the sheet before it starts absorbing into the fibers. Paper's absorbency also varies depending upon the nature of the solution being applied to it. Viscosity surface tension and chemical composition influence the sensitizer's ability to swell cellulose and open micropores in the cell wall, so admitting further liquid into the interfibrillar space.[10] With some papers, a sensitizer may tend to lie on the surface of the paper or in the open macroporous capillaries, rather than diffusing through the pores into the interfibrillar space. In other cases, the ingredients and sheet formation may lead to uneven absorption across a sensitized area. To help overcome these problems, and control the absorption and distribution of sensitizer, a wetting agent, also known as a "surfactant," or a humidifying agent, or "humectant," may be added.

Indications for use of either surfactant (Tween) or humectant (glycerin) are given in the previous section, *Testing Papers for Pt/Pd*. Generally, use Tween to resolve crystallization, retarded absorption problems when coating, and fibrousness in the highlight values of a print; use glycerin to reduce uneven coating and blotching, and mid-tone fibrousness in the print. At times, a combination of both is necessary.

Out of the various available types, non-ionic surfactants are most preferable, such as Tween 20™. A *final* concentration in the sensitizer solution of about 0.2−0.5%, depending on the paper, seems compatible with all the usual sensitizer chemicals. This is inexpensive and readily obtainable, though no doubt there may be many others that are also suitable, but with greater risk of foaming, including the longer chain siblings of Tween 40™ etc. Some trial and adjustments may be necessary, but a suitable starting formulation would be to prepare no more than 20 ml (very small amounts are needed) of a 15% solution of Tween 20 with distilled or purified water. Then, add an amount equivalent to ca. $^{1}/_{30}$th of the sensitizer volume to the sensitizer, e.g. add one drop (about 0.05 ml) of 15% Tween to 1.50 ml of sensitizer, then coat. Since Tween can, in diluted form, host micro-organisms, it is best to only add it to sensitizer on a coat-by-coat basis. Glycerin is easily prepared with distilled or purified water, to a 15% concentration and used in the same amounts as Tween.

Too much surfactant will cause the sensitizer to diffuse too deeply within the paper, the image will appear "flat" and "sunken." Too much humectant will prevent the sensitizer from drying out, and will damage the negative as well as render a flat and blemished print. The control of the balance of paper absorbency towards the sensitizer solution is crucial to the image quality, and the key to success in hand-coated photographic print-making.

Humidifying

Papers that have been stored in excessively dry conditions, as sometimes happens during the winter months in heated spaces, or when freshly received from warehouse storage, can sometimes render a crystallized coating. The simplest way to resolve this is to humidify the paper. One approach, if working conditions, time, and space allow, is to hang sheets in a space ca. 40% to 60% RH at room temperature, for 24 hours prior to coating. Alternatively, a sheet may be hydrated for 5 to 10 minutes in a hydration box of at least 50% RH, then coated with sensitizer—see Chapter 4. Regardless of the method, consistency is important.

Sizing

The entire platinum/palladium printing process requires acknowledgment of the relationship between the chemicals: those that make up the sensitizer, paper that it is coated on, and the substances used to clear a print after exposure to ultraviolet light. A consistency of ingredients and methods takes us a long way towards getting on with the key matter of making prints. Internally sized papers, as described above, will render the most predictable and repeatable results. Manually sizing paper, externally, introduces chemical and physical inconsistencies. I do not recommend this practice. If there is a need to size paper, for instance after decalcifying a machine-made stock, it may be best to use starch-based or alum-rosin agents—much depends on the paper. Avoid agents and surface treatments that include hydrophobic polymers, fumed or pyrogenic silica (strictly speaking this is not a sizing agent, but a thickener) and gelatin. The latter works well with palladium or palladium-platinum (where platinum makes up less than 20% of the sensitizer), but does not interact well with platinum or predominantly platinum sensitizers.[11] Size may be externally applied in a number of ways, including by brush, spray, or immersion. Attention must be paid to the concentration, viscosity, and temperature of the sizing agent, as well as the amount of applied solution. Of critical import is the way size is allowed to dry: the sizing agent should remain evenly distributed across the sheet as it dries. Simply hanging the paper by one corner until it is dry will create a gradient of varying absorptivity, and this will manifest in the tonal distribution of the print. It may be best to wrap a stack of freshly sized sheets in impervious material, and under slight pressure, for several hours—allowing the sizing agent to spread homogenously across and slightly within the sheets. Then, hang the sheet by two adjacent corners, rotating regularly from corner to corner until the sheets are dry. Again, it will help to strive for consistency through every step of this considerable effort.

Recommended Paper

This list only includes papers that are available as of the beginning of 2020, and work for platinum, palladium, and platinum/palladium printing. Not all are easily available in all regions. For a longer list of papers and suppliers, please refer to Appendix IV Resources, and the book's online resource to access up-to-date information. Note that naming conventions for any given paper may stay the same but its formulae, and even mill-source may change. Be sure to test each batch.

Arches Platine

Made by a French mill dating back to the fifteenth century, Arches Platine was developed in close consultation with the photographer/printer Martin Axon and others. This is a 100% cotton rag furnish, no buffer—or, as the manufacturer puts it, without an alkaline reserve, pH neutral, no optical brightening agents, and made on a cylinder mold. It comes in two weights and two sizes, with two deckled edges. Each side presents some differences in image quality. Arches Platine's dimensional stability is high, with negligible change from the original package-dimensions to the processed and dry print. It has now gone through several productions, each with slight variations in the formula. Consequently, each batch requires platinum/palladium printers to make small adjustments. The current batch, made without any watermarks, works well without Tween, and benefits from the addition of glycerin to the sensitizer. This paper is widely used in other siderotype processes.

Buxton Rag, Herschel Platinotype

Both of these handmade papers are produced by Ruscombe Mill. Buxton paper was developed in 1993 by the Mill's original master papermaker, Chris Bingham, when its studios were in Gloucestershire, England, out of a collaboration with Mike Ware. The name that Chris chose for it—Buxton—was a reference to Mike's home town, which has been noted historically for its spa waters since Roman times, when it was known as Aquae Arnemetiae. In the late 1990s, Ruscombe Mill was relocated to Margaux in France, in the heart of the Haut Médoc, where it shares the excellent local water with the winemakers. Buxton was made from the finest quality long cotton fiber, nearly 100% α-cellulose, which is stronger than papers made from cotton linters. The mill had problems with the quality of cotton furnishings supply in 2011 (which seem to have now been resolved as the mill has reintroduced a Buxton Rag paper), which eventually caused Chris Bingham to design a new paper. This new paper differs from all other alternative process papers in being handmade from 100% linen cellulose fiber (made entirely from the best quality flax, not cotton): he decided to name it "Herschel Platinotype" paper. Linen fiber is derived from the stem of the Common Flax plant, *Linum usitatissimum*, and has a similar size to cotton, but is cylindrical with a thick cell wall and a lumen (the tubular fiber's internal space or opening) of relatively small diameter. It has periodic knots or nodules, resembling bamboo. When beaten, the flax fibrillates more extensively than cotton and forms a stronger, less elastic paper sheet. These differences play out ultimately in the siderotype process—it is possible that the structure of the flax fiber may enhance its ability to retain the metal nanoparticles that lead to image formation.

Both Buxton Rag and Herschel Platine are cold-pressed, with an attractive "tooth" to give life to the surface, which does not change significantly after wet processing. The papers are internally sized with neutral alkyl ketene dimer (AKD), have a pH of approximately 7.5 and are very close to neutral, there are no additives, and they display a natural bright white color. There is no difference between wire and felt surfaces with regard to image formation, though each side may respond slightly differently when being coated. Use Tween with the sensitizer.

Revere Platinum

Related to the Magnani paper mill in Italy, but commissioned and distributed exclusively by Legion Paper Company based in New York, Revere Platinum is currently one of the most compatible papers for platinum/palladium and other siderotype processes. Based on a production by Magnani around 2011, variations in the formula since then have left printers veering from getting very good to disappointing results with the paper. Legion Paper is now committed to making a paper for alternative and siderotype processes, and the current production is proving to be very consistent. Revere Platinum is a 100% cotton paper, with an internal sizing agent, a smooth hot-pressed surface, and no buffering agent. It comes in one weight and two sheet sizes. Both sides are very similar, and yield good results. It is best to use a combination of Tween and glycerin when coating this paper. It has good wet strength and dimensional stability.

Reich CT

Commissioned and distributed by the family-owned Reich Paper company, Reich (pronounced "reesh") CT is a naturally translucent paper with unusual properties. It may be the only paper in the platinum/palladium materials list that has a food-safe certification! The paper is manufactured in England, using technology that was initially developed as a part of a strategic effort to prepare for World War II—all of Britain's tracing, or "engineering," paper was being imported during the 1930s, and mostly from Germany. Now made from 100% cellulose derived from a combination of northern hardwoods and southern softwoods, the fibers of which undergo heavy cellular defibrillation, Reich CT is a superfine, dense, and impervious paper. Due to the special processes employed in the manufacture of translucent papers, this paper does not receive heavy calendering (contrary to the impression one gets from its smooth surface). However, it is highly dimensionally unstable. It shrinks by as much as 25% by the end of the production process. From this state, its dimensional dynamics are a bit like a yoyo. The paper, relative to the original dry "packet" dimensions, expands by about 2% in the grain direction, and an astonishing 13% in the cross-grain direction when fully wet. It then, perplexingly, and depending on how much pressure is applied while drying the sheet, shrinks by about 1% in the grain direction and *increases* by about 5% in the cross-grain of the original dry dimensions. These dynamics can be compensated for to some extent when preparing digital negatives—see Chapter 7. Reich CT has a very smooth and fine surface on either side, and is available in a number of weights which are described as "bond" units: the thinner papers, 17# 24# and 30# have to be handled with the greatest of care during processing.

Wyndstone Vellum

This paper has many of the same characteristics as Reich CT, but is made in Germany. The main difference is that it is a bit more difficult and expensive to acquire in the USA.

Notes

1 Ware, *Platinomicon*, pp. 106–107. "Alfred Clements and his Work," *Photographic Times*, October 27, 1895, 216–220.

2 George Dawson, "The Platinotype Printing Process," *The British Journal of Photography*, August 20, 27 1880, 399–400.

3 http://darwin-online.org.uk/Francis_Darwins_Origin_of_species_1859.html, accessed 11.29.2019.

4 Dard Hunter, *Papermaking: The History and Technique of an Ancient Craft*. New York: Alfred A. Knopf, 1947, p. 50.

5 Ibid., p. 453.

6 Ware, *Platinomicon*, pp. 165–166.

7 Numerous calculators are available online for converting paper weights from lbs to g/m^{2}. Application of a conversion formula is possible, if the extra data are known, to give an equation to convert weights in lbs to g/m2.

The relationship between the weight, G, in g/m2 and the weight, W, in lbs (pounds) per ream is:

$$G = WF/AS$$

Where:

A is the area of one sheet in square inches

S is the number of sheets per ream (which can be 480, 500, or 516)

F is a numerical factor converting the Imperial units to metric: the numbers of grams in a pound times the number of square inches in a square meter:

$$F = 453.59237 \times 1550.0031 = 703069.58 = 703070 \text{ (approximately!)}$$

So, if we take a ream of S = 500 sheets of Imperial size (22 x 30 inches), the sheet area A = 660 square inches, and we get the conversion factor:

$$G = W \times 703070/(660 \times 500) = 2.13\ W$$

i.e. a "90 lbs" Imperial paper is 2.13 x 90 = 192 g/m2, etc.

8 Such as Canson Lavis Technique, Fabriano Artistico, Whatman Watercolour, 300gsm Canson Montval, 420gsm Gvarro, 180gsm Schoellerschammer dura matt, 200gsm Fabriano Artistico Traditional White HP, 300gsm Fabriano Artistico Extra White SP.

9 Serdar Bilici, and Loris Medici, *Sulfamic Acid for Paper Acidification*, https://photrio.com/forum/threads/sulfamic-acid-for-paper-acidification.102747/, accessed 12.03. 2019.

10 Mike Ware, *The Chrysotype Manual*. Brighton: ffotofilm publishing, 2006, p. 52.

11 Mike Ware, *Platinomicon*, pp. 218–219.

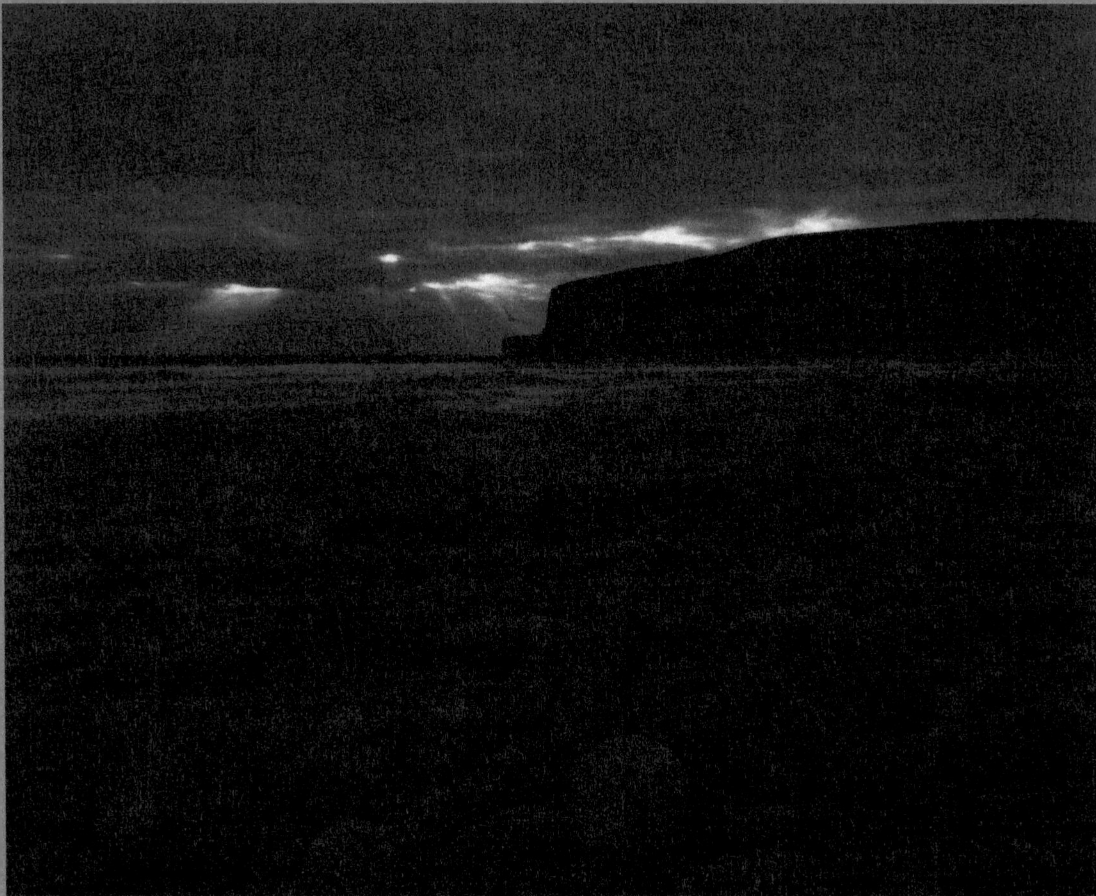

Figure 7.0 Mike Ware, *Rackwick, Hoy, Orkney*, 1984. Palladium print on Fabriano 5, from enlarged copy negative (Kodak SO

Chapter 7

Negatives

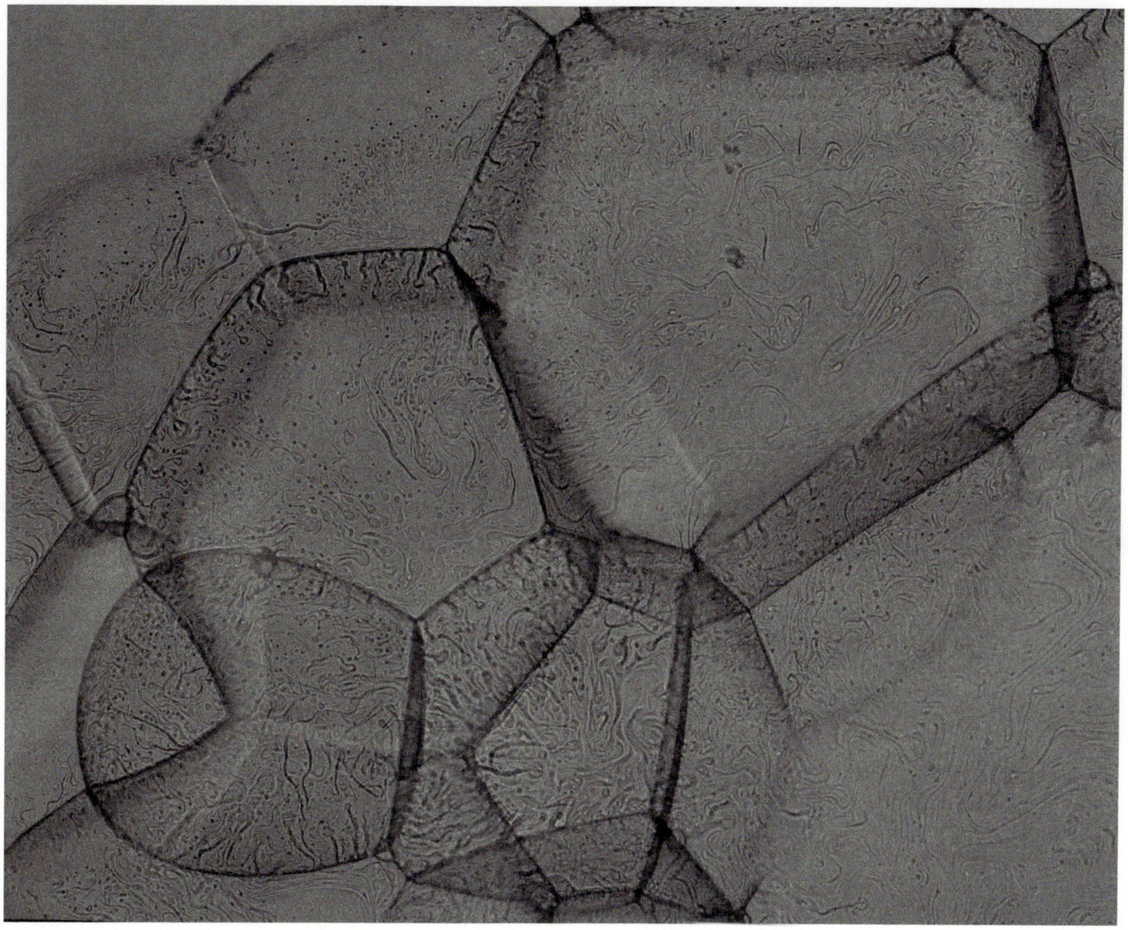

Figure 7.1 Jamie Steele, *Untitled (detergent bubbles)*, 2007. Platinum-palladium print on Crane paper, from original 4 x 5 inch negative, 12.5 x 10 cm, Fe3:Pt:Pd = 2:1:1, hydrated for 20 min in 100% RH / 20°C (water chamber).

At the beginning of this book, I described what first inspired me to make a platinum/palladium print. These aspirations still hold true and guide my work: I wish to make a print such that it defies separation into the many components of its making—a fusion of process and place; image and mark; tonal surfaces and paper. A good platinum/palladium print should let one look past, and thereby transcend, the process. This entire book comes from a desire to put good science and years of experience into the hands of artists, so that we may get on with the real work: to let beauty steady us as we stumble through the realm of human experience.[1] This chapter is about that journey, which begins with making an exposure and takes us to the point of having a negative ready to print. There is a lifetime of learning to absorb with each of these stages. A holistic approach may be most effective and productive: to cycle through all the stages of making a printed image, knowing that with each iteration, there will be a steady deepening of skill.

What follows are options for arriving at a printable negative. Each of these deserve more than a book's worth of text, and indeed, many volumes have been written about exposure, development, digital manipulation, and output. It may be helpful, however, to have a summary of workflows on hand, as they are presented here.

It is assumed that there is an understanding of the fundamental aspects of exposure, development (wet lab or digital image processing) and contrast controls (through exposure/development adjustments for film, or Photoshop for digital). It is also assumed that there is a familiarity with, and access to, the relevant facilities, such as a darkroom or computer and inkjet printer setup.

A black and white photographic process is not completely representative of, nor identical to, the human visual experience. To put it somewhat clumsily, the lens and sensor (or negative) do not "see" the way we see. Likewise, photo-sensitive paper does not respond to shades of light in the same way that we do. Less obviously, as the "image" progresses from lens to sensor to print, each of these media renders it in different ways. When our eyes may perceive a rich range of tonal variations of sunbeams over a boulder-strewn beach, the lens may bump up the scene's contrast, and the film may not differentiate beyond a certain range of tones. Hence, the captured image needs to be exposed and then processed, via wet lab or digital methods, to fall within the full tonal capacities of the negative or sensor. Further adjustments, by way of contrast controls and modulation of the negative's opacity, will need to be made to express fully the photographer's vision in the print. Each realm of control in the making of a fine print is affected by every other step of the entire process. So, it seems practical to envision what a print may look like while taking the photograph. One should also expect to discover meaning while preparing a negative and making a print.

Size, Contrast, and Density

Platinum/palladium and other iron-based sensitizers are about a million times less light-sensitive than silver-gelatin papers. Both practically and economically, current technology does not allow for projection, or enlargement, printing with this family of processes. Instead, contact-printing methods need to be used in order to have an adequate throughput of intense UV light. Thus, the basic requirement is a large negative that will eventually render a contact print of matching size. While diminutive contact prints as small as about 1″x 1.5″ / 24 mm x 36 mm may be made from 35 mm negatives, they require very finely textured paper and an image with clearly defined compositional elements in order to hold one's attention. The ideal minimum size seems to be about 5″x 7″ / 12 cm x 18 cm. The normal approach in producing large negatives, up to the beginning of this century, would have been to work with a large format camera. Contact prints from original, large negatives have a special allure because of their high resolution and tonal depth. The workflow for making negatives from photographic film is included for this reason, even though it may now be a rarefied practice. A more common approach is to make enlarged duplicate negatives via digital output onto inkjet-printed film. Along with adjustments in size, this also allows for unprecedented contrast and density modulations, albeit at the cost of a slight reduction in resolution.

"Contrast" and "density" have special meaning in photography. These terms apply to the systematic study of light-sensitive materials, called "sensitometry." The field, as an evaluation of relationships between an emulsion's response to

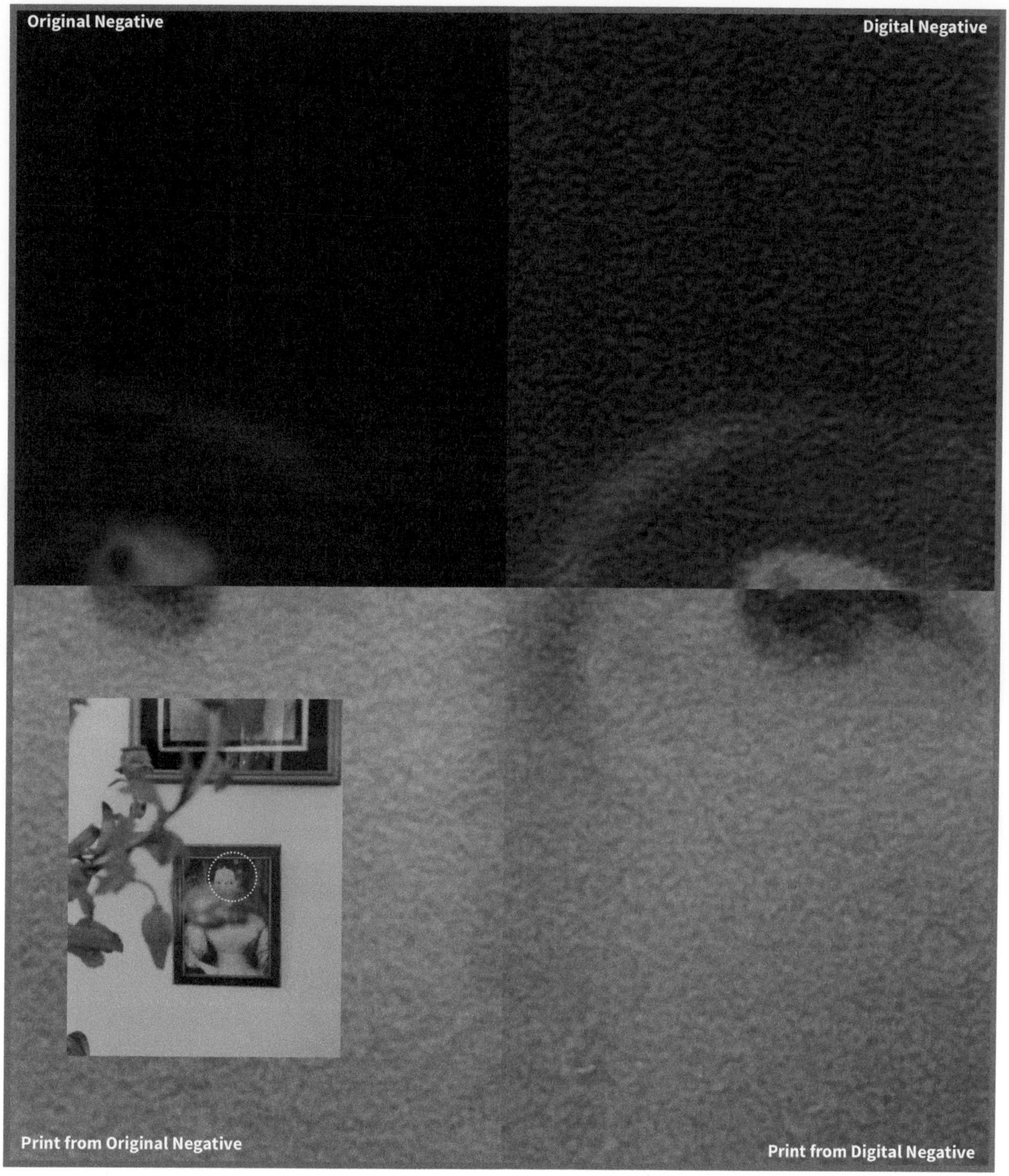

Figure 7.2 Comparison between an original and a copy negative: magnified details comparing the resolution of a print from an original 8 x 10 negative (inset) with a digital replica of the same negative. Both are negatives printed with Fe3 at 2:1:1, on Revere Platinum. The original negative is with Ilford HP5 developed in Pyro PMK. The digital negative was from a 600dpi scan with an Epson 10000XL, and output on Pictorico UP OHP, with an Epson 600 using PiezoDN Pro inks. There is no noticeable difference to the naked eye.

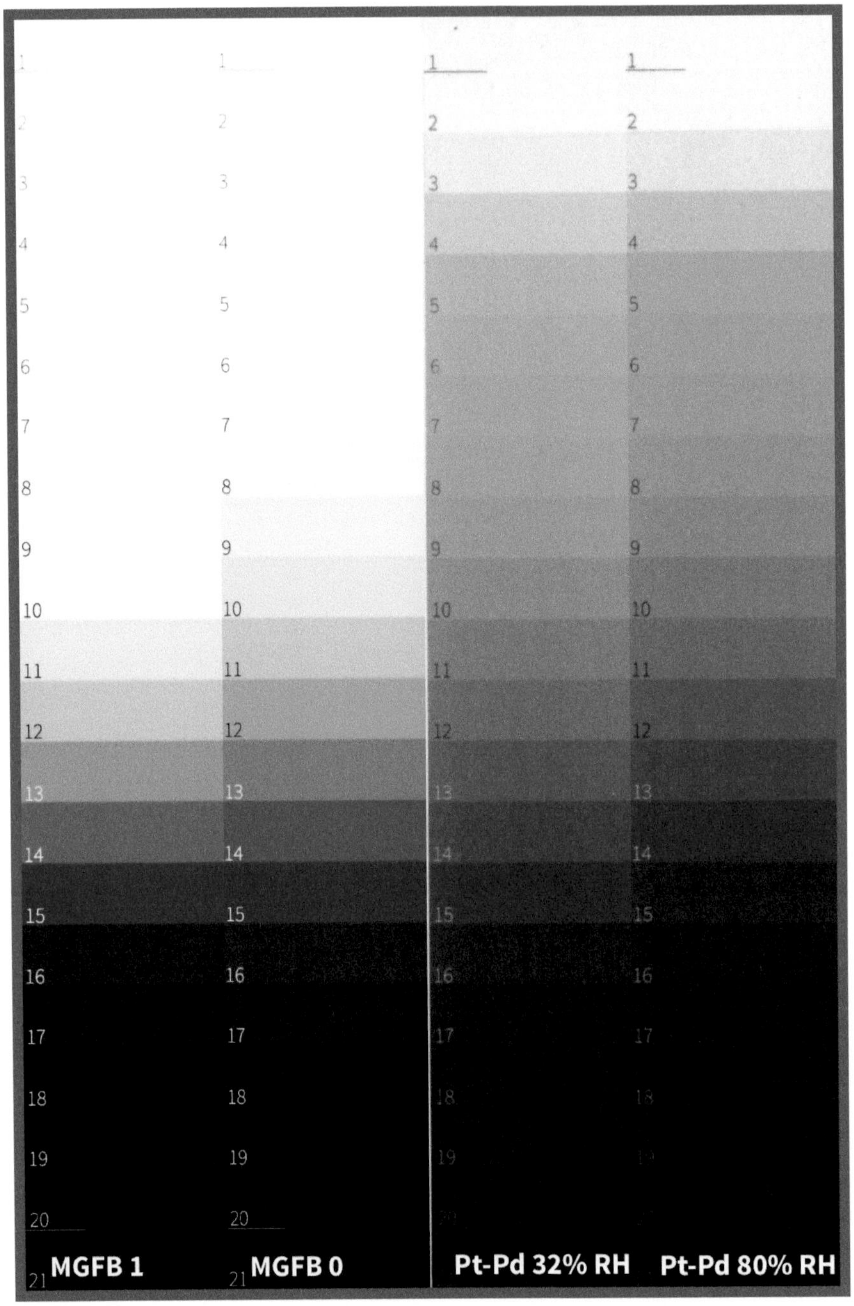

Figure 7.3 Differences of exposure scale and contrast between silver-gelatin and platinum-palladium processes, illustrated by step tests:

A silver-gelatin multigrade paper of grades #1 and #0 have exposure scales of 4 and 5 stops, respectively.

A Fe1:Pt:Pd = 2:1:1 paper, hydrated at 32% RH and at 80% RH has an exposure scale of 8 to 9 stops.

However, the density range of the silver-gelatin paper is perceptibly greater than that of the platinum-palladium paper; it is accordingly described as more contrasty.

All exposures were established for each process. To make comparison easier, the scans are shown without any color values.

light and the actual dosage of received light, was established in 1876 by two scientists and amateur photographers, Ferdinand Hurter (1844–1898) and Vero Charles Driffield (1848–1915).[2] Contrast is understood as the range of tonal grays extending between the two extreme points of black and white (or clear), and density describes how much light is absorbed or blocked by a material. By quantifying these two properties, we establish a way of matching and converting values between one system and another. Here, we need to align a negative's contrast and density range to the exposure scale of the platinum/palladium process. These ranges are measured as units of density with a device called a densitometer, and all the measurements are most practically rendered as a graph, or "curve." Negatives, being transparent, are measured by transmitted light passing through the material, and prints are measured by reflected light. As we consider the various approaches for making negatives, it is useful to know that platinum/palladium prints have an exposure scale of between 1.5 and 2.4, depending on whether the sensitizer is pure platinum, or pure palladium, or a mixture of the two; the paper; and the degree of hydration. These numbers need context, so let us relate the process to silver-gelatin printing papers: grade 0 (zero) papers, being of the lowest contrast, generally have an exposure scale of about 1.5. Platinum/palladium paper, and especially pure palladium or fully hydrated palladium-platinum, can render a very long, low contrast tonal scale that goes far beyond any silver-gelatin paper. One further cross-reference may be useful: these exposure scales may also be considered in terms of exposure values or "stops," ranging very roughly from about 5 to 8 stops, respectively. Compared to silver-gelatin printing, the print-out platinum/palladium method offers a much wider, or longer, exposure scale. Consequently, negatives have to be such that their densest areas do not "print through" and degrade the printed highlights and whites to gray. Specifically, a negative should display a maximum density of about 2.4 or more. For all of these reasons, negatives for platinum/palladium need to be adjusted for contrast and density.

Film, or Wet-lab Negatives

A negative's contrast and density are determined by the choice of film, exposure, the subject's contrast or brightness range, and development. Conventional, panchromatic black and white film is best for contact printing by ultraviolet. The most versatile films for contrast adjustments through development are the so-called medium to high speed materials like Bergger Pancro 400; Ilford FP4 and HP5; and Kodak TMAX 400.

Exposure and Subject Brightness

With the high sensitivity, or "dynamic" ranges of contemporary full frame digital sensors, we have much greater flexibility to make exposure adjustments during processing than with photographic film. With its narrower range, film exposure must be calculated with care, even though it is subjective in that the photographer decides which tone in the scene is going to match the mid-point gray in the print's tonal range. Exposure is also fine-tuned by the desired outcomes resulting from subject contrast and development. We think of a "normal" subject brightness range as having roughly a 6 to 7-stop difference from the darkest to the brightest areas in which we want to see textural detail in the print. Subject brightness ranges much above or below this average need to be "compressed" or "expanded," respectively, depending on the photographer's intentions, by the degree of development. Note that development is the sum of four components: the developer (chemical type and concentration), temperature, time, and agitation. A subject scene with 8 to 12 stops may require reduced development in order to bring the negative's contrast down to match the platinum/palladium sensitizer's exposure scale. Conversely, a foggy, 2-stop scene may require greater development.

Film Development and Pyro PMK

Generally, a workable negative for platinum/palladium printing may usually be produced from a film by giving a normally exposed camera negative about 70–100% extra development based on the normal (N) times suggested for standard film developers such as Agfa Rodinal (now made by Adox) and Kodak HC-110. For simplicity, let us refer to the manufacturer's recommended, "normal" development time as N, so N+70% is the optimal time for platinum/palladium negatives, and call this "P." Then, the following general guideline may provide a starting point from which film, exposure and development tests should be carried out, and intermediate points be established.

Particularly suitable for siderotype processes however, and when combined with Ilford FP4 or HP5 films, are the pyrogallol developers such as Pyro PMK. Reformulated by Gordon Hutchings and detailed in "The Book of Pyro,"[3] Pyro PMK renders negatives that are well suited for both platinum/palladium and silver gelatin printing. The process of development surrounds each silver grain with a yellow-green stain. Being opaque to ultraviolet, the stain presents as an area of tonal density to platinum/palladium sensitizer. Since the degree of staining increases in the dense areas of the negative, more ultraviolet will be blocked by the yellowish stain there than in the more transparent (subject shadow) areas. The net result amounts to a negative with a higher density range, or greater contrast index, in the ultraviolet spectrum than in blue light—this latter source being the one used for exposing silver-gelatin paper. Of particular note is that the same negative will not render as high a contrast when printed on variable contrast silver-gelatin paper, for which the more yellow densities will print "softer." The reduced stain in the less dense areas will render a higher-contrast set of tones because more blue light is passing through here. Thus, one negative can be produced to suit a wide range of printing methods.

The main active ingredient in Pyro PMK-1 is pyrogallic acid (also known as pyrogallol), which is one of the oldest developing agents used in photography. It was popularized by Frederick Scott Archer in 1851, when he introduced the wet plate collodion process. Pyrogallol is a very toxic substance in its pure, and especially crystalline form. When purchased in liquid form rather than mixed from the raw ingredients, and used with the same precautions as other photographic developers, Pyro PMK is much more benign. As with all darkroom chemicals, avoid contact by wearing impermeable gloves and have proper ventilation to prevent inhalation of vapors. If slight skin contact is made, flush with water.

Pyro PMK liquid comes as a two-part stock solution. Part A consists of pyrogallol, metol, and sodium bisulfite (also known as sodium hydrogen sulfite). Part B consists of sodium metaborate. The two bottles have an indefinite shelf life, but once mixed up, the developer will oxidize rapidly, and should be used within 10–15 minutes. For greatest consistency, I recommend that the processing

Table 7.1 Suggested development adjustments for controlling contrast in negatives.

Subject Brightness Range	*Development for Pt/Pd*	*'Zone System' notation*
2 to 5 stops	P + 30%	N +
6 to 7 stops	P	N
8 to 10 stops	P - 25%	N -

Note: Subject Brightness Range is the difference in stops (or exposure values, EV) between the darkest and lightest areas that you want to have detail or texture showing. "P" is the normal film development time to produce an optimal negative for platinum/palladium printing. The Zone System notation, formulated by Ansel Adams (1902–1984) and Fred Archer (1889–1963) in the 1940s, and finessed by many others since, built on the work of Hurter and Driffield. It uses a shorthand for film development, where "N" is understood as being normal development. Film requiring contrast modifications during development were noted with an "N," followed by plus or minus signs, sometimes with a numeral. See Table 7.2.

Table 7.2 Suggested Pyro PMK development adjustments for controlling contrast in negatives.

Subject brightness range, in stops	Development	Film Type Ilford Pan F (ISO 80)	Ilford FP4+ (ISO 100)	Ilford Delta 400 (ISO 320), HP5+	Kodak TMax 400 (ISO 320)
3 or less	P+2	Not recommended	Not recommended	21	23
4 to 5	P+1	12	14	17	19
6 to 7	P	10	12	14	15
8 to 9	P-1	9	10	11	12
10 or more	P-2	8	8	9	10

Note: The indicated development times are in minutes, at 68°F / 20°C, with agitation as recommended in Pyro PMK Processing Sequence

sequence be started within a minute of the developer being mixed up.

Suggested Pyro PMK Development

As suggested in Tables 7.1 and 7.2, development with Pyro PMK has been increased from times recommended for printing on silver-gelatin papers, but to a much lesser degree than for developers such as Rodinal and HC-110. These adjusted times are also referred to here as "P."

Pyro PMK Working Solution

Wear rubber gloves and work in a well-ventilated area. Pyro developers indelibly stain clothing and surfaces; prepare and contain all solutions in a sink.

1. 10 parts A + 20 parts B + 1000 parts water = working solution e.g. 10 ml of A + 20 ml of B + 1000 ml of water
2. Measure the water and add A and B solutions to it. Stir well. The color will change to pale and even dark amber after stirring. If this does not occur, there may be something wrong with your solutions or the ratio.
3. Use a non-hardening fixer such as Ilford Universal or Rapid Fixer, Kodak Rapid Fixer without the hardener, or Kodak F-24 formula. The last one is recommended by Gordon Hutchings.

Pyro PMK Processing Sequence

The following is offered as a guideline, since exact times and methods depend on a number of variables. It is widely accepted that for film development, only the time should be adjusted for contrast and density control. All solutions, including the intermediate and final water baths, should remain within ±2°F/°C.

Procedures will also be determined by the equipment being used. For tank development, use rubber gloves, but for tray development it may be better to use thinner, disposable medical gloves providing finer touch sensitivity. A closed tank is recommended for 35 mm and 120 roll film. Two sudden and vigorous inversions should be carried out every 15 seconds throughout the process and especially during the development time. Each complete operation should take 3 to 5 seconds. Trays are recommended for all sizes of sheet film. Use a tray one size larger than the film, e.g. 11″x 14″ trays for 8″x 10″ film. Have enough solution to cover each tray to a depth of at least 0.5″ / 1.25 cm. Process with the emulsion side up, and no more than four sheets of film per tray. Carry out constant agitation, while rotating the stacking order. It is important to establish a random agitation pattern to avoid flow patterns, and therefore best to rock the tray from corner to corner, then side to side, then top to bottom, etc.

1. **Presoak**: 1–5 minutes
   Water
   After use, discard or keep for step 2: Stop Bath
2. **Developer**: See Table 7.2 for time in minutes
   Working solution
   After use, keep for step 5: Stain Bath

3. **Stop Bath**: 1 minute
   Water or Presoak bath
   After use, discard
4. **Fixer**: 5–6 minutes
   Fixer F-24 yields the best stain, otherwise use any non-hardening fixing bath such as Ilford Universal Rapid Fix
   After use, keep or dispose as per manufacturer's recommendations
5. **Stain Bath**: 2 minutes
   Re-immerse in the developer solution. Alternative after-bath solution: 0.5 teaspoon of sodium metaborate to 1-liter of water
   Agitate gently every 30 seconds
   After use, do not reuse the developer—discard. The developer is a water-soluble, organic compound. It decomposes rapidly, and may be disposed of safely into domestic waste-water or septic tanks.
6. **Wash**: 20–30 minutes in gently running water
   or for tanks, 5 fresh water baths where all must be at the same temperature as process solutions ±2°F/ ±1°C, with each bath getting 5, then 10, 20, 10 and 5 inversions respectively with each subsequent change[4]
   or for trays, 5 changes of water at process temperature, for 5 minutes each with constant agitation for the first wash bath, then gentle agitation every minute for subsequent changes of water. More than 4 sheets of films may be washed at time.
7. **Final Rinse**: 1 minute water, gentle agitation (use purified or distilled water if tap water is unfiltered)
8. **Wetting Agent**: 1–2 minutes
   Kodak Photo-Flo™ or similar
   Very gentle agitation
9. **Dry, clean up**.

Digital Negatives

There will always be a slight loss of acutance and resolution when an original negative is duplicated. Known as generational loss, this phenomenon has been exacerbated by the advent of scanning, image-processing software, and inkjet printing. Yet these same technologies offer controls that are easier to implement than when working with traditional, wet-lab negatives. The trade-offs in both directions can be carefully balanced to render ideal printing negatives. The range of equipment, software, methods and materials available for preparing image files and printing digital negatives is growing, and constantly shifting. What follows, from the limitations of this book and my experience, is a summary of approaches using minimal resources. The objective is to make a digital, inkjet negative suitable for platinum/palladium printing. It is important to recognize that inkjet negatives have a different optical absorption spectrum from Pyro-processed silver-gelatin negatives as described in the previous section. Inkjet negatives are printed with a variety of dyes and pigments, from a number of channels, with each one having varying absorbances in the near ultraviolet which are quite different than their values in the visible range. This means that if we wish to make a platinum/palladium print after processing an image in Adobe Photoshop™, for instance, the file has to undergo some further modifications before it is printed as a negative. This may seem confusing, even bewildering. Yet, it is not that different from what traditional processes require: one places a negative in the enlarger, makes tests strips, examines them in exhibition light, and only after several iterations of this, is a final print made. We will do something similar here.

Acquisition (Digital Capture or Scanning)

A digital image may be rendered either via a scanner or a digital camera. The former is convenient for digitizing traditional photographic transparencies, be they negatives or slides. Digital cameras, along with generating actual photographic files, are increasingly being designed with "pixel shift" technology that allows for the capture of a super-high resolution, and low-noise images. This requires four exposures which are then blended into one file, making many of the latest full-frame cameras a viable replacement for flatbed scanners. An additional setup will be needed for illuminating the negative at close range. Regardless of the method, it is wise to capture images at the highest possible resolution, and with the minimum of processing at the time of acquisition. With scanners and their driver software, files should be saved in a lossless TIFF format, while cameras are best set to save files in a lossless RAW format. Manufacturers tend to create proprietary versions of the RAW standard—once files are transferred to a computer, it is advisable to convert these into the open-source standard of RAW, which is known as "DNG."

Scanning Workflow

Scanners are best controlled by software, or a "driver," running from a computer. Epson Scan software comes with all Epson™ scanners, for instance, but other companies also make drivers that can control a wide range of scanners. The following steps may be loosely applied to any setup, but is based on Epson™ and Apple™ equipment. It is assumed that all necessary software is installed and the monitor is calibrated (see the following section on Calibration). While it is generally good practice to calibrate a scanner, it is unnecessary in this workflow because all color and tone processing options are switched off.

1. Clean surfaces
   Wear lint-free gloves, and use lint-free optical wipes. Do a two-step cleaning: use a blower with very gentle wiping to remove surface detritus. Then, pour a small amount of 98% isopropyl alcohol onto a fresh wipe and carefully clean all glass surfaces in the scanner, as well as both sides of the original that is to be scanned.
2. Position original
   Some scanners require a special spacer or holder to properly activate the transparency scanning head. If placing large format negatives directly on the glass platen, Newton ring patterns may appear, especially if the upper, transparency scanning lid is also in close contact. The problem can be avoided by placing the original film emulsion-side down on the platen, and then a sheet of anti-Newton glass on top, before closing the scanning lid. Anti-Newton glass is expensive, but a worthwhile outlay if one is doing a lot of film scans. Alternatively, film may be scanned by wet mounting, which completely removes Newton rings.
   Square up the original as accurately as possible.
3. Preview
   Switch off any option to preview in Thumbnail form. Click the "Preview" button. This displays the original image. Select an area just slightly larger than the image. It is assumed that you know how to fine-tune the Preview window and the focus settings.
4. Settings
   - If a Mode option is available in the Settings panel, select "Professional" or the one providing the greatest number of manual settings
   - Set the document type to Film
   - Select Positive Film—we want to minimize any image processing carried out by the scanner software
   - Destination image type is to be 48-bit color
   - Resolution to the highest possible setting, but the original format, computer graphics processor and its storage limitations will also define this setting—ideally, no less than 300 dpi
   - Document size will be initially determined by the area selected during thc Preview step, but may be finessed here
   - Target size settings will depend on your original size and the intended size of your final print. It is best to overestimate, but here too, much will depend on your processor capacity and storage space. At the very least, set the target size to final print size or at least double the original size, whichever is greater. This setting will be reflected in the Scale setting.
   - Toggle Trimming off or on, it does not matter

- Click on the "Configurations" button, or access the Color scanning configurations and select No Color Correction
- Click on File Save Settings—in Epson Scan, this is only visible as an icon of a blue folder with a head-shot, to the right of the "Scan" button. While there will be an option to do this when a scan is started, it is helpful to complete this panel when scanning several images in one session
- Select the location for saving your file, enter file name text, and a start number suffix. See the step on Saving and Organization below
- Image formats should be saved as TIFF (*.tiff), then in Options, for Byte order select either Macintosh or Windows. Compression is None, and check Embed ICC profile
- Check "Show this dialogue box before next scan." Do not check "Add page dialogue after scanning." Select any other options as you prefer

5. Save settings as a Preset
The driver software usually provides an option to save all of these customized settings, thereby speeding up the next scanning session. However, it is useful to do screen grabs and take notes to assist when migrating from one machine to another, with software updates, or in creating other setups.
6. Scan
Be patient, and be still, because extraneous vibrations can affect scan quality.
7. Save the Scan and Organize
There are numerous strategies for naming and organizing files and their containing folders. If you already have a system in place, then work with it, and save your scans. For suggestions, see Chapter 9, Archival Processing, Presentation, and Documentation. Generally, name a scan to correspond with its original, plus one or two digits indicating the scan version. At some future point, it is possible that the original may be scanned again, either at a higher resolution, or with other equipment. File names sequenced with incremental numbers or letters will avoid inadvertently destroying or overwriting data. Keep the original scan intact, and do not edit it. Always create a copy, while appending characters to the original file name which indicates its generation or version. Keep the parent and all child files in the same directory.

Calibration

The following describes what one should do at the very minimum to create a calibrated work space, without requiring hardware other than what is on hand or readily available. This will not stand as a full substitute for the precision offered by a hardware-based tool. In moving from a digitally captured image to a physical print, we encounter a number of quite different rendering environments. The image is akin to a bank note being used time and again, across different markets in a variety of cultures: there has to be some understanding and agreement so that the note's value remains fairly intact, if not having even accrued in worth. Similarly, an image displayed on-screen should anticipate the final, printed output. While color is less critical with this particular workflow, on-screen adjustments to contrast and tonal distribution require that monitor, inkjet printer, inks, and the film on which a negative is printed are all calibrated with the platinum/palladium print as an end point. Ideally, one should use hardware/software calibration tools for setting up and maintaining the monitor, as well as a print viewing space in controlled light. These tools, made by companies such as X-Rite, come with detailed instructions.

Monitor Setup

The monitor, and its surrounding ergonomics, should be carefully set up: the screen is at shoulder to head height and about an arm's length away, with the keyboard and mouse/pad level with your elbows. Block or dim any light source pointing at your face from within a 45° cone of vision as you look at the monitor—this reduces eyestrain and increases your ability to accurately assess subtle tonal information. Use a monitor hood to enhance this. Any lighting around the monitor should be subdued and carefully shielded to illuminate keyboards and desk areas.

Monitor Calibration

All operating systems provide software to carry out basic monitor calibration procedures. Work in the space described above, Monitor Setup, and set your screen to a mid-point brightness level. As you proceed through the display calibration options, select the following when available:

- Target gamma set to 2.2
- Target white point set as close as possible to white point of your print viewing area, around 4,500K. See *Print Viewing Area*, below

For Windows:

1. Open the Calibrate Display Color wizard, by launching Settings > Display > Advanced Display Settings. Then scroll down to "Display adapter properties [your monitor]" and open this, then select the Color Management tab, and then the button with the same name. In the Color Management pane, select the Advanced tab and at the bottom of the window, click "Calibrate Display." You are now ready to calibrate the monitor.
2. Follow the options until calibration is complete.

For macOS:

1. Open the Displays setting, by going to System Preferences > Display. Select the Color tab, then Option + click "Calibrate…". Doing this accesses the Display Calibrator Assistant in Expert Mode (it will be checked in the pane).
2. Follow the instructions until calibration is complete and you have been asked to save the setup.

Print Viewing Area

It is helpful to have a print-viewing stand close to the monitor. Devices are made to serve this purpose, but are generally bulky and expensive. A cheap alternative would be to use a small book stand, in an area illuminated by a dimmable (important) hooded lamp installed with a 4000–5000K bulb. The objective is to match white balance, contrast (loosely, this is also referred to as gamma) and brightness, as closely as possible between the monitor and print viewing conditions. After having calibrated the monitor and prepared the viewing area, open a blank Photoshop document with a white ground and view it in Full Screen mode. Hold a sheet of paper—ideally a blank piece of opaque stock which is used for platinum/palladium printing— in the viewing light, and adjust the dimmer control until the brightness levels of paper and screen match as closely as possible. If printing with a translucent paper, such as Reich CT, then place a blank sheet of it on the same surface as your final mounting board before examining it in the viewing light.

Printing a Digital Negative

As the text of this book was being finalized, the prolific series editor, Christina Z. Anderson, completed and sent to press a definitive book about making digital negatives with QuadTone RIP software and ink-jet printers. If digital negatives are part of the reader's workflow, then refer to *Digital Negatives with QuadTone RIP, Demystifying QTR for Photographers and Printmakers*[5] for advanced approaches to digital negative-making. Other approaches, such as the Piezography Digital Negative system (PiezoDN), include the use of customized printer ink configurations and offer nuanced possibilities. To take an image from digital to analogue requires a combination of image processing software such as Lightroom, Photoshop, or Capture One, with software that interfaces between screen image and printer output such as QuadTone RIP. In that mix is the matter of manipulating the image itself, for which plug-in tools such as Negative Lab Pro, which finesse a scanned negative, become very useful.

If the platinum/palladium printing process is unfamiliar, it may be advisable to simply acquire a commercially produced step tablet, such as those made by Stouffer, and skip to the instructions in Chapter 8, Making the Print, then return to these guidelines once a basic familiarity has been established with the printing process.

There are generally two levels of making digital negatives in your own studio, each determined by cost, desired quality, degree of expertise, and creative investment in the final print. The first, basic approach provides a workflow using the most basic of configurations: a photo-quality inkjet printer installed with its manufacturer's inks. It requires some experience with the entire sequence of operations from digital file to platinum/palladium print, using an image editing program such as Adobe Photoshop CC and printer driver settings. What you will gain from using "OEM" (original equipment manufacturer) inks is convenience and cost savings, but by trading off some quality and fine control. The second or advanced approach is at the other end of complexity and expense: it will require some investment of time and money in any one of several combinations involving special software, ink, and printer. While the choice of printer, ink, and even software may be broad, I strongly feel that the only suitable material for inkjet negatives is Pictorico Ultra Premium OHP (UP OHP). There are other products available, and cheaper ones, but UP OHP has a greater capacity to absorb inks than other films and renders a wider tonal range with high resolution. Note that Pictorico, confusingly, makes another similarly named product called "Premium OHP." This film may work well for silver-gelatin printing, but it holds about 25% less ink than UP OHP, and I do not recommend it for platinum/palladium printing.

Printing Negatives—Basic Setup

Base Exposure Time

It is important to know how much exposure is needed to create the darkest possible tone, or Dmax, in any given process, and this is referred to as the base exposure time. Exposure is determined by the intensity of the exposing source and duration. Most ultraviolet exposure units require a short warm-up period to bring the source to a steady output, from which point fairly consistent exposures may be measured by time. Since most sources fluctuate in intensity, and even suffer from reduced output as the bulbs age, a better measurement for exposure may be made by the use of an "integrator" or "dose-meter" that displays the accumulation of ultraviolet dosage.[6]

Sensitize and hydrate a sheet of paper by following the standard procedures described in Chapter 8. When ready, lay a clear piece of the same substrate from which your negatives are made, e.g. Pictorico UP OHP, or Ilford HP5, over about two-thirds of the sensitized area. Place this in your contact printing setup. Lay a sheet of 4-ply black card across the entire coated area, and then slide it so that a portion of both the paper and paper+film is uncovered. If using a flip-over vacuum frame, you'll need to tape the card down. Space this so that at least five stepped exposures may be made, with each strip being no less than ca. 0.5″ / 1 cm wide. Make a series of 2-minute / 100-dose exposures, and then process and dry the print. On inspecting the dry print, the base exposure is indicated by the darkest step after which no further change can be discerned between the areas covered and not covered by film (there may always be a slight difference). It may be necessary to repeat the test, either because exposures were not long enough, or to fine-tune the base exposure with smaller exposure increments. Make a note of the final base exposure, and use it for all subsequent exposures in this guide.

Creating an Adjustment Curve

This is the stage at which we create a "translator" or, referring to the aforementioned analogy of money in marketplaces, a "converter." We want the monitor to give us a reasonable indication of the final print. Since monitor and print are essentially two quite different realms of imaging, as we shall see, then the monitor image must pass through a final adjustment before it goes to be printed as a negative. The following steps provide a rough, but fairly effective, substitute for the more advanced approaches described latterly. While I refer to current versions of Photoshop CC, older versions of the software, and many other image editing tools, provide similar options.

1. Photoshop color settings

It is important to work with grayscale files in a consistent manner. This may not seem important at first, but as you move on to more refined printer setups, defining color spaces and settings can be very helpful. Go to Photoshop > Edit > Color Settings, and make the following selections:

Working Spaces
- RGB: Adobe RGB (1998)
- CMYK: leave as is (usually U.S. Web Coated (SWOP) v2
- Gray: Gray Gamma 2.2
- Spot: leave as is (usually Dot Gain 20%)

Color Management Policies
- RGB: Convert to Working RGB
- CMYK: Off
- Gray: Convert to Working Gray
- Profile Mismatches: check Ask When Opening and Ask When Pasting
- Missing Profiles: check Ask When Opening

Conversion Options
- Engine: Adobe (ACE)
- Intent: Perceptual
- Use Black Point Compensation: check
- Use Dither (8-bit/channel image): check
- Compensate for Scene-referred Profiles: uncheck

Advanced Controls
- Uncheck all options

Save the settings by clicking the "Save" button, entering a suitable name (e.g. "PtPdNegative ColorSettings" and a description, such as "Settings for Gray Gamma 2.2 for negatives for Pt/Pd." Note the characters, or lack of, for file naming here. It is good practice to avoid all special characters in file names except a hyphen or underscore. Words may be distinguished from each other by capitalizing the first letter. Avoid spaces (which becomes particularly problematic if posting files to web sites), and do not ever use period, exclamation, hash, percentage, quotation, ampersand, or asterisk characters within a filename.

2. Make a 21-step target

A step target, or "wedge," may be downloaded from *pradipmalde.com/step-wedge.*

A 21-step image file, sometimes called a test target, may be downloaded from the book's online resources listed in Appendix 4. This, or any similar target, arranged in 5% increases of opacity, will provide an effective gauge for fine-tuning Photoshop adjustments.

Open the file in Photoshop, and immediately save it as a "child" of the original, e.g. rename "21-step target.tif" to "21-step target-1-1.tif." This ensures that the original remains unchanged, and any copies of it, or copies of copies, can be tracked and easily paired with prints made from that negative. Ultimately, this naming procedure is more important when working with actual photographic files than test targets such as this one.

Go to Photoshop > Edit > Convert to Profile..., and in the pane, make sure that both Source and Destination Spaces are set as Gray Gamma 2.2 (the latter will be prefixed by "Working Gray –"), Intent is "Perceptual" and Black Point Compensation is checked. Uncheck Flatten Image to Preserve Appearance.

You will now prepare the image for printing as a negative. It is best to do this two-step adjustment just before sending a file to print, and from within Photoshop rather than the printer driver. Save the file. In the Layers panel, click on the top-most layer and insert an Invert Adjustment Layer. This reverses all the values of the image to a negative. Then go to Image > Mode > Image Rotation > Flip Canvas Horizontal.

Go to File > Save As… and in the dialogue box, rename the file by appending "flat" to its name, e.g. 21-step target-1-1_flat.tif, uncheck layers, check embed color profile: Gray Gamma 2.2, locate it in the proper directory (i.e. next to its parent) and click Save. Flattening a file ensures that only the visible layers are printed, and reduces the chance of software-based artifacts showing up as glitches in the output negative.

3. Set the printer driver

In Photoshop > File > Print…, adjust the following settings:

Printer Settings

- Printer: select your printer from the drop-down list
- Click Print Settings…, where you will make a number of adjustments and then select Save for subsequent printing sessions.
  - Paper Size: Select the sheet size being printed, usually this is US Letter > US Letter for 8.5″x 11″ (sheet) UP OHP.
  - Layout:
    Border set to None
    uncheck "Reverse page orientation" and 'Flip horizontally'
  - Color Matching, select "EPSON Color Controls"
  - Printer Features (the layout and naming of these will vary, depending on your operating system and printer model, but the most essential settings are here):
    Print Settings Selection ID: Advanced Color Settings
    Media Type: Premium Glossy Photo Paper (170)
    Ink: Photo Black
    Print Mode: Advanced B&W Photo
    Output Resolution: SuperPhoto – 2880 dpi
    High Speed: Off
    Super MicroWeave: On
    Finest Detail: On
    Color Toning: Fine Adjustment
    Tone: Normal
  - Advanced Media Controls (or keep scrolling in Printer Features until you see these)
    Paper Thickness: 2 (UP OHP is actually 145 microns / 0.145 mm thick, and Epson's unit here is in 0.1 mm. Going down to 1 may cause smearing)
    Drying time per print head pass: 5 (x 0.1 sec)
  - To save all of the above settings, click on presets > scroll all the way down in the drop-down list, and select Save Current Settings as Preset…, give it an appropriate name such as "UP OHP Negatives," select "Only this printer" and click OK. From this point on, you will always be able to select this setting from the Presets dropdown list.
  - Click Save to close the window.
- Color Management
  - Color Handling: Printer Manages Color
  - Check "Send 16-bit Data"
  - Select "Normal Printing"
  - Rendering Intent: Perceptual

4. Print the target negative

Load the printer with a sheet of film and send the 21-step target file to print. Let the printed negative air dry for at least 1 hour in dust-free room conditions before printing with it.

5. Print target on platinum/palladium

Following the instructions in Chapter 8, make a platinum/palladium print with the 21-step target. The paper, sensitizer recipe, and hydration controls together affect the print. Take careful notes, since it is important to work to these same conditions for subsequent prints as you design an adjustment curve. Process and dry the print before proceeding to the next step.

6. The adjustment curve

Place the platinum/palladium print from step 5 beside your monitor, viewing both in calibrated conditions.

Open the 21-step target file in Photoshop, and save a copy of it as a second generation of the printed one, e.g. if the (layered) printed file was 21-step target-1-1.tif, then open it and save it as 21-step target-1-2.tif.

Hide the Invert Adjustment layer if it is visible. The screen image is now a positive, in the same way as the platinum/palladium print. You may also need to do a horizontal canvas flip if it is not reading correctly. At this point, it is normal to see a significant difference between the screen and printed image.

We are now going to make the screen image match the platinum/palladium print, just by making visual comparisons. Just below the Invert Adjustment, create a new Curves Adjustment, and call it "Preview," making sure that the new layer's Properties panel is displaying a histogram. Select the curve adjustment tool in the Properties panel—it looks like a left-pointing hand, with up/down arrows. Hover the cursor, now appearing as an eye dropper, over the 10% step, click and drag up or down until the tones are approximately matching between screen and print. Repeat this process with the 20%, 30% etc. You will notice the graph change, and that it is now populated by a trail of anchors corresponding to every other step on the target. Click on any one of these anchors to fine-tune the tones, not just in comparison to the print now, but also to each other in the same way the print tones relate. Leave the 0% (white) patch alone, but as a final tuning, you may need to adjust the 100% step once more. Overall, the curve should look like a lazy "S" that has been rotated a few degrees clockwise. If there are any inverted "S" curves, or valleys and peaks within the overall shape, re-examine the tones with care, and smooth out the shape. Otherwise some tones will invert and the resulting print will appear solarized. Save your file.

7. The printing curve

Having modified the screen image to look like the printed version of the original file, the next step is to create a curve which adjusts an unmodified (ideal) screen image to match the biases of the paper-sensitizer-hydration combination of our print.

Create one more Curves Adjustment layer, just below the Invert layer (which is still hidden). Call it something like, "Printing Curve." Select the curve adjustment tool. Open the Info panel (it looks like an "I" inside a circle, or go to Window > Info). As you move the eyedropper cursor around, notice how numbers shimmer through the Info panel display. Pay attention to the "K" (which represents black values) readouts in percentage points. Click and drag on the 10% patch, moving the cursor up or down until the "K" value reads 10%. Repeat this for the 20%, 30%, etc. The final value for 100% may need to be numerically modified: click on its anchor, and the Output value reads 100, reduce it to 98 or 99.

Save your file.

8. Print the adjusted target negative

A reminder may be helpful: the objective is to view and adjust a digital image on our calibrated monitor, apply the "Printing Curve," invert and flatten the file, and output a negative that produces a print roughly matching the screen.

Working with the same file as in step 7, hide both the "Preview" and "Printing Curve" layers. This is what we want the final print to look like. Make visible the "Printing Curve" and Invert Adjustment layers, apply a horizontal canvas flip, then save the file as... e.g. "21-step target-1-2_flat.tif" and send to the printer. Repeat Step 5. Compare the monitor image to the dry printed image from this negative. You should be fairly close to a good match. Repeat Steps 6 onwards if necessary, and certainly go through the entire process when changing papers, sensitizer mix, and hydration levels. When ready with a final printing curve, save the 21-step target, with the "Printing Curve" layer as a master reference file for future use with other images. Name it so that the curve's printing parameters are identifiable. The paper, sensitizer, and hydration levels should be indicated, along with a version number, e.g. "ReverePlatinum_Fe2PtPd_2-1-1_70RH-1" would denote the curve was prepared for Revere Platinum paper, coated with an ammonium ferric oxalate (prepared by Method 2): platinum: palladium ratio of 2:1:1, and hydrated at 70% RH (assuming the room/ambient tempt is standard at 20°C).

9. Digital to print

From this point on, follow these steps each time a digital file is to be prepared and printed as a negative for platinum/palladium printing:

1. Manipulate an image in Photoshop, paying attention to file names and version numbers.
2. When ready to print, open the master reference file prepared in step 8. Select the "Printing Curve" layer, then go to menu Layer > Duplicate Layer..., and in the Duplicate Layer window, for Destination Document, select the image file you wish to print. Click OK. In the image file, move the Printing Curve layer to the top of the layer list.
3. Apply an Invert Adjustment, flip horizontally, save, then Save as... a flattened version of the file, and send to print, with the settings in Step 3.

Printing Negatives—Advanced Setups

Far more advanced approaches than the one described above, with some increase in expense and the investment of time, may be used to great advantage. There is not much to say here, because there are books or manuals that accompany each of these methods. At their core, they provide fine-tuned software to control the inkjet printer heads, and require users to learn how to build sophisticated adjustment curves based on parameters defined by any given photographic process. Programs and utilities such as Mark Nelson's Precision Digital Negatives or Roy Harrington's/Ron Reeder's QuadTone RIP (QTR), and Richard Boutwell's suite of QTR add-ons, are highly effective software-based tools. Additionally, customized inks may be installed in printers, and when combined with software, these can produce the most remarkable negatives. I have used, and am particularly taken by, the PiezoDN package of inks and software provided by Cone Editions in Vermont. The one thing common to all of these, however, is that they take some time to learn and assimilate into the workflow. It is not an approach to take casually, but certainly one that must be grappled with if digital negatives need to be made that compare well to their silver-gelatin counterparts.

Masks

The matter of modulating tonal relationships within the printed image is different from simply controlling contrast or arriving at linear translations of tones from positive to negative to positive. Modulating, or varying exposure across the printed area, is traditionally done by dodging and burning techniques and is a highly subjective and nuanced practice. It requires the printer to both feel the truth of the image, as well as imprint pathways, if you will, for a viewer's eyes to be coaxed along. In more ways than one, the practice of modulating tones, and being guided by these modulations as one looks at a print, is both a choreography and a dance. By these means for instance, a good printer can make a photograph *of* an energetic, high-action event seem to be *about* a serene, transcendent moment. Or the opposite, and so on. Projection-printing via an enlarger permits the use of tools and cut-outs for selectively reducing (dodging) or adding (burning) exposure to part of an image. With contact printing, and under ultraviolet, the matter of dodging and burning gains a few levels of impracticality and difficulty. I do not recommend the ski-mask/UV blocking goggles/silver-foiled oven mitt approach. The following is a technique based on Emmet Gowin's (1941) contact-printing methods,[7] and with a bit of practice, is both more practical and precise.

1. Tape down a negative on a lightbox, just enough to prevent it from sliding around. The image should be oriented correctly, i.e. emulsion side down.

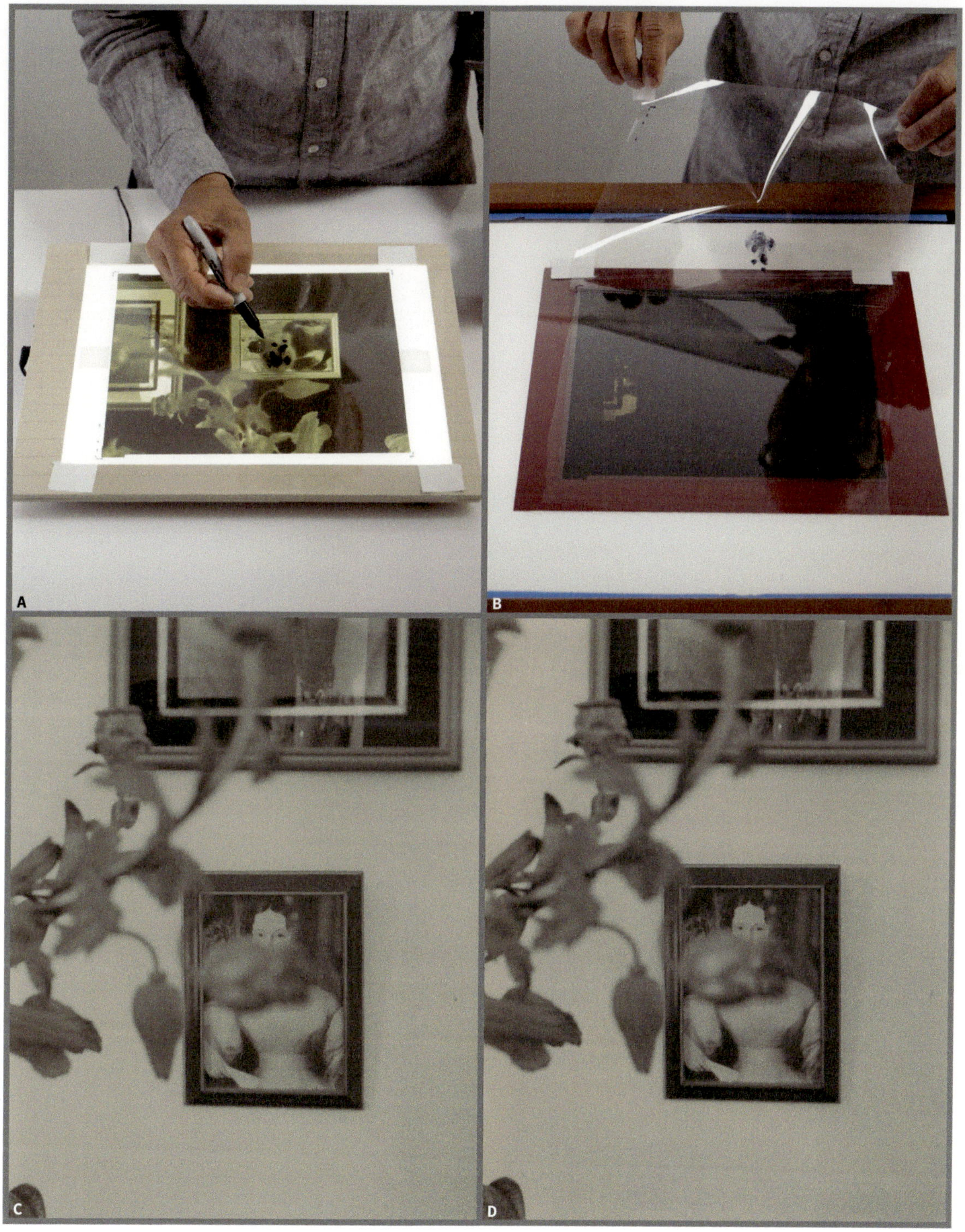

Figure 7.4 How to draw and use acetate masks:
A. drawing a mask by tracing a negative,
B. attaching the acetate sheet to a printing frame, and two prints:
C. made without mask
D. made with the mask.

2. Place a slightly larger, clear acetate sheet over the negative, and tape it down by the corners.
3. Using a fine-tipped indelible marker on the acetate, create registration marks by tracing a dot or fine line over each corner of the negative. Also trace the notch marks, or provide some other indication, so that you will be able to correctly orientate this mask later.
4. On one edge, write down an identifier for the negative, plus a mask number, as it is likely that you will work with multiple masks for any one image.
5. Working with a variety of indelible markers, each with different tip sizes, begin to paint over areas that need to be dodged. Marker colors, solvents, and other approaches may be used to blend and control transparency. Multiple acetate masks permit nuance and further control.
6. After loading sensitized paper and negative into the contact printing frame, attach the mask to the outside, and in registration with the negative. If using more than one mask, arrange them so that they may be peeled off in sequence, as exposure progresses.

A similar approach may be applied to digital files being manipulated in Photoshop by using multiple Curve Adjustment layers.

1. Between the Background (image) and the topmost Invert Adjustment layers, create a Curve Adjustment layer.
2. Using the curve modifier tool, hover, then click and drag up or down over an area you wish to lighten or darken, or even alter contrast. You may create multiple anchor points. Ignore the effect this has on surrounding areas that you do not wish to change.
3. Click on the Layer Mask that automatically appears to the right of the graph icon in the Curve layer. In the Properties panel, which has now changed and no longer displays a histogram, click the "Invert" button. This effectively prevents any adjustments made in step 2 from affecting the image, by filling the mask with an opaque black tone.
4. Set the foreground color to white, select and adjust the Brush tool, and start "painting in" the adjustments made in step 2.
5. Repeat as necessary until ready to output the image as a negative.

Caring for Negatives

In contact printing there is always a risk that the sensitizer chemicals may damage the negative, which will be particularly vulnerable if the paper has been very fully humidified. Original negatives may be protected during contact printing by interposing a very thin sheet of archival polyester film—such as Mylar™ or Melinex™—between paper and negative. Digital negatives, being re-printable, may not require such attention. If the thickness of plastic protective sheet, which is generally expressed in "mil," is too great then it will diminish the edge sharpness or acutance of the print, and some resolution in the image may be lost, depending on the size and nature of the light source. Diffuse light sources, such as arrays of multiple fluorescent tubes, will be more prone to rendering a loss of edge sharpness than single-lamp point sources.

For these reasons, the protective sheet should ideally be no thicker than 0.4 mil / 10 μm, but it is hard to obtain such thin films. More practically, the loss of sharpness is imperceptible with a film of 0.5 mil / 12.7 μm to 1 mil / 25.4 μm plastic, even under a magnifying lens. 2 mil / 50.8 μm film, when used under diffuse light sources will cause a "softening" of the edges that is visible under close inspection, but may be fine with point sources.

Notes

1 Elaine Scarry, *On Beauty And Being Just*. Princeton, N.J.: Princeton University Press, 1999.

2 R.M. Callender, "The Photographic Researches of Hurter and Driffield," *Journal of Photographic Science*, Vol. 27, 1979, pp. 174–180.

3 Gordon Hutchings, *The Book of Pyro*. Granite Bay: Bitter Dog Press, 1992.

4 This method of conserving water was formulated by my photo-tech instructor, Philip Hypher, and taught to students while I was at the Bournemouth and Poole College of Art

and Design. Phil was a reconnaissance photographer in the Royal Air Force during the Second World War, and his crew had to go out in balsa wood planes, before and after bombing raids, each time having exposed and processed sheet film while in the air, and return to base with dry (dipped in methylated spirits) film to hand over to RAF Intelligence. An unsung hero.

5 Ron Reeder and Christina Z. Anderson. *Digital Negatives with QuadToneRIP, Demystifying QTR for Photographers and Printmakers*. New York: Routledge, 2021.

6 Integrators or dosemeters display exposure via an arbitrary unit or "dose," with sensor types and software resulting in a non-standardized display of measurements. In other words, the dose reading from one design of integrator may not be the same as that from another of the same exposure.

7 Jain Kelly (ed.), *Darkroom 2*. New York: Lustrum Press, 1978, pp. 43–44. Emmet Gowin (1941) refers to a "contour printing map," a technique, that resulted from a collaboration between Frederick Sommer (1905–1999) and his apprentice, Alex Jamison (?), whereby multiple layers of slightly opaque material vary exposure across a contact print.

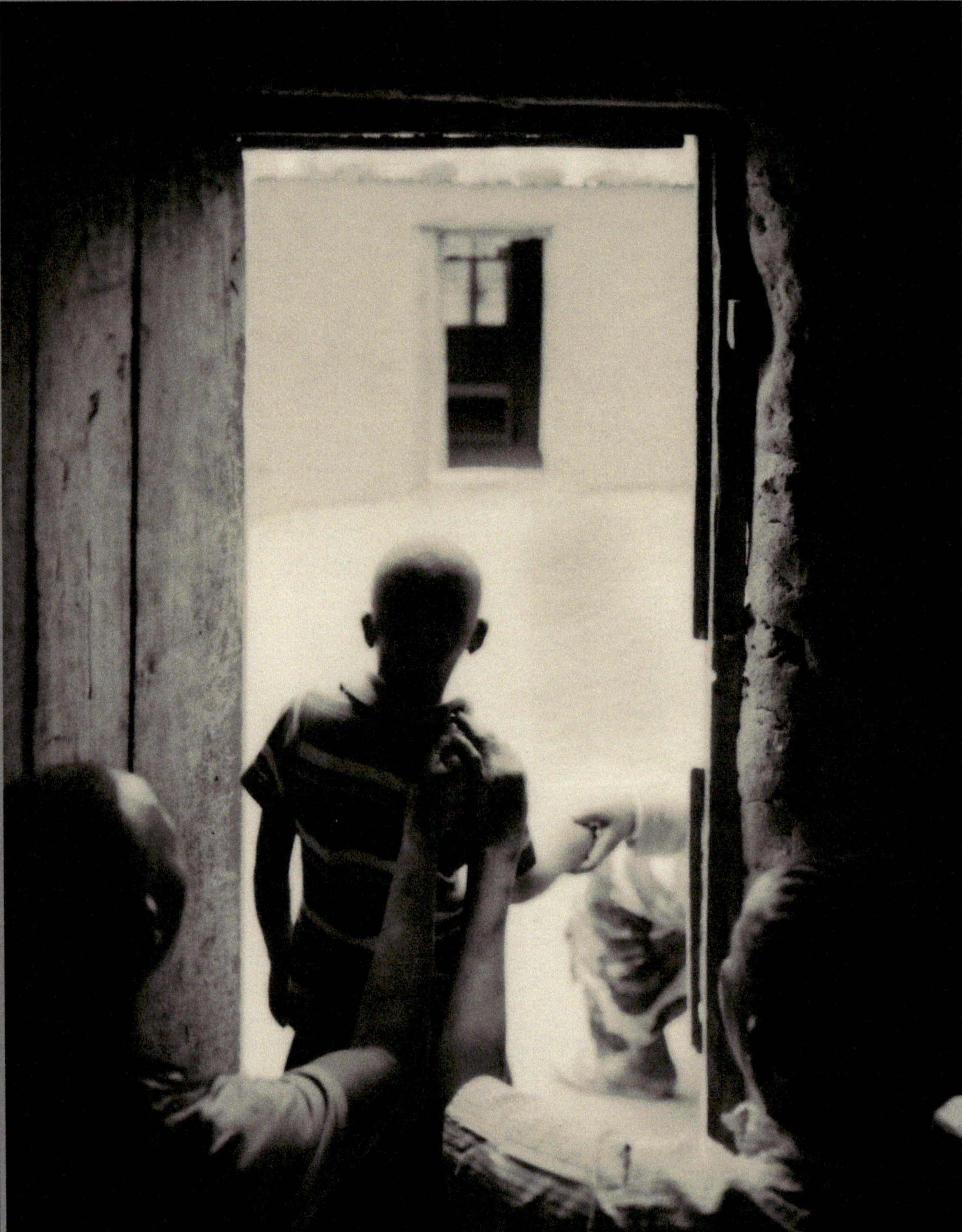

Figure 8.0 Pradip Malde, *Julia Ivan, a retired ngariba, young Sarah Christoms reaching out to her mother, Rose Petro, and Raheli Msagala, a retired ngariba and AFNET activist, all of whom reject the practice of female genital cutting, near Kongwa, Tanzania*, 2017. Platinum-palladium print on Reich CT 24#, from digitized 8 x 10 inch original, 24.5 x 19.5 cm. Fe2:Pt:Pd = 2:1:1, 20 min hydration in 100% RH / 20°C.

Chapter 8
Making the Print

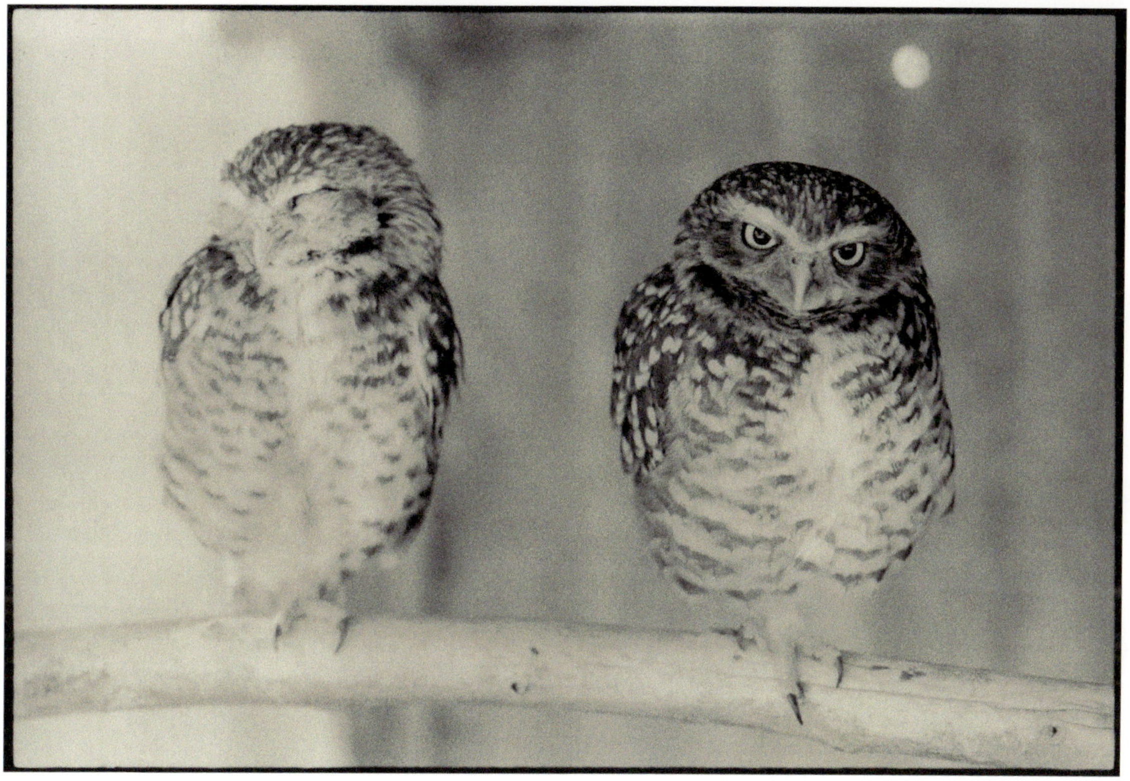

Figure 8.1 Paula Luttringer, from the series *Bestiario Las Lechuzas*, 2005. Platinum-palladium print on Revere Platinum paper, from PiezoDN negative of 35mm film original, 19.5 x 29 cm. Courtesy Paula Luttringer. Fe:Pt:Pd = 2:1:1, hydrated for 90 min. at 76% RH / 20°C.

It may seem mischievous, if not downright willful, to embed the core set of processes in the middle of a book such as this. Yet, it has been suggested several times, and may be clear by now, that a variety of components need to be in place, and dynamics explained, before we can bring them all together and start to make prints. I have established the historical and chemical reasons that result in the printing-out platinum/palladium process. Guidelines have been provided for a safe workspace, prepared and equipped with adequate surfaces, hydration chambers and an exposing unit. Sensitizer and processing solutions, and a wet area for clearing prints, should be ready, along with the ambient conditions for storing, coating. and hydrating paper as described in the preceding chapters.

Hydration Reminder

It is important to revisit the matter of hydration: the presence of water molecules in the paper during exposure has a significant effect on image formation and quality. The key to the printing-out version of platinum/palladium lies in controlling the humidity of the sensitized paper just before exposure. The effect of ambient relative humidity (RH) on the extent of printing-out, color and contrast is summarized in Table 8.1, from which we see that optimum results are obtained when sensitized paper is equilibrated with an ambient RH between 50% and 80% at ca. 68°F / 20°C. Below 50% RH, the printing-out effect is partial and tonal information continues to form during the first clearing bath; above 80% RH the maximum density of the image may tend to weaken because the sensitizer diffuses too deeply into the paper. There are two convenient methods for introducing a controlled amount of humidity into the sensitized paper, a process which we refer to as "hydration." Chapter 4 provides more details about the preferred two methods, namely ambient and salt hydration. A hygrometer helps to monitor the ambient method, whereby you can simply make use of the prevailing RH (if suitable). Hanging the paper in a dark ambience at room temperature (ca. 20°C) for an hour or two before exposure can render fairly predictable results, provided that conditions do not fluctuate by more than 5% for either the RH or temperature. The greatest control is provided by a salt hydration tank using a saturated aqueous solution which provides an atmosphere of constant, known, relative humidity across the entire "comfort zone" (68–72°F / 20-22°C). With this method, ambient temperatures may vary and are not critical, but it is good practice to sustain a steady room temperature. Hydration with either method should not be less than an hour in order to ensure evenness. The upper time limit is not critical and can be as much as two hours, but here too, consistency makes for greater control, which in turn makes room for intention and expression.

Before proceeding with step-by-step instructions for making a print, we need to go into a few more details about the variety of results that can arise from different combinations of sensitizer (i.e. ammonium ferric oxalate or ammonium ferric oxalate with excess ammonium nitrate, plus palladium, platinum, or a mix of both) and hydration levels. One of two ammonium ferric oxalate formulae may be used for this process: "Method 1" is a 60% w/v solution of pure ammonium ferric oxalate, while "Method 2" is a solution of 60% w/v ammonium ferric oxalate + 34% w/v ammonium nitrate. Method 1 renders warmer and slightly higher contrast prints than Method 2. The warming effect of Method 1 is much greater at lower hydration levels, shifting to an almost dirt-red color with Aquapel-sized papers. Method 2 sensitizers lean towards a very neutral tone at higher hydrations, especially when platinum is the predominant or sole imaging metal. Generally, palladium yields brown or sepia tones and a softer image, i.e. a longer exposure scale, with great delicacy in the high values. This is because palladium is more reactive than platinum,[1] and given this rapidity, can form finer nanoparticles, which in turn appear as warmer hues. The effect is greatest with gelatin-sized papers than others, but near-neutral tones may be achieved in well-humidified preparations. Platinum yields neutral grey tones and tends to provide a slightly higher contrast (shorter exposure scale) than palladium, and with greater maximum density. However, the overall speed of printing may be slower, needing greater exposures. Being less reactive, the reduction of platinum complexes can also be inhibited by impurities in the paper, especially when gelatin is the primary sizing agent. Hence, pure platinum sensitizers are generally harder to pair with a compatible paper. Platinum-palladium mixtures combine their characteristics proportionally, and usually offer, in sum, much more than the parts on their own. From all of this, it may be clear that paper and hydration levels affect image color, contrast, and maximum density, as well as sensitivity—what is also referred to as printing speed. There is one further quality, and a very important one, that results from the dynamics between sensitizer, hydration, and paper: resolution, or grain, or smoothness of tone. At a microscopic level, this has to do with the way the imaging metal particles distribute and aggregate within the paper. We use Tween and glycerin to fine-tune this process of precipitation, but always be mindful that other factors beyond our control, such as the presence of incompatible substances in the paper, may be guilty of undermining image quality. Most of these dynamics are summed up in Table 8.1.

Note that the Relative Speed is arithmetic, referring to mid-tone formation; the Exposure Scale is in ΔlogH, from fog+0.04 to 0.9Dmax; and Development is in logH units (0.3=1 stop), where 0 is total printing-out. All of these, and color, will vary with the choice of paper. These observations are derived from a sensitizer prepared with pure ammonium ferric oxalate, or Method 1. In comparison, Method 2, due to an excess of the humectant ammonium nitrate, at mid to higher hydrations will render cooler print colors and have slightly greater printing exposure ranges.

The following general points should be kept in mind:

- Contrast: The printing exposure scale (ΔlogH) values indicate that contrast may be controlled by hydration and the combination of platinum and palladium in various ratios. As the sensitized paper's humidity is increased prior to exposure, there is a corresponding increase in its exposure range, or a lowering of contrast. A similar effect is rendered by the metals, with palladium-only sensitizer resulting in a lower contrast print than one prepared with just platinum. Mixing the platinum and palladium solutions in the ratio of 3:1, respectively, gives a sensitizer with a contrast and speed that are fairly constant over wide variations in humidity (RH 40–70%), and with a long range of well-graduated neutral tones, and good Dmax.

Table 8.1 Characteristics of printing-out platinum/palladium sensitizers: the highlights.

Sensitizer	*RH / Salt in hydration chamber*	*Relative Speed*	*Printing Exposure Scale*	*"Development" (0.3 = 1 stop)*	*Color*
Platinum	32% / Calcium chloride	1.8	1.5	0.9	warm black
	56% / Calcium nitrate	1.7	1.5	0.3	warm black
	80% / Ammonium chloride	1.0	1.8	0	neutral
Palladium	32	0.5	2.0	0.4	sepia
	56	1.3	2.2	0.2	brown/red
	80	2.5	2.4	0	warm black
Platinum-palladium (3:1)	32	1.2	1.6	0.6	warm black
	56	1.0	2.0	0.1	neutral
	80	1.0	2.2	0	neutral
Platinum-palladium (1:1)	32	1.2	1.6	0.6	warm black
	56	1.0	1.8	0.2	neutral
	80	1.0	2.0	0	neutral

Note: **Sensitizer** describes the metal(s) which have been combined with ammonium ferric oxalate. **RH/Salt in Hydration Chamber** is the amount of hydration given to the coated paper before exposure. **Relative Speed** is the coated paper's sensitivity to UV. "**1.0**" is the base exposure. If my exposure for a sheet of paper coated with platinum-palladium sensitizer (3:1 or 1:1) at a hydration of 56% RH is 10 minutes, then I should expect another sheet of the same paper and sensitizer, but hydrated to 32% RH, to need 10 x 1.2 = 12 minutes exposure. **Printing Exposure Scale** is the tonal range of the print, from the deepest black to the brightest white and gives an indication of contrast; the smaller the number, the higher the contrast. **Development** (strictly speaking, not development—the term is used to give some indication of tonal values that finish precipitating when immersed in the first clearing bath) or the extent one may expect printing-out of the image. At the lowest hydrations, expect to see around 1 stops-worth of tones to "come in" in the highlights.

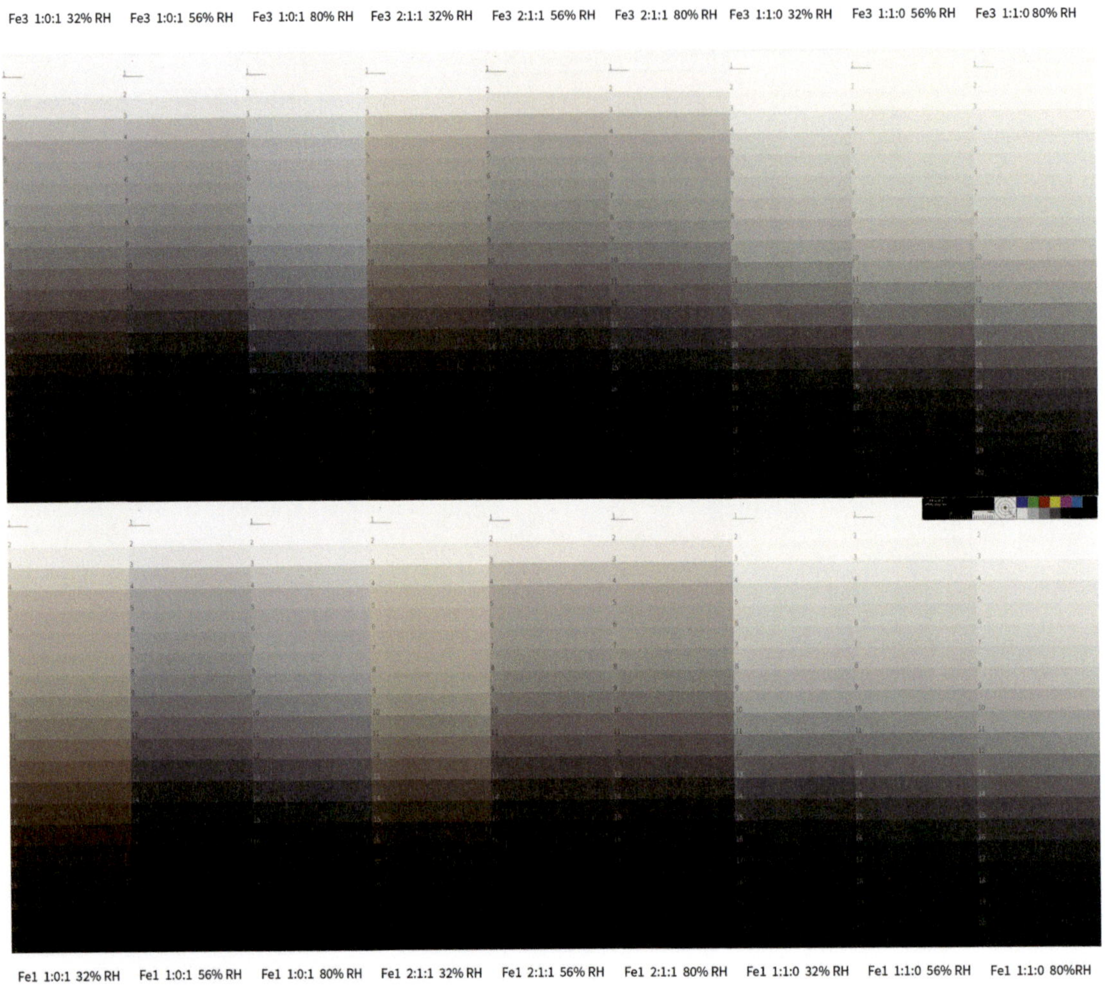

Figure 8.2 The combined effects of sensitizer composition and hydration on print color.
Key to test labels:
Fe1 Pure ammonium ferric oxalate, Method 1, (bottom row)
Fe3 Ammonium ferric oxalate with excess ammonium nitrate, Method 3, (top row);
Three volume ratios:
Fe:Pt:Pd = 1:0:1 (pure Pd)
2:1:1 (Pt/Pd)
1:1:0 (pure Pt)
Hydrated at RH values:
32%, 56%, 80% at 20°C

- Color: Hydration affects color up to around RH 56%, shifting the tones from warmer to cooler with increasing levels of humidity in the paper at the time of exposure.
- Relative Speed: As hydration increases, sensitized paper becomes more responsive to ultraviolet, and levels out for platinum/palladium at 1:1 or greater for paper hydrated past 56%. Note

the exception to this with palladium-only sensitizers, where the opposite effect plays out.

- Printing-out: The printing-out effect, where the image is formed during exposure, is almost or fully complete with hydration levels beyond 56%. Described here as "development," at lower hydration levels, some image formation occurs after exposure and during the first minute or so of the print's immersion in the initial processing solution.
- Resolution: Not indicated in the table, but a result observable to the naked eye, is that increasing hydration renders smoother and more highly resolved tonal gradations.

Workspace

As described in Chapter 3, set up in a work area that has a comfortable ambient temperature, relative humidity not less than 30%, and suitable lighting. Assemble the following:

Negatives
Timer
Notebook
Pencil
Hydration chamber
Glass sheet leveled for coating
Coating area template (cardboard sheet the same size as negative)
Paper to be sensitized
Masking tape
Blotting paper strips, 0.5″ / 1 cm, same length as coating road
Sensitizer
Additives (Tween/glycerin)
Syringes / pipettes
Mixing glass
Ultraviolet exposure unit
Processing solutions
Processing trays
Washing system
Drying rack

Arrange sensitizer, coating rod, timers, blotting paper, tape etc. around the coating platen. Protect the coating rod from fine scratches and damage by always placing it on a pad of paper towel.

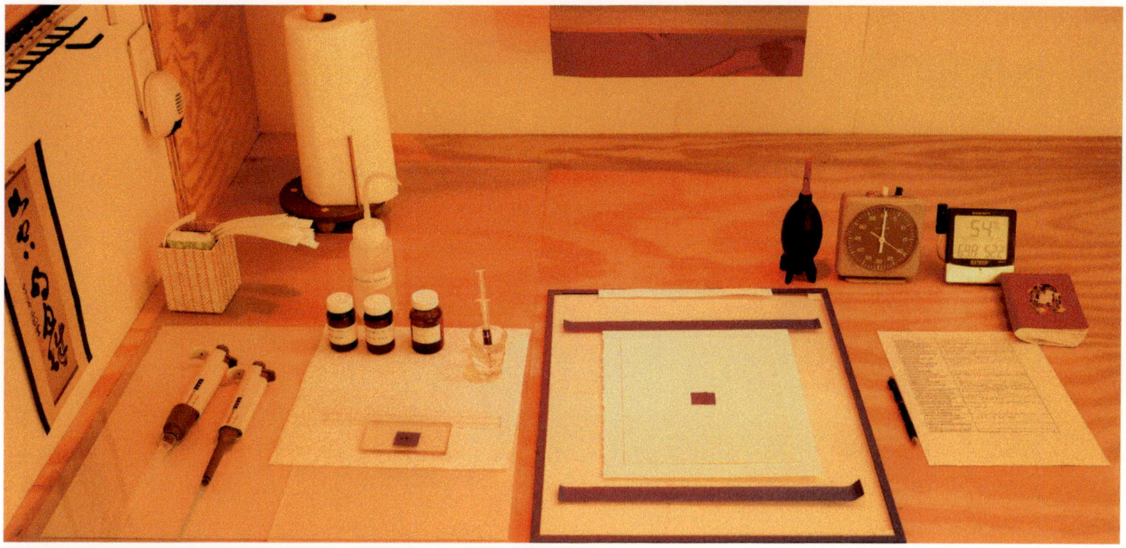

Figure 8.3 Workspace organization: an efficient workspace for coating.

Table 8.2 **Paper types and additives.**

Paper type	*Tween 15% v/v*	*Glycerin 15% v/v*
Arches Platine		0.05 ml
Reich CT Vellum	0.05 ml	
Revere Platinum	0.05 ml	0.05 ml
Wyndstone Vellum	0.05 ml	

Note: Volumes are per 1.5 ml of sensitizer which is sufficient to coat an 8″ x 10″ / 20 cm x 25 cm area on most papers—this is derived from an average specific coating volume of 30 milliliters per square meter of surface area (30 ml/m^2). These very small volumes are best measured with micropipettes; a more practical but approximate method is to use a 1 ml syringe and count out drops: 1 drop = ca. 0.05 ml. Since absorbency varies amongst papers and surfaces, the final sensitizer and additive volumes may need to be modified after some testing. See Chapter 6.

Printing Workflow

Preparing to Coat

1. *Pre-coat hydration for 5 to 10 minutes.* Hydrate the paper (better referred to as "pre-coat hydration"). This is always helpful, but essential to carry out with papers that have been recently purchased or stored in dry conditions of less than 30% relative humidity. Use an aqueous salt hydration chamber so long as it has a relative humidity atmosphere greater than 50%. Alternatively, sheets of paper may be hung up for at least 24 hours in any working environment of ca. 68°F / 20°C with RH 50–80%.
2. *Make notes* of all necessary details about the print, including an identifier for it and the negative. See Chapter 9.
3. *Mark* the paper with the area to be coated using a coating area template. The coating dimensions should be ca. 0.5″ / 1 cm larger than the negative to allow for irregularities. Write a print identifying number/title in one corner.

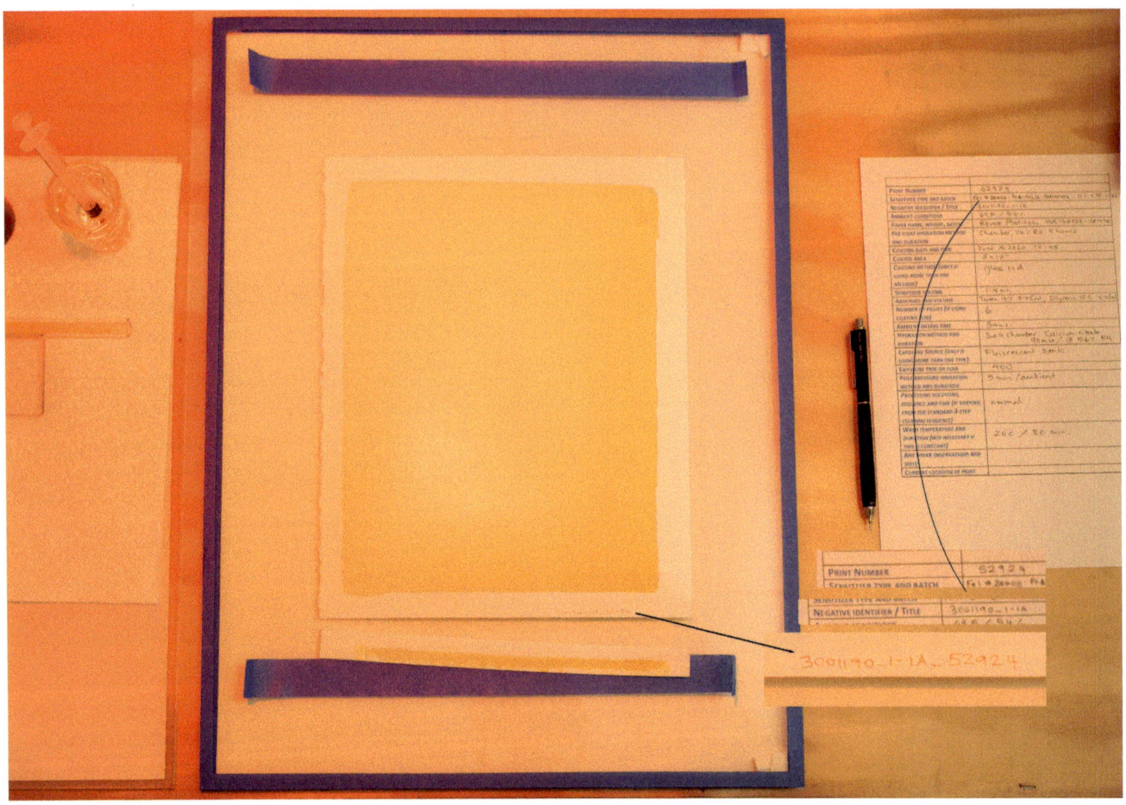

Figure 8.4 Mark-up the area to be coated, lightly with pencil, and number the print. A good coating should have a silky sheen to the surface for the first couple of minutes after the last pass.

4. *Additive* (may be optional): measure out an appropriate number of drops of additive into a mixing glass. You will already have prepared Tween and glycerin solutions if needed. The choice of paper will determine the additive. See Table 8.2.
5. *Sensitizer*: measure a volume of sensitizer (see Chapter 5: Sensitizer solutions), reading from the bottom of the plunger, and gently syringe it into the mixing glass. Then thoroughly mix it with additive, if used, by slowly syringing the liquids in and out about three or four times. Finally, pull the sensitizer into the syringe. The amount will depend on the coating method, type of paper, and area to be coated. When using a glass coating rod, a trial volume may be calculated from the average specific coating volume of 30 milliliters per square meter of surface area (30 ml/m^2). E.g. use 1.50 ml for an 8″ x 10″ area as a starting guideline. Brushes absorb sensitizer, and lead to wastage. Exact brushing volumes should be at least that used with coating rods, and may need to be increased by as much as double the amount. See step 9 below.
6. *Tape the paper* or clip to a heavy glass sheet which has been arranged to serve as a coating platen. Use a low-tack pressure sensitive tape such as Scotch Removable 3M811. Check that the glass is level left to right; this is critical, so adjust it if necessary by inserting spacers below the glass plate. It is best to coat in an up-down direction rather than side to side.
7. *Dust off* the paper surface with a bulb blower.

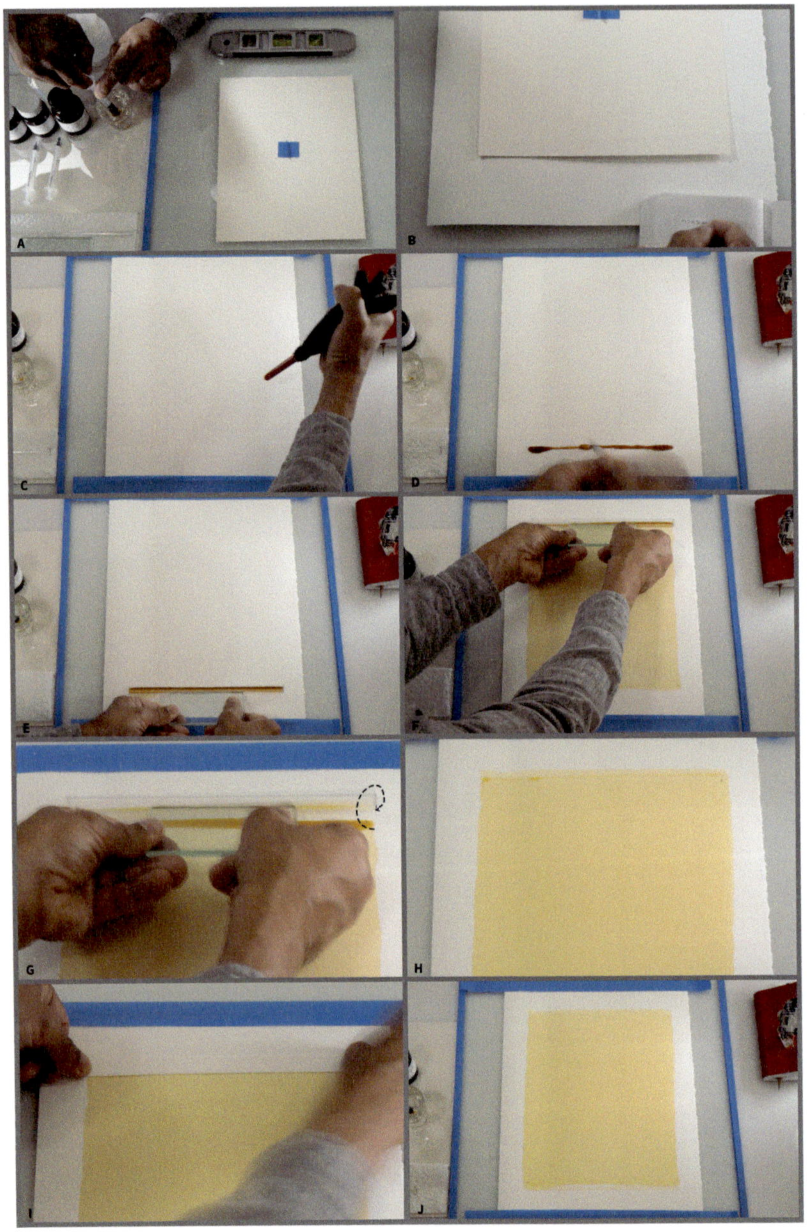

Figure 8.5 Coating the paper with sensitizer:

A. Mix sensitizer and check glass plate is level.

B. Attach paper onto clean glass plate with tape, number the sheet and take notes.

C & D. Dust off with a blower. Syringe sensitizer in a line at bottom of paper area to be coated.

E. Introduce coating rod to line of sensitizer.

F. Firmly make first pass, pushing line of sensitizer to the top of the area.

G. Hop coating rod over line of sensitizer and draw it back down to the area bottom.

H. Continue passes of rod 4 to 6 times.

I. Blot off any remaining thin line of sensitizer at top or bottom after the last pass.

J. Resulting uniform coat.

Coating

8. *Syringe the sensitizer onto the paper.* Point the nozzle of the syringe at the bottom left mark, and expel the liquid slowly and gently (use two hands) in an unbroken strip of sensitizer as you move it steadily from left to right across the width of the coating area. For steadiness, touch the paper lightly with the syringe nozzle. *Do not be too slow or fussy*—irregularities will even out in the coating.
9. *Begin coating the paper.* When using a glass rod, hold it with both hands, and place the straight, central portion onto the paper, a little below and parallel to the line of sensitizer. Then, with moderate pressure, push the coating rod up into the strip of solution. Pause briefly (about two seconds at most) while the liquid distributes itself uniformly along the length of the rod, then steadily push the strip of solution up the paper, like a tiny tidal wave running in front of the rod.

   If using a brush, a number of factors need to be considered. Dry bristles will increase the amount of sensitizer being absorbed into the brush, and consequently wasted. Dampening brush fibers will reduce this dynamic, but the presence of water within the brush may affect sensitizer concentrations in the paper—regulate this process by using the same amount of (distilled) water and blotting up excess water with a paper towel. Japanese hake-style brushes such as the Richeson 9000™ or Kobayashi Soft Synthetic™ series, designed for applying sensitizer washes, are ideal because they are made with synthetic fibers held together by relatively inert ferrules.

   It is preferable to mask the borders of the printed area from exposure, rather than darken them to show the rough edges left by passes of the glass rod or by brush strokes. There are three reasons for this preference: aesthetic, technical, and practical.

   Aesthetically, it seems unnecessary to "show the brush marks" in order to prove that it's a handmade print—connoisseurs will already know that anyway. Moreover, an erratic black border provides a strong peripheral distraction from the image content. The formal, geometrical qualities of the rectangular frame have the time-honored virtue of being unpretentious. The technical reason is most important: the masked border that has been coated with sensitizer, but remains unexposed, provides a direct visual test for clearing to paper-base white. If the borders are not masked, but exposed and darkened, one can never tell if the print has been properly cleared. This can be a cruelly demanding test, but is very desirable as an indicator of archival processing. In addition, if there is a large non-image area of redundant sensitizer which is heavily exposed, the dense photoproduct may "bleed" during wet processing, into light areas near the edge of an image, such as a sky scene, and ruin the print. Moreover, an unnecessary excess of CO_2 gas will be produced. Masking the negative with ink (if outputting a digital negative), Rubylith®, or any other UV-opaque material, is easy.
10. *Make four to six passes.* Working with a glass rod, when you reach the pencil marks at the top of the coated area, "hop" the rod over the strip of solution and pull it back to the bottom of the coating; then hop over the sensitizer strip once more and push it up to repeat the spreading. Just four passes over the paper should suffice for its surface layer to become saturated with a uniform coating of sensitizer. The first pass should be made quite rapidly (three or four seconds for an 8″ x 10″ print) applying moderate pressure to ensure complete wetting of the surface. The last three passes should be done as slowly as possible (10 to 15 seconds each) with very

little pressure applied, to allow maximum absorption. At the end of the fourth pass, drag the spreader below the picture area with any excess sensitizer. If much excess solution remains you can apply two more passes.

11. *Blot up the excess sensitizer.* At the end of the last pass, gently lift off the spreader rod, and soak up any residual liquid left at the bottom of the coating with the edge of a clean strip of blotting paper to prevent any excess sensitizer from crystallization as this can damage the negative. Make a mental note of the remaining sensitizer. There should be just a very thin but connected line of solution remaining at the end of the fourth pass. If not, in future coatings, try to fine-tune the exact coating volume on the basis of experience with your chosen paper, in order to minimize waste. See Chapter 6.
12. *Air-dry the coated paper for five minutes.* Either place the paper on a dark shelf or in a drawer, or under a covering such as a box lid. Do not put anything in contact with the moist surface. This step allows sensitizer to distribute into the paper's surface. I do not recommend drying with heat unless it is carried out with a regulated air flow at 100°F / 40°C ±1°, for exactly five minutes.

Hydrating

13. *Clean up the coating area and tools.* Do this at the same time as step 12. Wipe down the coating rod with a clean, damp, soft, paper towel or cloth and return it to a clean padded surface. Use the same paper towel to clean the glass platen. Flush out syringes with distilled water and clean the mixing vessel. Place any blotting paper and wipes with residual sensitizer into a trash container. Make sure that the sensitizer bottle is firmly closed.
14. *Hydrate the sensitized sheet, one to two hours, or 5 to 30 minutes if using the Timed Method.* See Chapter 4, Hydration chambers. Examine the coated surface—the sensitizer should be evenly distributed. Place the paper in a hydration chamber or in a darkened area with suitable ambient conditions. If any fine crystals are present on the surface (visible in glancing light), and if you must proceed with printing, then be sure to separate the negative and paper with a sheet of clear acetate film—but you should also revisit Chapter 6.

Exposing

15. *Arrange negative and paper*. Since this is substantially a printing-out process, a hinged-back contact printing frame enables inspection of the intermediate result. Carbon dioxide is always evolved during the exposure, and the resulting gaseous pressure can create pockets of reduced contact between negative and paper, which in turn affects image sharpness. A porous sheet of felt or thick blotting paper placed in the contact-printing frame, between an impervious layer (to maintain the hydration dynamics of the paper) and the coated paper is recommended. This provides a "buffer zone" into which the carbon dioxide can transit. This arrangement may not be necessary if working with vacuum frames—however, some form of registration, such as with tape, may be necessary here if you intend to inspect (with care) the print during exposure. Close the frame, and, if using them, attach any masks to the exterior-facing glass.
16. *Exposure* is between two and 10 minutes under an average 120 watt fluorescent UVA light source, with peak emission around a wavelength of 365 nm, but can be as short as 40 seconds with high-intensity discharge lamps; such a short exposure may produce warmer tones with palladium. Expose until the highlight tones are just visible. Under conditions of full printing-out (e.g. platinum-palladium prints hydrated at 50% RH or greater) you can continue exposing until the highlight detail is fully resolved; the shadows will not "block up" totally, like development Pt/Pd prints, because the printing-out process has a self-masking action in regions of high print density. Inspection during exposure should be carried out in a well illuminated area which is shielded from ultraviolet. In order to maintain paper hydration levels (which will affect color, contrast, and sensitivity), make these inspections as short as possible.
17. *Post-exposure hydration*, for two to five minutes. To enhance the gradation in the high values, especially if printing-out is not complete, return the print to a hydration chamber (with an RH of at least 50%) for 2–5 minutes, or expose the print surface uniformly to water vapor over a tray of water at ca. 40°C. *Do not immerse the print in the tray!* This increased presence of water molecules in the sensitized areas allows highlight values to precipitate finely and bond to the paper.

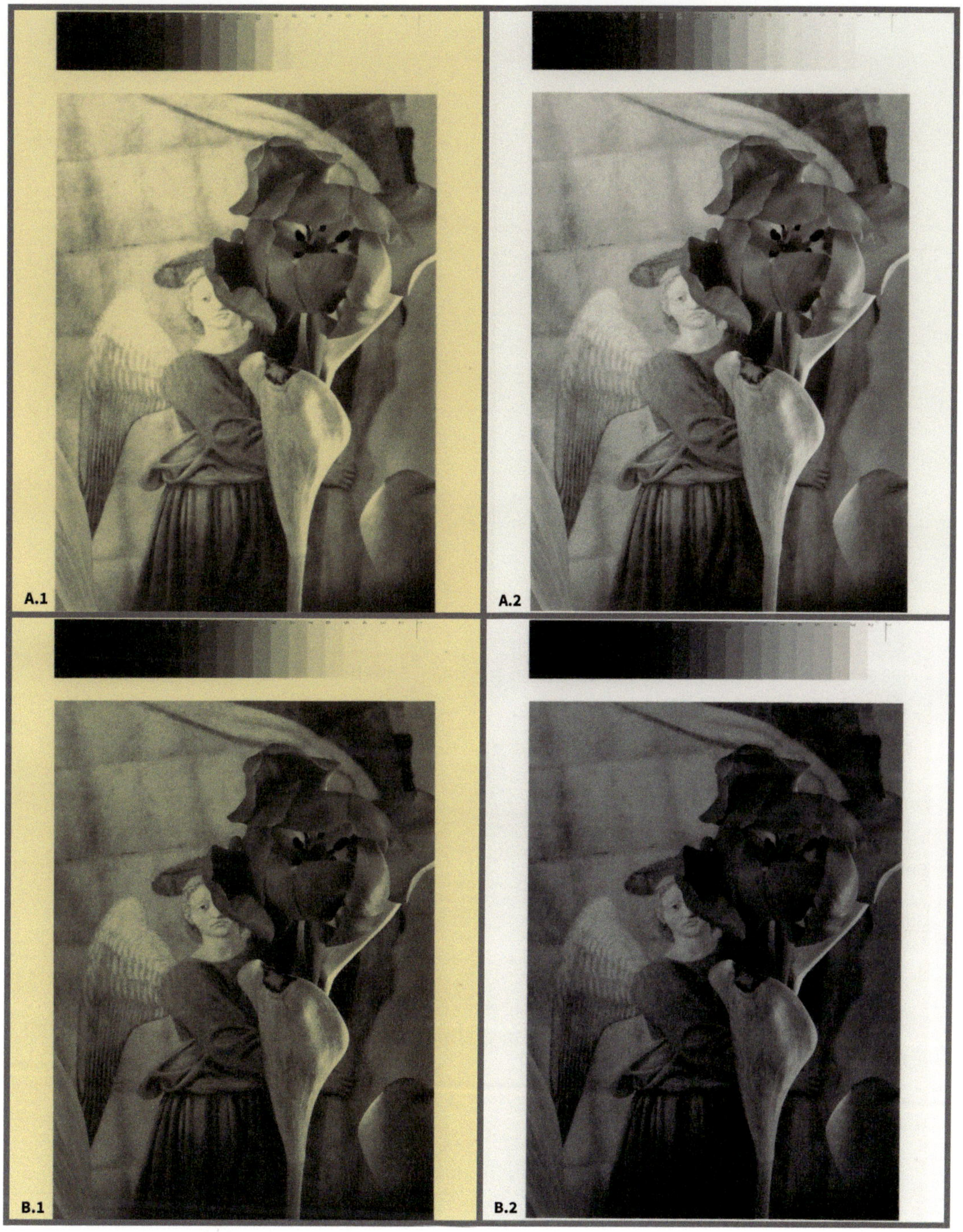

Figure 8.6 The printing-out effect:
A. under-exposed and
B. over-exposed
1. before processing and
2. after processing.

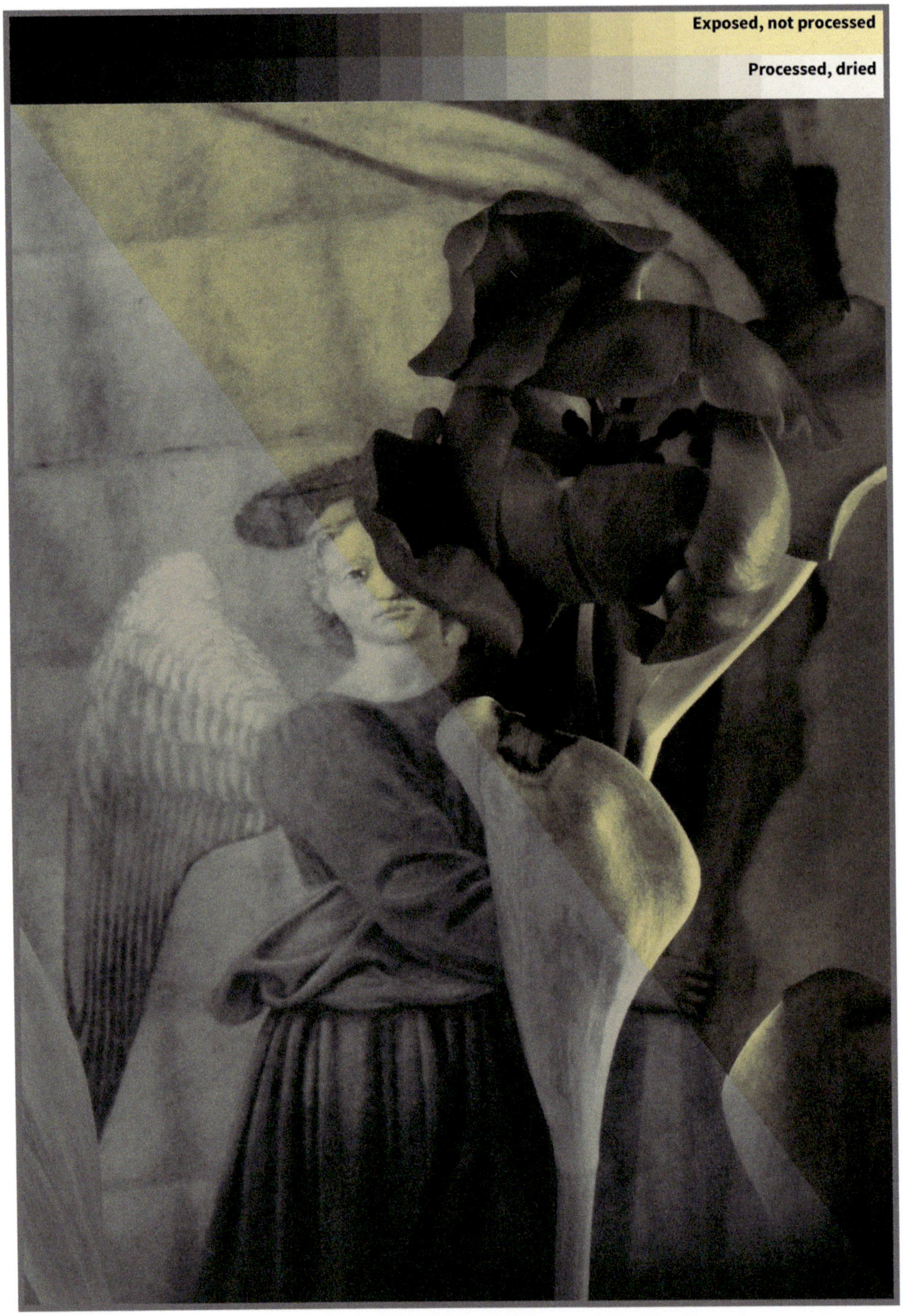

Figure 8.7 The appearance of a correctly exposed print just prior to being inserted into disodium EDTA, and after full processing and drying.

Processing

18. *Develop/clear in disodium EDTA* (5% w/v) for 10 minutes. Wear gloves or use tongs. Immerse the print face up, evenly and swiftly covering its surface with solution. Apply constant and gentle agitation for first one to two minutes, then occasional gentle agitation. With thin and low wet-strength papers, the potential for damage may be reduced by first placing the print, face up, in a clean, dry, flat-bottomed tray, and then pouring the solution over the paper. After the initial immersion and agitation, prints may be processed face up if they tend to sink, and face down if they tend to float. It is important that this first bath should be acidic, around pH 4. *Do not use tetrasodium EDTA.* The capacity of this bath is ca. sixty 10″ × 8″ prints, or until the volume diminishes to a level where more solution is needed to adequately immerse the print.
19. *Rinse in fresh water* for 30 seconds, with gentle agitation.
20. *Clear in sodium sulfite* (2.5% w/v), for 15 minutes, with occasional gentle agitation.
    This sulfite bath does not keep, so it should not be stored and re-used, but made up fresh for each printing session. The capacity of this solution is about five 10″ × 8″prints.
    Kodak Hypo Clearing Agent may be substituted for sodium sulfite.
    Replace with fresh solution as soon as it begins to look very slightly yellowish.
21. *Rinse in fresh water* for 30 seconds, with gentle agitation.
22. *Clear in tetrasodium EDTA* (5% w/v) for 15 minutes, with occasional gentle agitation.
    The capacity of this bath is at least sixty 10″ × 8″ prints. It should always look clear—replace if it discolors.
23. *Wash in running water* (65–75°F / 18–24°C) for a minimum of 30 to 40 minutes. Far less water may be used by washing the print in three 10-minute baths of fresh water, with occasional, gentle agitation.
24. *Drain face out*, on a clean, near-vertical sheet of glass or acrylic for 5–10 minutes.
25. *Dry* at room temperature, for ca. two to four hours, on horizontal plastic or fiberglass screen, blotters, or any arrangement that does not undermine the physical and archival integrity of the paper.

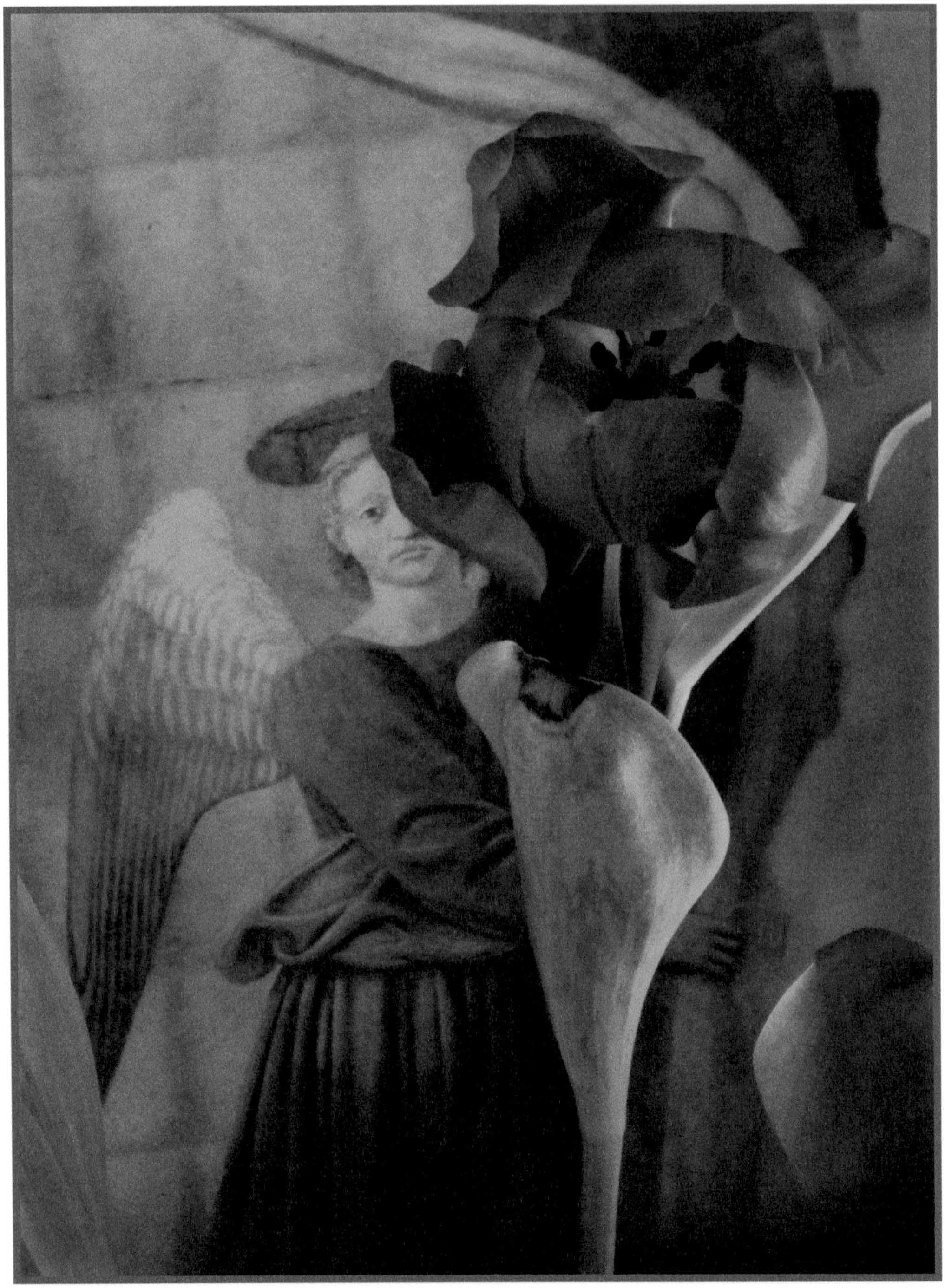

Figure 8.8 The complete, dried print from 8.7 . Pradip Malde, *Tulip with Madonna del Parto, the Left Angel (Piero della Francesca)*, 2020. Platinum-palladium print on Revere Platinum paper, from PiezoDN negative of full frame digital original, 19.5 x 25 cm. Courtesy Pradip Malde. Fe2:Pt:Pd = 2:1:1, hydrated for 90 min. at 76% RH / 20 ° C

Summary Printing Workflow

From pre-coat hydration to a finished exposure takes about two hours. The wet processing sequence runs for about one hour and 15 minutes, followed by the final drying stage. The entire cycle may be repeated every 30 minutes or so, such that a good number of prints may be made within a five to eight-hour work session. Keep in mind that final decisions about exposure, and even adjustments to digital negatives, should take into consideration the darkening effect, or "dry-down," resulting from greater scattering of light as it reflects off the dry, fibrous paper surface. When working with translucent papers, also consider the effect that their mount support will have on print color and tone.

I. *Pre-coat hydration*: in 70–80% RH, 5–10 minutes. Choose side, wire, or felt, and mark up coating area
II. *Prepare sensitizer*: measure volumes of iron and platinum+palladium solutions—these may have already been pre-mixed at least one hour prior to coating; add Tween or glycerin if needed
III. *Coat paper*: ca. 1.5 ml per 10″ x 8″ area, four to six passes with a coating rod
IV. *Air dry*: five minutes, in dark, ambient conditions
V. *Hydration*: one to two hours in controlled RH, in hydration chamber or in dark ambient area
VI. *Load contact printing frame* with negative and paper, mask if necessary
VII. *Expose*: to UVA source until highlights are just visible, by inspecting the print as necessary
VIII. *Post-exposure hydration*: optional, humidify the print over 40°C water for two minutes
IX. *Develop and clear #1*: Disodium EDTA 5% for 10 minutes
X. *Rinse* in fresh water 30 seconds
XI. *Reduce and clear #2*: Sodium sulfite 2.5% for 15 minutes
XII. *Rinse* in water 30 seconds
XIII. *Final clear #3*: Tetrasodium EDTA 5% for 15 minutes
XIV. *Wash*: at least 30 minutes or three 10-minute static baths
XV. *Drain* and *dry*

Note

1 Ware, *Platinomicon*, 11.4 passim.

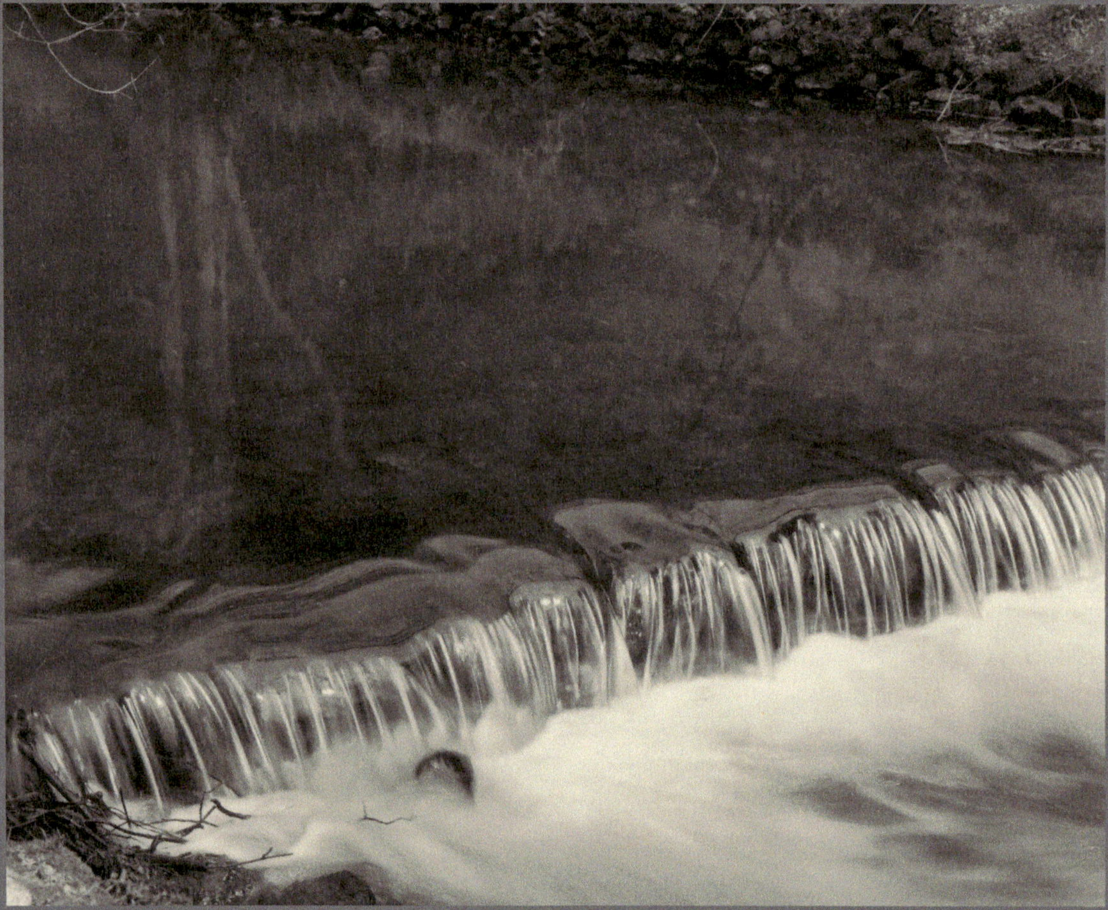

Figure 9.0 Mike Ware, *River Lathkill,* 1982. Palladium print on Fabriano 5, from enlarged copy negative (Kodak SO 015) of 6 x 7 cm. original, ca. 19 x 24 cm. Courtesy Pradip Malde. Fe1:Pt:Pd = 1:0:1

Chapter 9

Archival Processing, Presentation, and Documentation

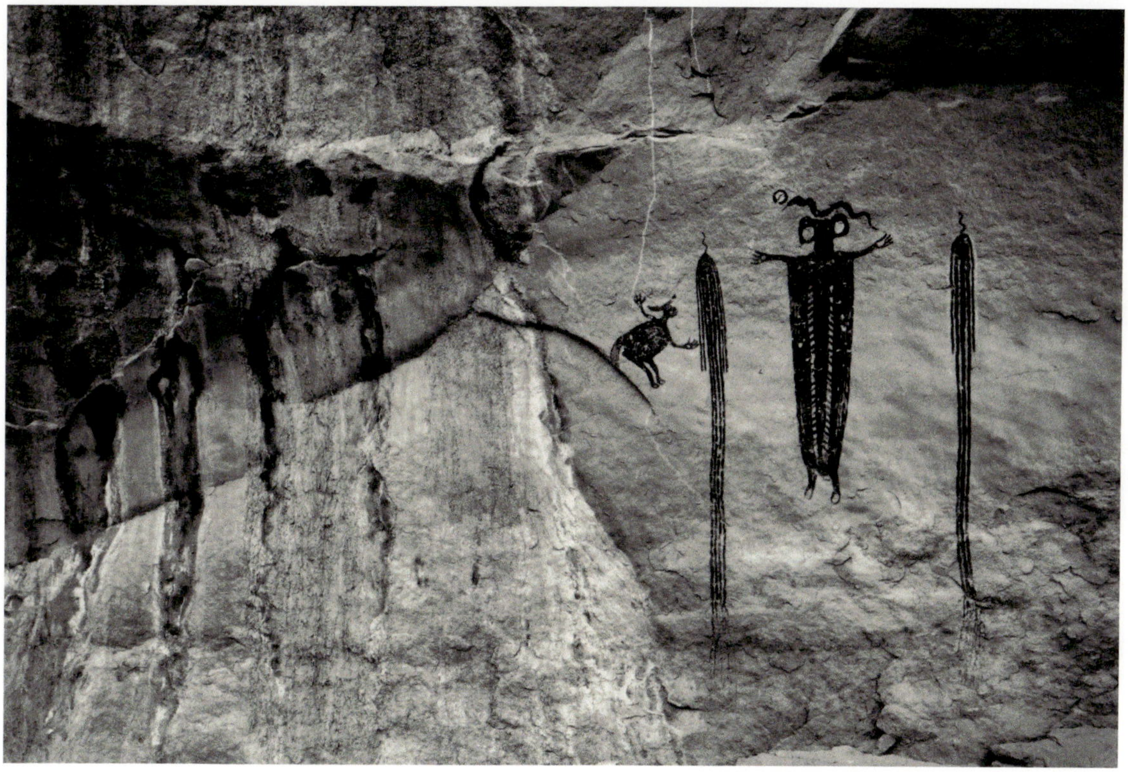

Figure 9.1 Stephen Alvarez, *Head of Sinbad*, 2019. Platinum-palladium print on Arches Platine, from an enlarged digital negative. 23.5 x 35.5 cm. Printed as an edition of 12 for The Ancient Art Archive by Pradip Malde, 2019. Courtesy Stephen Alvarez. Fe3:Pt:Pd = 2:1:1, hydrated at 52% RH / 20°C for 60 minutes.

In the basement of the National Gallery of Art, Washington D.C., are the well-organized facilities of its photograph conservation department. With ultraviolet-blocked daylight decanting in from high windows, telescoping air-vents hovering over work tables from ceiling-bound gantries, and radiant pools of illumination, this hybrid of surgical theater and chemistry laboratory would set any photographer's heart pumping. Then you notice the labeled folio boxes, neatly arranged prints, and carefully written notes. Small clusters of grey patch after patch accompanied by data sheets—step tablets, and books, scalpels, brushes, trays, cameras. Amidst all of this order, there was a moment in 2001 when a conservation scientist, a curator, and two photograph conservators "exchanged bewildered stares."[1] Connie McCabe, Head of Photograph Conservation at the National Gallery of Art, was summarizing the puzzle before them: they were looking at platinum prints, and analytic data clearly supported this; yet, contrary to all that was believed to be true about this process, the prints were faded and fading. How could it be, the experts wondered, that the most permanent of photographic processes was susceptible to decay? By 2011, the question led to a partnership of more than 40 scientists, curators, conservators, and artists, working internationally across a range of museums, galleries, libraries, archives, and universities. What had been assumed previously about platinotypes, along with the matter of permanence, was that they were warm to neutral in tone, always on mat paper, and looked freshly printed. As it turned out, by 2014, this group had established that platinum/palladium prints could display far more complex and varied visual qualities, and presented a dynamic chemical challenge to conservators. Their research clarified obscure and ill-documented historic practice, while providing answers to the triggering questions from 2001:

- Why do platinum/palladium prints deteriorate?
- How do we restore and conserve historic platinotypes, palladiotypes and their many variants?
- What best practices could contemporary printers employ to ensure the permanence of platinum/palladium prints?

Researchers found that platinotypes, and later palladiotypes, far from being the result of one process, actually derive from a variety of formulae and development techniques. Along with the familiar matte finish, prints were made on a range of textured and even almost glossy surfaces, while the print color could vary from orange-brown to neutral, and even to bluish-black depending on the addition of chemicals to the sensitizer or developer. This latter area of inquiry provided clues to why platinum and palladium prints fade, and their paper supports discolor and crumble. McCabe's group, dubbed the "Platinistas" by Mike Ware, reported its work through a series of presentations and panels at the landmark 2014 conference in Washington D.C., *Platinum and Palladium Photographs: Technical History, Connoisseurship, and Preservation* and the proceedings were later published as a book by the same name.[2] Besides standing as an essential text in any dedicated printer's library, the book is richly illustrated with plates, each serving as a valuable visual gauge to perhaps the most important question of all: what is a good platinum/palladium print? This chapter presents a condensed response to the why, how, and what questions listed above.

From the outset, permanence was considered a distinguishing characteristic of platinum and (later) palladium prints. At the same time, however, the plethora of articles about reviving and restoring the initial, lustrous appearance of prints made with these noble metals indicates that something was awry. Instructions about how to clean a print and remove surface dirt are understandable, since the paper is porous and relatively unprotected by standard photographic colloids such as gelatin, and is therefore susceptible to accumulating particulate matter from the environment.[3] However, there were also instructions and articles about print toning and intensification, and about the removal of "yellowing" by bleaching recipes, for instance,[4] indicating that platinum/palladium prints, for one reason or another, were not as stable as had been assumed. The Platinistas found that, out of over 30 known additives[5] to the sensitizer, some had a detrimental effect on the image over time.[6] Mercuric chloride was often used in sensitizers and developers to shift image color towards a sepia tone and smooth out granularity or "noise" (in itself induced by the use of contrast-controlling agents such as ammonium or potassium dichromate and potassium chlorate). However, analysis of historic prints[7] and aging tests suggest that if metallic mercury contributes significantly to the image density, then its loss —because it is volatile—will cause an apparent fading. The presence of mercury in combination with other components actually made platinum and palladium prints fade or discolor. There was another, more troubling "bewilderment"—discoloration, which was also sometimes perceived as the fading of a platinum/palladium image. While it was apparent that the reasons for this were related to residual salts in the paper, few, especially until the 1980s, realized the full reasons for and implications of this. The problem was not, actually, about the loss of image-forming metal but rather the result of a chemical change in the residual iron compounds, existing as impurities in the paper substrate itself—it is not (usually) the cellulose that is yellowing. As paper yellowed, the tonal contrast diminished and it seemed as though a platinum/palladium print was fading.

Archival Processing

The exposed sensitizer layer (made by the printing-out or development Pt/Pd methods) of a platinum/palladium print will contain, in addition to the nanoparticle noble metal image, the following ions: iron(III) complexes, free oxalate, ammonium or potassium, tetrachloroplatinate or tetrachloropalladate, and chloride ions. These ions need to be cleared from the paper as efficiently and completely as possible. Traditionally, to remove excess iron and platinum salts from platinotypes, it was generally recommended that they be cleared for five to ten minutes in each of three successive baths of hydrochloric acid, diluted 1:60 from concentrated (36% *w/w* HCl) acid, followed by thorough washing, for as much as two hours.[8] Palladium-containing sensitizers, however, cannot be treated in such an acidic solution without incurring intolerable image loss: the necessity to employ more dilute (1:200) hydrochloric acid in their processing provides one reason for their being less well cleared of iron salts, and therefore even *more* susceptible to yellowing than platinum-only prints. Since the 1960s, with the increasing use of predominantly palladium-based sensitizers, printers have cleared papers with less aggressive agents such as citric acid. More recently, a variety of clearing baths and mixtures of tetrasodium EDTA, sodium sulfite, and citric acid have been touted, erroneously, as being an effective means of clearing prints. All of these have assumed that if an unexposed coated area of the print has the same paper-base white as an uncoated area after clearing, then it is free of all sensitizer remnants and image formation by-products. Mike Ware, in the early 1980s, reviewed this practice and arrived at a scientifically derived set of procedures for the proper clearing of platinum and palladium prints, which were later recommended as a best practice by the Platinistas.[9] Ware understood that while most of the excess chemicals, being water-soluble, can be removed by thorough washing, the iron(III) complex is inclined to bind chemically to cellulose, i.e. paper, and can, with time, transform into a more highly-colored iron(III) complex.[10] Moreover, in clearing baths at pH values above 4, these chemicals tend to react further with water, or hydrolyse. In doing so, iron(III) hydroxide is formed, which could lodge permanently in the paper fiber and eventually impart a yellow stain. In addition, any calcium ions in the washing water ("hardness") or in the paper (chalk buffer) will tend to precipitate insoluble calcium oxalate and promote the hydrolysis of more iron(III) complexes. While freshly formed iron(III) hydroxide can be re-dissolved in dilute acids initially, if it is not removed promptly it slowly transforms irreversibly into the mineral called Goethite—which is quite insoluble in dilute acids. It is therefore essential to remove all iron(III) ions at the wet processing stage before the print dries. This is best accomplished by the three-bath clearing sequence described in Chapter 8 and below. Note that this chemistry of clearing applies to all versions of the platinum/palladium process, be they printing-out or development Pt/Pd methods. The three-bath sequence (disodium EDTA / sodium sulfite / tetrasodium EDTA) described in this book should be used with all potassium-based sensitizers and other variants, once development is complete.

The two clearing baths, namely disodium EDTA and tetrasodium EDTA, are chelating agents—they bond with the iron compounds and form a water-soluble complex. Along with this important property, each of the three agents serves

Figure 9.2 Effects of various proposed methods of processing and clearing platinum-palladium prints. Note that the sequence used in this work produces the least staining on artificial ageing of the test print. Courtesy Matthew L. Clarke.

a specific purpose. Disodium EDTA is acidic at ca. pH 4 to 5, suppresses hydrolysis, chelates iron(III) well, and removes most of it. Interspersed by a water rinse, the next bath of sodium sulfite or disulfite reduces any residual or bound iron(III) to iron(II). Importantly, iron(II) is less strongly bound to cellulose and less extensively hydrolysed than iron(III), so is more completely removed in the final complexing bath that follows. After one more intermediate water rinse, the bath of tetrasodium EDTA, which is alkaline at ca. pH 9 and efficiently chelates residual iron(II), removes the last traces of iron and leaves the paper sheet in an alkaline condition—a desirable state for the print's longevity.

It is important not to use tetrasodium EDTA, or, as has also been practiced, trisodium EDTA, for the first processing bath—both have high pH at which level the iron(III) is not effectively complexed, but rather hydrolysed, which will eventually result in a yellow or brown staining. Nor is the use of just tetrasodium EDTA, or some combination of this with sodium sulfite, an effective means of clearing. Dr. Matthew L. Clarke, one of the Platinistas and now conservation scientist at the Smithsonian Institution in Washington, D.C., carried out an extensive study to test for the most effective way of clearing platinum/palladium prints. His team prepared scores of step-tablet prints, clearing each by means of the standard as well as some unusual methods. Along with the traditional method of using just hydrochloric acid, they tested for effective clearing with only tetrasodium EDTA; tetrasodium EDTA with sodium sulfite (one of the most widely recommended current approaches for development Pt/Pd); only sodium sulfite; two unusual solutions: Lime-A-Way™ which is a hard-water stain remover and the soda, 7Up; and the three-bath sequence described in this volume. Many of these methods have been recommended and documented in numerous contemporary texts for development Pt/Pd. Clarke's results, which measured paper-base whites after drying and after a standardized aging test, concluded that "After aging, it is apparent that many clearing methods are not adequate. The 3-step method (disodium EDTA, followed by sodium sulfite, followed by tetrasodium EDTA) gives the cleanest appearing print after aging."[11] His findings go on to state that the "3-step bath is therefore highly recommended regardless of the paper support (the) sensitizer employed, as it consistently and efficiently reduced the residual iron."[4] Along with confirming the effectiveness of this 3-step clearing bath, Clarke's team also highlighted the importance of sufficient washing as a final step in the process, noting that while the most dramatic staining arose from poor clearing, poorly washed prints also contributed, but to a lesser degree, to discoloration and damage over time.

To determine, by the use of instruments alone, if a print has received sufficient clearing during processing is resource-intensive, expensive, and time-consuming. Most of us do not have the skills or access to facilities for this. Certainly, it is impossible to assess for full clearing by the naked eye alone. So, the decision to take remedial conservation measures by non-conservators should be triggered by a clear acknowledgment of erroneous practice: if the printer knows that the 3-bath system was not used, then the corrective treatment should be considered. For those who may already have works processed by means other than what is recommended here, it is possible to retroactively "correct" platinum/palladium prints. This may not completely protect against a deterioration of the image, but it will certainly reduce the effect. As a "best practice" then, it is advisable to have the final stages of *any* platinum/palladium printing process be the clearing sequence described from steps 18 to 24 in Chapter 8.

Presentation and Storage

Another important way to minimize discoloration and fading can be accomplished by the proper presentation and storage of platinum/palladium prints. Besides the archival storing and handling of prints, their presentation for viewing impacts longevity. And yet, we want to look, to experience the print, and the way we experience them is heavily influenced by presentation—keep in mind that platinum/palladium prints have a distinctive tone, color, and sub-surface appearance. The technical reasons for this fingerprint are apparent from earlier chapters, but it does no harm to consider them in a slightly different context here, and then to approach the matter of how best to look at and exhibit a print. The following printing parameters affect the color produced in platinum/palladium images:

- *Ambient temperature*
  For a given relative humidity, a higher temperature determines a higher water content of the atmosphere, *i.e.* the *absolute* humidity is greater, so it promotes more neutral tones. e.g. 56% RH @ 25°C is the same absolute humidity as 75% RH @ 20°C, and vice versa
- *Paper base composition*
  The differing fiber morphology, e.g. of celluloses derived from flax (linen) versus those from cotton or gampi, may favor different sized metal nanoparticles. Palladium prints are more neutral-toned on linen paper than on cotton paper. The presence of gelatin sizing agent in the paper may interact, like surfactants, by "protecting" nanoparticle metals, favoring the formation of smaller particles and therefore warmer colors. Gelatin also inhibits the precipitation of platinum. Sizing by starch, alum-rosin, or alkyl ketene dimer (AKD or "Aquapel") gives more neutral tones with palladium. Paper surfaces also have some bearing on color and the scattering of reflected light.
- *Pre-humidification or relative humidity of ambient atmosphere*
  Higher relative humidity in the atmosphere increases the amount of absorbed water in the sensitized sheet, which enables mobility of the platinum/palladium ions for printing-out, and growth of metal nanoparticle size, giving more neutral color. Hydration levels of the paper before and after sensitization is important in printing-out platinum/palladium, and can be achieved by constant humidity enclosures,

containing saturated salt solutions, or even water for a timed humidification.

- *Proportion of platinum to palladium in the sensitizer*
  Platinum tends to yield neutral gray tones whereas palladium tends to give warm brown images.
- *Surfactants in the sensitizer*
  Tween 20 may tend to stabilize smaller nanoparticles by surface adsorption, inhibiting particle growth and therefore causing warmer color.
- *Humectants in the sensitizer*
  Their presence promotes an increased amount of water to be absorbed by the cellulose fibers. Humectants like glycerine and ammonium nitrate are hygroscopic and give more neutral color.
- *Humidity of the sensitized paper*
  The amount of water molecules in the paper prior to and during exposure will affect the size of platinum/palladium nanoparticles, and determine color, where smaller particles appear warmer and larger ones appear more neutral.
- *Atmospheric pressure during exposure*
  i.e. whether using a pressure frame or a vacuum easel for the contact exposure: vacuum pumping can reduce the amount of absorbed water and tends to produce warmer toned prints with palladium. Remember that carbon dioxide is always evolved during the exposure, so will seek a pathway to escape from the paper: porous felt backing is recommended for the contact printing frame.
- *Exposure duration*
  Lengthy exposures of palladium papers at lower light intensity are more likely to give neutral tones than brief exposures at higher intensity. Iron(II) begins to randomly migrate during exposure, "seeking" target Pd(II) to reduce, and it is the actual speed of this process that determines nanoparticle growth. Longer exposures give more time for palladium particle growth.
- *Post-exposure hydration by "steaming"*
  Suspending a print over warm water (typically @ 40°C for two minutes) after exposure, but before wet processing, shifts the colour of palladium, or palladium-platinum prints, towards more neutral tones. For all platinum/palladium prints made with (pre-exposure) hydrations less than 50% RH, it also allows the highlight values to form more slowly than if the print is immersed in the first clearing bath immediately after exposure—the post-exposure hydration renders smoother highlight tones, and anticipates any "development" (see Chapter 8).
- *Chemistry of developer, if used*
  With the development Pt/Pd process, the chemistry of the first clearing bath can affect the colour of platinum/palladium prints, which are warmer with oxalate than with citrate developers. The Platinotype Company formulated a phosphate-containing developer said to provide colder tones with platinum.

The experience of looking at a platinum/palladium print is shaped by the viewing light, the areas surrounding the image, and framing. All of these are in play whether a print is simply viewed on a table, pinned up on a wall, or framed in a gallery. You will notice (at times dramatic) differences between looking at the same print in raking sunlight at 5000K or higher, and in diffuse, warm 2700–3000K domestic light. The color of window mount boards can enhance or severely undermine the image, all the more so when a board has to be placed under translucent papers. The effect of showing or masking out coated areas (and process marks such as brush strokes) outside of the image frame needs to be considered. All matte-surfaced prints, but especially those with a preponderance of dark tonal values, change when placed under glass or acrylic sheet. I offer the following as a set of personal preferences:

- *Low ultraviolet and infrared radiation*
  These are invisible, do not deteriorate the image-forming metal particles but can have

a detrimental effect on paper if there are uncleared iron(III) complexes. While less important for occasional viewing, it will be necessary to consider this when prints are on extensive or permanent display. Conservators refer to a print's exposure to light over the duration of an exhibition as the "total light dosage" (TLD).[12]

- *Illumination levels of ca. 150 to 400 lux*
  The tonal range of most platinum/palladium print and paper combinations is fully perceptible in these conditions.
- *Color temperature, optimal around 3000K*
  A range from standard domestic to gallery lighting, between 2500K and 5000K respectively, is also fine.
- *Viewing light is diffused with soft-edged shadows*
  The slightly directional light helps pick up surface texture without having it interfere with the image.
- *Borders around the image are even and distraction-free*
  Avoid displaying brush or coating marks—they generally create a distracting conversation when seen by those who do not know much about the process, and appear superfluous to those who are familiar with it. See Chapter 8, step 9.
- *Mounts are carefully matched to print and image color*
  Do the matching under controlled lighting. Platinum/palladium prints can tolerate gray and black mounts along with the more usual white to ivory range.
- *When possible, avoid glass or acrylic*
  Cheaper "picture" glass has a slight green cast, and this can interfere with the image. Acrylic is much more neutral, but is more susceptible to scratching than glass. When prints must be glazed use museum-quality glass or acrylic that is UV-blocking and reduces reflections by means of special surface coatings. Do not opt for the cheaper "anti-reflective" materials that have a stippled surface.

As with all valuable photographic works on paper, archival materials should be used for the presentation and storage of prints, and each print interleaved or separated by an acid-free sheet. Buffered paper may be used for this. Ideal ambient storage conditions are about 40–50% RH and 68–72°F / 20–22°C.

Retouching

Platinum/palladium prints are highly lightfast and robustly resistant to all contaminants likely to arise in a normal environment. For retouching white spots, use permanent, high-quality watercolor pigments. A blend of Lamp Black, Prussian Blue, and Burnt Umber will usually result in a suitable retouch color for most papers and tones. Matching a hue for translucent paper is best done by laying it on the final backing board, since the base white color is a combination of both.

The removal of black spots resulting from small transparent dots on the negative is not easy. (Black spots can also manifest with certain development Pt/Pd processes, but this has not been a problem with the printing-out version.) There is no convenient reagent for dissolving platinum black spots. Palladium is slightly attacked by dilute hydrochloric acid in air, but dilute hydrobromic acid is much more effective as a print "reducer" (*i.e.* oxidising agent): palladium images may be etched by dilute hydrobromic acid (1% to 10% *v/v* depending on the rate desired). The "reducing" action appears to be "proportional" in its effect across the density scale. None of this is very convenient, and at times, a more drastic approach may be the most convenient; it requires a steady hand, a very sharp-tipped scalpel, and some practice. Remove the spot with precise scrapes, in tiny movements, and by minimal gouging of the paper surface. Then, lightly smooth out the abraded surface with the tip of a folding bone; laying a sheet of tissue paper over the print and scratched area prevents burnishing marks from appearing. Finally, consider retouching what may now be a white spot.

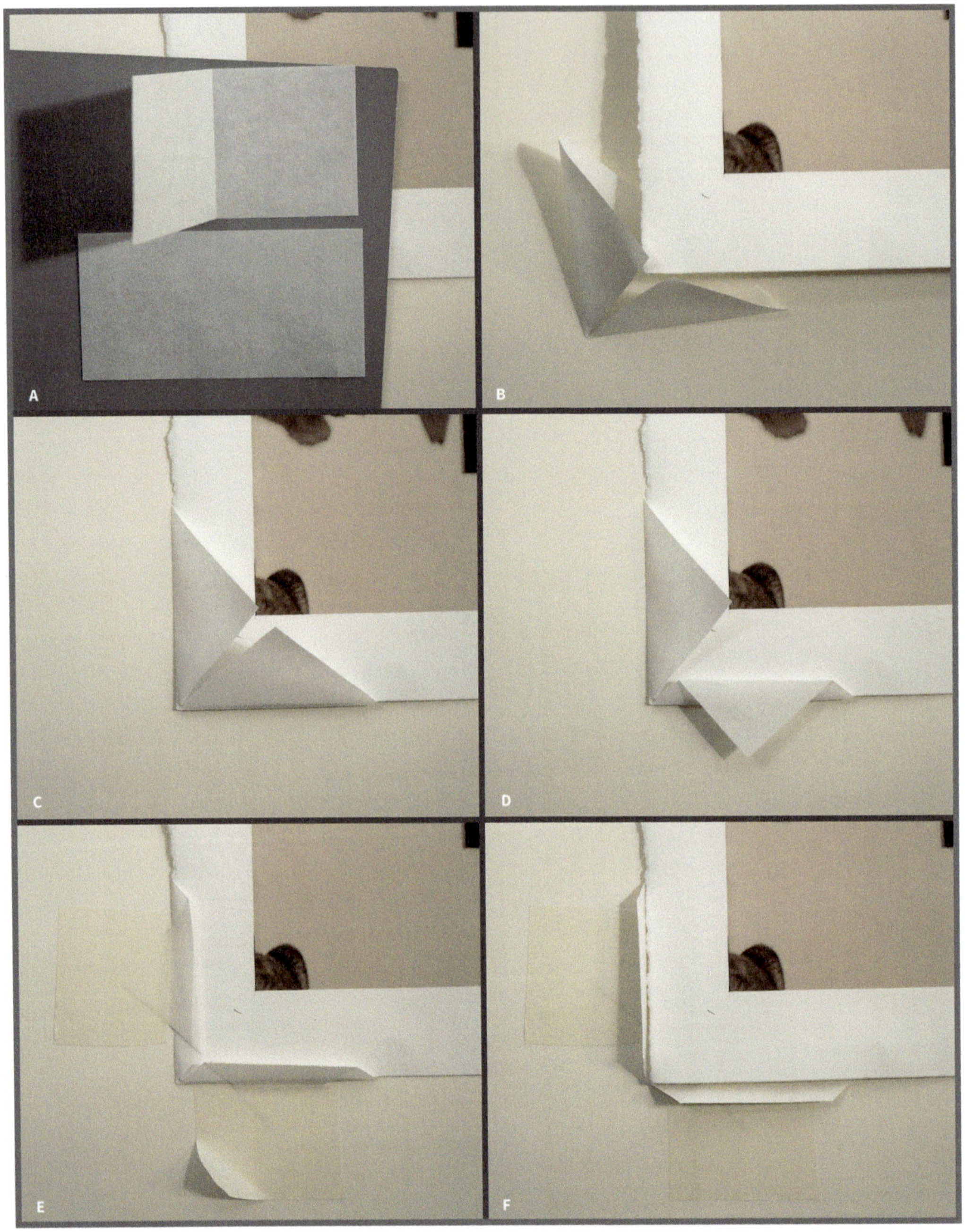

Figure 9.3 A window-mounted print using a non-adhesive method for attaching a print to a backboard.[13]
A. fold 2:1 aspect ratio strip into half, then turn each corner down to center fold to make a triangle.
B. position and weigh print down on board. Tuck the triangle under print.
C. position snugly.
D. fold out each corner, leaving a portion of fold covering the print.
E. tape the outer fold down to board, and press down the folds.
F. print may be easily removed or replaced on board.

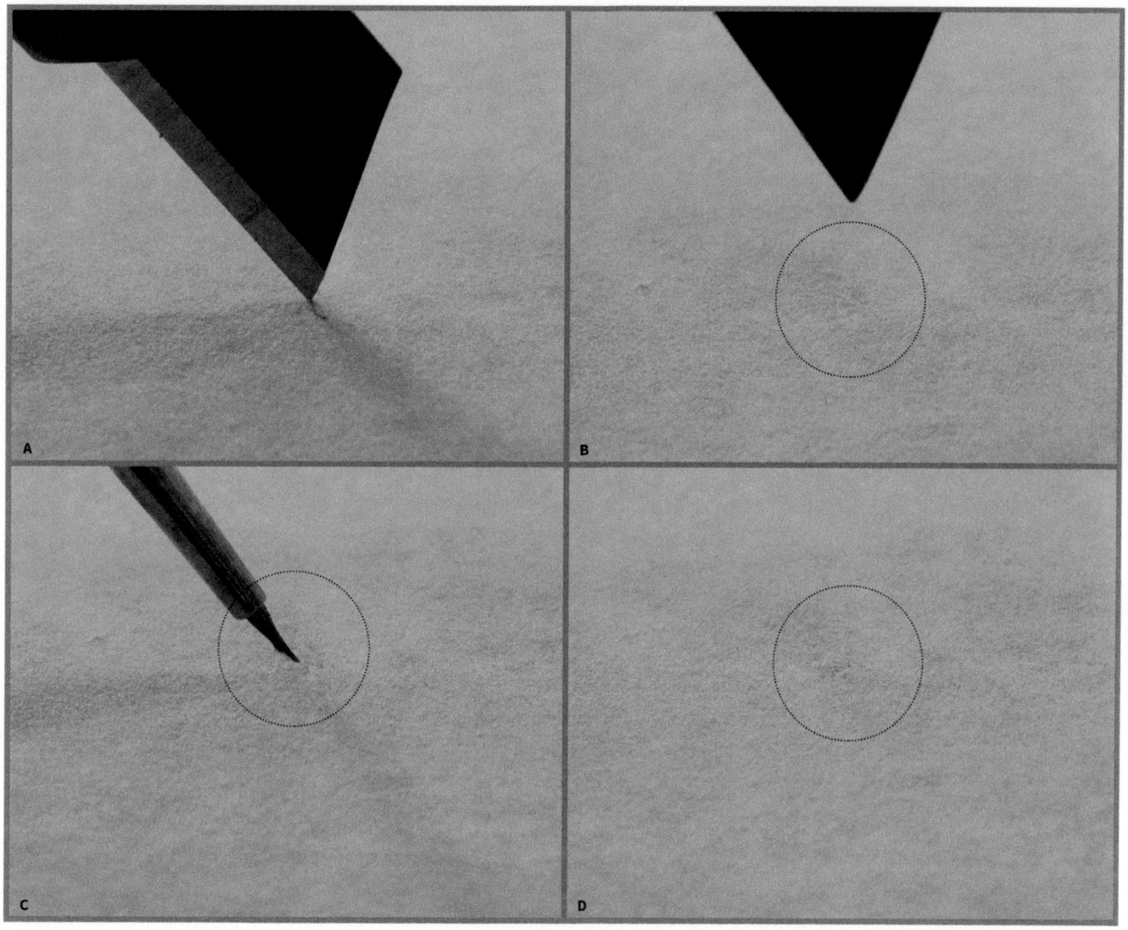

Figure 9.4 Retouching: Illustrations for removing a black spot A. with a very sharp blade tip, gently scrape blemished surface. B. continue in small movements until mark is gone. C. retouch with mat gouache using a #00000 brush. D. the scraped area is matched to surrounding tones.

Documentation

Maintain records of all matters relating to the prints you make. Most photographers and printers follow a typological form to organize images: by subject, job, client, etc. There are strong arguments now to break from this, with the increasing use of digital asset management (DAM) programs such as Adobe's Lightroom™, and to shift towards a more logical approach. I recommend organizing all image files by date, and utilizing meta or keyword tags to automatically generate typological substructures; and name them numerically. Regardless of the specific organizational and file-naming approach one takes, there will be a need to set up correspondences between the original (negative or digital file), any multiple versions of this (if using a digital negative), and each print made from a given negative. When making digital negatives, while a trace of all work is kept via Photoshop's Layers, and its History tool (you will need to adjust Preferences for this), some data will need to be noted down, such as the printer profile, output settings, printer, and inkset. Then, there is a need to maintain platinum/palladium print data. The main reason for keeping process notes is to facilitate subsequent printing sessions for any given image. A photograph, printed months or years ago, may need to be matched. Or you, the photographer/printer, may just want to pick up from a previous iteration, and form a new expression without having to start from scratch. In these cases, it helps to have a record of the exact ratio of platinum to palladium, type of ammonium ferric oxalate, paper, and the hydration conditions, and then to be able to proceed from these parameters. There is another type of case to be made for well-maintained printing notes: when problems need to be solved or the process needs to be modified, having detailed notes keeps chaos at bay. Despite all that has gone into the preparation for material in this book, the printing-out platinum/palladium *process* is soft-edged and human. It will evolve, will need to be adapted as materials change, and will require fine-tuning in order to be fully assimilated into your creative practice. Keeping notes is an essential part of a cyclic and deepening practice: good science can lead to efficient methods, which in turn fosters systematic record-keeping, which leads to evidence-based problem solving and reflection, which takes us to modified and evolving practice, and we arrive at virtuosity, which, returning, inspires good science and art. With all of this in mind, I now list and describe each of the variables you may encounter with the printing-out process. There is a certain logic to the organization of this list, and I hope that is apparent. I conclude with what may serve as a useful template.

Original Images and Negatives

Assign negatives and digital files with sensible names. A sequence of numbers is a manageable way of allocating an identifying name, especially if it is accompanied by a descriptive title or caption in a digital asset management database, such as Lightroom™, or paper notebook. For instance, a negative's unique number may be marked on the sleeve; the corresponding entry in a database or notebook will record a date, title, description, and any other information. Digital negatives should be named such that they can be clearly traced to parent files, and a notebook entry for each should provide all necessary information if a similar print needs to be made at some future time.

Printing Data

For maintaining records about sensitizer and processing solutions, see Chapter 5. Of all the notes you keep, the most detailed will be for the printing segment of the workflow. A notebook, or some digital form of it, should record information for every print. An almost comprehensive list of variables and parameters are listed below, with the most essential indicated by an asterisk (*).

- **Print Number*
  Always unique and even applied to test prints, should also be penciled onto a corner of the print. E.g. 00001
- **Sensitizer batch and type*
  See Chapter 5
- **Negative identifier*
  Along with noting this in the printing notes, it may be helpful to also append this to the Print Number on the print. E.g. 00001_52763
- *Negative title*
  It helps to have the title or some description of the image in the printing notes
- *Ambient conditions*
  Record the room's temperature and relative humidity at the time paper is coated
- **Paper name and weight*
  See Chapter 6
- **Paper batch*
  See Chapter 6
- *Pre-coat hydration*
  Note the method and duration, e.g. "hung on line, ambient, 20°C at RH 60% for 24 hours"
- *Coated date and time*
  Note the date and time that the coating was actually carried out
- *Coated area*
  e.g. 8″ x 10″
- *Coating method*
  e.g. "glass rod" or "Richeson-9000 3-in"

PRINT NUMBER	52924
SENSITIZER TYPE AND BATCH	Fel #200608: Pt#190606: Pd#190606 0·7: 0·35: 0:3
NEGATIVE IDENTIFIER / TITLE	3001190_1-1A
AMBIENT CONDITIONS	69 F / 54%
PAPER NAME, WEIGHT, BATCH	Revere Platinum VOR 160425-15014781
PRE-COAT HYDRATION METHOD AND DURATION	Chamber, 76% RH, 4 hours
COATING DATE AND TIME	June 15, 2020, 12:45
COATED AREA	8 x 10"
COATING METHOD (ONLY IF USING MORE THAN ONE METHOD)	glass rod
SENSITIZER VOLUME	1·4 ml
ADDITIVES AND VOLUME	Tween 15% 0·05ml, Glycerin 15% 0·05ml
NUMBER OF PASSES (IF USING COATING ROD)	6
AMBIENT DRYING TIME	5mins.
HYDRATION METHOD AND DURATION	Salt Chamber. Calcium nitrate 90mins / @ 56% RH
EXPOSURE SOURCE (ONLY IF USING MORE THAN ONE TYPE)	Fluorescent bank
EXPOSURE TIME OR DOSE	400
POST-EXPOSURE HYDRATION METHOD AND DURATION	5 min. / ambient
PROCESSING SOLUTIONS, SEQUENCE AND TIME (IF VARYING FROM THE STANDARD 3-STEP CLEARING SEQUENCE)	normal.
WASH TEMPERATURE AND DURATION (NOT NECESSARY IF THIS IS CONSTANT)	20°C / 30 min.
ANY OTHER OBSERVATIONS AND NOTES	
CURRENT LOCATION OF PRINT	

Figure 9.5 A sample printing data sheet, with annotations, of the print in Figure 7.2. Also refer to the printing data sheet template in Appendix II—Workflow Summaries.

- *Sensitizer volume*
  Amount of sensitizer used to coat the image area, e.g. 1.5 ml
- *Sensitizer additive(s) and volume*
  Indicate the additive(s), concentration, and volume, e.g. "Tween 15%, 0.05 ml" or "Tween 15%, 1 drop"
- *Number of passes*
  Only practical if using a coating rod, e.g. 4
- **Ambient drying time*
  Duration is important to note if there are any changes from previously recorded ambient conditions. This is not critical but can be very helpful when producing an edition of prints
- **Hydration method*
  e.g. calcium nitrate, RH 56%
- **Hydration time*
  Duration, e.g. 90 minutes
- *Exposure source*
  Important if working with more than one source
- **Exposure time or dose*
  e.g. 3 mins or 200 units
- *Post-exposure hydration method and time*
  e.g. 80°F water in tray, suspended print for 2 mins, or ammonium chloride hydration chamber (80%) for 2 minutes
- *Processing chemicals, sequence and times*
  Not necessary if always using the recommended 3-step clearing sequence, but any variation from this should be noted
- *Wash time and temperature*
- *Other printing notes and observations*
- *Current location*
  Useful to note the location of a print, especially if it is in a collection away from your studio

Notes

1 Constance McCabe,. (ed.), *Platinum and Palladium Photographs: Technical History, Connoisseurship, and Preservation*. Washington D.C.: Photographic Materials Group of the American Institute for Conservation of Historic and Artistic Works, 2017, p. 8.

2 Constance McCabe, *Platinum and Palladium Photographs* 2017.

3 Erin L. Murphy, "A Summary of Early Chemical and Physical Treatment of Platinum Prints," in McCabe, *Platinum and Palladium Photographs*, 2017, pp. 251–257.

4 Murphy, "Summary."

5 Alice Carver-Kubik, Saori Kawasumi Lewis, Andrew Robb with contributions by Erin L. Murphy, "Additives Used in the Platinum Process," in McCabe, *Platinum and Palladium Photographs*, 2017, pp. 92–97.

6 Matthew L. Clarke, "Characterization, Degradation, and Analysis of Platinum and Palladium Prints," in McCabe, *Platinum and Palladium Photographs*, 2017, pp. 108–114.

7 Margaret Wessling, "Preservation Concerns for Edward Weston's Platinum and Palladium Photographs" in McCabe, *Platinum and Palladium Photographs*, 2017, pp. 274–275.

8 W.J. Barker, "Yellowed Platinotypes," *The British Journal of Photography*, 64, September 7, 1917, 467.

9 Clarke, "Characterization."

10 Constance McCabe, Christopher A. Maines, Mike Ware, and Matthew I. Clarke, "Alfred Stieglitz's Palladium Prints: Treated by Steichen," in McCabe, *Platinum and Palladium Photographs*, 2017, pp. 368–369.

11 Clarke, "Characterization."

12 Jennifer J. Gutierrez, "Caring for Platinum and Palladium Prints: Storage and Display" in McCabe, *Platinum and Palladium Photographs*, 2017, p. 264.

13 McGlinchey Sexton, Jennifer and Jennifer Jae Gutierrez. "Nonadhesive Mounting Methods for Photographic Prints", in *Platinum and Palladium Photographs: Technical History, Connoisseurship, and Preservation*, edited by Constance McCabe, pp. 268–273. Washington D.C.: Photographic Materials Group of the American Institute for Conservation of Historic and Artistic Works, 2017.

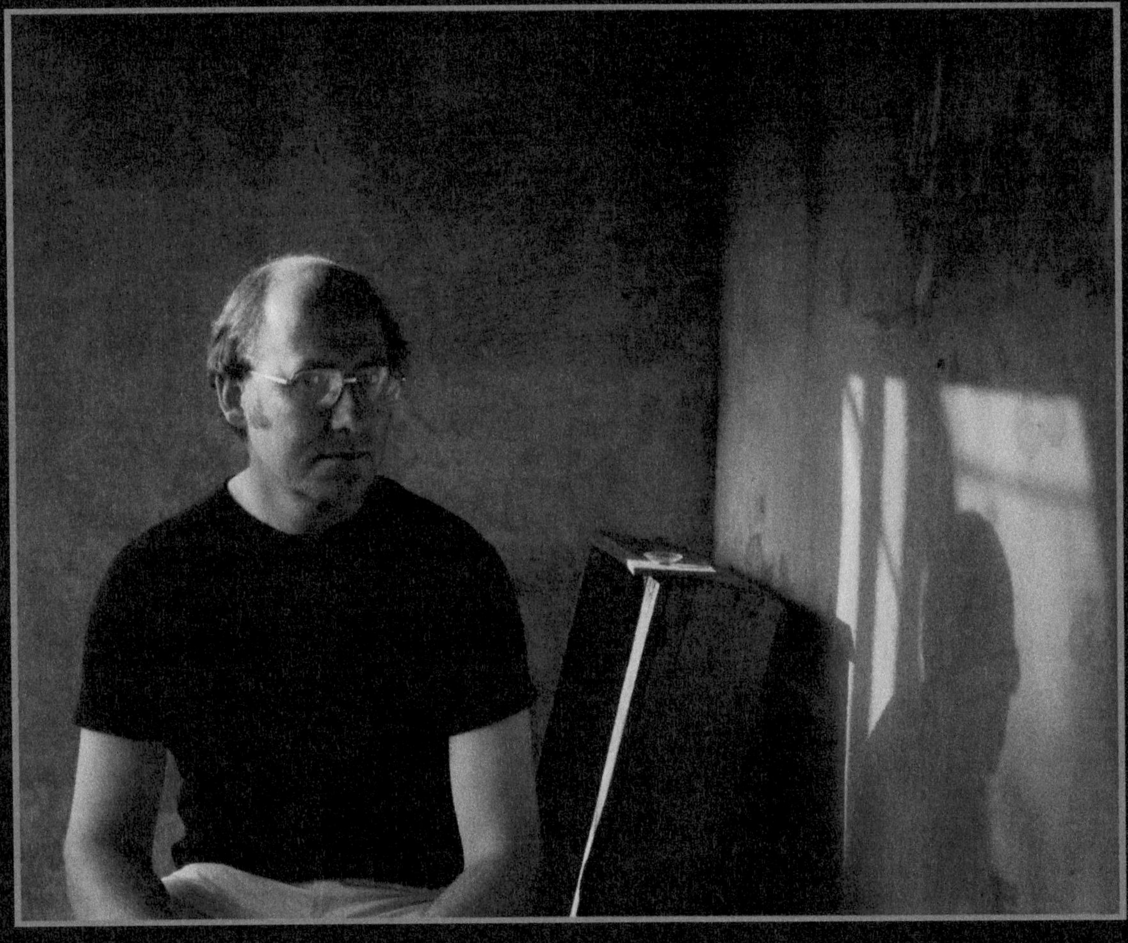

Figure 10.0 Pradip Malde, *Mike Ware, Orkney*, 1984. Platinum-palladium print on Van Gelder Simili Japon from 18 x 25 cm original, 17.75 x 24.5 cm. Courtesy Pradip Malde. Fe1:Pt:Pd = 2:1:1, 20 min hydration in 100% RH / 20°C.

Chapter 10

Looking and Seeing

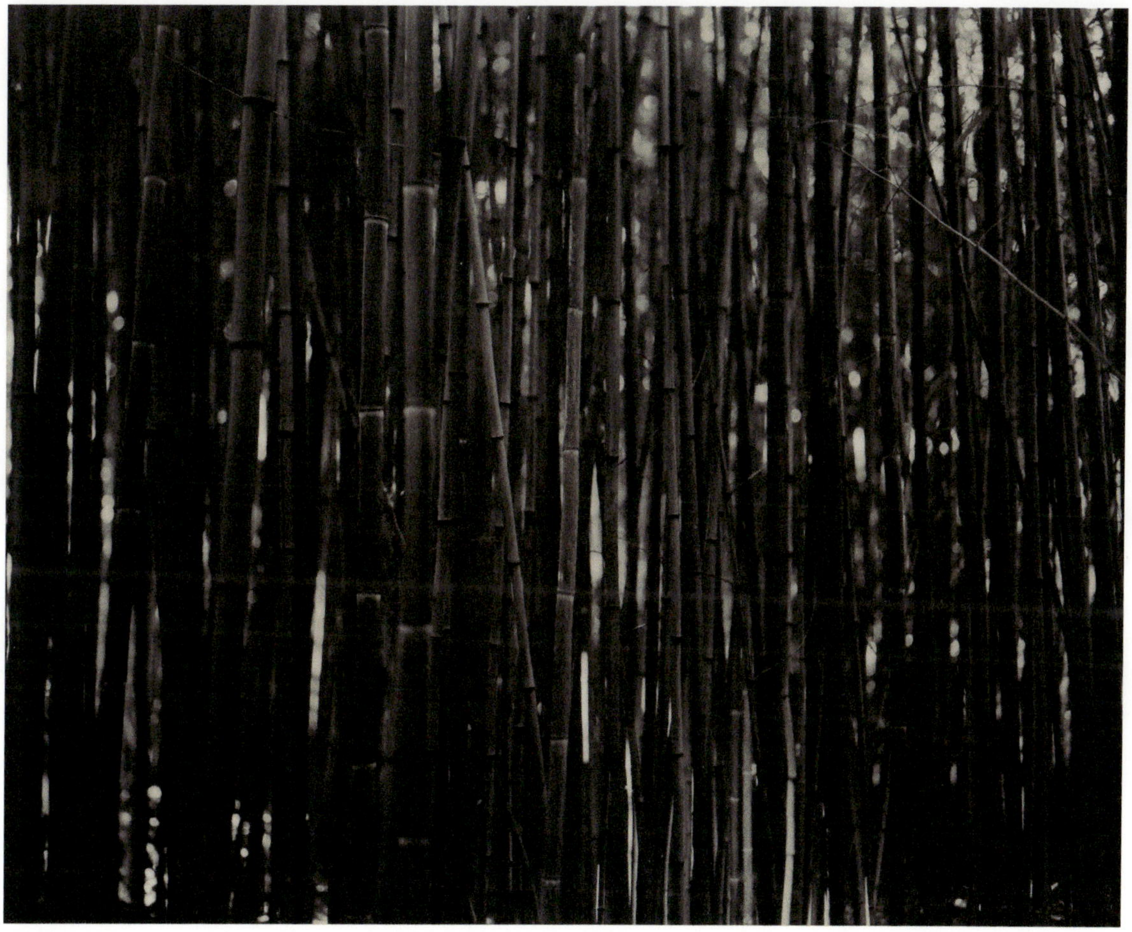

Figure 10.1 Allie Horick, *Bamboo*, 2016. Platinum-palladium print on Reich CT 24# paper, from a PiezoDN negative of 8 X 10 original developed in pyro PMK, 25.5 x 20.5 cm. Courtesy Allie Horick. Fe1:Pt:Pd = 2:1:1.

Every photographic print has a particular look. Its clusters of opaque, microscopic image particles, bound to a piece of paper, are the result of many, interdependent and finite steps that lead finally to the moment of looking at a print: from the passage of light reflecting off the subject's surfaces, through lenses, to film and sensors, and ultimately traveling from the printed image and into our eyes. As printers, aficionados of platinum/palladium images, we are going to take time to scrutinize prints. However, looking closely at a photographic print in the hope of accessing its mechanics actually has a contrary effect to what we would expect from scrutiny: instead of becoming increasingly aware of the materials of a print, we tend to acquire a more intimate and palpable sense of "image." It seems the more we do this, the less separation there is between oneself, the print, and the photographic "window." There is a feeling of not only being a part of the world, but of being apart from the world. The more we look at and consider a photographic print, the more it seems that we are gazing out at another world. That is the avid wonder and magic induced by the platinum/palladium print. The desire for wonder and magic drives our decisions, as photographers/printers, that ultimately lead to a fine print. Should one choose to print digitally, with pigments in absorbent papers, or use a conventional silver process to form the image in a gelatin emulsion, supported by paper or film? Or should the print be made using a historic siderotype process, in which the metal particles are embedded within the intertwined and flattened cellulose fibers of paper? Each of these possibilities will result in a distinguishing set of aesthetic experiences that become part of the print.

What distinguishes platinum and palladium prints from their chemical relatives? Platinum and palladium prints are made of precious metals precipitated so finely that they render tones with highly nuanced values, creating highlights that slide to the very edge of the paper's base white. At the other end of the scale, persistently whispering shadows are often not as intense as in some other processes, and generally do not have the deep black values of emulsion-based silver prints. As with the way we actually see, the darkest tones in platinum/palladium prints are rarely an undifferentiated mass. For the British photographer Peter Henry Emerson (1856–1936), the platinum process was capable of producing images that aligned with his understanding of nature and human vision. The process mirrored, by its nuanced tonal range, his experience of life on the Norfolk Broads,[1] a system of wetlands in eastern England largely made up of flooded peat pits dating back to medieval times. The Broads have a special luminosity, a subtle radiance. Emerson, a medical doctor, was concerned about the health and economic conditions of those who lived on the Broads, but he was also moved by the tension between the soft landscape of this special place and its tenacious inhabitants. He must have intuitively understood that the platinum process could unveil his empirical and intimate experience of the Norfolk Broads. To some extent, Emerson's photographs of the Broads were conditioned by the in-camera manipulation of selective focus. But more important is the way Emerson looked to the process to map his experience of this place.

Emerson was intrigued by the possibility that platinum prints could articulate his idea of "natural" vision.[2] He was enthralled by its scientific potential, in the sense that images could be rendered in ways that mimicked human vision: brightness and darkness are rarely perceived in absolute values by the human eye, so why should tone be rendered as such in a photograph?

Emerson's naturalistic photography aligned well with the neutral tones and long exposure scale of the platinum print. His photograph, "Gunner Working up to Fowl" was taken with the lens pointing *into* the primary light source, yet the print reconciles the wide gamut of bright tones and deep shadow. The platinum process itself shaped Emerson's imagery, with subjects envisioned and even photographed in a certain way because they would become platinum prints. In doing so, he set out the possibilities of the platinum process, and his prints went on to shape a particular kind of photographic vernacular.

Emerson's use of process to recreate the experience of place also resonates deeply in the work of Frederick H. Evans (1853–1943), who regarded places as worthy of being photographed largely in terms of the platinum print's tonal range.

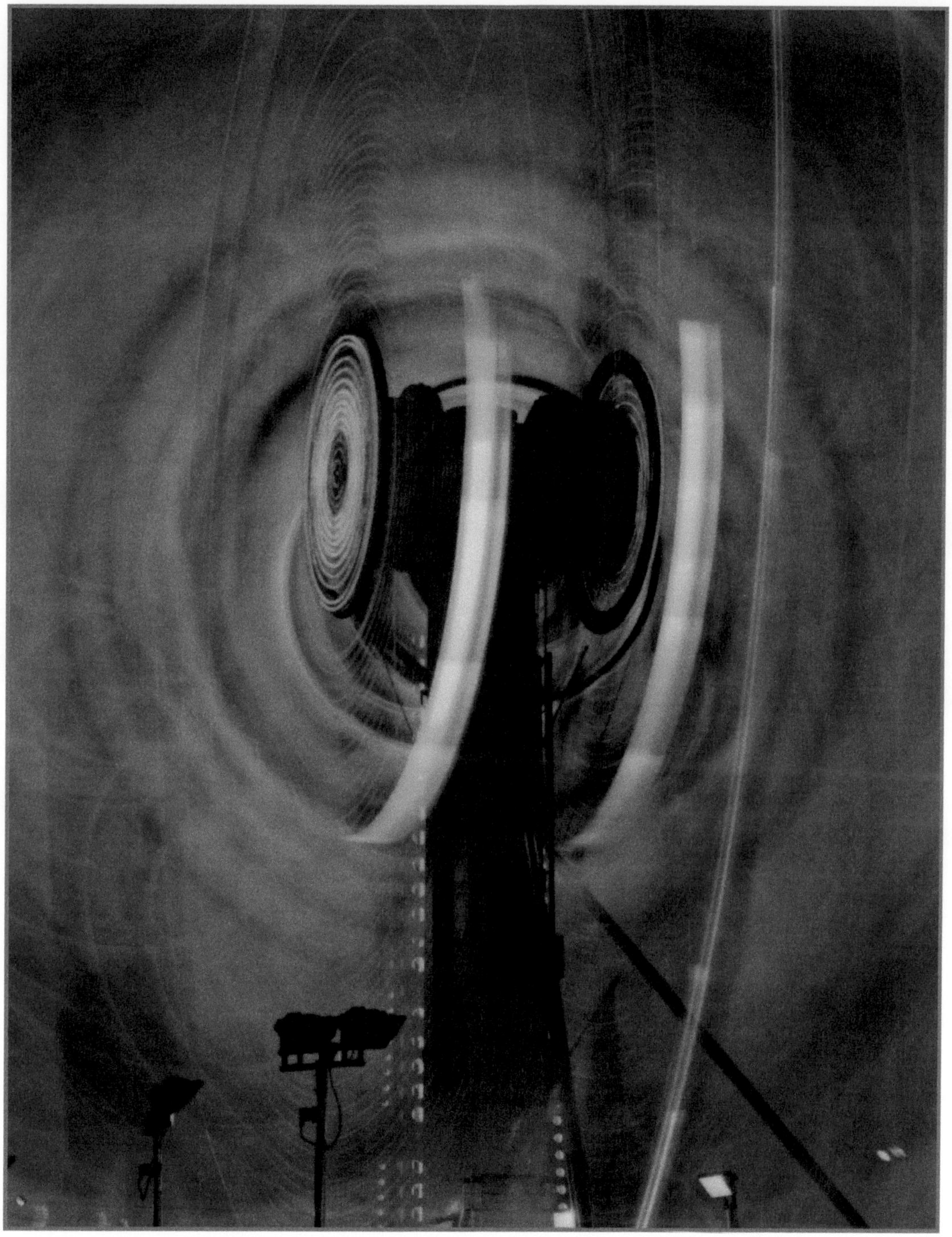

Figure 10.2 Roger Vail, *Kamikaze #2*, 1996. Platinum-palladium print from original 8 x 10 inch negative, 24 x 19 cm. Courtesy Roger Vail. The image has flowing edges that are more like territorial transitions or, conceptually, the excited states of energy. Just as with our understanding of quantum behavior, we sense there is an edge, but we are never certain of where exactly it is— even with the naked eye. Scrutinizing these rounded edges almost induce a sense of vertigo.

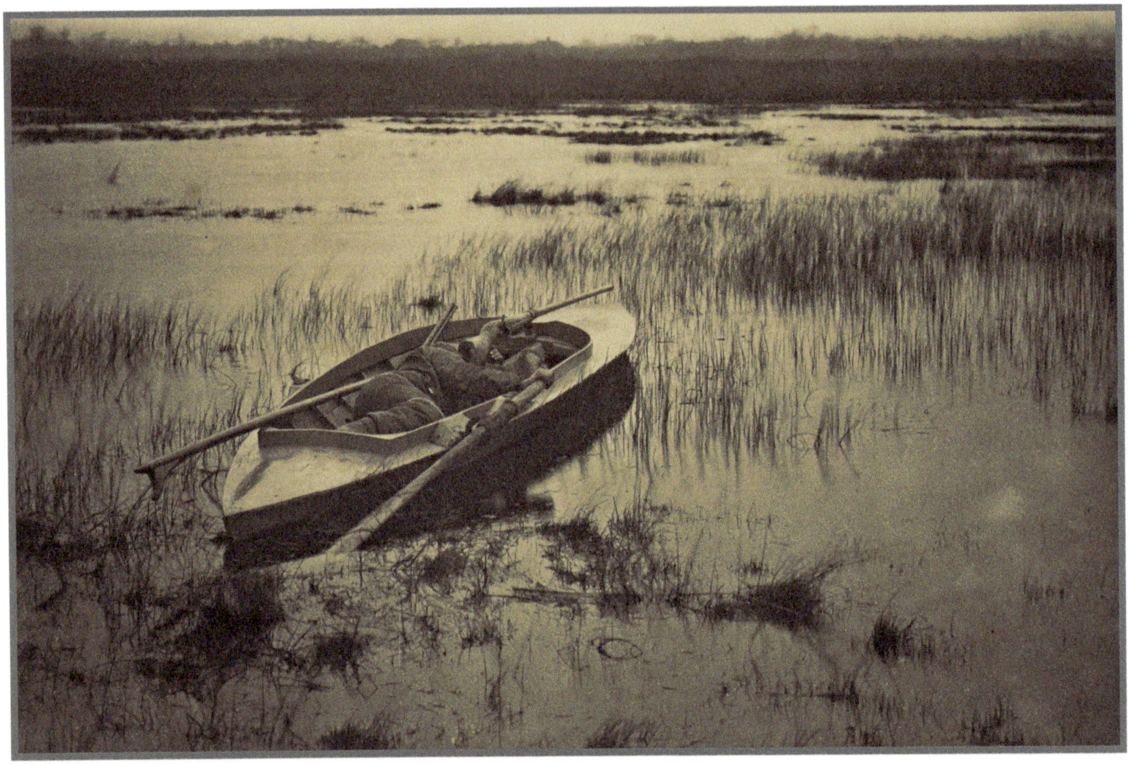

Figure 10.3 P. H. Emerson: *Gunner Working up to Fowl*, 1886. 18.9 × 28.7 cm. Courtesy of the Getty's Open Content Program. Development platinum print, printed by Valentine and Sons, Dundee. Images could be rendered in ways that mimicked human vision: brightness and darkness are rarely perceived in absolute values by the human eye. The platinum process itself shaped Emerson's imagery, with subjects envisioned and even photographed in a certain way because they would become platinum prints.

Consider Evans as he stood within the magnificent cathedral to photograph York Minster's North Transept: "In Sure and Certain Hope." The scene before him could have been translated into any one of many silver-rich, emulsion-based papers available in his day, each capable of rendering deep, thundering blacks that crescendo rapidly to shrieking whites. But Evans chose the platinum process, with its capacity to portray misty and velvety shadows, which then float, along with myriad dust motes, towards a creamy shimmer of reflecting stone and window light. This print of the North Transept is remarkably nuanced, quiet, and harmonious. The spirit of the place, but also of the photographer, is in the print, which is in turn revealed as the nature of the place, re-imagined through the photograph.

Vernacular: Then and Now

With each thing that is crafted, there is an accompanying conversation between maker and material, and this special pairing, almost regardless of what is actually made, profoundly informs the experience of production. In the case of looking at a print and constructing meaning, the tools we use to understand the world also form our world. Tools and technology create their own contexts, or syntaxes, conditioned by shifts in culture and changing materials. This is a complex, recursive process that can have distinguishing features which may be described as "vernaculars." For instance, the way by which a print is made, from the materials used to prepare the light sensitive paper, to the more specific techniques used in printing, such as dodging, burning, and contrast controls, forms the vernacular

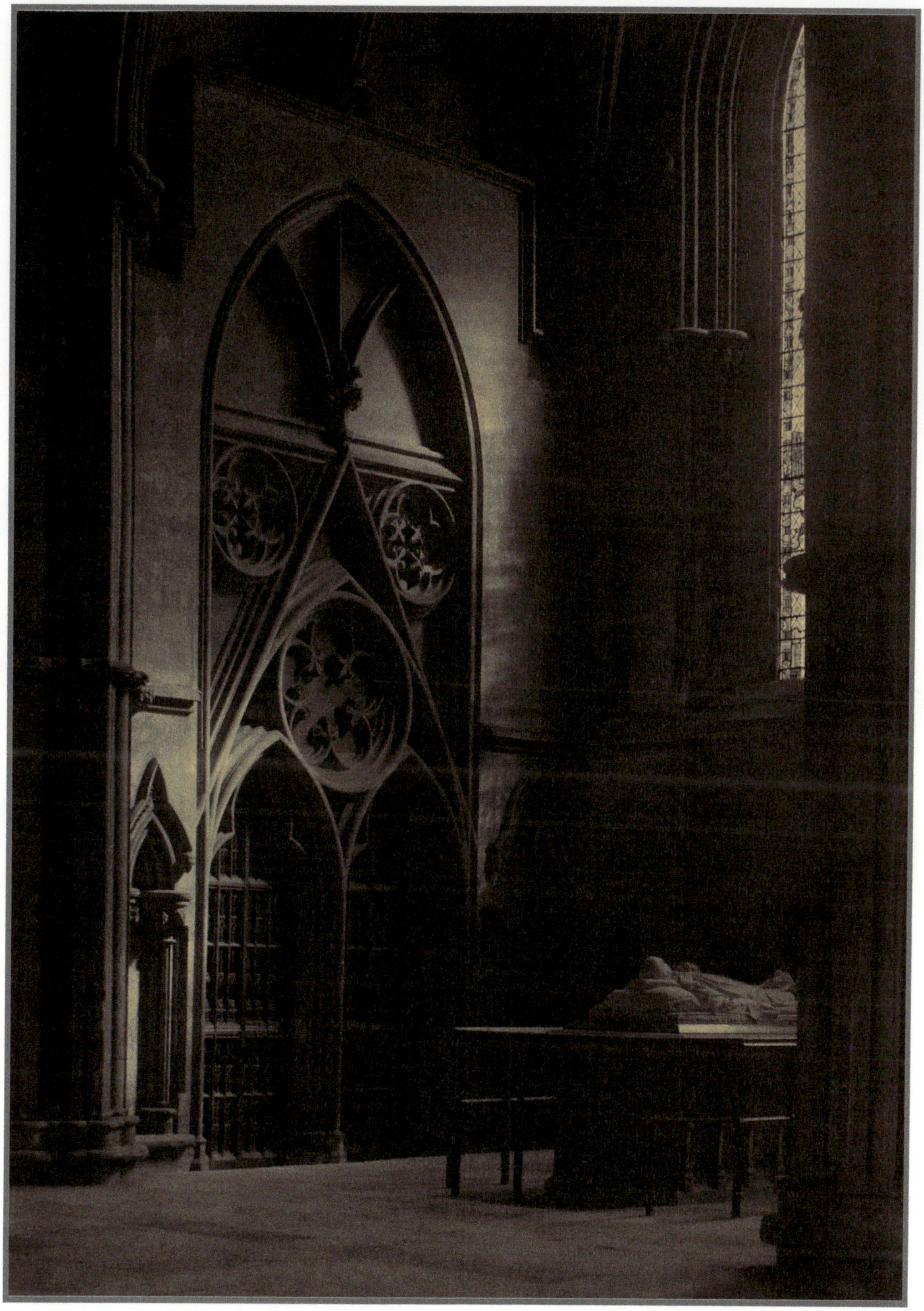

Figure 10.4 Frederick H. Evans: *York Minster, North Transept, "In Sure and Certain Hope,"* 1902. Development platinum print. 27.46 x 19.69 cm. National Gallery of Art, Carolyn Brody Fund and the Milmore Memorial Fund, 2011.18.1. Velvety shadow tones and detailed highlights create a particular "vernacular."

of an image. Vernaculars are often understood to be regional, cultural, and temporal variations of language and architecture. In terms of a photographic vernacular, what was understood and expected of the platinum print in the 1890s to the 1910s may be quite different from the platinum/palladium print of today. Understanding these differences and similarities in terms of the vernacular can better frame a contemporary approach to appreciating these remarkable photographs.

The language of the photograph affects the way we experience and understand the world around us. While words point to objects and experiences, the inverse is also true. Experiences shape language and how language is understood. The more specific an experience, and the more frequently a certain kind of experience occurs, the more likely it is that a vernacular language will form around the experience. (Scots Gaelic has over four hundred words to describe wintry weather conditions.[3]) A vernacular form of language grows from nearness: its ripples are tighter and more local, more attenuated but more precise. And in that precision, vernacular expressions are capable of conveying a more personal and immediate sense of place. While photographing in Norfolk, Emerson sensed the delicate balance of the local people's lives and their fragile environs. It is no coincidence that "delicate" is a word used often to describe fine platinum/palladium prints.

Cultural and technological shifts are closely linked with vernacular expressions. How a photographed moment was envisioned and expressed as a print in the late 1800s is far different from how a photograph might be taken today for an instant post on social media. Since the language of the platinum/palladium print has shifted over time, it may be useful to highlight the most distinguishing characteristics and similarities of the process, and some differences between historic and contemporary practices, before considering the vernacular itself. For the purposes of comparison, the "initial period" will be defined as between 1882 to around 1914, the heyday of industrially manufactured platinum papers. The years since 1990 or thereabouts will describe "contemporary" platinum/palladium photography, a period during which there has been an extraordinary growth of interest in the process, expanding significantly in the early twenty-first century.

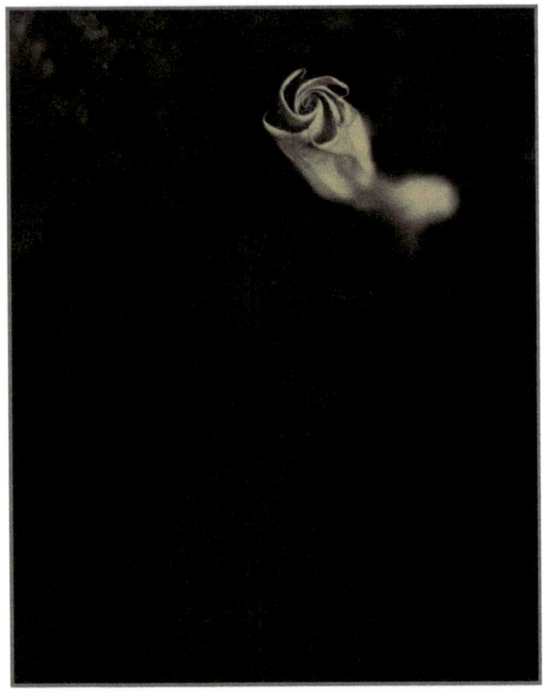

Figure 10.5 Emily J. Gómez, *Opening Moonflower, 2nd in a series of 4*, 2000. Platinum-palladium print on Crane's Platinotype, from original 5 x 5 inch negative developed in Pyro PMK, 12 x 9.5 cm. Courtesy Pradip Malde. Gómez's print of the moonflower jumps between extremes, with its blazing whites and swirling blacks. The image, but also the print, remains animated. There are no homogeneous zones or encampments of blandness. The platinum-palladium print is capable of rendering tonal nuance at the extreme ends of highlight and shadow values. Midtones transition from dark to light in whispers, rather than with the megaphonic sparkle so typical of dual- and triple-layer processes such as glazed albumen or silver gelatin.

What Has Changed?

During the initial period, photographs were profoundly shaped by the slowness of negative emulsions and the need to mount cameras on

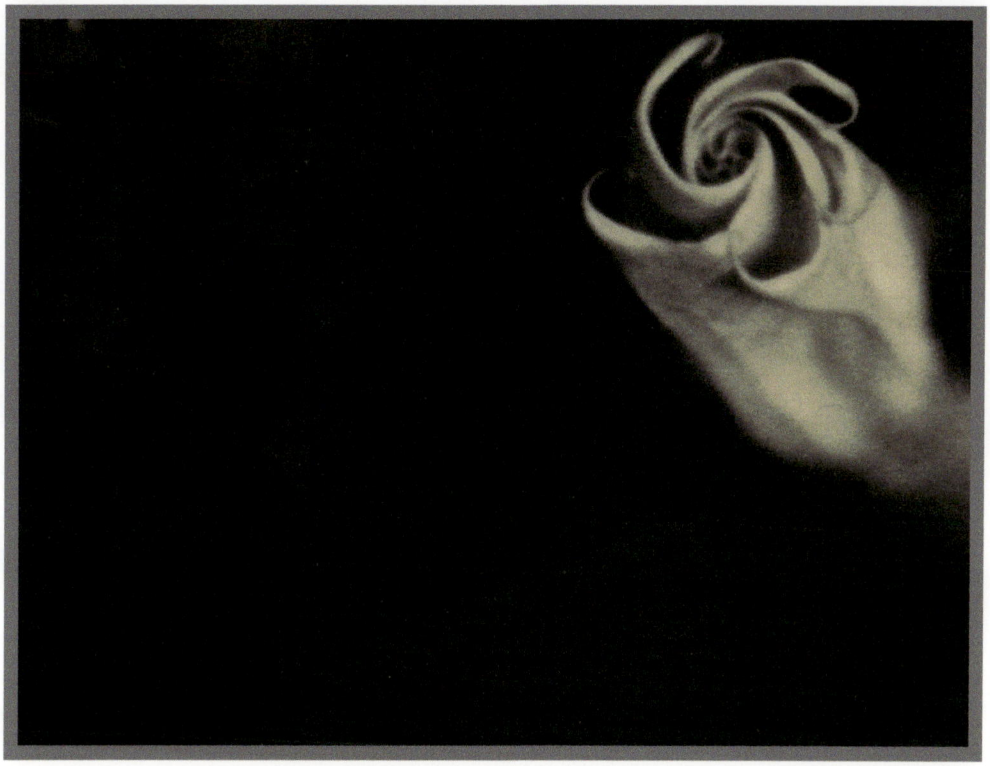

Figure 10.6 Detail: Emily J. Gómez, *Opening Moonflower*, 2000.

tripods. Carefully composed—and usually static—subjects characterized the imagery. The contrast range of negatives was long, and it suited the tonal response of the printing-out papers of the day. Lenses, while capable of rendering high acutance and resolution, were often designed to project a fashionably diffused focus. Prints were most often prepared on carefully formulated machine-made papers, surfaced to optimal specifications, which were then mechanically sensitized to produce a predictable look and feel. Industrialization facilitated and established an understanding and expectation of how a platinum print should look.[4] Contemporary platinum/palladium printers, now working as they must with handcrafted techniques, may use traditional large-format methods to make film negatives, or they may generate enlarged negatives from any kind of analogue or digitally originated image. The combination of highly sensitive digital sensors, contemporary films, and unprecedented lens acutance leads to images that have a very different, often more spontaneous, feel when compared with anything found in the initial period. Hand-sensitized prints may be made on many different paper supports, few of which are custom-made for the platinum/palladium process, and they vary tremendously in appearance.

It may not be possible to define the look and feel of the quintessential contemporary platinum/palladium print. The initial period was driven by a kind of scientism and the corollary belief that there is an underlying order to everything, waiting to be discovered and expressed, or, in Emerson's case, prescribed for the rest of us. Contemporary platinum/palladium photographers pay little attention to linking process with philosophy or to the conceptual premises of Emerson's naturalistic photography.[5]

What Has Not Changed?

From the very outset, platinum/palladium prints were, and continue to be, lauded for their permanence. Along with their archival robustness, an image rendered with noble metals has marketing cachet, garnering the interest of collectors and curators. The *impression* that the look and feel of a platinum/palladium print was and is somehow different from all other photographic prints also persists. For all these reasons, whether made in the initial period or by a contemporary artist, platinum/palladium prints are among the most prized in today's collections.

The distinctive "hush" often felt when looking at fine platinum/palladium prints invites a sustained examination of the image. Over time, even the most innocuous-seeming photograph begins to unveil its glory. This approach invites the photographer to make work arising from a sustained relationship with the subject, and the viewer to sustain a relationship with the print.

Breathless and Forgetting: Qualities that Set Platinum/Palladium Prints Apart

There was a moment in the late 1990s when I found myself in the print room of the Art Institute of Chicago looking at a display of unframed platinum prints by Frederick H. Evans. My breath, quite physically, left me, as did my sense of where I was. My hush became part of a larger silence that filled the room. These were platinum prints of a quality rarely seen.

Good silver-gelatin prints are capable of evoking an awed response. But that sense of awe, even for the most admirable silver-gelatin prints, may more aptly be described as a gasp rather than a hush. Printers who have shaped their vision through the more conventional silver-gelatin emulsion papers are, often inadvertently, challenged by this difference. Contemporary platinum/palladium printers work in the realm of hushes and sighs, not in the gasping atmosphere of silver-gelatin printing. The platinum/palladium process has formal qualities distinct from other photographic processes, the vernacular of which calls for description.

The evocative "hush" of the platinum/palladium print may be understood as *intensity, contrast, tone, edge, resolution,* and *volume*, terms that are traditionally used by printers and connoisseurs. These aspects are explained below in emotive terms, using words to approximate another realm of experience rather than using the more traditional photographic terminology based on optical and sensitometric values. The singular and highly nuanced qualities of a platinum/palladium print, and the ways in which it is distinct from other photographic printing processes, may be further understood by considering its *form, pattern, overtone, texture,* and *delicacy*. Combined, this terminology more clearly establishes how a platinum/palladium print evokes its distinctive atmosphere of wonder and magic. Each of these facets may provide a path to connoisseurship: identifying and appreciating a good platinum/palladium print. Much like a single line of notes in a fugue by Johann Sebastian Bach (1685–1750), these facets overlay each other and together form a grand harmony, or, as Paul Caponigro (1932) described it, the distinctive "voice of the print."[6]

The signature components of platinum/palladium prints (intensity, contrast, tone, edge, resolution, and volume) activate forms that are visible only because they are expressed in this medium.

Intensity and Contrast

The intensity of a platinum/palladium print is a potent quality, forceful and full of presence. It has an ability to persist in just the dark values, or just the light ones, or across the entire tonal range, resulting in the capacity to hold a viewer's attention. Other printing processes rarely sustain this kind of dynamic. Overall print contrast, whether low or high, is expressed by a scale of tones that may vary from soft and subtle gradations

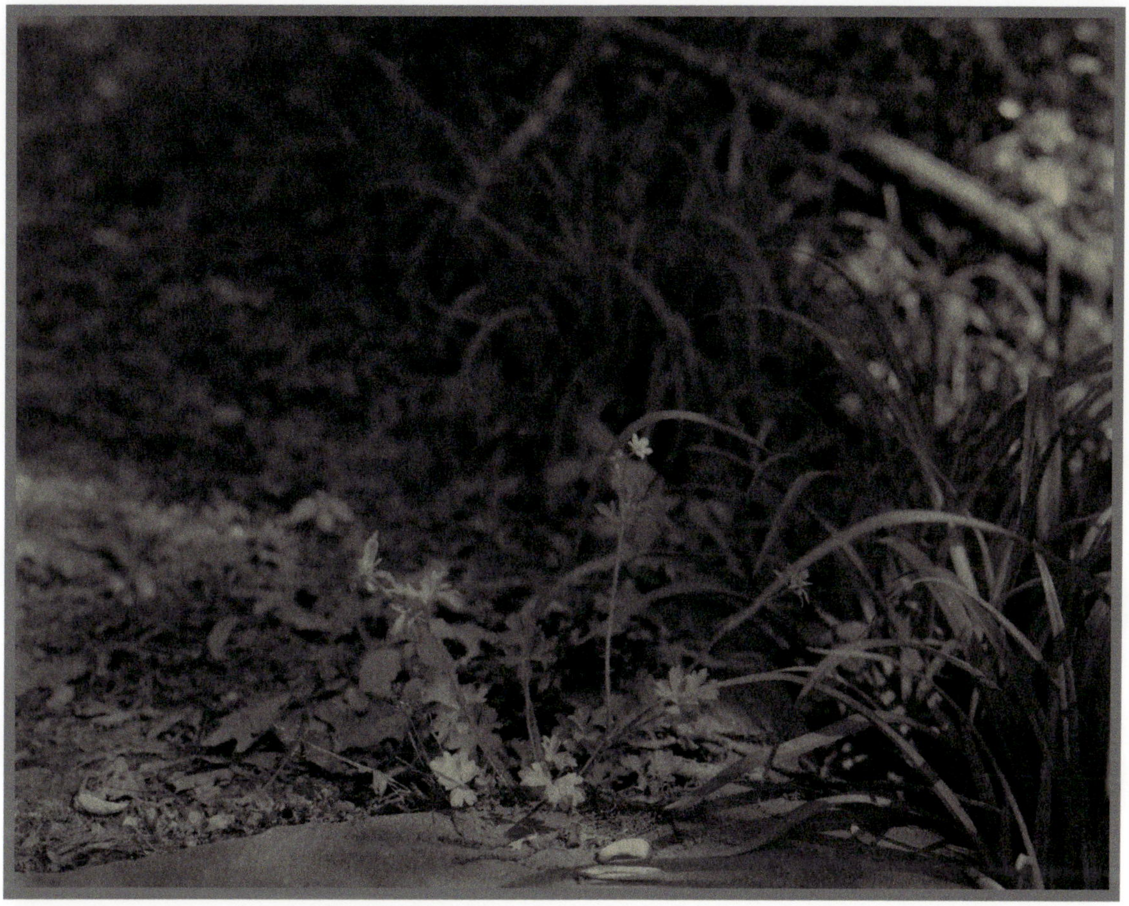

Figure 10.7 Ashley Block: *Betty Lou*, April 24, 2012. Platinum-palladium print on Weston Parchment, from original 4 x 5 inch negative, 9.5 x 12 cm. Courtesy Pradip Malde. The platinum/palladium photograph is about making: about making something so well that self-expression evaporates in the presence of the thing made, that interpretation becomes just another act in the drama of experience, and technique simply becomes the flame in the lamp.

to abrupt switches from white to black and back again. With platinum/palladium prints, this distinctive tension between intensity and contrast generates rippling tones and eddies that are like emotional turbulence.

Tone

Tonal qualities in platinum/palladium prints are influenced by the intricate physical structures to which these metals are bound. Color is influenced by the way paper, along with the other components of the image, absorbs and reflects light. It is directly related to the nanoparticle size of platinum and palladium and the shape, quality, and tone of the cellulose structure in which they reside. The microscopic structure of the paper consists of translucent matrices of cellulose that absorb, reflect, and scatter light to varying degrees. The color of the paper may range from bright white to warmer shades of off-white and buff. Its surface quality, which may range from very smooth to highly textured, further influences the appearance of a print. All of this plays out in a dimensional space that is unlike the optically homogeneous layer of emulsion-based

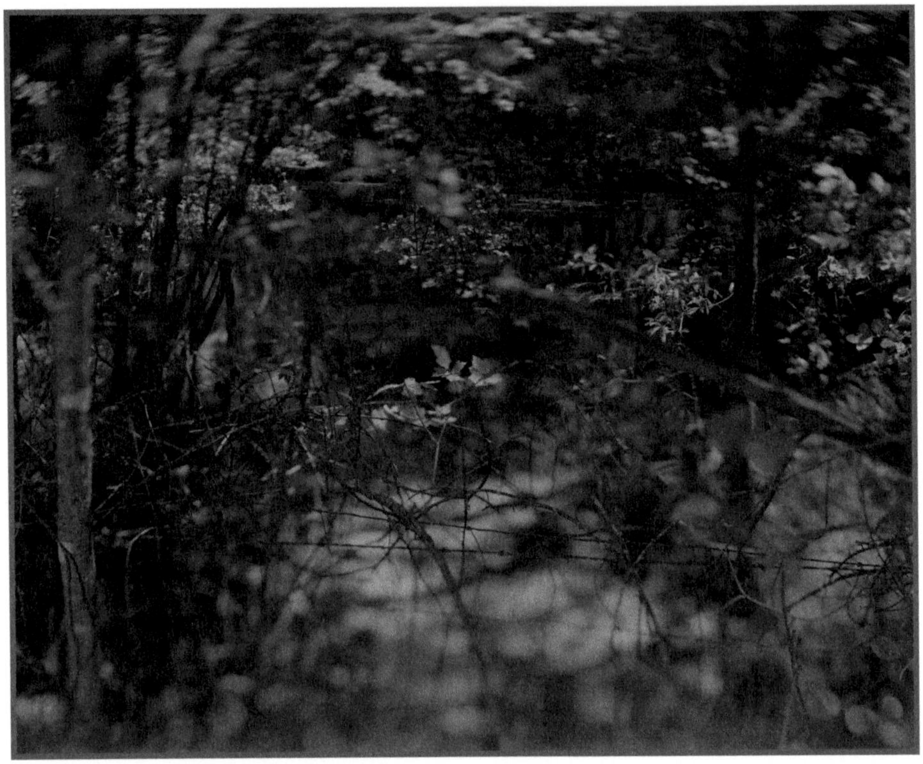

Figure 10.8 Chandler Sowden, *Prairie Chapel Road, Decherd, Tennessee*, 2012. Platinum-palladium print on Weston Parchment, from original 8 x 10 inch negative, 19.5 x 24.5 cm. Courtesy Pradip Malde.

prints. It is therefore misleading to think about a platinum/palladium print as being monochromatic. It can be replete with a nuanced range of color which serves as a delightful expression of how light is absorbed and reflected by metal atoms and cellulose fibers.

Edge

Gazing at a fine platinum/palladium print is a patient activity, and one that is needed to access its velvety qualities. A Middle Eastern dessert known as *qata'if* is described as the food for those who are patient.[7] *Qata'if*, derived from the Arabic root meaning "velvet," is a term often used to describe the way edges are rendered in platinum/palladium prints. An edge defines the way one tonal zone, or territory, is distinguished from another. Rather than describing the degree to which an edge is "sharp" (the more traditional understanding of edge), this process yields edges as rounded and flowing, velvety, and ethereal, even when the image is clearly delineated and sharply focused. The ability to recognize this kind of edge promises rewards for those who spend time with a platinum/palladium print and leads to an appreciation of the process's peculiar display of resolution.

Resolution

Resolution is a quality that is controlled by the chemical and physical aspects of diffusion of substances within the substrate. It is not to be mistaken for sharpness of focus or film grain. Rather, it describes how the flow of tones in the negative splice with the dimensionality of the metallic and organic structures in the print. Resolution of detail is also related to the size of the original negative and the need for an exposure to be made with the negative in contact with the printing paper. When a print is closely and patiently scrutinized, these highly resolved aspects of the image begin to unveil themselves.

Volume

Many, if not all, of these formal components and qualities exist in a three-dimensional space, or volume. The platinum/palladium print is produced by the presence of noble metals, in clusters of varying size, amid the interwoven cellulosic mat of the paper. Cellulose is normally quite transparent but appears white because of the scattering effect of light as it passes through the surface layers and reflects back out. The metals that form the image are opaque, and absorb and reflect varying wavelengths of light. Just as the print is not monochromatic, it is not two-dimensional. All plain-paper siderotype prints have a depth or volume to their tonal spaces, but platinum/palladium prints seem to have an uncommon spatial fingerprint. When considering the print in terms of volume, it also helps to think of it in terms that are similarly multidimensional, as in the following section.

Form, Pattern, Harmony, Texture, and Delicacy

Form, in visual terms, is frequently understood as being three-dimensional and provides a critical framework for considering what is seen, along with how the components of an image relate to each other. As with all expressive media, the form of a platinum/palladium print is distinguished by its technical characteristics, which may be manipulated to direct the viewing experience towards an aesthetic. The signature components of platinum/palladium prints (intensity, contrast, tone, edge, resolution, and volume) activate forms that are visible only because they are expressed in this medium.

The formal relationships within an image are seen within a context, or set of patterns, that help describe and express experience in poetic terms. *Pattern* is seen in particular kinds of repetitive symmetry that are almost independent of what was actually photographed. In other words, patterns provide evidence that something exists, even if it is not seen. The way one establishes patterns in a photographic print can be so assertive that we see what may not have been photographed, and these actions determine, in distinct and wondrous ways, what *is* envisioned through the print.

A vocabulary of terms is required to express the phenomenon of how visible and invisible layers interact with perceptual and physical layers. *Harmony* is a term borrowed from sound and music. In platinum and palladium printing it is understood as the whole being greater than the sum of its parts, with the parts, or layers, modulating and modifying each other. The visible layers interact with the perceptual and physical layers: what is in the negative and what is rendered in the print are the result of these parts. These overlays may or may not be compatible with each other, but, just as in the improvisatory music of Thelonious Monk (1917–1982) or Ludwig van Beethoven (1770–1827), the fact of having been brought together *in terms of the medium* creates a harmonious whole.

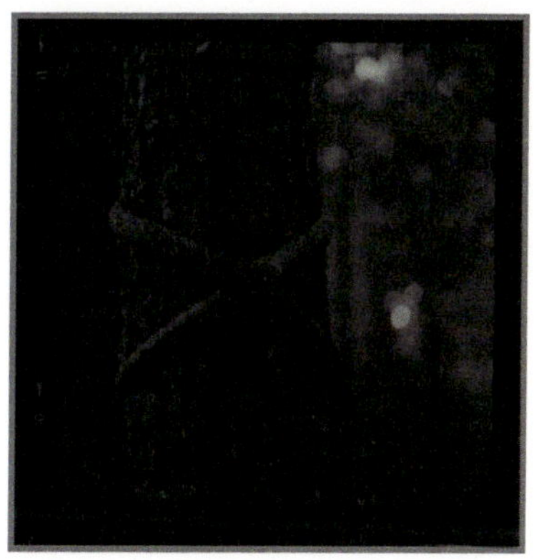

Figure 10.9 Brooke Irvine, *Crossed Vines, Sewanee, Tennessee*, September, 2014. Platinum-palladium print on Weston Parchment, from original 6 x 6 cm negative, 6 x 6 cm. Courtesy Pradip Malde.

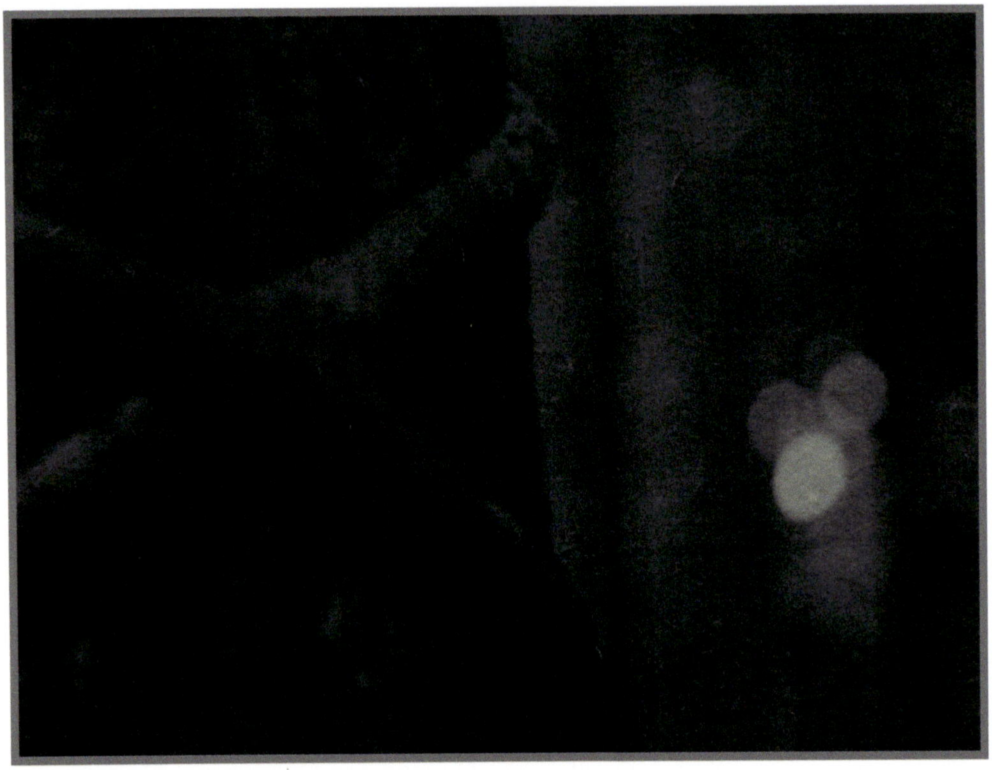

Figure 10.10 Detail: Brooke Irvine, *Crossed Vines, Sewanee, Tennessee*, September, 2014. Irvine's print is mostly dark, with just three or four luminous circles on the right providing some relief. The surface and swarming cellulose fibers form the paper as well as the tree trunk. The heavy black parts of the image are unyielding in the sense that they do not easily let the viewer's eyes wander away. Still, the lighter, glowing circles, hovering just above the print's surface seem to lead us under and behind the surrounding darkness. One becomes enraptured by looking. This tiny contact print is composed of mostly darker midtones, with few truly black or white values. Despite its tonal limits and diminutive size, the print provides an impression of a full tonal gamut. The oily black details unveil themselves with a steady continuity of tone and texture all the way through to the brightest areas of the print.

As the visual experience not only shifts across the surface, but also enters into the dimensional expression of the print, we encounter *texture*. Texture includes some of the most singular parts of photography: what the image is of, how the tonal information interplays with the paper, and how it ultimately leads us to consider what is the image about. Paper presents itself not just as surface but also as a kind of interior. Together, the print renders as singular texture, one that is at the same time ordered and chaotic, an unchanging formlessness.[8]

The dance between order and chaos is a delicate one. *Delicacy*, as a visual experience of the platinum/palladium print, is largely shaped by resolution, described above. Delicacy is seen in fine nuances of tone, where there are few absolute values, where the continuum is valued more than the end points, where the flow from one state to another seems proper and according to an innate sequence.

Making and Seeing the Platinum/Palladium Print

I write as a photographer. As I gaze at the tiny, 5.5 cm x 5.5 cm palladium print beside me, by Brooke Irvine, "Crossed Vines, Sewanee, Tennessee," and then out through a window to some shimmering leaves in a cluster of cottonwoods, I flip to a memory: it is 1964, and

I am skipping over iridescent oil-scummed water puddles in a rubble-filled alley in my hometown of Arusha, Tanzania. This combination of luminous shadow tones in the print and the rippling leaves outside my window connect with a childhood memory about iridescence.

Irvine's print is mostly dark, with just three or four luminous circles on the right providing some relief, some place to move toward. I look at the print through the filter of my childhood memory. This much is familiar to me when regarding a photograph: I am seeing through the photographer, the lens, and the print; I know my eyes are dancing between one part of the image and another. I see varying densities and cellulose fibers forming on the paper as well as the tree trunk. The heavy black tones of the print demand my attention, and still my eyes slide across to the lighter, glowing circles, then back to the darkness, glancing not just over the print's surface, but under and into the surface. I see things that were not photographed but are certainly and clearly in the print. It is only because of the distinct physical nature of the object, the print, that I form this particular sense of what the image is about. I cannot look away. The process of looking has enraptured me.

Ultimately, the platinum/palladium photograph is about making: about making something so well that self-expression evaporates in the presence of the thing made, that interpretation becomes just another act in the drama of experience, and technique simply becomes the flame in the lamp. The platinum/palladium print is more a poem. Unlike most other photographic images and processes, the platinum/palladium print invites a constantly unfolding scrutiny, a conversation without words—and recursively, the making of a photograph is conditioned by knowing it may be printed in platinum/palladium. The print's vernacular deserves careful attention so that we may better partake of the "gaze that binds us to the world."[9]

Notes

1 Peter H. Emerson and T. F. Goodall. *Life and Landscape on the Norfolk Broads*. London: Sampson Low, Marston, Searle and Rivington, 1886. See also Philippa Wright and John Taylor, "Peter Henry Emerson's Platinum Prints and Photogravures," in *Platinum and Palladium Photographs: Technical History, Connoisseurship, and Preservation*, edited by Constance McCabe, Washington D.C.: Photographic Materials Group of the American Institute for Conservation of Historic and Artistic Works, 2017, pp. 290–301.

2 Nancy Newhall, *P. H. Emerson: The Fight for Photography as a Fine Art*. New York: Aperture, 1975, p. 28.

3 Susan Rennie (ed.), *Historical Thesaurus of Scots*. https://scotsthesaurus.org, accessed 12/30/2019.

4 Sarah S. Wagner, "Manufactured Platinum and *Faux Platinum* Papers, 1880s–1920s", in McCabe, *Platinum and Palladium Photographs*, 2017, pp. 144–183.

5 See Tatiana Cole, "The Platinum Renaissance: Oral Histories of Platinum-Palladium Printers and Artists," in McCabe, *Platinum and Palladium Photographs*, 2017, pp. 432–449.

6 Paul Caponigro, *The Voice of the Print*, Williamsburg, Mass.: Allethaire Press, 1990.

7 Habib Salloum, Muna Salloum, and Leila Salloum Elias, *Sweet Delights from a Thousand and One Nights: The Story of Traditional Arab Sweets*. London: I. B. Tauris, 2013.

8 Sōetsu Yanagi, *The Unknown Craftsman*. Tokyo: Kodansha International, 1972.

9 Mark Doty, *Still Life with Oysters and Lemon*. Boston: Beacon Press, 2001, p. 50.

PART 2

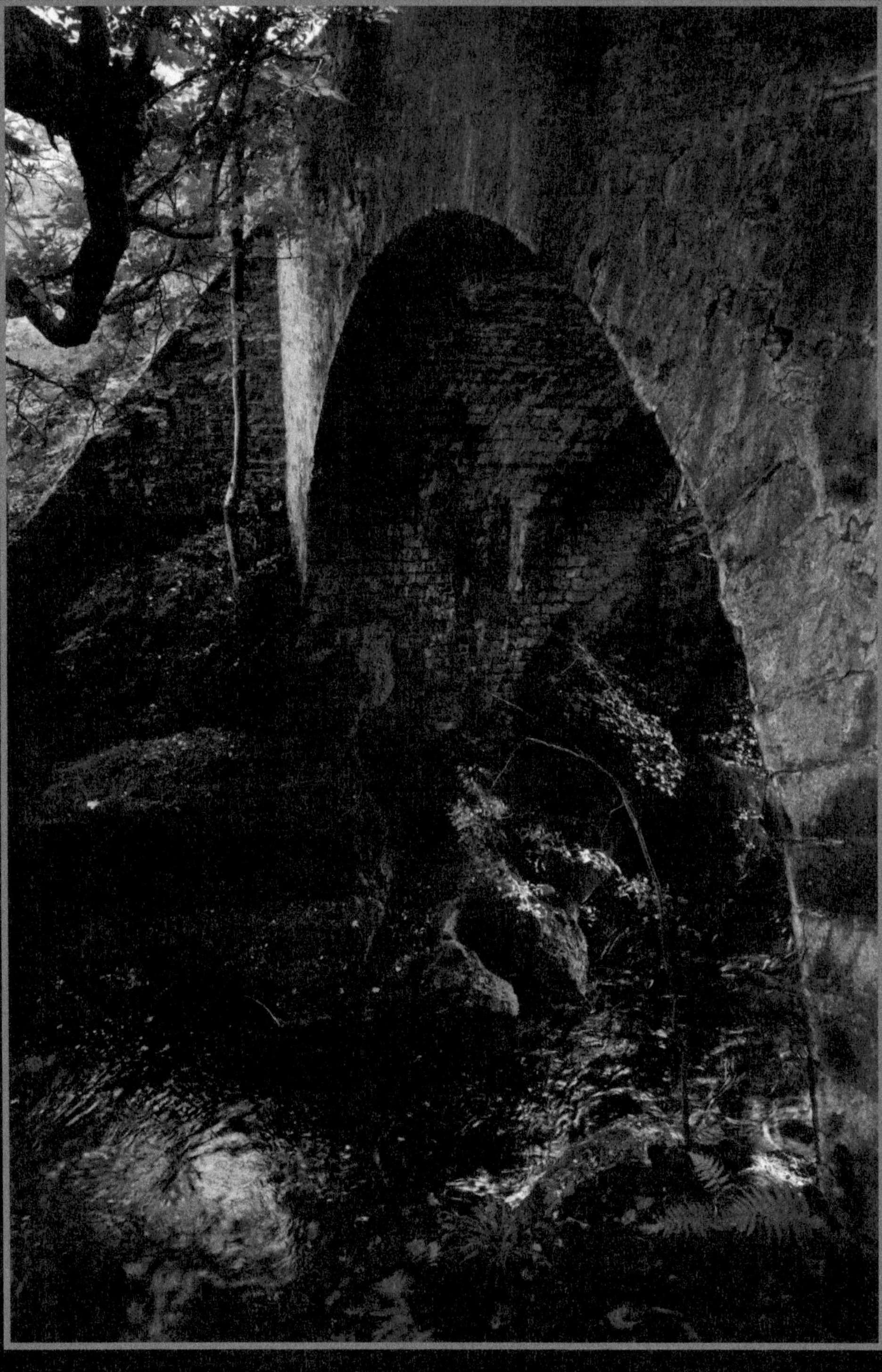

Figure 11.0 Mike Ware, *Bridge at Buchanty, Glenalmond, Perthshire*, 1985. Platinum-palladium print on Revere Platinum from enlarged digital PiezoDN negative of 20.2 x 28.8 cm. Original. 6 x 9 cm. Fe3:Pt:Pd = 2:1:1, hydration for 90 min at 76% RH / 20°C Courtesy Mike Ware

Chapter 11

Contemporary Practices

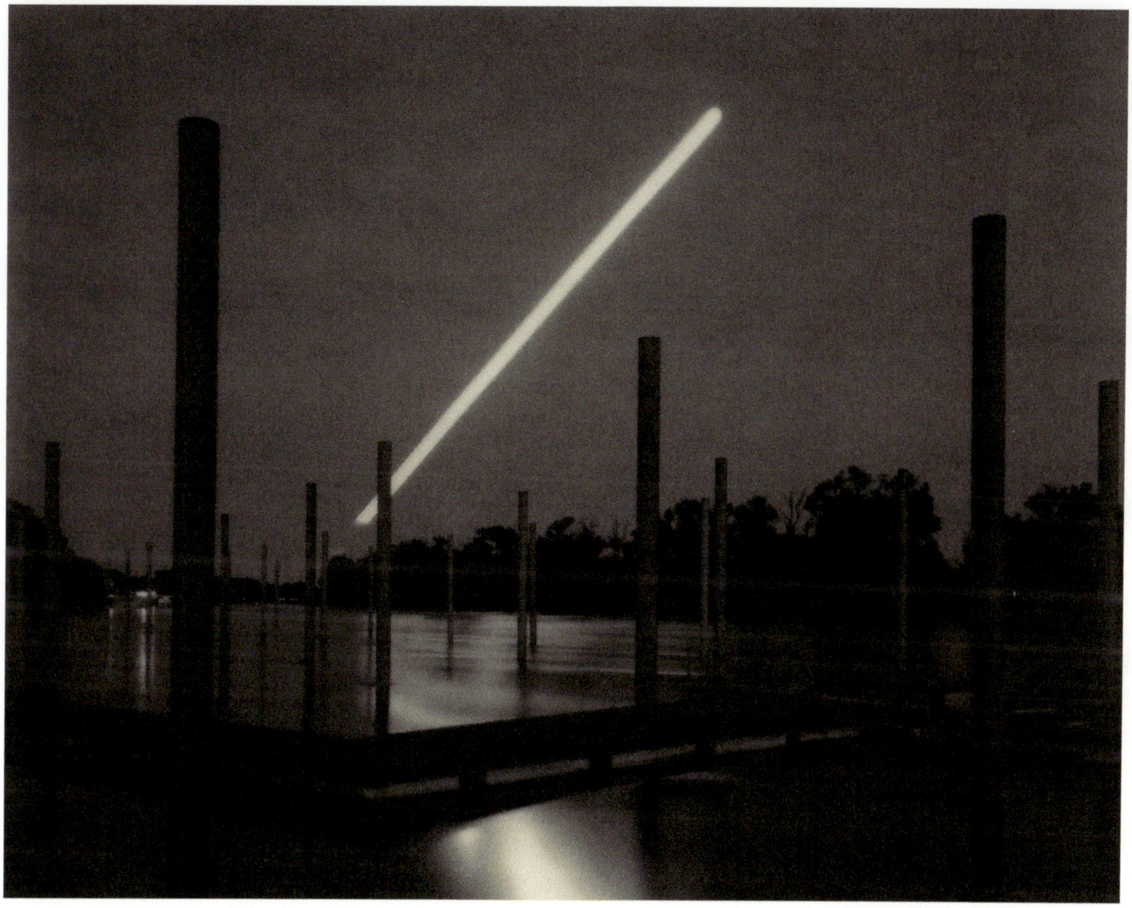

Figure 11.1 Roger Vail, *Sacramento River with Moon and Poles,* 1979. Platinum-palladium print on Weston Parchment paper, from original 8 x 10 inch negative developed in PMK Pyro, 20 x 24.5 cm. Courtesy Roger Vail. Fe1:Pt:Pd = 2:1:1, hydrated for 20 min. at 100% RH / 20°C.

June 7, 1982, Orkney, Scotland: Grey. My first platinum print, tenderly coated with a hake brush, and exposed under a humming mercury vapor bulb, now floated up in the potassium oxalate developer with barely an image. There had been weeks of preparation: conversations with my bank manager (negotiating an overdraft so I could buy chemicals and equipment); a grant from the Scottish Arts Council; my dear friend and photographer, Alf Magne Andreassen from

Norway beside me thanks to a hard-earned grant from the Norwegian Arts Council—and it came to this! A dark-gray print. Over the following months, Alf and I struggled to make a decent print, compelled only by the memory of platinum prints we had seen at the Hayward Gallery in London four years previously. Now, here we were, frustrated by and feeling ignorant about platinum printing.

Knowing what is possible and being inspired by the achievement of others is a kind of energy that pushes you forward at moments of despondency. It is a special form of hope, as John Berger suggests, that appears secretly, quietly, like contraband.[1] As we struggled to make those first prints, Alf and I—and a few months later, Mike Ware—persisted largely because we mentally recalled the work of Frederick Evans, Gertrude Käsebier, P.H. Emerson, Laura Gilpin and others. There was no internet, of course, and Orkney, even though it has the oldest public lending library in Scotland, offered limited access to print collections. Seeing platinum/palladium prints reproduced in books, leave alone first-hand, was a rare and precious experience.

Some time around August of 1982, I made the long trek to London for an appointment with the Print Collection at the Victoria and Albert Museum, to study their copy of Peter Henry Emerson's *Life and Landscape on the Norfolk Broads*, the legendary book with hand-tipped platinum prints. A puzzled assistant greeted me with news that the book could not be found, but was probably with the conservation department; could I come back another time perchance? Heartbroken, I promised myself that if I ever successfully made platinum prints, I would share, along with the recipes, well-reproduced prints that serve as a guide. Much has changed since that trip from Scotland to London. *Life and Landscape* may now be viewed in high resolution online,[2] and there is an abundant list of other resources and well-illustrated books related to platinum/palladium printing. *Platinotype: Making Photographs in Platinum and Palladium With a Contemporary Printing-out Process*, the result of Mike Ware's and my collaboration since late 1982, has been guided by that early pledge. So, we conclude with a chapter of examples of printing-out platinotypes. I have selected a broad range of work, hoping that the collection serves as a cynosure on several fronts. Platinotype, loosely defined, has not fully shed its Pictorialist "look": the propensity for soft focus, for human figure-work, landscape, portraits, and still life. Certainly, there are justifiable reasons for this. The medium lends itself to a nuanced tonal range, and therefore to renderings of human skin and fluid landscape. Platinum/palladium prints, especially printing-out versions, have a velvety quality that can mesmerize a viewer; and being formed from precious metals, they create a compelling cachet for collectors. In sum, these properties can change the way the medium is used in contemporary photographic practice and vice versa, and I hope these shifts are apparent in the following pages. Documentary work, printed in platinum/palladium, brings us to consider both the "vent" and the conveyance (the print), to consciously traverse the difficult territory between privilege and access as in works by Virgil DiBiase and Lucy Wimmer. Highly conceptual and political approaches, as taken by Angela Berry, Shaun O'Dell, and Paula Luttringer, become subjective and visceral through the materiality of the process. The painterly and magical realms also find their voice here, unveiled in prints such as those by Cathy Cone, Pamela Heemskerk, and Gilles Lorin. These approaches, though they seem to be on the periphery of platinotype orthodoxy, don't ever lose their relevance to more familiar expressive uses of the medium, as we see in Imogen Cunningham's or Ellie Young's work. My point is that the physical characteristics of a well-seen and printed image, and the ensuing experience of looking at a platinum/palladium print are captivating. This experience has the capacity—like any effective artistic experience—to transform us, and it is for this reason that I feel we, as photographer/printers,

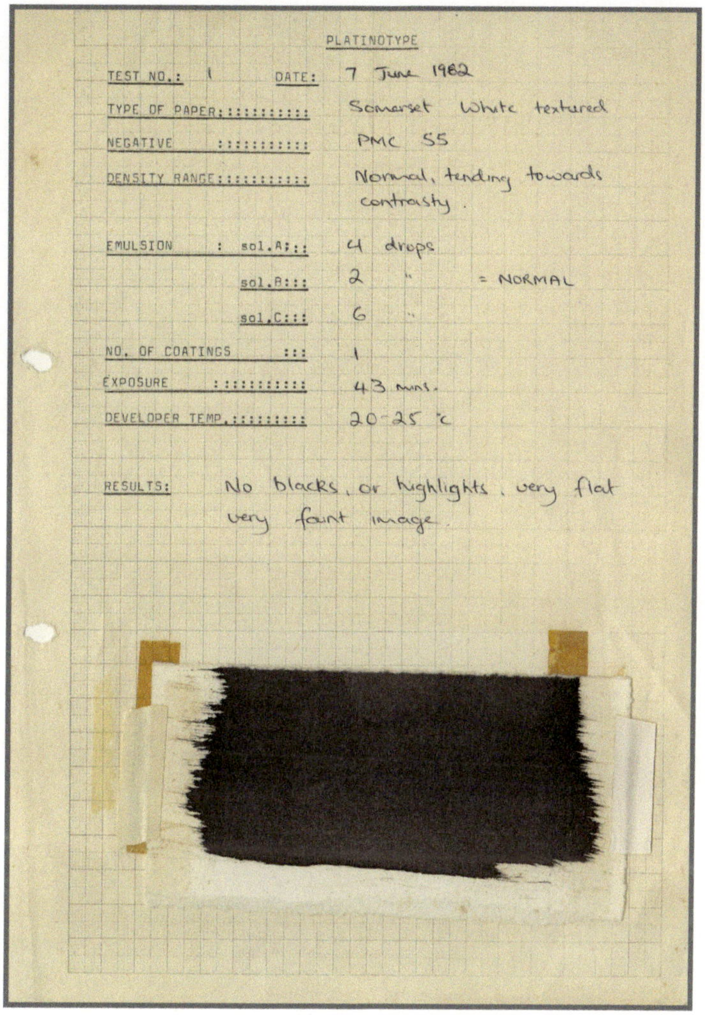

PLATINOTYPE

TEST NO.: 1 DATE: 7 June 1982

TYPE OF PAPER:::::::::: Somerset White textured

NEGATIVE :::::::::: PMC 55

DENSITY RANGE:::::::::: Normal, tending towards contrasty.

EMULSION : sol.A::: 4 drops

sol.B::: 2 " = NORMAL

sol.C::: 6 "

NO. OF COATINGS ::: 1

EXPOSURE :::::::::: 43 mins.

DEVELOPER TEMP.:::::::: 20-25 °c

RESULTS: No blacks, or highlights, very flat very faint image.

Figure 11.2 Pradip Malde, the first platinum print, June 7, 1982. Data sheet and platinum print on Somerset paper, made with Alf Magne Andreassen. Development Pt print, with ferric oxalate and potassium tetrachloroplatinate(II) on Somerset White Textured paper, and developed in potassium oxalate.

have a moral imperative. What follows here are examples of not just technical prowess, but of process being harmoniously absorbed into the creative act of seeking and discovering. The experience of the beautiful is unifying, and is the keel that keeps us balanced, that stabilizes us through darkness and dilemma. Roger Vail speaks of this imperative in another way, "what we photographers do in the moment is very much like a form of meditation, that is, [we] enter into states of heightened awareness... And in those moments of discovery and making there is no distinction between form and content..."[3] What follows is a collection of fine work—mostly contemporary. Eighty printing-out platinum/palladium prints represent the work, ideas and methods of accomplished practitioners. Each folio includes the artist's statement, advice for the reader, a biography, and process notes.

Alf Magne Andreassen

I grew up in Angola, and after 20 years, moved back to Norway with my parents. I hardly understood my mother tongue, and spoke better English than Norwegian. I have always seen the world around me as a stranger. My photographic studies at college in Bournemouth (1979), where I also met Pradip Malde, gave me a way of expressing this. A few years after the end of our studies, in 1982, Pradip began to talk about platinum printing and for a period of time we tried to discover its secrets. Sadly for me, this did not last long. I made platinum/palladium prints between 1982 and 1985, and I have never made prints since then. But maybe this will change.

Advice about making printing-out Pt/Pd prints:

If someone wants to work with platinum/palladium, remember, it is expensive, but it is very, very beautiful. It breathes.

Alf Magne Andreassen continues to live and work from Oslo, Norway, as a photographer. He has been commercially and artistically recognized for his distinctive imagery which borders on magical realism. He has taught at the Oslo National Academy of Arts and the Norwegian National Academy of Craft and Industry.

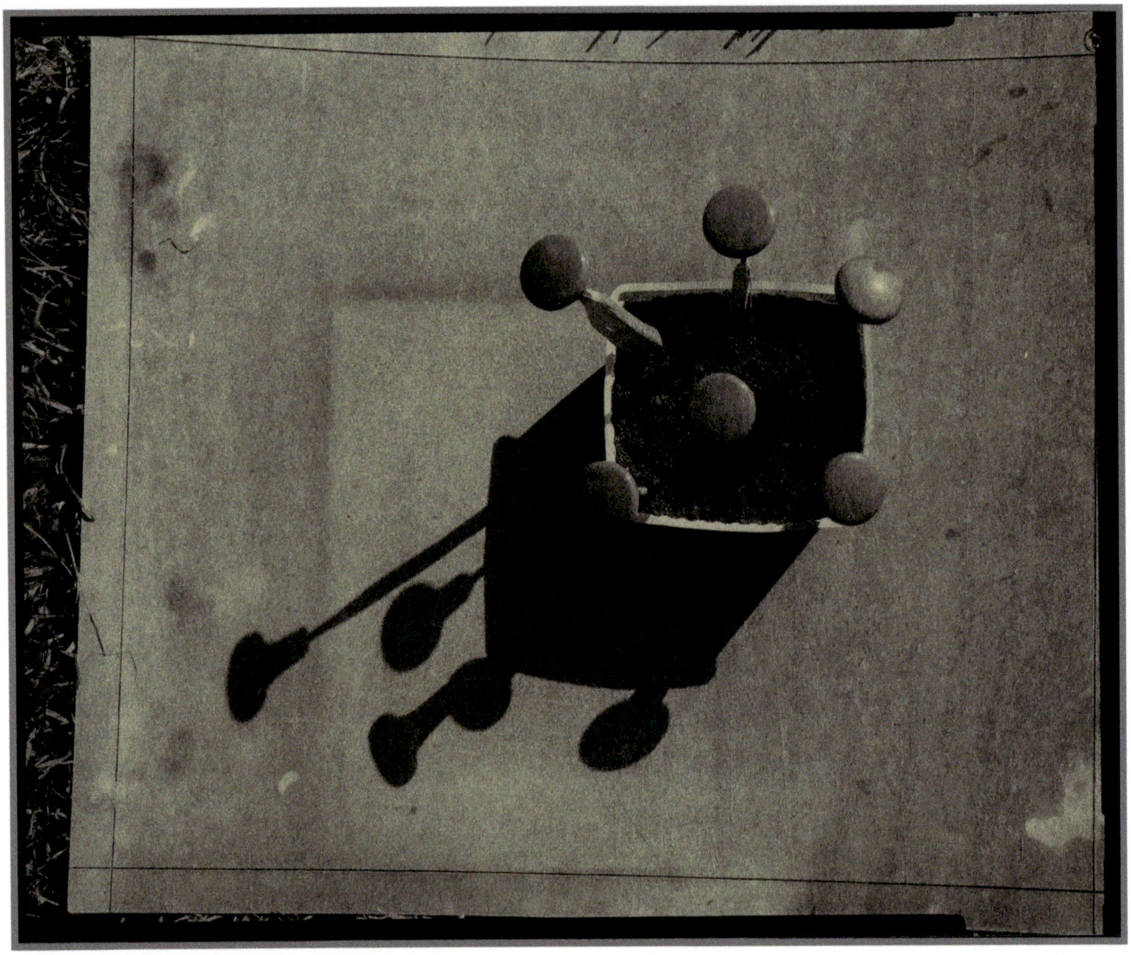

Figure 11.3 Alf Magne Andreassen, *Untitled*, 1981, printed 1984. Platinum-palladium print on Van Gelder Simili Japon paper, from original 10 x 12.5 cm negative, 10 x 12.5 cm. Courtesy the artist.

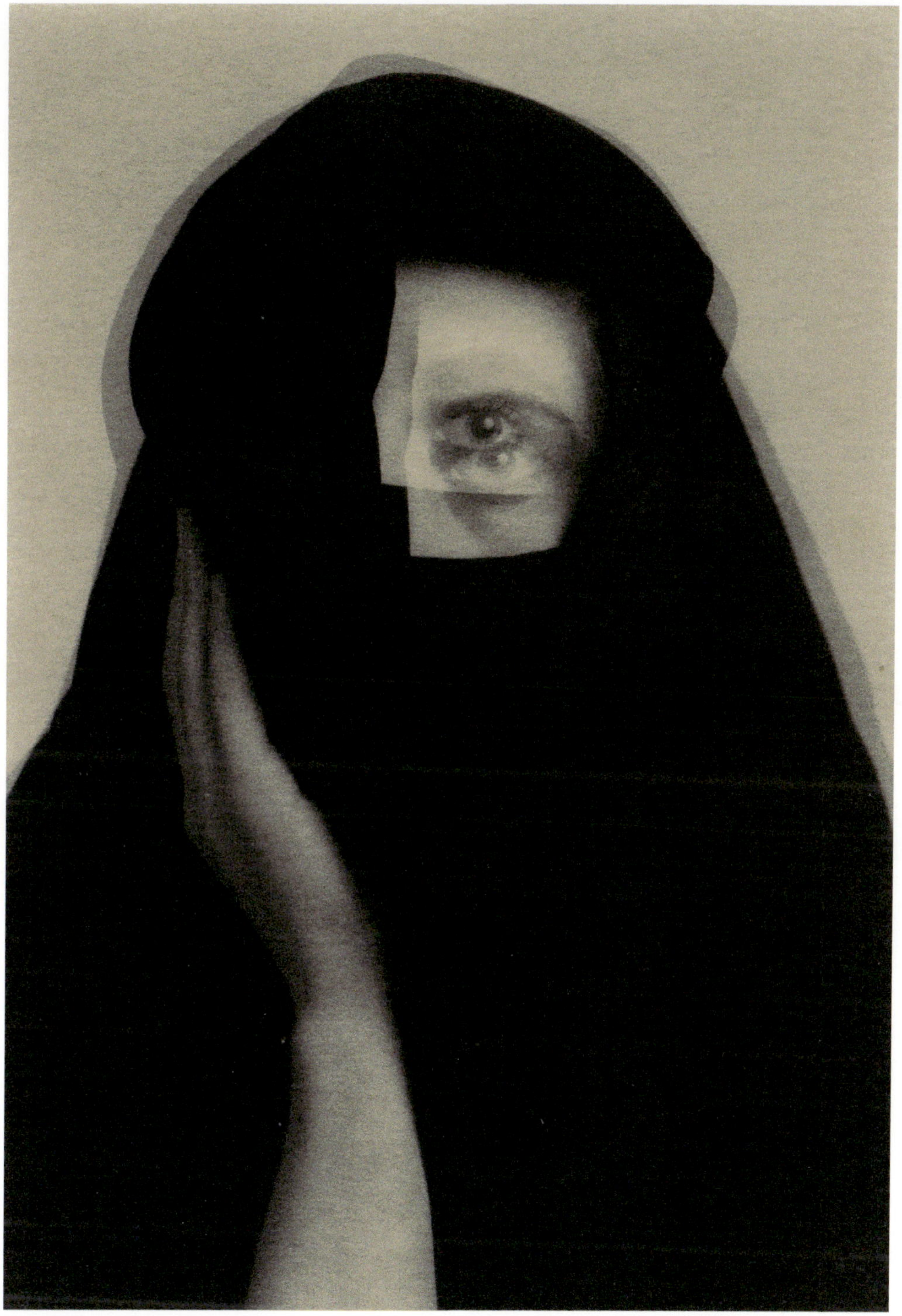

Figure 11.4 Alf Magne Andreassen, *Untitled*, 1978. Printed 1984. Platinum-palladium print on Van Gelder Simili Japon paper, from original 10 x 12.5 cm negative, 10 x 12.5 cm. Courtesy the artist.

Angela Berry

My work examines the attribution of human traits, needs, intentions, and feelings to non-human entities. I am influenced by pioneering feminist Mary Wollstonecraft's critique of the naturalistic system of representation championed by her male contemporaries like Edmund Burke; which she argued problematically omitted the social, material, and physical conditions of the view like labor, industry, trauma, and history to project monolithic ideas of power and land ownership. In my work historical elements of the "picturesque" like animals, plants, meadows, rocks, monuments, natural phenomenon, and fortresses are reimagined formally and conceptually to focus on the collapse of the anthropomorphic mystic and the drama of co-regulating oneself with other life forms. Recent work has investigated fault lines and fissures on either boundary of the North American Plate in San Diego County and Iceland, eruptions from paintings collected by volcanologist and geochemist Haraldur Sigurðsson exhibited at his Volcano Museum, violent interactions between animals and drones, thermonastic leaf movements of rhododendron maximum, *and the real and fictionalized life of the 1925 US court case The Scopes Monkey Trial. The series represented in this volume,* East of the Sun *explores fortification, vulnerability, and captivity, and is titled from the Norwegian fairy tale* East of the Sun and West of the Moon. *The tonal range and physical character of the subject matter along with the rhythms of waiting and working in fortified time—the formal properties and content—led me to produce printing-out Pt-Pd prints for the series.*

Advice about making printing-out Pt/Pd prints:

Darkroom state of mind meets the ease of a Light-room process that produces an impressive range of tones and absorbent surface. The cost of platinum and palladium fluctuates and can be expensive, especially for larger sized prints or editions. Control and keep track of humidity during print runs, and consider working from a finely tuned digital negative.

Angela Berry is a San Francisco based artist and educator. Her work has been exhibited at The Ganza Art Lab, Tokyo; Eldfjallasafn / Volcano Museum, Stykkishólmur, Iceland; CTRL+SHFT, Oakland; Jewish Community Center of San Francisco; Ogden Museum of Southern Art, New Orleans; Contemporary Arts Center, New Orleans; Jonathan Ferrara Gallery, New Orleans; and Vox Populi, Philadelphia among others. She has received artist and research residencies at the Vermont Studio Center (2013) and Vatnasafn / Library of Water (2017). Berry is a former artist-member of The Front collective in New Orleans (2012–2015). She received an MFA in Fine Art (2016) and an MFA in Visual & Critical Studies from California College of the Arts (2017), where she was awarded the Bernard Osher Scholarship (2014–2017), and a BA from The University of the South (2008). She has taught at the California College of the Arts (2017–2018), The University of the South (2018), and the San Francisco Arts Institute (2020). Her work and curatorial collaborations have been reviewed in *Art In America*, *Artforum*, and *Art Papers*.

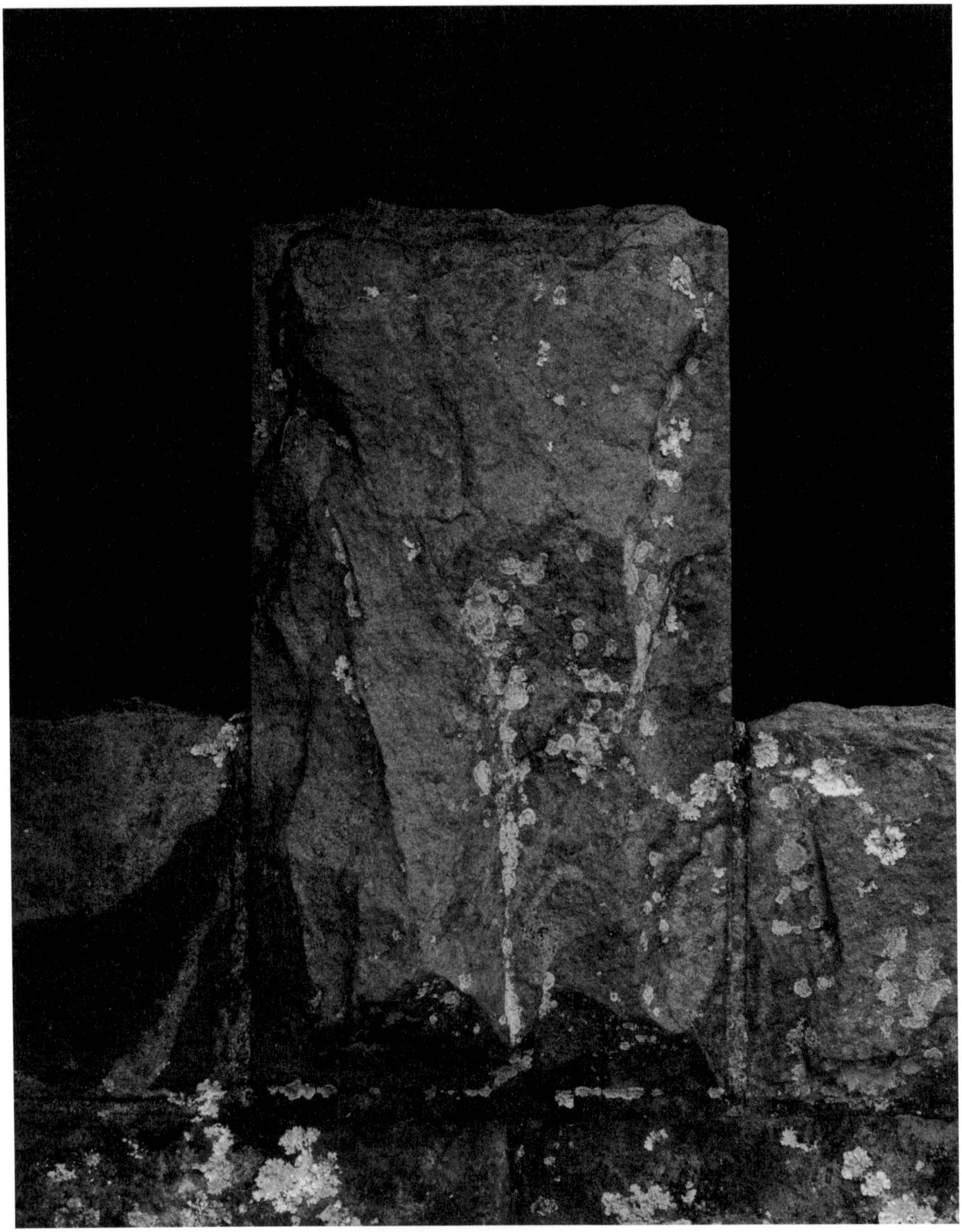

Figure 11.5 Angela Berry, *East of the Sun #32*, 2020. Platinum-palladium print on Revere Platinum paper, from PiezoDn negative of full frame digital original 25 x 20 cm. Courtesy Angela Berry.
Fe:Pt:Pd = 2:1:1, hydrated for 90 min. at 76% RH / 20°C.

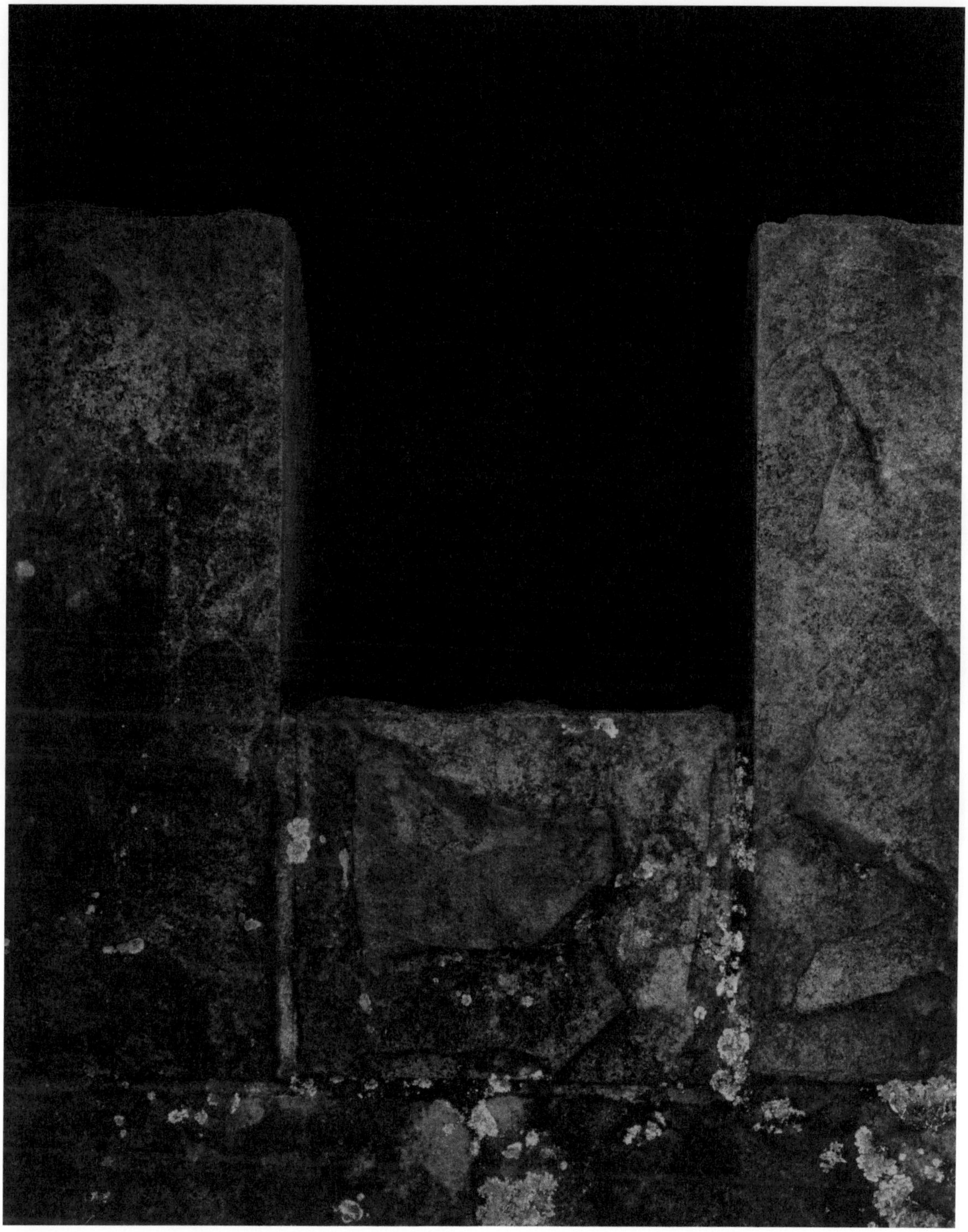

Figure 11.6 Angela Berry, *East of the Sun #26*, 2020. Platinum-palladium print on Revere Platinum paper, from PiezoDn digital negative of full frame digital original 25 x 20 cm. Courtesy Angela Berry.
Fe:Pt:Pd = 2:1:1, hydrated for 90 min. at 76% RH / 20°C.

Stuart Clook

My photography and print making is influenced by the Pictoralist and Tonalist movements of the late nineteenth century. These historical processes—renowned for their subtle tonal range, luminosity, and inherent permanence—are labor and time-intensive yet rewarding and addictive. I'm exploring analogue and alternative photographic processes to help me express my thoughts and feelings about the landscape that we live, work, and play in. These processes help me make my prints in a uniquely personal way and where I can use my hands in today's digital and machine centric world. My goal is to inspire those who see my work to look more carefully at the world around them and to discover beauty in the familiar and in the lesser known places around us

Advice about making printing-out Pt/Pd prints:

When printing a new image for the first time and my initial choice of paper has not given me the result I was looking for, I find that I can use the same (digital) negative to explore my other cotton rag, vellum, and gampi papers by inspecting the print during exposure and adjusting the time accordingly. This saves me time and materials, and I often arrive at a final result just by inspection. Secondly, compared to the development Pt/Pd process, I have much better control especially over the hue or color of the print. I achieve this by using one of a number of fixed ratios of Pt:Pd (100:0, 50:50, 20:80, 10:90 and 0:100) and standardize my humidification process by first drying the coated paper for two minutes under a gentle flow of low heat from a hair dryer and then hydrating it for 20 minutes over a tray of water at 20 °C. Don't spend too long trying to perfect step wedge prints. They are important in helping you dial in your digital negative, but my advice is to get close and then go ahead and make a print. There are many variables in platinum printing and some of these will only become apparent the more you print. You don't need to spend a huge amount of money on equipment to get started. There are several low-cost UV lamps on the internet and a print frame can be made from a piece of custom board, window glass, and bulldogs clips. Use plastic storage containers for process trays and save your money for buying the chemistry.

Stuart Clook is a self-taught landscape photographer and printmaker who lives on the east coast of New Zealand's South Island. Working almost exclusively with medium and large format cameras, he prints his work using digital negatives and the photographic printing processes of the late 19th century. Stuart has held solo exhibitions in New Zealand and Australia and was the featured photographer in *On Landscape* magazine in 2019. Stuart's prints are held in private collections throughout New Zealand and Australia.

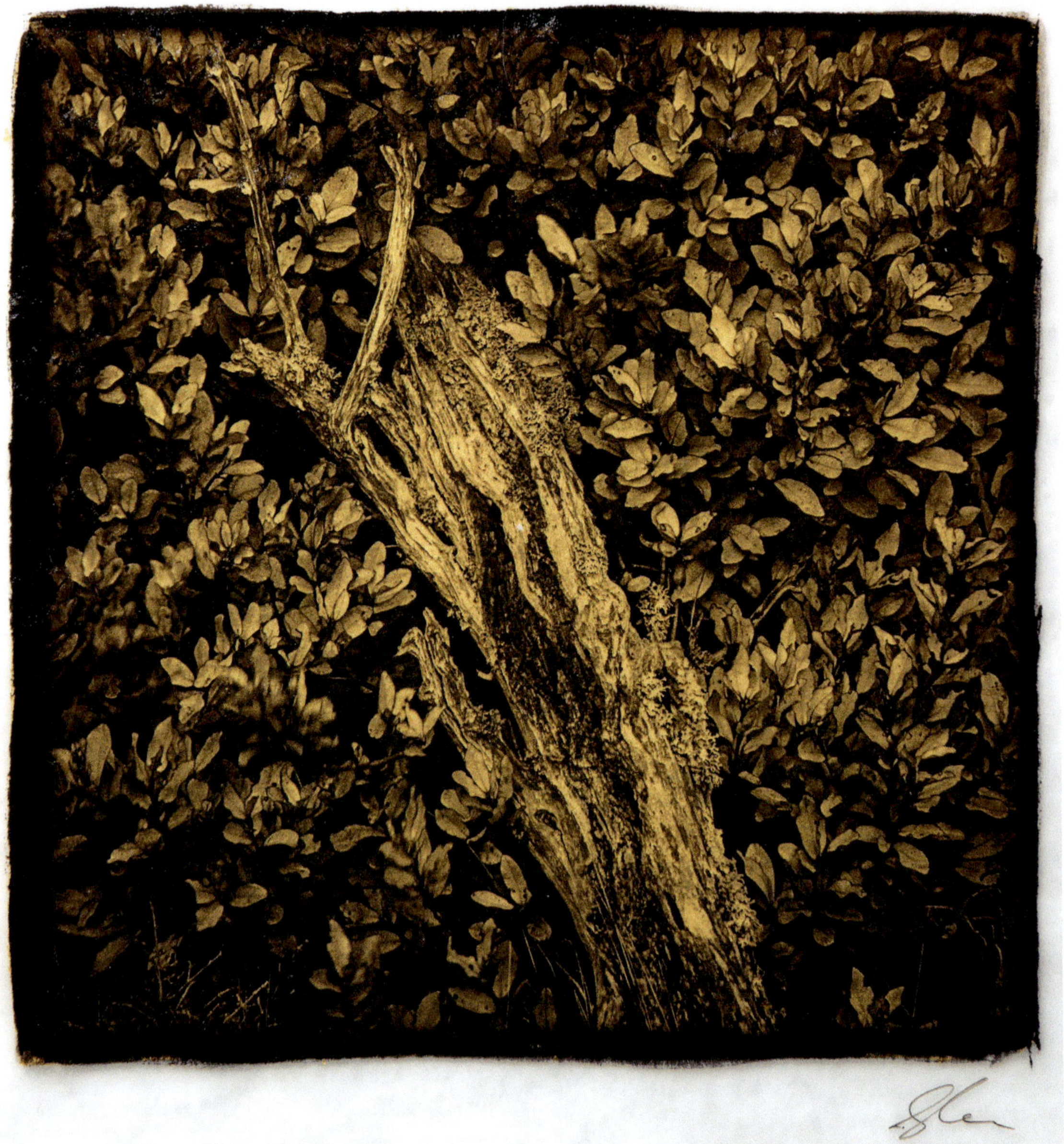

Figure 11.7 Stuart Clook, *Companions I*,
2019. Platinum-palladium print on Reich CT paper, from PiezoDN of a 6 x 6 cm original,
22 x 22 cm. Courtesy Stewart Clook. Fe:Pt:Pd = 2:1:1. After drying 24 ct gold leaf is gilded on to the reverse and then varnished.

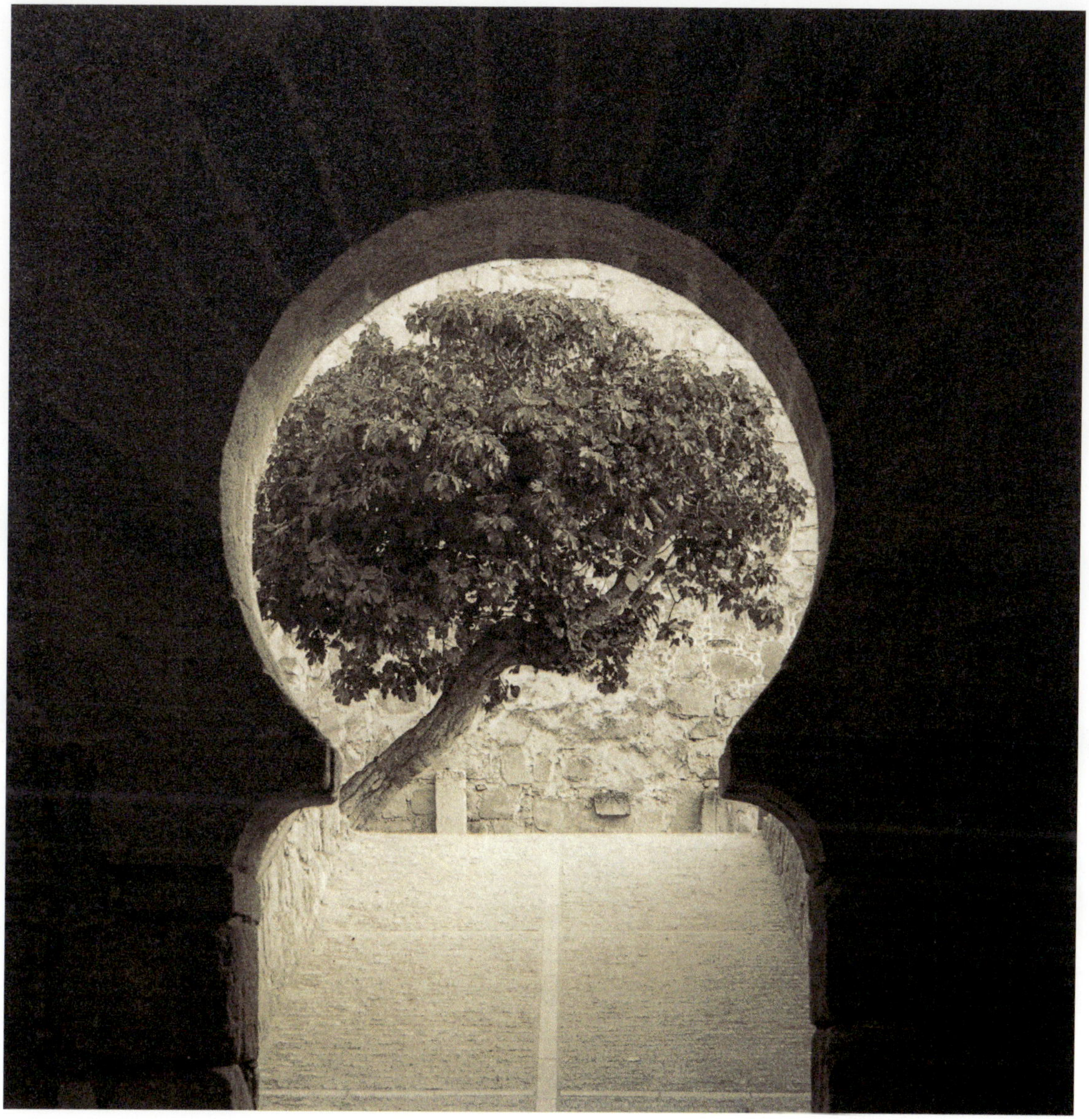

Figure 11.8 Stuart Clook, *Trujillo Castle*, 2018. Palladium print on Bergger COT 320 paper, from an enlarged digital negative of full frame digital original, 22 x 22 cm. Courtesy Stewart Clook.
Fe:Pt:Pd = 1:0:1.

Cathy Cone

I begin by finding an image in the world. The image emerges and appears as a rather strange, unexpected surprise. This is the evidence that I take back into my studio as "something" found and not unlike collecting specimens. I photograph so I can see what something looks like. The finding and constructing of what I understand and know, links up with what I don't know or can't see. If I'm lucky, they meld into a new image that lives on the threshold.

In terms of advice, I don't usually give it and prefer to share things I love. Heartsong. The printing-out process transforms the heart center of my images into suspended gems. The medium's patina reminds me of a delicate shell. Jon Cone and I have the privilege with our business to offer workshops and materials. Walker Blackwell, our colleague, and Jon are skilled technicians and I'm thankful for the osmosis they provide! I make photographs as a way to listen to my heart's song, and then I practice like hell to sing it.

Cathy Cone is a photographer and painter. Educated in programs at Ohio University, and at the Vermont Studio Center, she received her MFA from Maine Media College. Her work explores the liminal space inhabited by the relationship between what is known about the world and the parts of nature that exist just beyond the limits of human understanding. Her subjects are chosen by what speaks to her visually, and the process that unfolds is one of layered revelation. She describes her own artistic practice in a way that resembles a scientist's approach to collecting evidence or specimens from the natural world: "I take a slice of life through my camera lens in attempt to see what I missed." Cone, with her husband, master printer Jon Cone, founded Cone Editions Press (CEP), a collaborative printmaking workshop in 1980. She currently lives and works as director of the Workshops and Studio for CEP in East Topsham, Vermont. She has exhibited her work at the Weisman Art Museum, University of Alabama, DeCordova Museum, and the Vermont Center for Photography. Her work currently resides in the collections of IBM, MCI, Hallmark Fine Art Collections, American Express, and the Beekman (a Thompson Hotel in New York), among notable others.

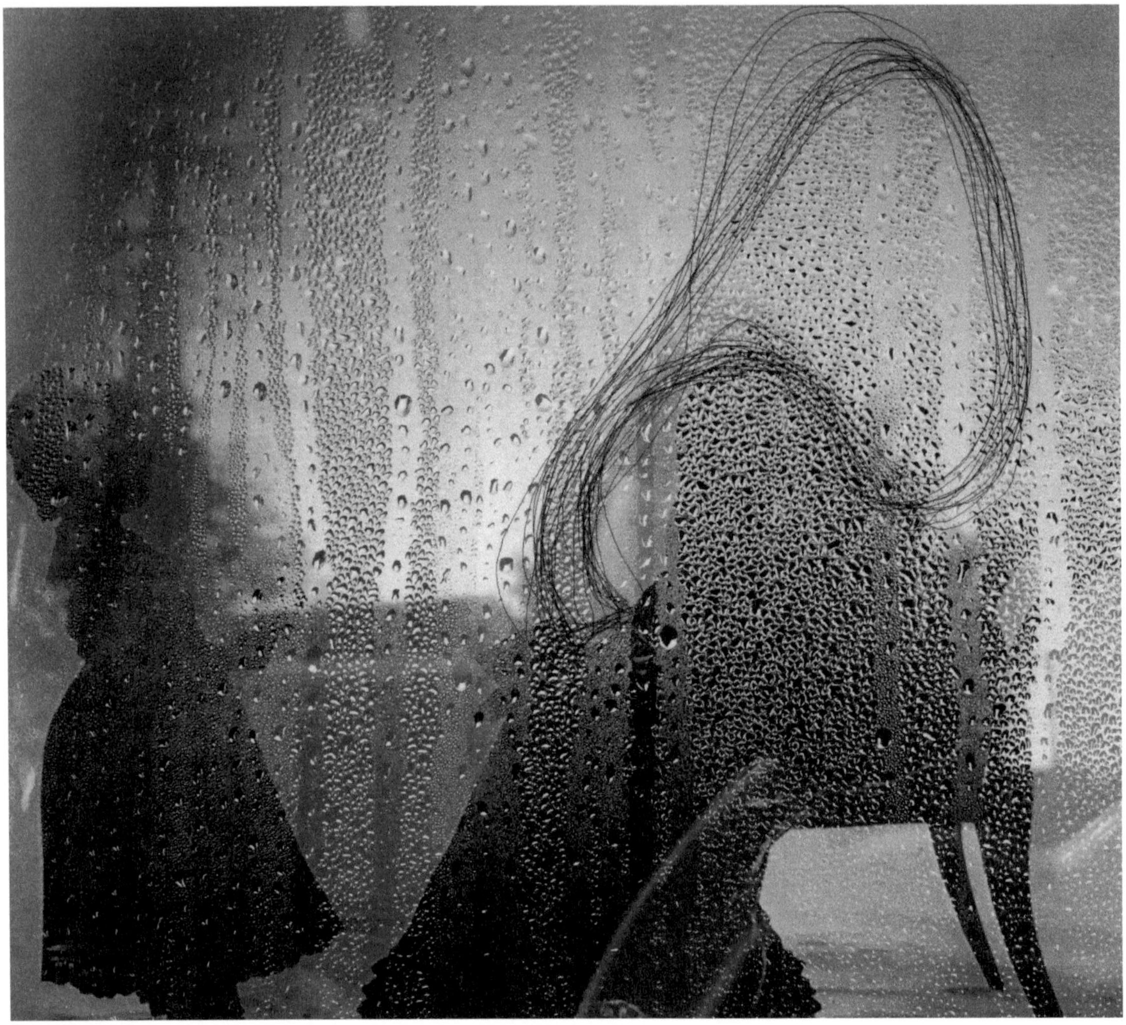

Figure 11.9 Cathy Cone, *Rewinding Forward, "Situation,"* 2019. Platinum-palladium print on Revere Platinum paper, from PiezoDN negative of full frame digital original,
23 x 19 cm. Courtesy Cathy Cone.
Fe3:Pt:Pd = 2:1:1, pre-coating humidity 50% for 2 hours at 68°F, rod coated, hydration 50% RH at 68°F, 60 minutes. Exposure using NuArc FT32. Post-exposure hydration 10 minutes 50% RH at 68°F. Cleared using the M-W clearing method.

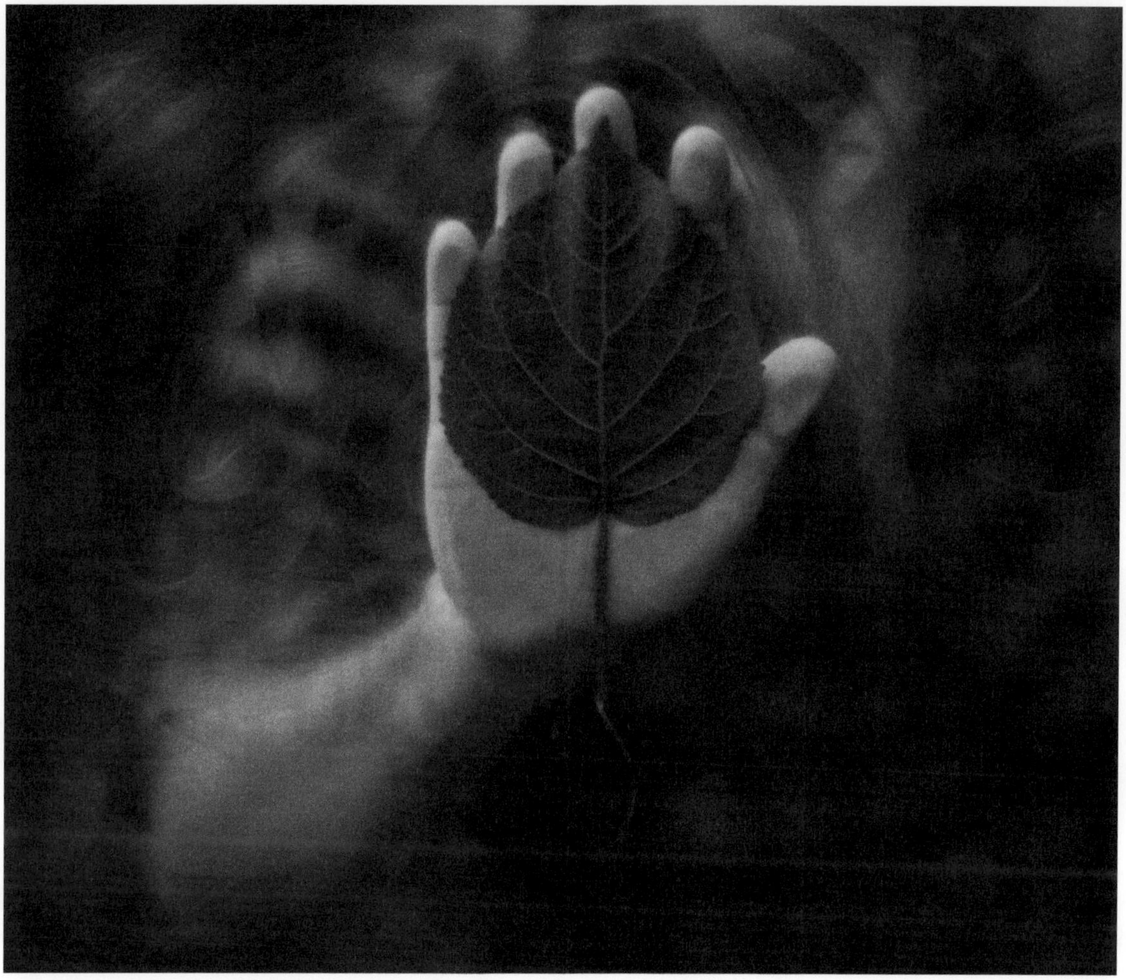

Figure 11.10 Cathy Cone, *Pledge*, 2019. Platinum-palladium print on Revere Platinum paper, from PiezoDN negative of full frame digital original,
23 x 20 cm. Courtesy Cathy Cone.
Fe3:Pt:Pd = 2:1:1, pre-coating humidity 50% for 2 hours at 68°F, rod coated, hydration 50% RH at 68°F, 60 minutes. Exposure using NuArc FT32. Post-exposure hydration 10 minutes 50% RH at 68°F. Cleared using the M-W clearing method.

Jon Cone

There are so many pros of the printing-out process that they are almost too numerous to list. However, I most appreciate not having to mix and work with ferric oxalate, or worrying about spots from stray iron in the development stage, or dealing with hot potassium oxalate developer in the first place. With printing-out I am able to more perfectly carry my calibrations in digital negative making because humidity is a guiding force in printing-out aesthetics and as a byproduct more perfectly controls speed and contrast. I enjoy the higher Dmax of printing out. And mostly, I love the smoothness gained by adding platinum, not to mention a much higher acuity.

Advice about making printing-out Pt/Pd prints:

The printing-out process requires precision in timing. Really all Pt/Pd printing does, but it is not taught. With the printing-out process as explained in this book, timing and humidity become everything. I would recommend, rather than using humidity tanks, that you find a way to precisely control the humidity in your processing room. With that level of control, you can make quite large prints by simply hanging them to humidify. The precision of the ammonium printing out process when coupled with PiezoDN digital negative making is so precise that we, as an editioning service, can create and share Soft Proof ICC profiles so that the photographers we print for can preview how their images will print. These Soft Proof ICC profiles are loaded into Photoshop or Lightroom. We can share Soft Proof ICC profiles of different combinations of metal, iron solution, humidity, and paper. One other benefit of the precision of the ammonium printing-out system is that we can also make replicate proofing setups using the Piezography Pro ink set to proof platinum/palladium by inkjet. This considerably lowers the cost of trialing images prior to the expense of employing actual platinum and palladium.

Jon and Cathy Cone founded the first professional digital printmaking studio in 1984. "The Digital Ark" was a part of Cone Editions Press, founded in 1980 in Port Chester, New York. Jon Cone is a Master Printer in screenprint, relief print, photogravure, intaglio and many other conventional and digital printmaking methods. Relocated to Vermont, Cone began pioneering inkjet technologies in 1990 and spent five years as IRIS Graphics Marketing and Development Partner in Fine Arts before developing systems for the Epson printer platform. Cone developed pigment inks for Epson printers prior to Epson. Many of these inventions included light blacks, quad blacks, and monochromatic software, eventually releasing the Piezography system of carbon-based inks and software for Epson printers in 1999, which are still in continual development. A subset of the Piezography system is the PiezoDN digital negative system. As a printmaker, Cone has collaborated with Richard Avedon, Norman Bluhm, David Bowie, Gregory Colbert, Lester Johnson, Wolf Kahn, Saul Leiter, Melissa Meyers, James Nachtwey, Gordon Parks, Kiki Smith, and many others. Cone teaches digital printmaking, direct to plate photopolymer gravure, printing-out Pt/Pd, and digital negative making at Cone Editions Press in Topsham, Vermont.

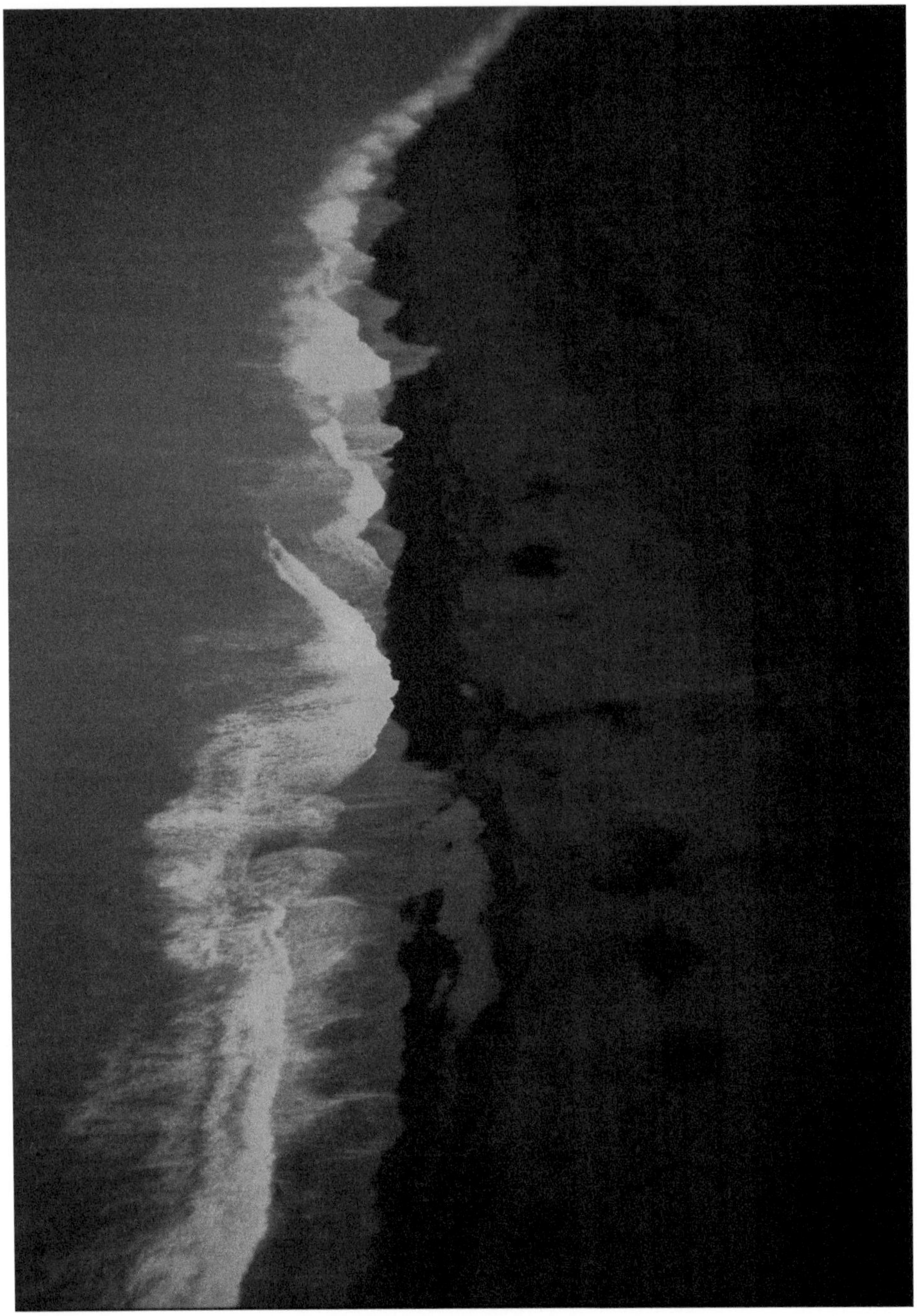

Figure 11.11 Jon Cone, *Reynisfjara Beach II*, Photo 2013 / Print 2019.
Platinum-palladium print on Revere Platinum paper, from PiezoDN negative of digital original with Sony Nex-7 using a Kern-Paillard Switar 75mm lens,
26 x 19 cm. Courtesy Jon Cone.
Fe3:Pt:Pd = 2:1:1, pre-coating humidity 50% for 2 hours at 68°F, rod coated, hydration 50% RH at 68°F, 60 minutes. Exposure using NuArc FT32. Post-exposure hydration 10 minutes 50% RH at 68°F. Cleared using the M-W clearing method.

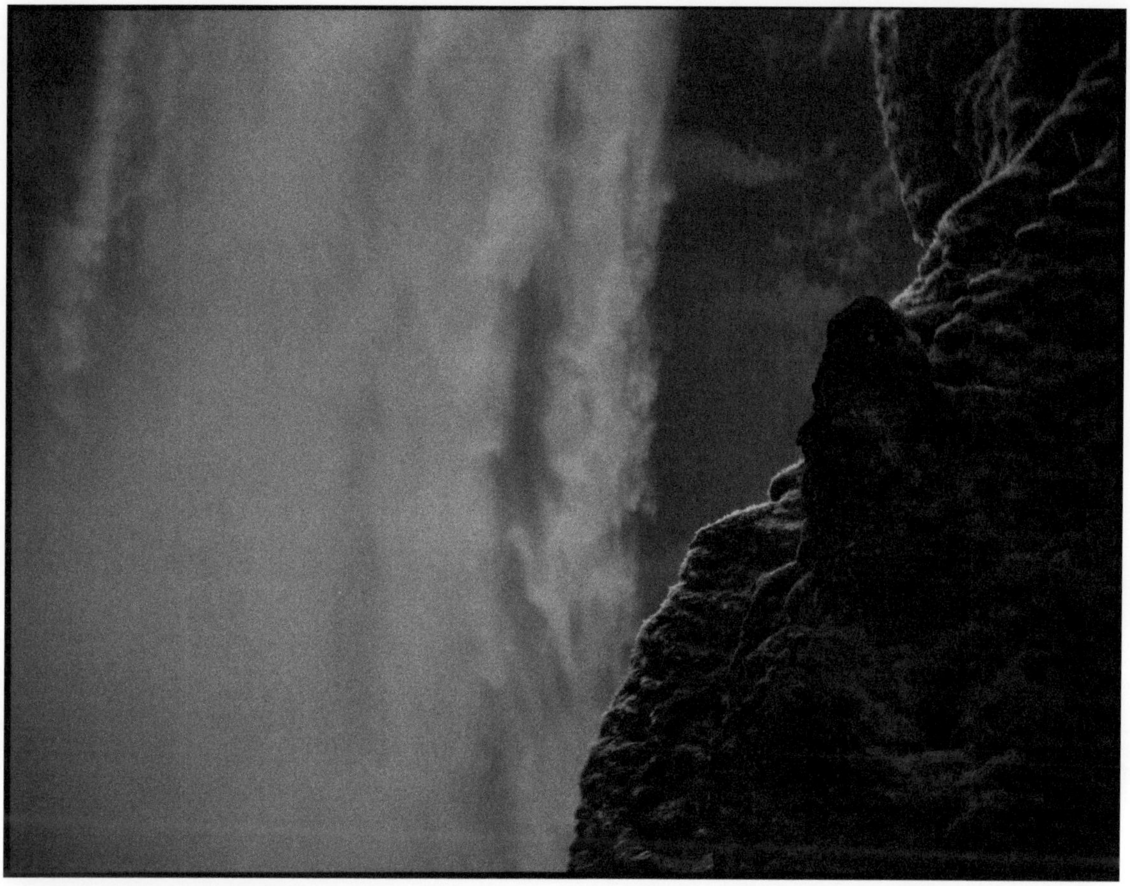

Figure 11.12 Jon Cone, *Skógafoss X*, Photo 2013 / Print 2019.
Platinum-palladium print on Revere Platinum paper, from PiezoDN negative of digital original with Sony Nex-7 using a Kern-Paillard Switar 75mm lens,
26 x 19 cm. Courtesy Jon Cone.
Fe3:Pt:Pd = 2:1:1, pre-coating humidity 50% for 2 hours at 68°F, rod coated, hydration 50% RH at 68°F, 60 minutes. Exposure using NuArc FT32. Post-exposure hydration 10 minutes 50% RH at 68°F. Cleared using the M-W clearing method.

Imogen Cunningham

Meg Partridge, Imogen Cunningham's granddaughter and director of the Imogen Cunningham Trust (ICT), writes:

> *From the ICT's perspective, because Imogen was always interested in, and had considerable knowledge of platinum printing, it is consistent with her vision for us to print Imogen's work in Pt/Pd. Along with the aesthetic link, there are several directives and practicalities that have brought the ICT to working with the ammonium-based printing-out process. All ICT prints have to stand the test of time, and remain as close to the original intent/initial resulting print, as possible. The printing-out Pt/Pd process is very consistent and scalable, and pairs exquisitely with Imogen's original, as well as calibrated digital enlargement, negatives. Printing-out Pt/Pd is ultimately a simpler process. Like anything, once past the learning curve, it is clear and direct. The constants— humidity, exposure time, etc.— help to clarify the process and make it easy to repeat, regardless of whether one is a beginner or advanced printer. Being able inspect a print during exposure is not only a pleasure but also offers excellent resulting prints. Then, there's the matter of money! With the price of metals today, using the printing-out process results in far less platinum/palladium chemicals being lost to prints that are thrown out. The printing-out process is less toxic than the processing chemicals needed for the development Pt/Pd version.*

Cunningham grew up in a big family in Portland, Oregon, and Seattle, Washington. Cunningham's interest in photography came early, and in 1901, she sent away $15 for her first camera, commencing what would become one of the longest photographic careers in the history of the medium. Cunningham's formal training in photography was self-constructed: she enrolled at the University of Washington as a chemistry student—because there were no photography classes or department. After graduation, she was a photo assistant at the Edward Curtis studio, learning the business of running a portrait studio and experiencing the technique of platinum printing. After two years at the Curtis studio, she spent a year in Dresden, Germany which enabled her to formally study the photographic process with the renowned Dr. Robert Luther at the Technische Hochschule where she became proficient in the technique of platinum printing. Returning to the United States, Cunningham opened her own photography studio in Seattle and soon became a popular portrait photographer while pursuing her own Pictorialist photographic work with a large format camera. Much of her studio work was produced in platinum. In 1917 Cunningham moved to northern California where she resided for the rest of her life. As a professional photographer, Cunningham continued to work as a portrait photographer while creating work for herself and managing gallery relationships, exhibitions of her work, and accepting part-time teaching positions in California. In the 1930s Cunningham was a member of the California-based Group f/ 64, whose photographic intent departed from Pictorialism with sharply focused, longer depth of field photographs.

The printing-out platinum-palladium prints reproduced here are from Cunningham's negatives, and printed by Pradip Malde under the supervision of the Imogen Cunningham Trust.

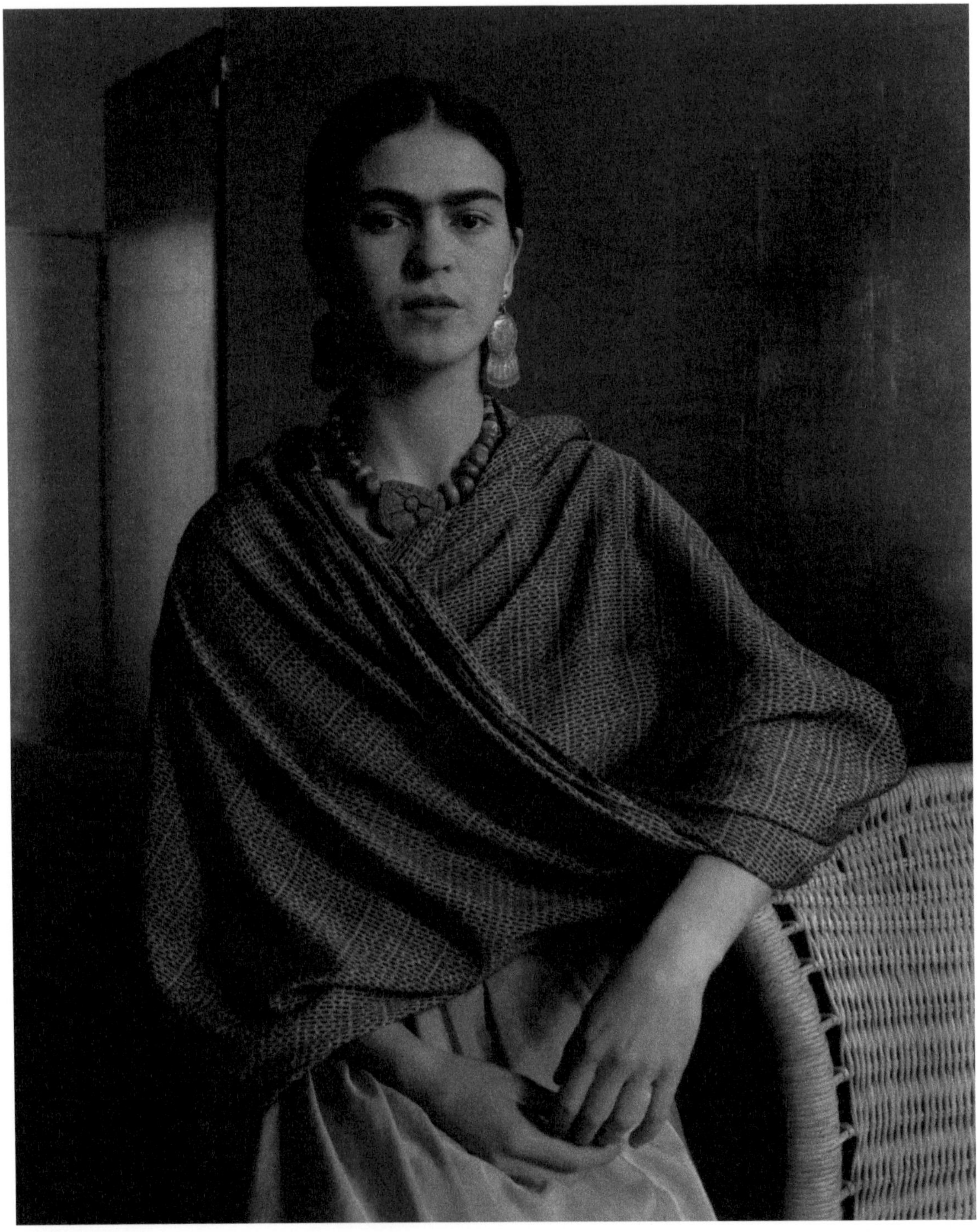

Figure 11.13 Imogen Cunningham, *Frida Kahlo Rivera, Painter and Wife of Diego Rivera,* 1931. Platinum-palladium print on Arches Platine paper, from PiezoDN version of the original 4 x 5 inch celluloid negative, 24.9 x 17.5 cm. Courtesy Imogen Cunningham Trust.
Fe1:Pt:Pd = 2:1:1, hydration for 90 minutes at 56% RH / 20°C.

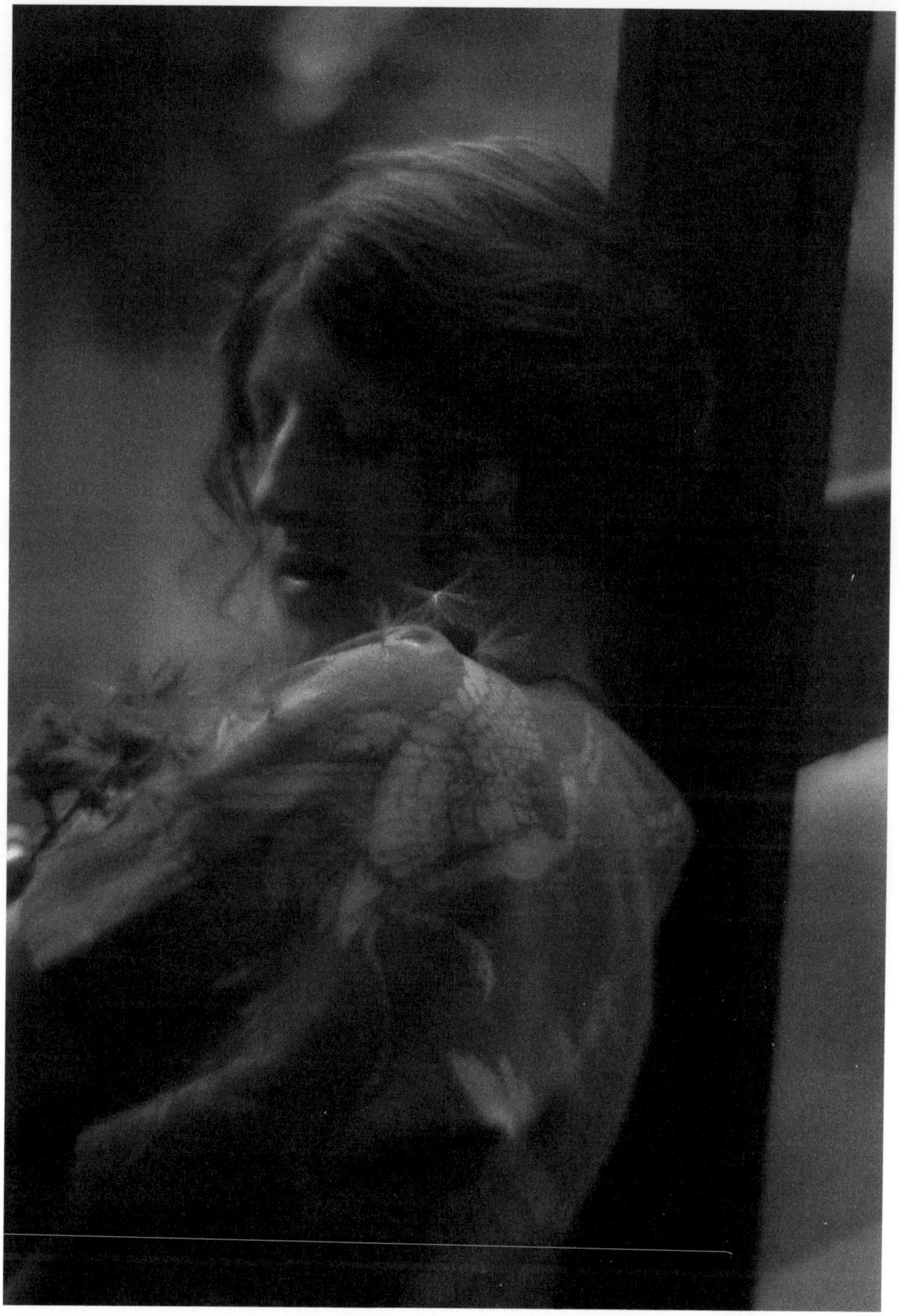

Figure 11.14 Imogen Cunningham, *Clare and the Floating Seeds*, 1910. Platinum-palladium print on Arches Platine paper, from PiezoDN version of the original 5 x 7 glass plate negative, 24.5 x 19.7 cm. Courtesy Imogen Cunningham Trust. Fe1:Pt:Pd = 2:1:1, hydration for 90 minutes at 56% RH / 20°C.

Ivey Dahlstrom

I grew up moving around the world from which I cultivated a willingness to be comfortable in new environments. Soon after losing a member of my family, I produced a family album for my undergraduate thesis. Most family albums document vacations or children growing up; they recall staged happiness. Family albums do little to reflect the true vastness of what has been emotionally experienced. Attempting to reclaim memory, my photographs illustrated tensions and collisions between how something looks and how it felt; they were an argument for a truth about how it feels to exist in the context of how a family looks. Regardless of what one tries to share, what one feels is seldom recognized by anyone other than oneself. My work bridged memory and expectation and acknowledged the inexactness of both memory and expectation. When one returns to a family album to rejoice in past experiences, they look through a curtained window, blinded by what was decidedly left out of frame. My work acknowledges the warped nature of memory and invites us to return home and to memory with more experienced eyes and a greater understanding of life, lived then, living now.

Ivey B. Dahlstrom is a photographer and a teacher, doing a post-baccalaureate fellowship based out of Tracy City, Tennessee. Her interests include matters of visual literacy, storytelling, and enabling fair access to education. Currently running a photo-based narrative program at Tracy City Elementary, Dahlstrom aims to foster self-confidence and resiliency in her students living in her rural community in southern Appalachia. She recently graduated cum laude and with Honors from The University of the South, at which time she also received the John McCrady Memorial Award for excellence in Art and the Shepard Award for Excellence in Photography.

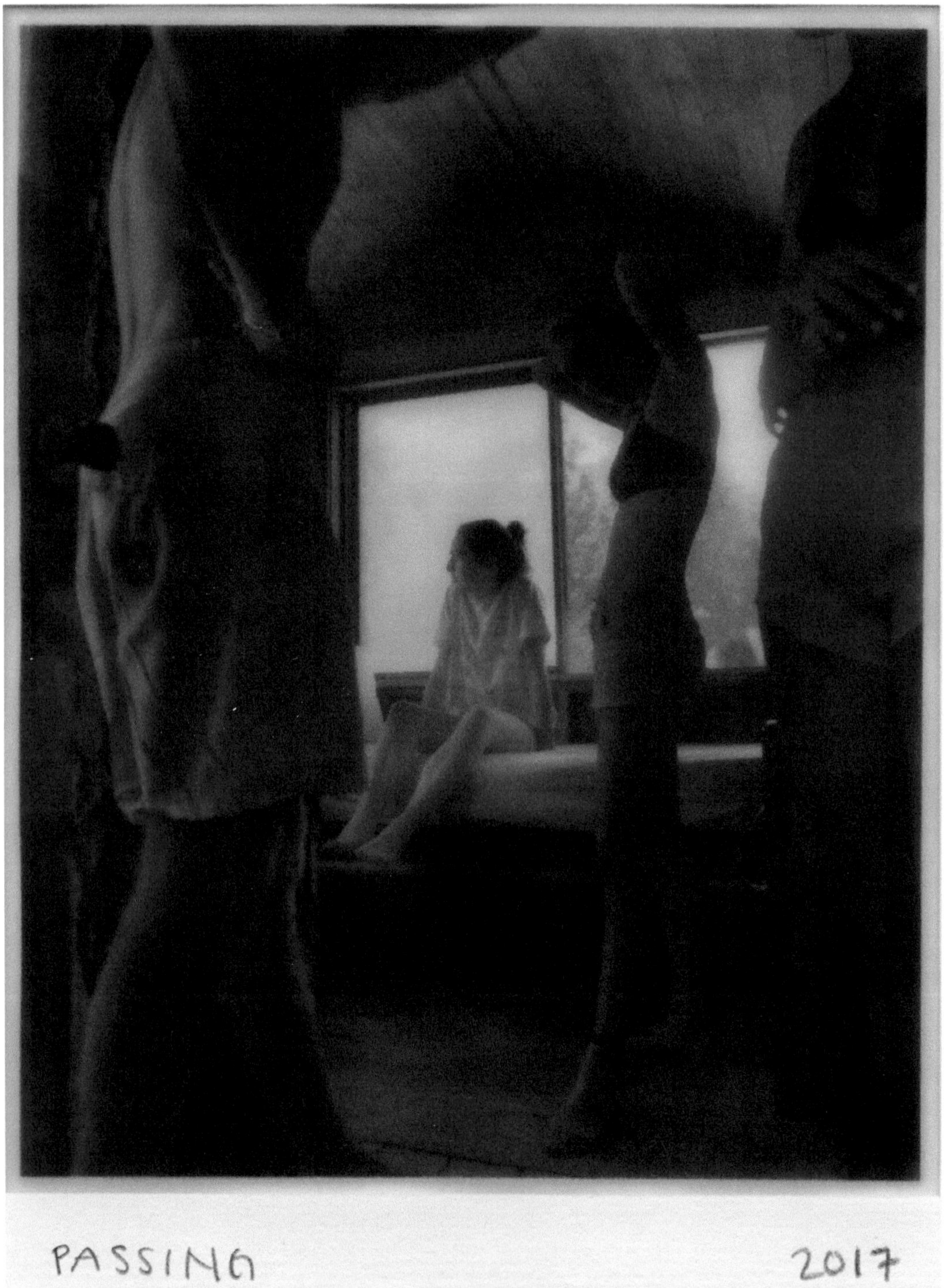

Figure 11.15 Ivey B. Dahlstrom, *Passing*, 2017. Platinum-palladium print on Reich CT 24# paper, from original 4x5 inch negative, 12 x 10 cm. Courtesy Ivey Dahlstrom.
Fe1:Pt:Pd = 2:1:1, hydration for 30 minutes at 100% RH / 20°C.

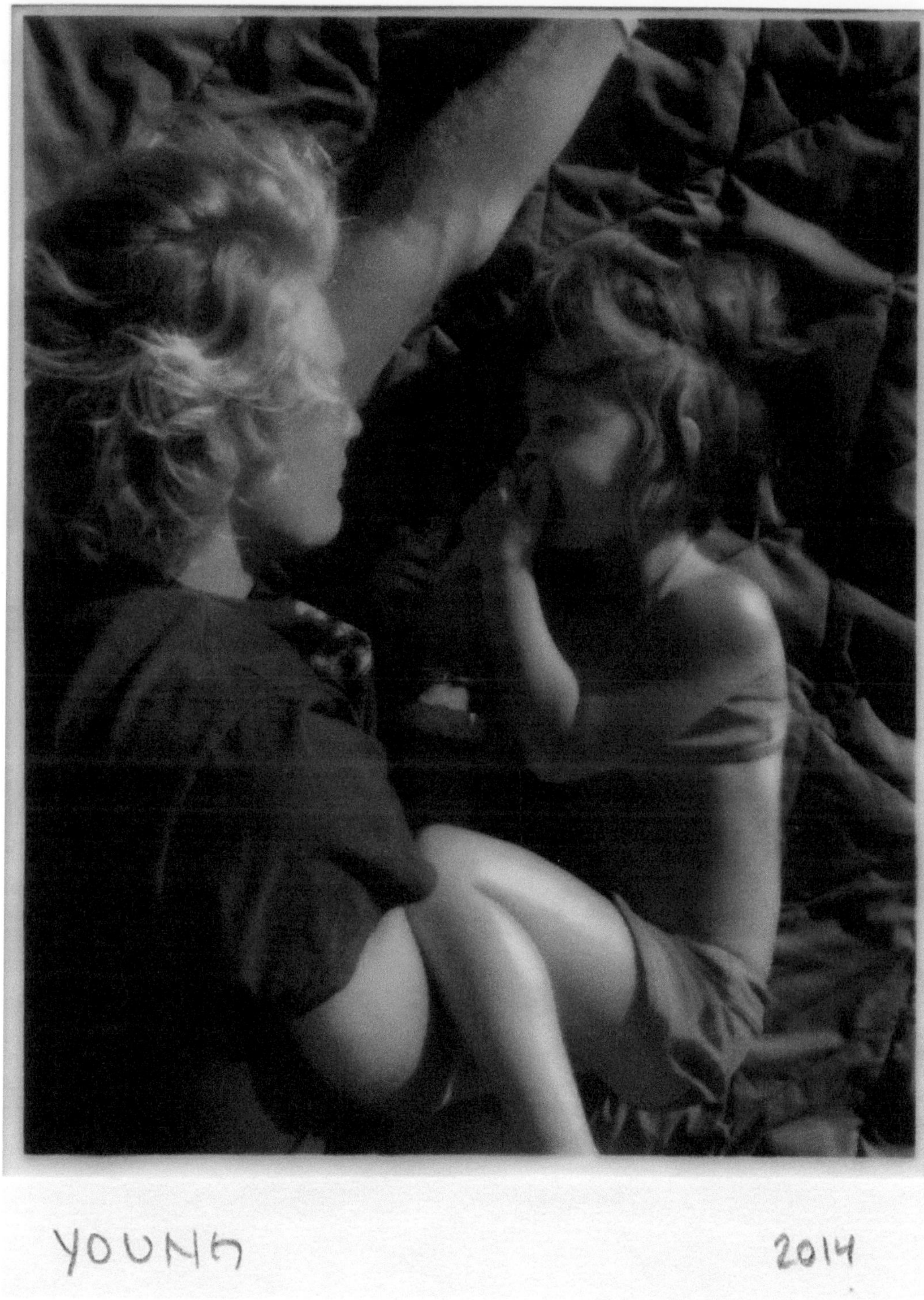

Figure 11.16 Ivey B. Dahlstrom, *Young*, 2014. Platinum-palladium print on Reich CT 24# paper, from original 4x5 inch negative, 12 x 10 cm. Courtesy Ivey Dahlstrom.
Fe1:Pt:Pd = 2:1:1, hydration for 30 minutes at 100% RH / 20°C.

David Dennard

In 1993, when I knew little about art or photography, I was exposed to the work of Paul Caponigro. I was profoundly touched by his ability to reach into the deeper realms of the human psyche and spirit through his images. Seeing that possibility, it has since been my goal to touch the hearts of my viewers in a way that adds something positive to their lives. I call that simply, "beauty." My hope is that my audience will be touched at a level meaningful to them, whether at the mental, emotional, or spiritual level or a combination of all three. After many years of digital printing, I have begun exploring Platinum/Palladium printing as my primary medium due to its depth of expression and its unique feeling when viewed.

Advice about making printing-out Pt/Pd prints:

An obvious advantage of the printing-out method is the ability to see the print as it is exposed. There are also advantages in terms of tonal range over the development Pt/Pd process. The downside is the need for humidity control if you want to make reproducible prints. I find a good piece of art is more about the feeling of the artist than the specific technique. Choose the method that works best for you and your needs. Paul Caponigro once said, "In order to be a good photographer, you need to work more on your emotions than you do on your technique." Sound advice.

David Dennard is a physician, photographer and father of three who lives in Dalton, Georgia with his wife and six rescue dogs. His photographic focus is on black and white imagery. His work crosses many genres including landscapes, architecture, portraits, and nudes. After many years of digital printing, he returned to the darkroom in 2017. Currently, his artistic focus is on platinum/palladium printing from digital negatives. Dr. Dennard has had solo exhibitions as well as individual images in exhibitions in galleries across the United States.

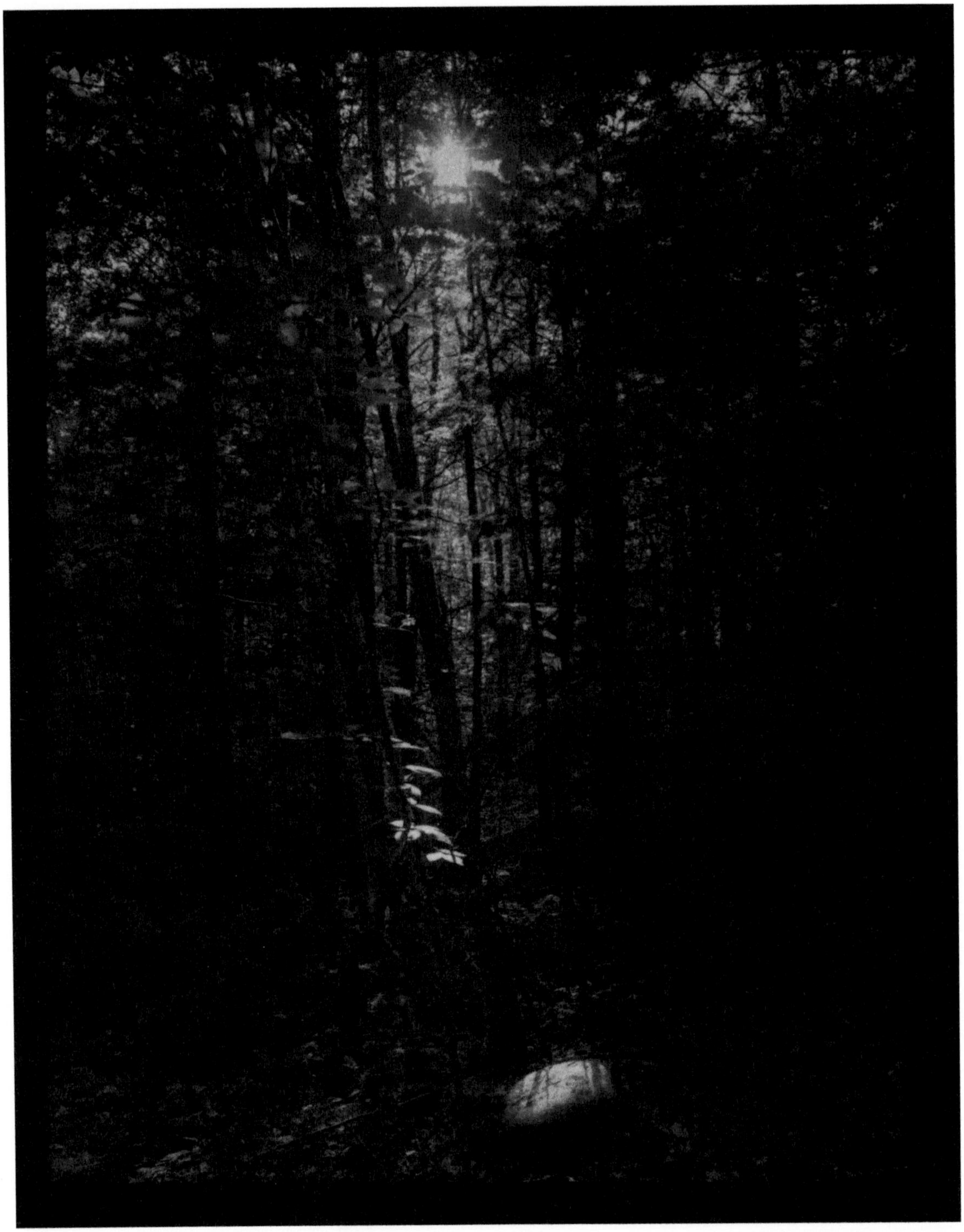

Figure 11.17 David Dennard, *Windsor, Maine*, 2019. Platinum-palladium print on Revere Platinum paper, from digital negative printed on Epson 3880 of full frame digital original,
12 x 10 cm. Courtesy David Dennard. Fe:Pt:Pd = 2:1:1.

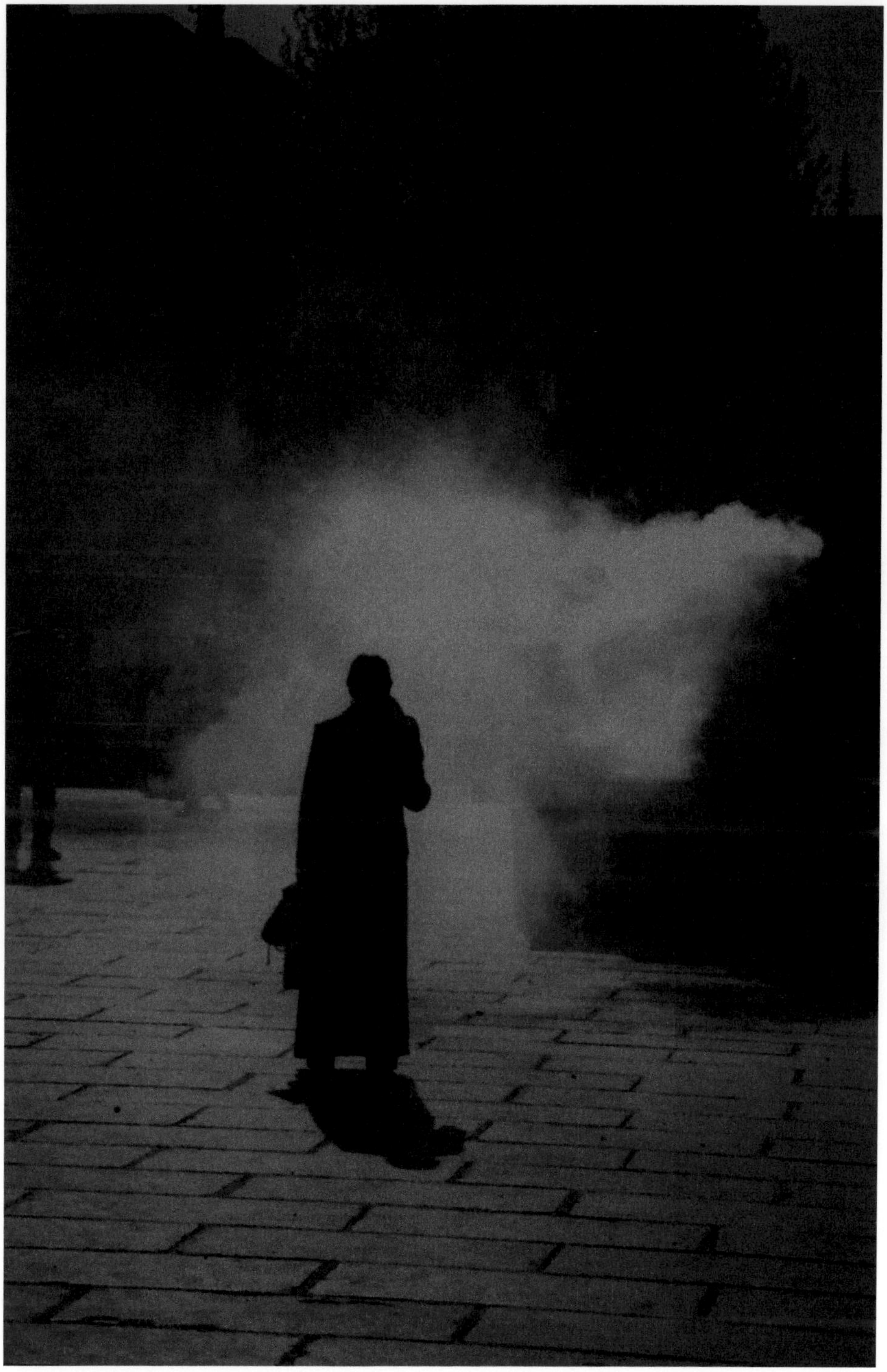

Figure 11.18 David Dennard, *Lhasa, Tibet*, 2013
Platinum-palladium print on Revere Platinum paper, from digital negative printed on Epson 3880 of full frame digital original, 31.5 x 21 cm. Courtesy David Dennard. Fe:Pt:Pd = 2:1:1.

Virgil DiBiase

In the 1980s when I was in medical school there were 500,000 people with dementia in this country. Now there are nearly 6 million people and that number will triple by mid-century. There is no cure for dementia but that doesn't mean we can't take care of the people in a humane way. I started to make portraits of my patients with dementia and spend time with them in their homes and make a social event out of it. When they came back to see me in my clinic they had an emotional memory of our interaction. Two years ago, my father developed dementia and he said to me: "get me out of here," a metaphor for being trapped cognitively by his disease. Since that time, I've been making Pt/Pd prints of perceptions, using quotes from my patients to visually shape my photos. I aim to create a visual language in order for the viewer to experience the inner state of dementia. We categorize people with dementia as suffering from the same set of conditions and having the same symptoms. They may have the same disease but they experience it in unique ways. They are unique individuals; just as non-demented people are. And just as Platinotype prints are unique even when printed from the same negative. Metaphorically this process parallels my relationship with my dementia patients (or any other relationships). These are unique individuals and anything goes. It takes patience. This project deserves a hand-crafted process rather than inkjet prints or even silver gelatin prints where every print looks the same. I would like to preserve these memories in photographs. Platinum prints are very stable, but memories are not.

Advice about making printing-out Pt/Pd prints:

I can adjust the exposure for the highlights. The printing-out formula uses 30% less volume than development Pt/Pd methods. There is no burning of the shadows during over-exposure because of the process's self-masking effect. Consequently, one can get higher Dmax than development Pt/Pd methods. I can't think of any disadvantages, but calibrate, calibrate, calibrate—the monitor with the digital negative especially. And pre-humidify the paper before coating it.

Virgil DiBiase is a photographer and neurologist based in Valparaiso, Indiana. He merges his photographic practice with his clinical practice in order to create a more social environment with his dementia patients. His father taught him darkroom skills as a child and now he dedicates his time making Pt/Pd prints from digital negatives. He has had three one-person shows and has been in numerous group shows, most recently at the Griffin Museum of Photography. He has been shortlisted three years in a row for the Royal Photographic Society International Photographic exhibition and the Critical Mass top 200. His works are included in the Fort Wayne Museum of Art and many private collections.

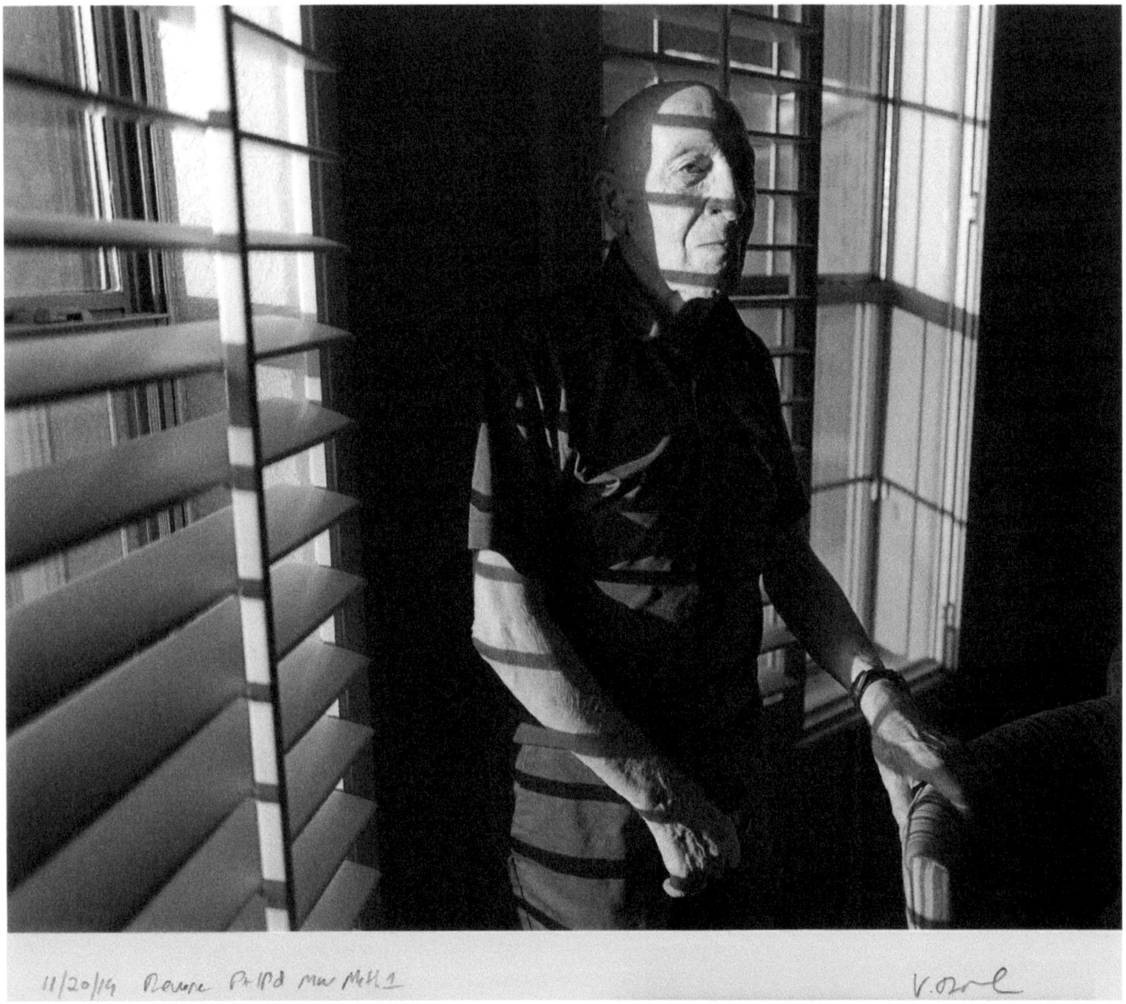

Figure 11.19 Virgil DiBiase, *Orfeo*, 2017. Platinum-palladium print on Revere Platinum paper, from PiezoDN negative printed on Epson P600 of full frame digital original, captured with Leica Q, 20 x 25 cm. Courtesy Virgil DiBiase.
Fe1:Pt:Pd = 2:1:1.

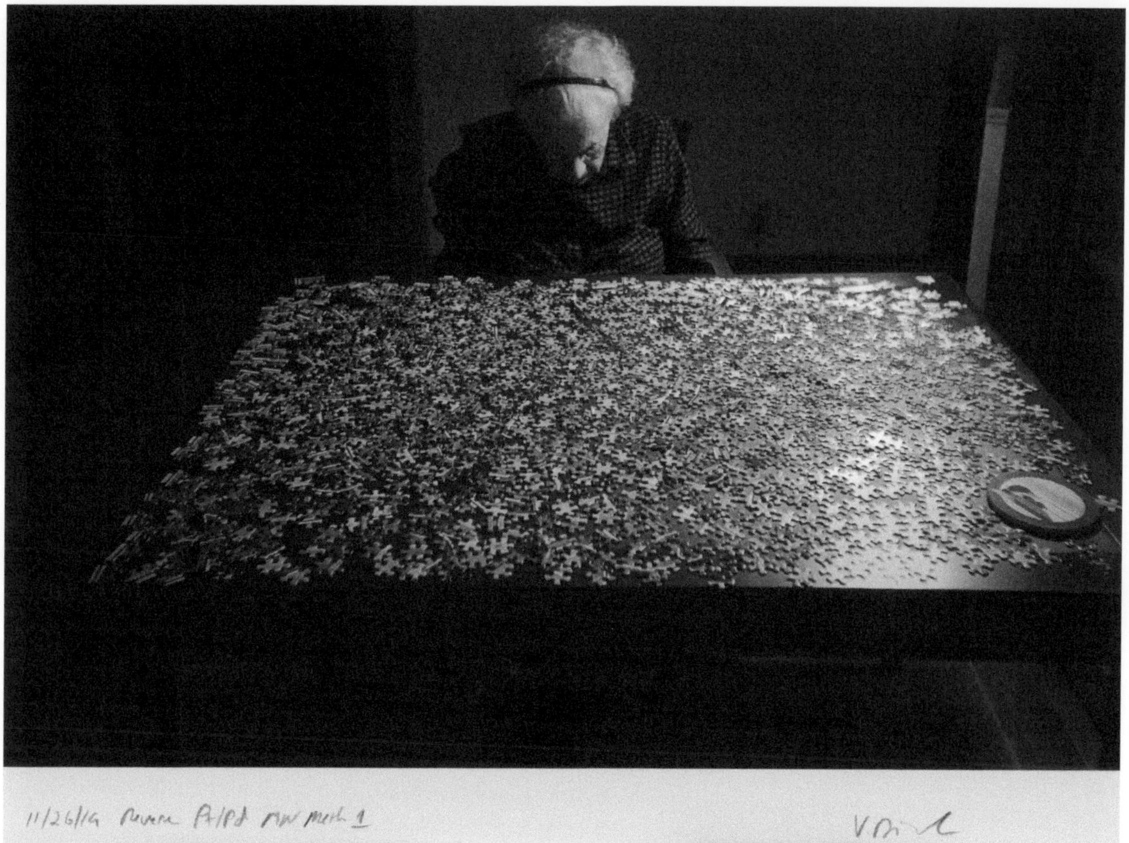

Figure 11.20 Virgil DiBiase, *1000 Pieces*, 2017.
Platinum-palladium print on Revere Platinum paper, from PiezoDN negative printed on Epson P600 of full frame digital original, captured with Leica Monochrom,
17 x25 cm. Courtesy Virgil DiBiase.
Fe1:Pt:Pd = 2:1:1

Danielle Edwards

I am passionate about trees and the environment. I often photograph in digital infrared as it brings another element to the images I capture with its broad tonal range. This translates well to platinum-palladium printing. The hybrid system of digital and analogue in tandem works well for me. Without the availability of this system I could not achieve this as infrared film is no longer available. The Pond *highlights the extreme tonal range that can be captured in infrared, but also that these extremes can be maintained in the platinum/palladium print.*

Advice about making printing-out Pt/Pd prints:

This is a process with proven longevity. The tonal values are wide and the paper combination with this process is alluring to the senses. You want to physically hold the prints and soak in the sensuality of this process. It takes imaging to another level. Take your time and enjoy the process. It has a similar effect to slow photography. It is time-consuming but worth waiting the time that it takes to complete a print. It is a type of meditation. It is both relaxing and rewarding. With the proper application this is quite an economical process. It is important to create negatives with the tones you wish to see in the print. If it's not in the negative you will not see it in the print. It is a thrill to inspect your image and watch it appear before your very eyes due to the printing-out effect. Creating and holding a platinum/palladium print evokes a feeling, with many intangibles, that cannot be equated to digital photography alone.

Danielle Edwards is an Australian photographer known for her images of plants, flowers, and decaying items. Danielle utilizes a variety of alternative and non-alternative photographic processes to visualize her imaging concepts, platinum/palladium, Vandyke brown and Chrysotype being some of the processes she employs. She is a regular contributor to national and international conferences. Her expertise is varied with a background in both science and art, and she has published both technical and creative works.

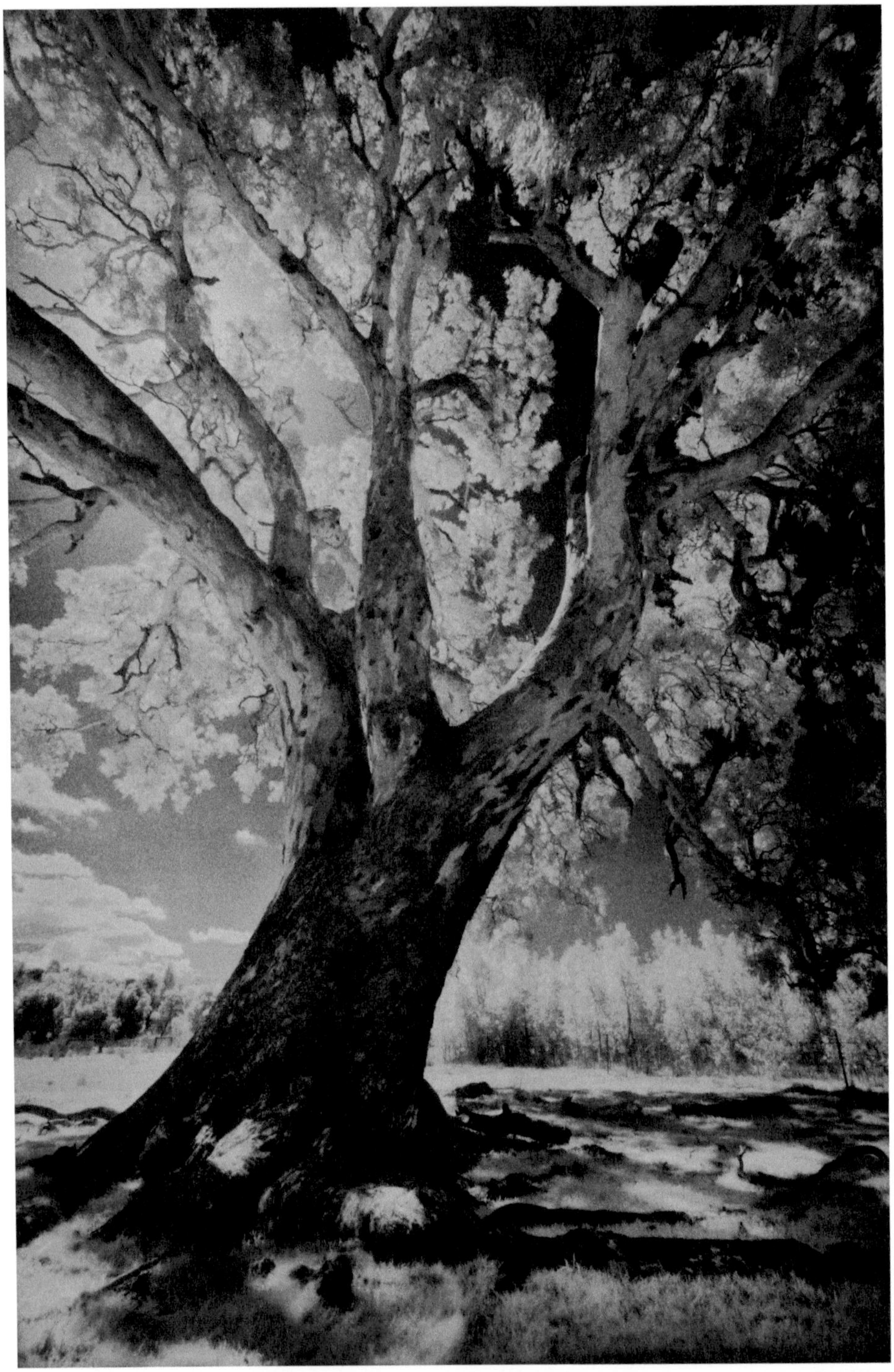

Figure 11.21 Danielle Edwards, *The Sentinel, Eucalyptus camaldulensis*, 2019. Platinum-palladium print on Bergger Cot 160 paper, from PiezoDN negative of full frame digital original, Nikon D3 Infrared conversion, 29 x 19 cm. Courtesy of Danielle Edwards. Fe:Pd:Pt = 4:3:1, hydrated for 40 min. in chamber at 66% RH / 19°C.

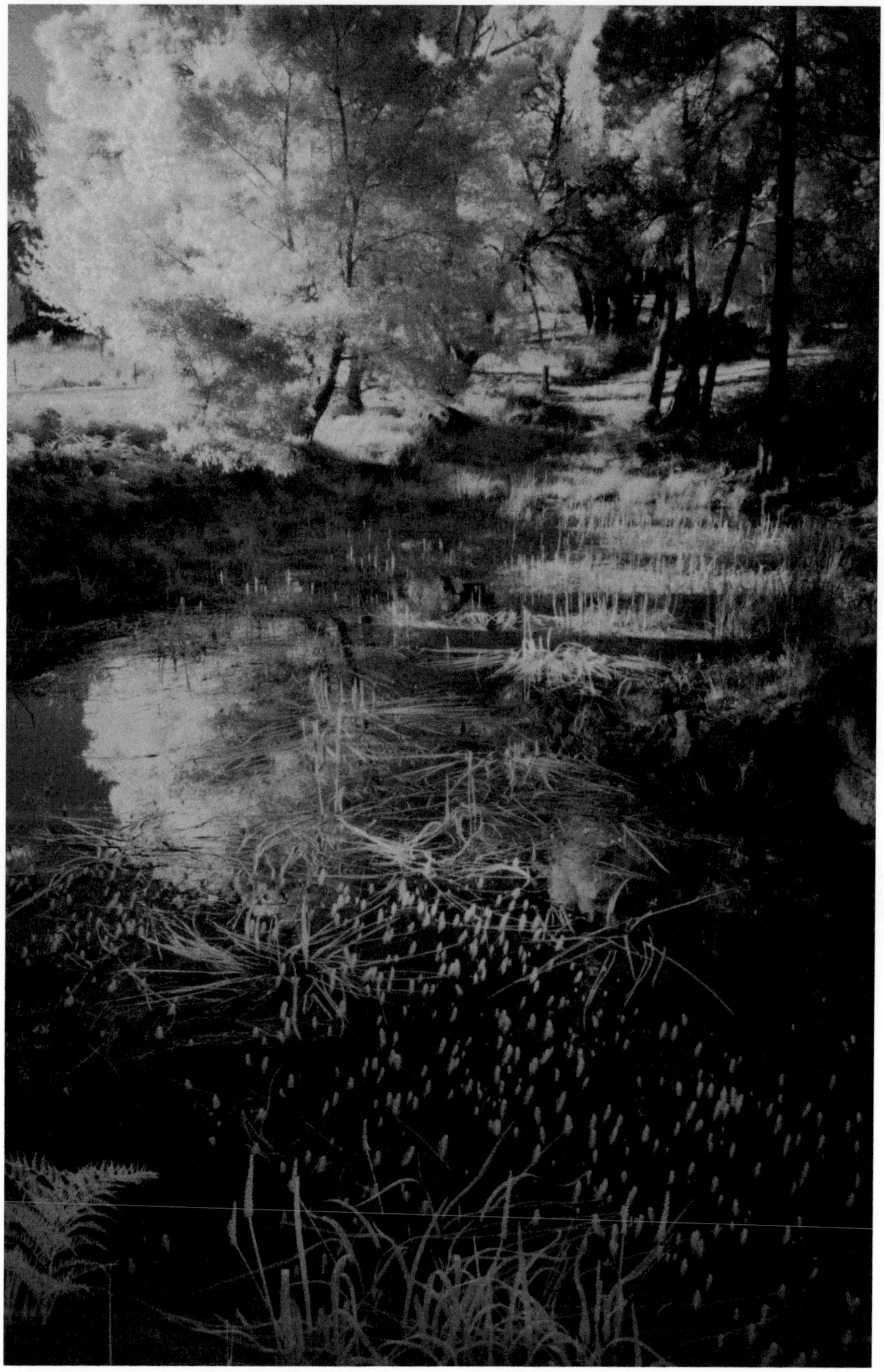

Figure 11.22 Danielle Edwards, *Into the Dell*, 2019. Platinum-palladium print on Bergger Cot 160 paper, from PiezoDN negative of full frame digital original, Nikon D3 Infrared conversion, 29 x 19 cm. Courtesy of Danielle Edwards. Fe:Pd:Pt = 4:3:1, hydrated for 40 min. in chamber at 66% RH / 19°C.

Jill Enfield

I tend to work on several different bodies of work at one time, one of which is made while commuting. Having lived in New York City for 35 years, a recent move to the Hudson Valley was a big change. Photographing out the window of the train that I take twice a week has helped make this adjustment a little easier. Looking out the window and seeing the Hudson River on one side and small river towns that are growing and expanding on the other, I feel that I am recording the change of seasons as well as changing America. The photographs are made with a smartphone, from which I produce positives to make contact prints onto glass using wet plate collodion. The resulting negatives are then used to make platinum and/or palladium prints.

Advice about making printing-out Pt/Pd prints:

As I was experimenting, I noticed a difference between waiting for the print for an hour in a humidified room or using it immediately. I loved seeing the color change between the two. I also loved seeing the image as you take it out of the ultraviolet light instead of waiting for the print to be processed. Take good notes. The humidifying is really important as is the waiting time so that you are consistent with your end-color. It is important that the negative be pretty perfect—with detail in the whites and the blacks. I seem to work a lot with flatter negatives, although I love seeing contrast in other photographers' work.

Jill Enfield is based out of the Hudson Valley in New York State. She is the author of three books on alternative photographic processes, including the recent edition of *Jill Enfield's Guide to Photographic Alternative Processes*, and teaches workshops throughout the world on the subject. Enfield's work is part of many museum and private collections including the Amon Carter Museum in Dallas, Texas, the Albrecht-Kemper Museum of Art in St. Joseph, Missouri and the Bibliothèque Nationale in Paris, France. Her wet collodion portraits and Glasshouse of Immigrants was featured at the Ellis Island Museum and Photoville in DUMBO, New York and the Annenberg Center of Photography in Los Angeles.

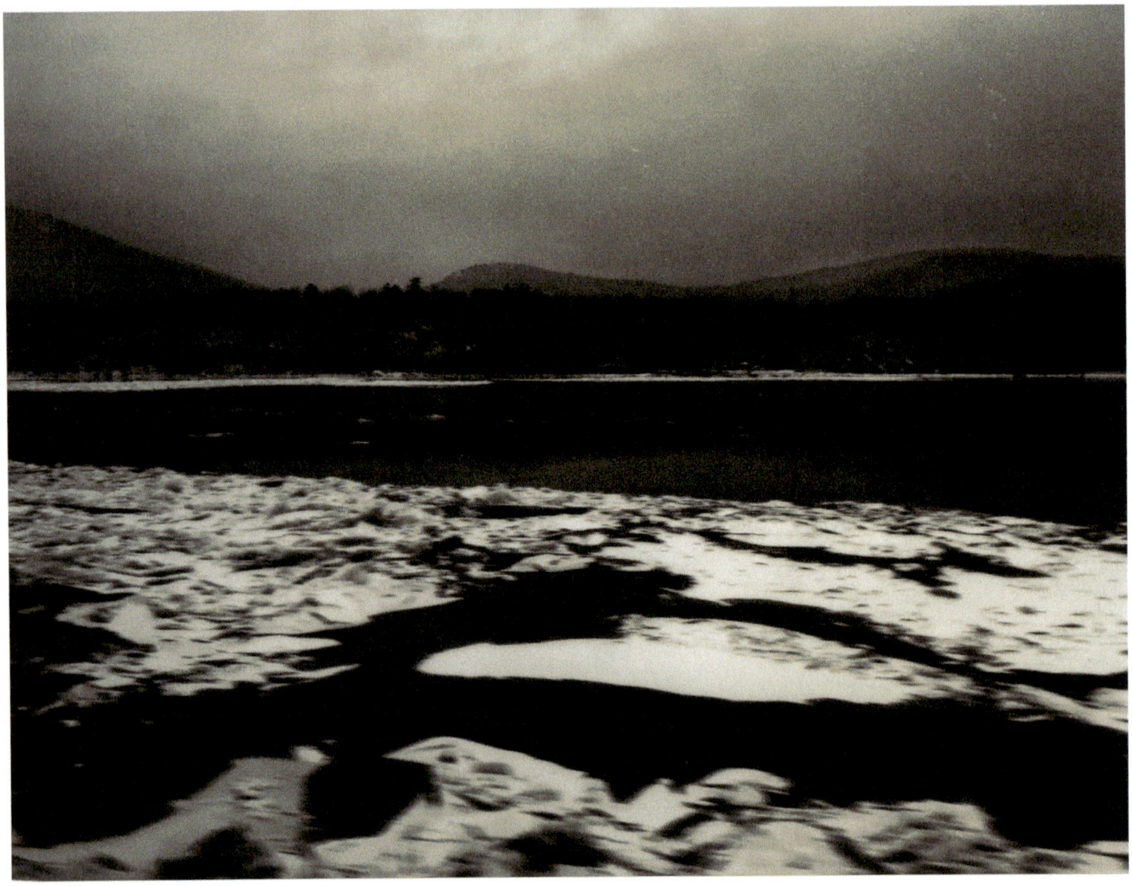

Figure 11.23 Jill Enfield, *Untitled*, 2019. Platinum-palladium print on Revere Platinum paper, 25.5 x 19 cm. Courtesy Jill Enfield. Fe1:Pt:Pd = 2:1:1.

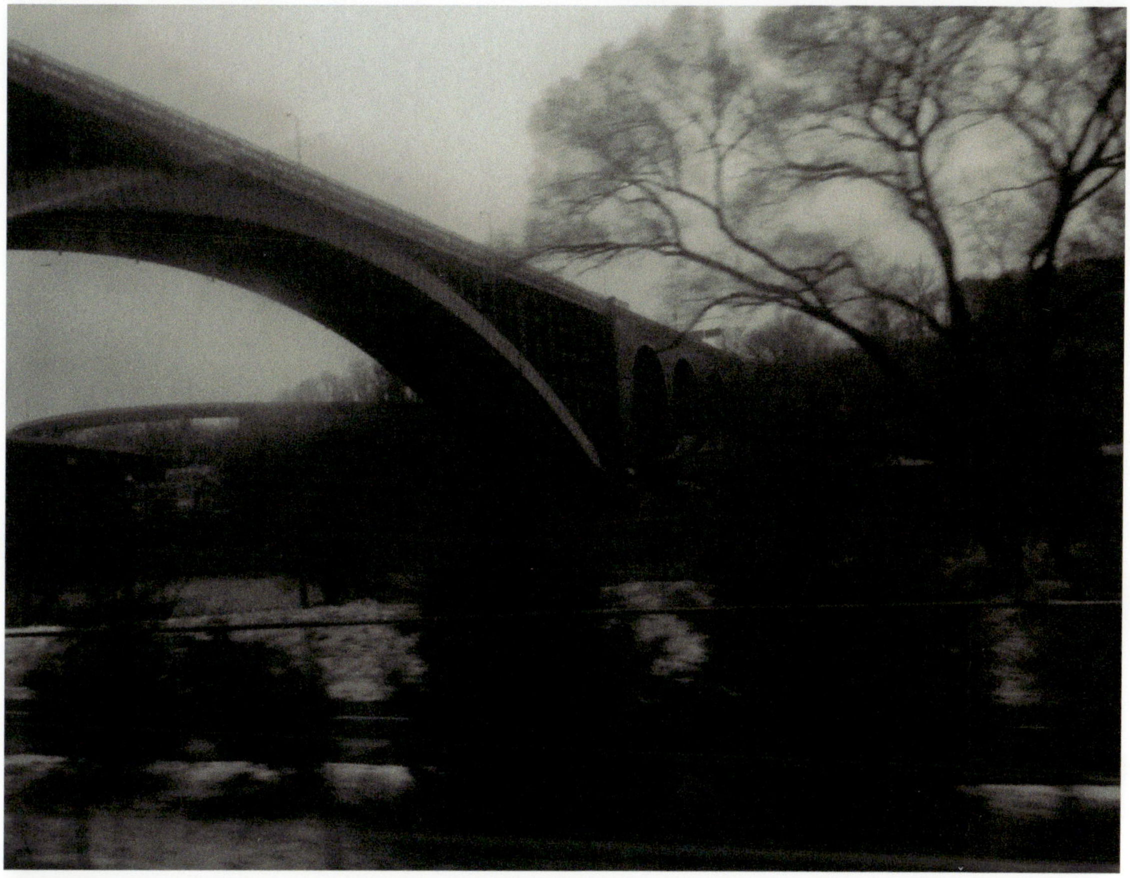

Figure 11.24 Jill Enfield, *Untitled*, 2019. Platinum-palladium print on Revere Platinum paper, 25.5 x 19 cm. Courtesy Jill Enfield. Fe1:Pt:Pd = 2:1:1.

Emily Gómez

Creating a platinum/palladium print is similar to watching a moonflower unfold at dusk, or finding a hidden waterfall in the woods—startling, visceral, magical. The two series that I present, though made years apart, retain the sense of awe I seek to preserve when setting out into nature with my large format camera. Whether photographing a blooming moonflower in my backyard or waterfalls in the mountains of North Georgia, my goal is to provide images of wonder that take us away from our everyday worries, and invite us to stare.

Advice about making printing-out Pt/Pd prints:

This process has advantages over other versions, including the ability to control contrast through hydration, the ease of making a correct exposure in the print, and the beauty of the vast tonal range. I used to have difficulty buying the bulk chemicals and mixing them myself. Since Inkjet Mall sells the pre-mixed sensitizers now, that is no longer an issue. Trust your instinct when it comes to making a correct exposure. With some alternative processes, you need to overexpose or underexpose before processing. With printing-out Pt/Pd, the exposure is correct when it looks right to you. Choose a negative with a wide tonal range to see the full capability of what Pt/Pd will reveal to you. It may look like a complicated process on the surface, but once you have your darkroom set up, printing is almost like a dance. Use the Multi-timer [4] app that Pradip recommends to keep you on track with hydration and processing time. Have fun!

Emily J. Gómez is a professor of photography at Georgia College in Milledgeville, Georgia. Her interests lie in finding awe-inspiring landscapes, often while contemplating their complicated histories of human habitation. Primarily an alternative process printer, she uses the printing-out platinum/palladium process, as well as the New Cyanotype[5] because of their superior abilities to express tonal variation and depth. Her work is in the permanent collection of the Zuckerman Museum of Art, and her photograph *Opening Moonflower (3rd in a series of 4)* was published in the book *Platinum and Palladium Photographs: Technical History, Connoisseurship, and Preservation*. Her work can be found at emilyjgomez.com.

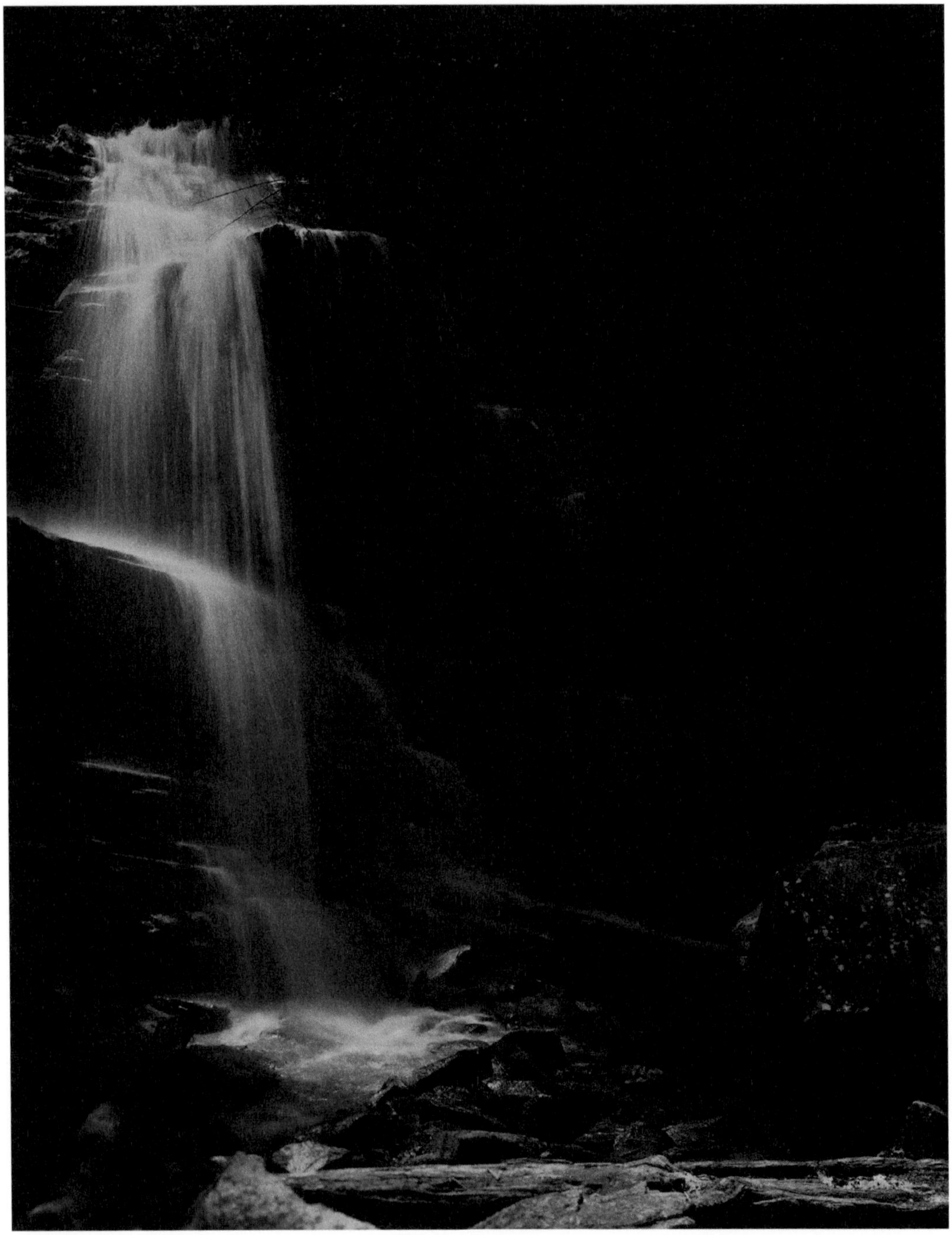

Figure 11.25 Emily J. Gómez, *Bad Branch Falls, Rabun County, Georgia,* 2014. Platinum-palladium on Revere Platinum paper, from original 8 x 10 inch negative developed in PyroPMK, 24 x 19.5 cm. Courtesy Emily J. Gómez.
Fe3:Pt:Pd = 2:1:1, hydrated for 20 min. in 100% RH / 20°C.

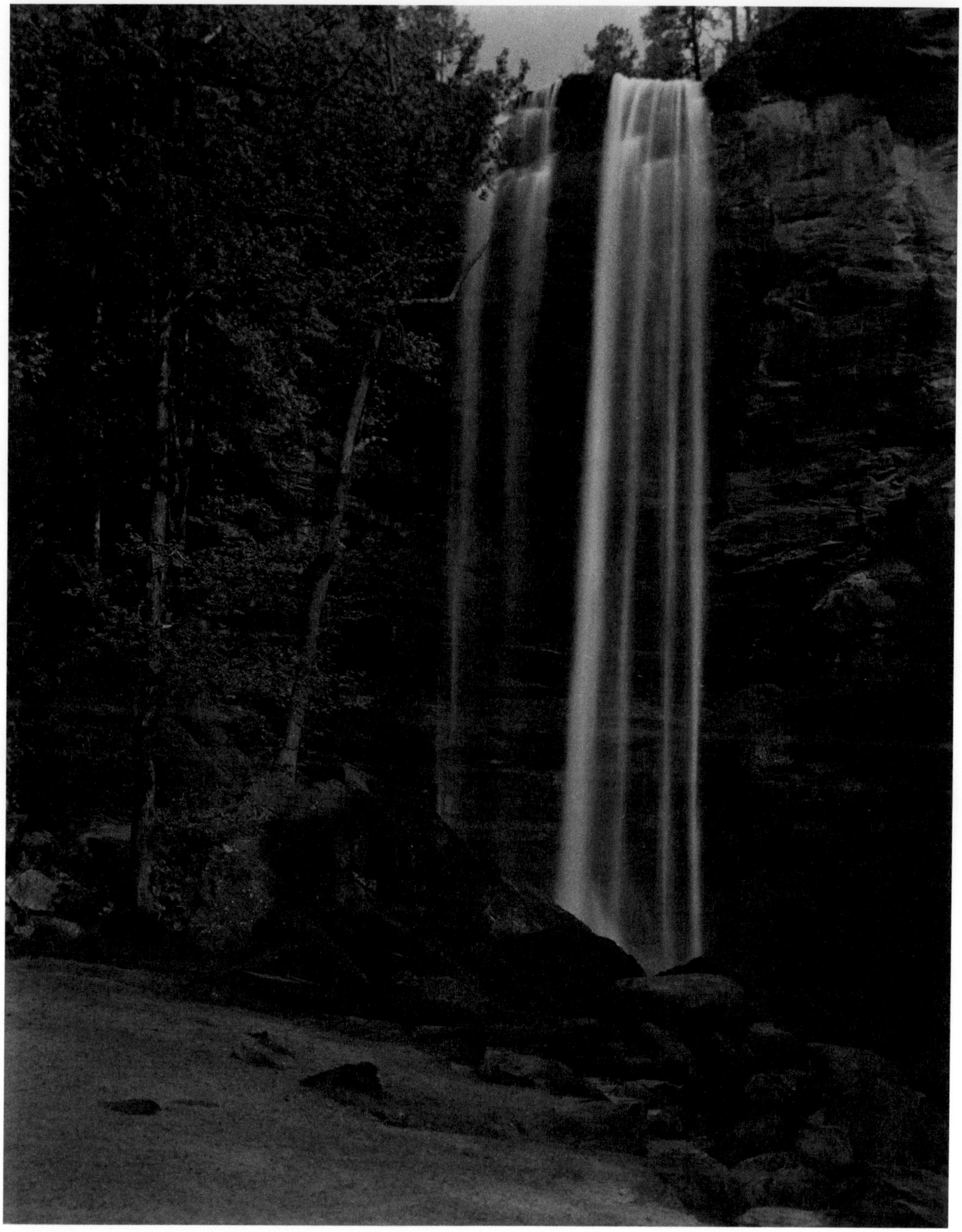

Figure 11.26 Emily Gómez, *Toccoa Falls, Toccoa, Georgia*, 2014.
Platinum-palladium on Revere Platinum paper, from original 8 x 10 inch negative developed in PyroPMK, 24 x 19.5 cm. Courtesy Emily J. Gómez.
Fe3:Pt:Pd = 2:1:1, hydrated for 20 min. in 100% RH / 20°C.

Pamela Heemskerk

Orchestra. *I was born a girl child in the plains of Montana, an achingly beautiful and yet terrifying landscape. The wind howled across the plains. Weather rolled in loud and fearsome, and the mountains, rivers, sky, and vast prairie were the backdrop. Into a transient military family, I arrived. I grew up with the understanding that wandering, walking, and being in nature was restorative. My playground was in the woods, on horseback, and on foot. I had a freedom most children do not experience today. Soon though, cultural, familial, military, gender, and religious pressures would eventually have their way with me, squeezing and molding me into an acceptable female role. It kept me safe and helped me to survive. Until it no longer worked, as the pain of living a prescribed life became unbearable. I set out to find freedom on my own terms. I walked out into the natural world and found a mirror there. I made pictures and wrote poetry. The sounds of the natural world infuse these works. I use elements of the primal world, light, and symbol as my vocabulary. These are images meant to be read, as they stand alone, or when viewed together, as stanzas in a poem, driving the narrative.* Orchestra *is my journey to reclaiming voice, identity and freedom through a connection to the natural world, domestic life and the tension between the two.*

I was drawn to the printing-out Pt/Pd process out of a deep sense of longing to feel my images with my hands. I work primarily in color. Yet, I could sense and feel that what I was making was asking me for something more. Essentially, I was looking for a translation from the original digital negative to something far more magical in the paper. And that is what I find to be so alluring about this process, watching the image become embedded in the fibers of the paper. The image transforms through chemistry, process, and the photographer's hand.

Advice about making printing-out Pt/Pd prints:

I feel like I am in a unique place to comment on this. I have come to this process as a complete novice. I am a photographer who came up primarily as a digital practitioner. In the past, the darkroom printing process had seemed too difficult and too toxic to approach. The "Malde-Ware" or printing-out process came highly recommended and I enrolled in a workshop taught by Pradip Malde at Cone Editions Press in Vermont. What I found was this process is very accessible to someone who has never worked in the darkroom. As it was mostly nontoxic, I had no reactions to any chemicals. I highly recommend attending a workshop, and then being prepared to be blown away by the beauty of the print.

Pam Heemskerk is an American photographer and poet living in rural Maryland where she writes, makes her work, and teaches. Using a variety of photographic tools, including digital, film, instant and found family images, she explores different ways of seeing the geography of human experience as reflected in the land, people, and objects that surround her. The narratives that often appear deal with her desire for connection to self and to the wild. Fascinated with thc handmade book she works to blend image and the written word into unique book structures. She teaches book arts and photography at the Frederick Book Arts Center. Pam is a graduate of Maine Media College MFA program and a 2018 Critical Mass 200 Finalist.

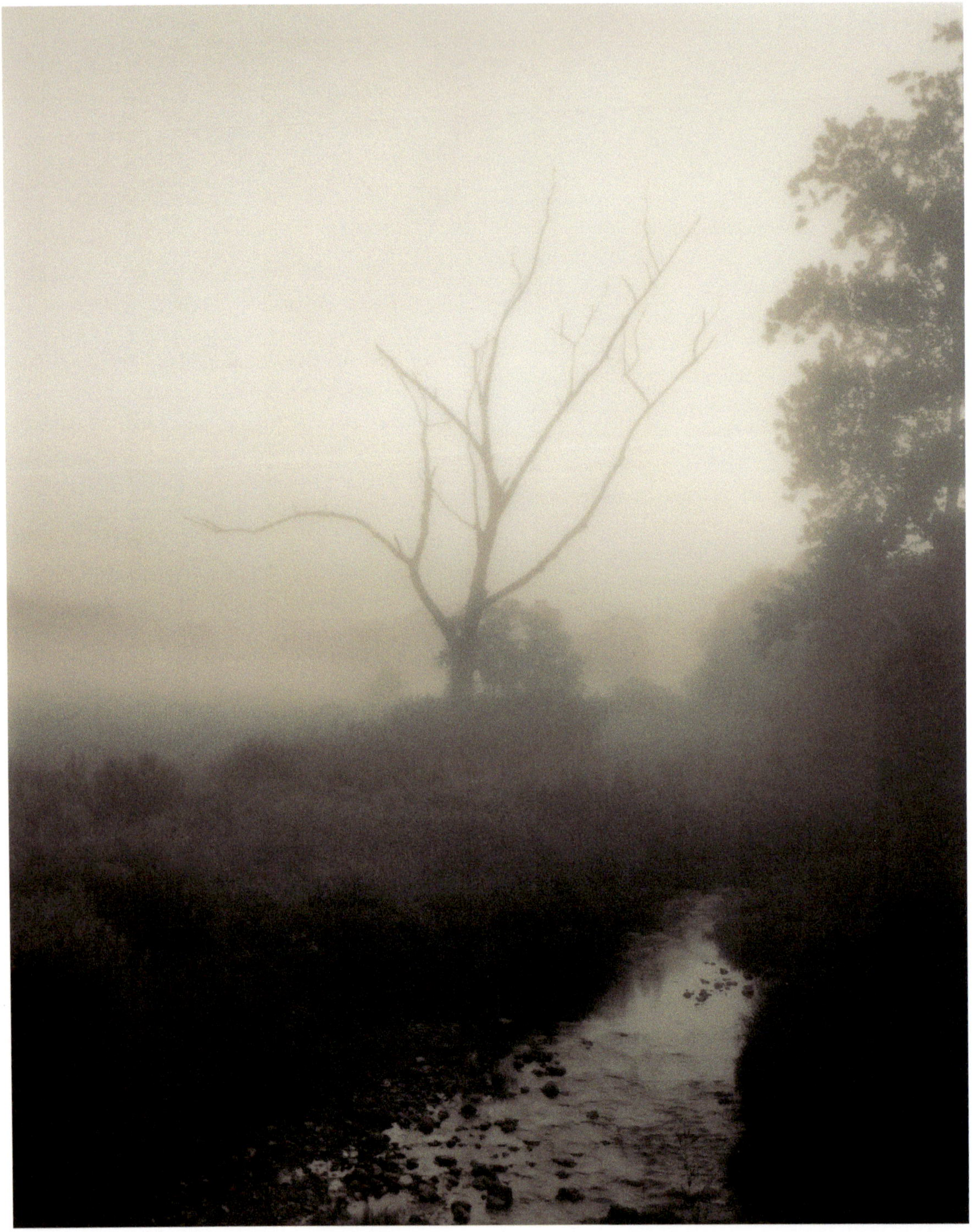

Figure 11.27 Pamela Heemskerk, *Little Bird*, 2017. Platinum-palladium print on Reich CT 24# paper, 22 x 18 cm. Courtesy of Pamela Heemskerk.
Fe1:Pt:Pd = 2:1:1, hydrated for 32% RH / 20°C.

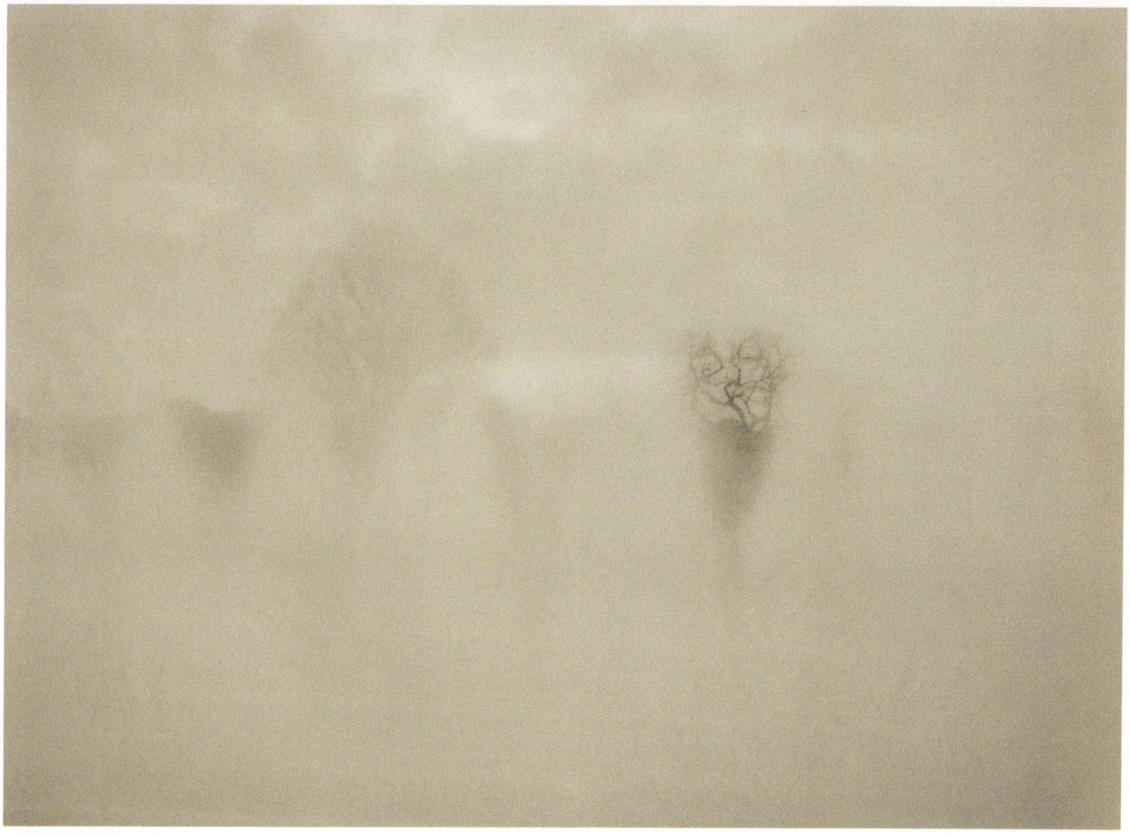

Figure 11.28 Pamela Heemskerk, *Yoni,* 2017. Platinum-palladium print on Reich CT 24# paper, 16 x 22 cm. Courtesy of Pamela Heemskerk.
Fe1:Pt:Pd = 2:1:1, hydrated for 32% RH / 20°C.

Allie Horick

My photographic work explores human-environment interactions, focusing specifically on the concept of wilderness. I am interested in how it has changed through history, the internalization of wildness, the ecological concept of rewilding, and ecofeminism. The environmental writings of authors such as Annie Dillard, Helen Macdonald, and Mary Oliver have encouraged me to seek out female voices and experiences and have helped to validate my own experiences as a woman who explores "wild" places. My interest in these interactions stems from my own relationship with the environment, which has been cultivated through years of extended backcountry travel in remote settings with small groups of women and female-identifying individuals. As a teenager, I participated in camping-canoeing trips and have gone on to lead similar trips for youngsters in Montana. Through this, I have seen how moving slowly through a landscape can help reconnect and ground you. Wilderness areas are tied to women in my mind because my experience of these places has taken place almost solely in their company. It is for this reason that much of my work focuses on female bodies in nature.

Advice about making printing-out Pt/Pd prints:

The printing-out process is the only one I have experience with, being taught by Pradip as an undergraduate student at college. I love its physicality, tonal range, and connection to environmental influences. A good platinum-palladium print takes time and finessing. Give yourself fully to the process and it will give back. If I had not found platinum-palladium printing, I may have never fully pursued photography. The process was so engaging and fulfilling that I was drawn completely in by the medium.

Allie Horick was born and raised in Nashville, Tennessee. She graduated from the University of the South in May 2015 where she studied environmental studies and photography. She has spent the past four years teaching photography and environmental education. She is currently pursuing an MFA at the University of Montana. Her work focuses on interactions between humans and the environment, drawing on the concepts of rewilding and ecofeminism as well as her personal experiences leading others and traveling in remote wilderness settings.

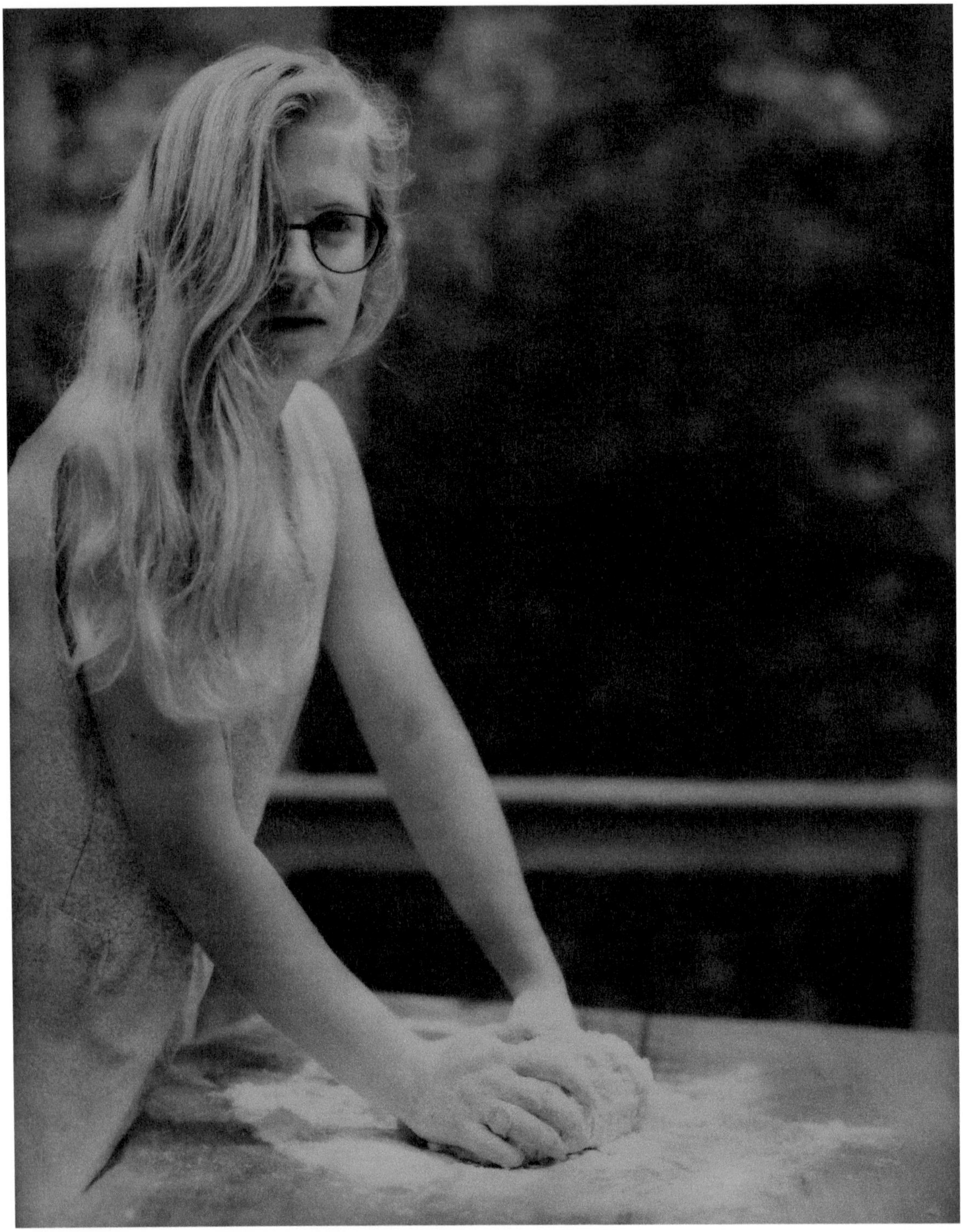

Figure 11.29 Allie Horrick, *Ansley Making Bread*, 2016.
Platinum-palladium print on Reich CT 24# paper, from a PiezoDN negative of 8 X 10 original developed in pyro PMK, 25.5 x 20.5 cm. Courtesy Allie Horick. Fe1:Pt:Pd = 2:1:1.

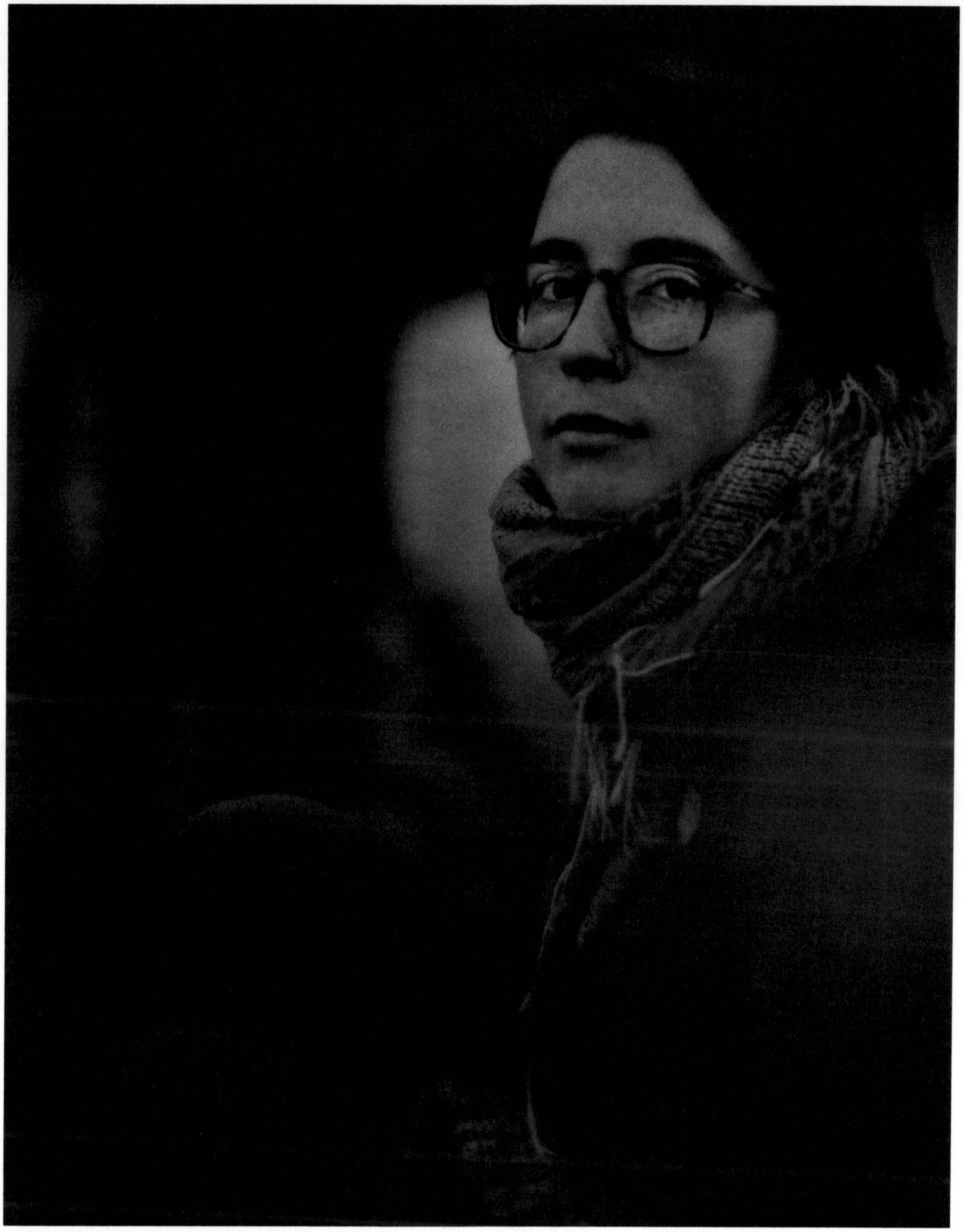

Figure 11.30 Allie Horrick, *Bea by the Creek*, 2018.
Platinum-palladium print on Reich CT 24# paper, from a PiezoDN negative of 8 X 10 original developed in pyro PMK, 25.5 x 20.5 cm. Courtesy Allie Horick. Fe1:Pt:Pd = 2:1:1.

Skip Klein

I am a photographer because the process of making images provides the opportunity to stare at people and examine the human condition for a much longer time than otherwise would be socially acceptable. For me, carrying a camera is a way to be fully alive in the world and a good excuse to engage with new people. I started as a voyeur with a long lens, and over time have moved to shorter focal lengths which requires me to get close and make contact. As I get closer, I can enter their world more fully and experience their human condition. Over time, I feel that I am seeing more with my heart and eyes, and less with my analytical, technical mind. By making exposures of others, I find myself exposing things deep inside me, first to myself, and ultimately to the viewer of my images. When it is successful, the artistic process infuses the objective, "just the facts" documentary with just a bit of lyricism, in which the emotions, aspirations, and interpretations of the photographer and the viewer come into play. I enjoy returning to my subjects to provide a physical print which captures the moment in time that we spent together.

Advice about making printing-out Pt/Pd prints:

The printing-out Pt/Pd process has elegance, ultimate simplicity, an efficiency of materials, and the ability to achieve a wide range of tone and color ... all of this in a scientific way that appears to be reproducible. There is the need to carefully control and monitor the humidity in the room and in the paper. Run, don't walk, to take a workshop at Cone Editions in Vermont. You will learn how to make digital negatives and then (Malde-Ware) printing-out Pt/Pd all in the same week, in a beautiful spot with wonderful food. If Irving Penn were alive, I have little doubt that he would be experimenting with the Malde-Ware printing-out process. He might discover that he no longer needs to double coat on paper mounted on aluminum. And, he might reinterpret some of his earlier prints in a beautiful new light.

Skip Klein is a Baltimore-based artist who uses photography to explore the human condition and his fascination with people and faces. A graduate of Yale University and the Stanford Graduate School of Business, Skip returned to his childhood passion for photography after managing a health sciences mutual fund led to three severe bouts of treatment-resistant depression. For Skip, carrying a camera is way to be fully alive in the world and a good excuse to meet new people. Skip has self-published eight artist books, including *Michaela's World*, *GingerSnaps*, and *Ventanas De Miguel*. He was the executive producer of the film *Return to Cuba: In the Footsteps of Walker Evans*, which premiered at the Camden International Film Festival (CIFF) in September, 2016.

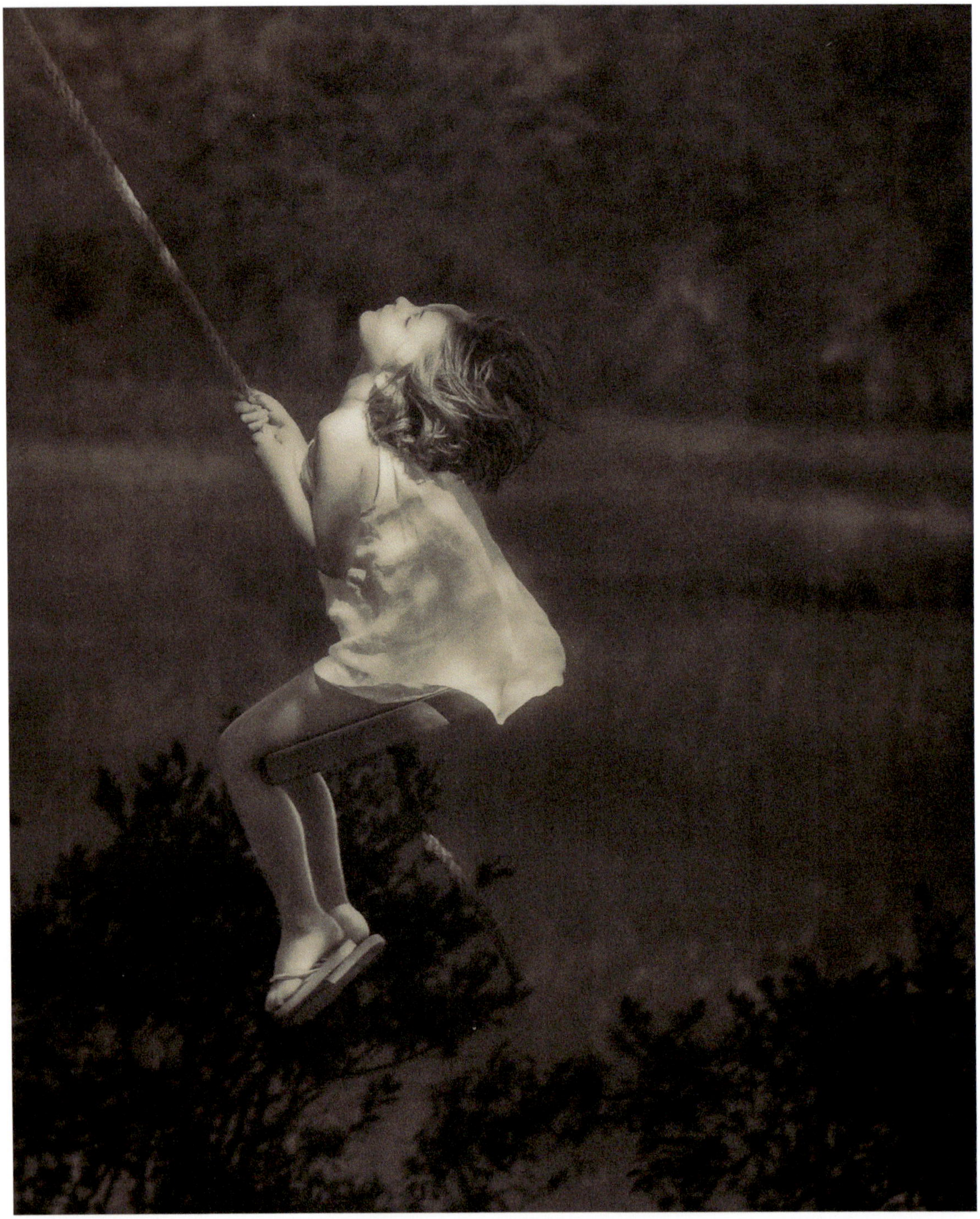

Figure 11.31 Skip Klein, *Swinging Into Another World*, 2018. Palladium-platinum print on Opalux paper, from PiezoDN negative of a full frame digital original, 22 x 18 cm. Courtesy Skip Klein.
Fe1:Pt:Pd = 4:1:3.

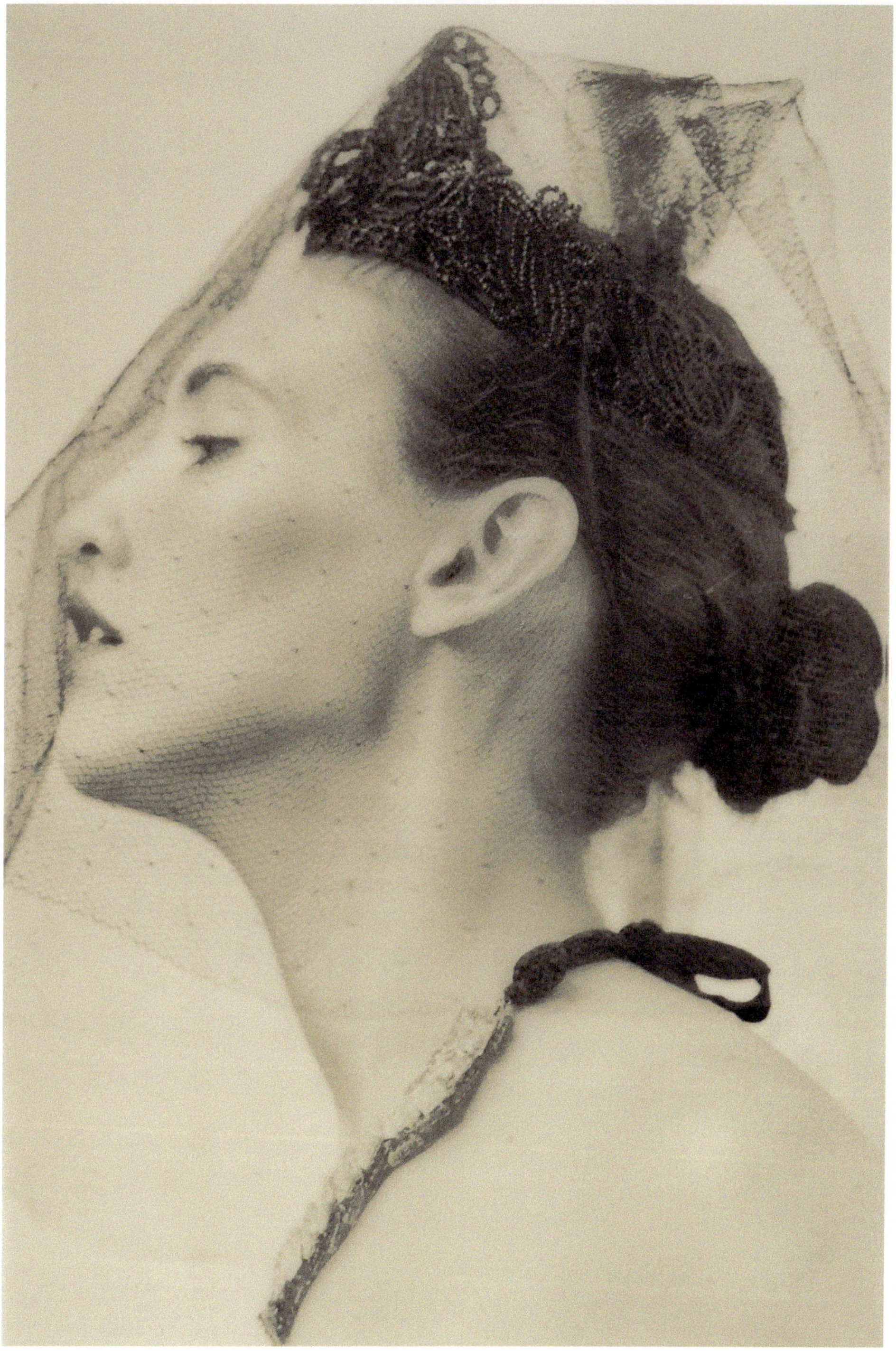

Figure 11.32 Skip Klein, *No Smoke and Veil*, 2018.
Palladium-platinum print on Opalux paper, from PiezoDN negative of a full frame digital original, 22 x 16 cm. Courtesy Skip Klein. Fe1:Pt:Pd = 4:1:3.

Pablo Kolodny

I express myself best when using historic photographic printing processes. They offer a velvety nuance which stands in contrast to life's harshness. The sharing of such prints lets me share an important goal: to create a peaceful ambience, to let myself and others meditate about improving the lives of others. I plunged into the field of photography because it made me think philosophically. I have studied and practiced photography daily for over 20 years, refining my technique and my thinking.

Advice about making printing-out Pt/Pd prints:

I find that the printing-out process, because I can preview, or inspect, during exposure, makes it easy to see the final print. It is really impressive, after having mostly made development Pt/Pd prints, to see the print coming out from the exposing unit as it "should" look. After exposure, the chemical process is quite simple, with no unexpected changes in density and/or contrast in the workflow. It has been tricky trying to set up ambient temperature and relative humidity conditions for my work room, but when I have it fine-tuned, the process is pretty consistent in terms of print color. The biggest disadvantage is how expensive the platinum and palladium salts have become lately.

Pablo Kolodny is a photographer based in Buenos Aires, Argentina. He mostly does historic printing techniques like Kallitype, Van Dyke brown and Platinum/Palladium. His interests range from social and political matters on one side to painting, sculpture, music, and other creative forms. He is currently working on an upcoming exhibition in Arboretum Lussich, Maldonado Department, Uruguay, for which 18 x 24 cm film negatives will be printed as platinum-toned kallitypes. He is a renowned photographer in Latin America and has been represented in Photo Patagonia 2019, an analogue Photo Festival where he exhibited some of his works and was in charge of a printing workshop.

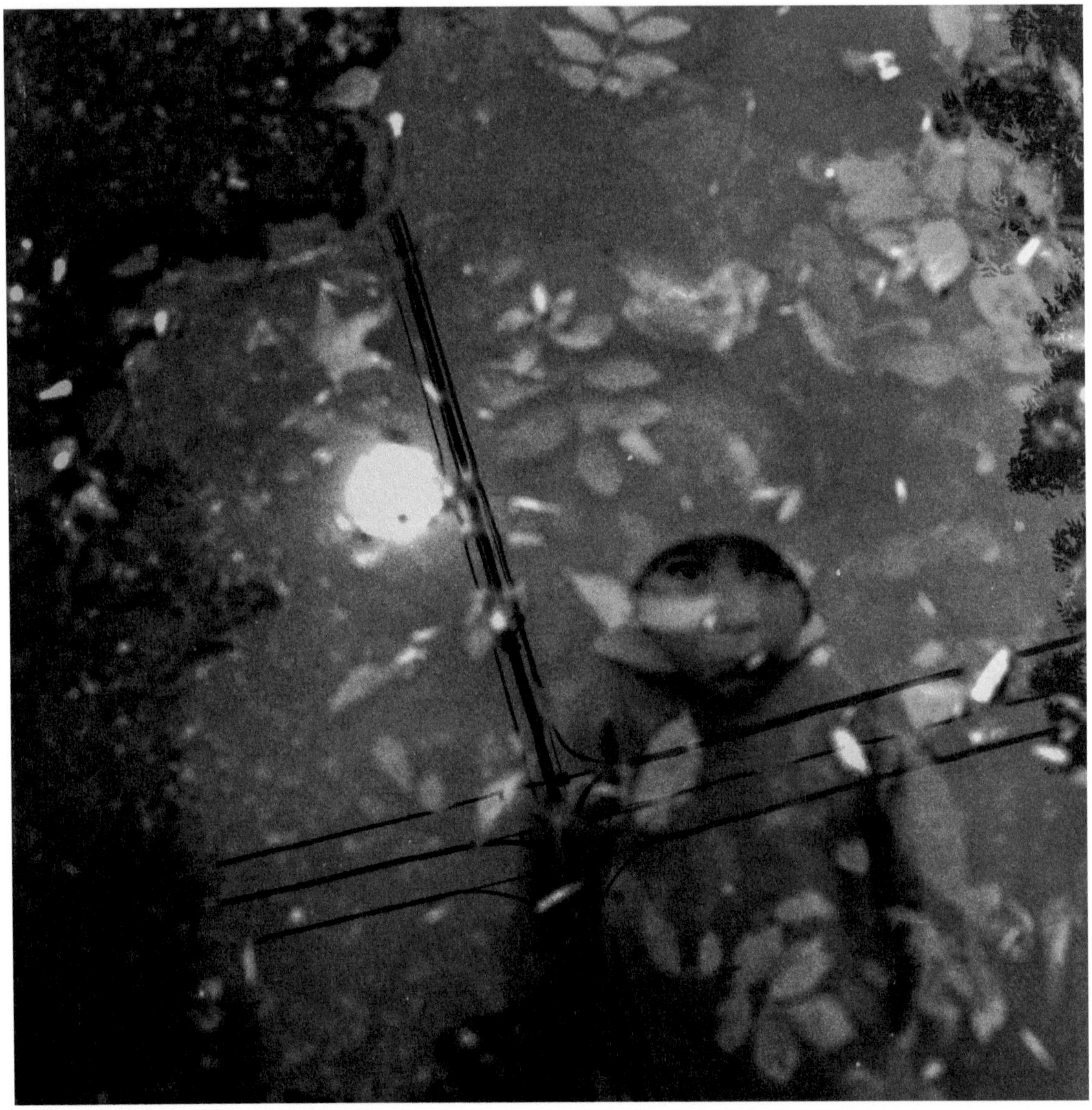

Figure 11.33 Pablo Kolodny, *Dream/Nightmare*, 1989. Platinum print on Arches Platine 300 gsm paper, from PiezoDN negative, 25 x 25 cm. Courtesy Pablo Kolodny. Brush-coated.
Fe:Pt:Pd = 1:1:0, hydrated at ca. 75% RH / 24°C.

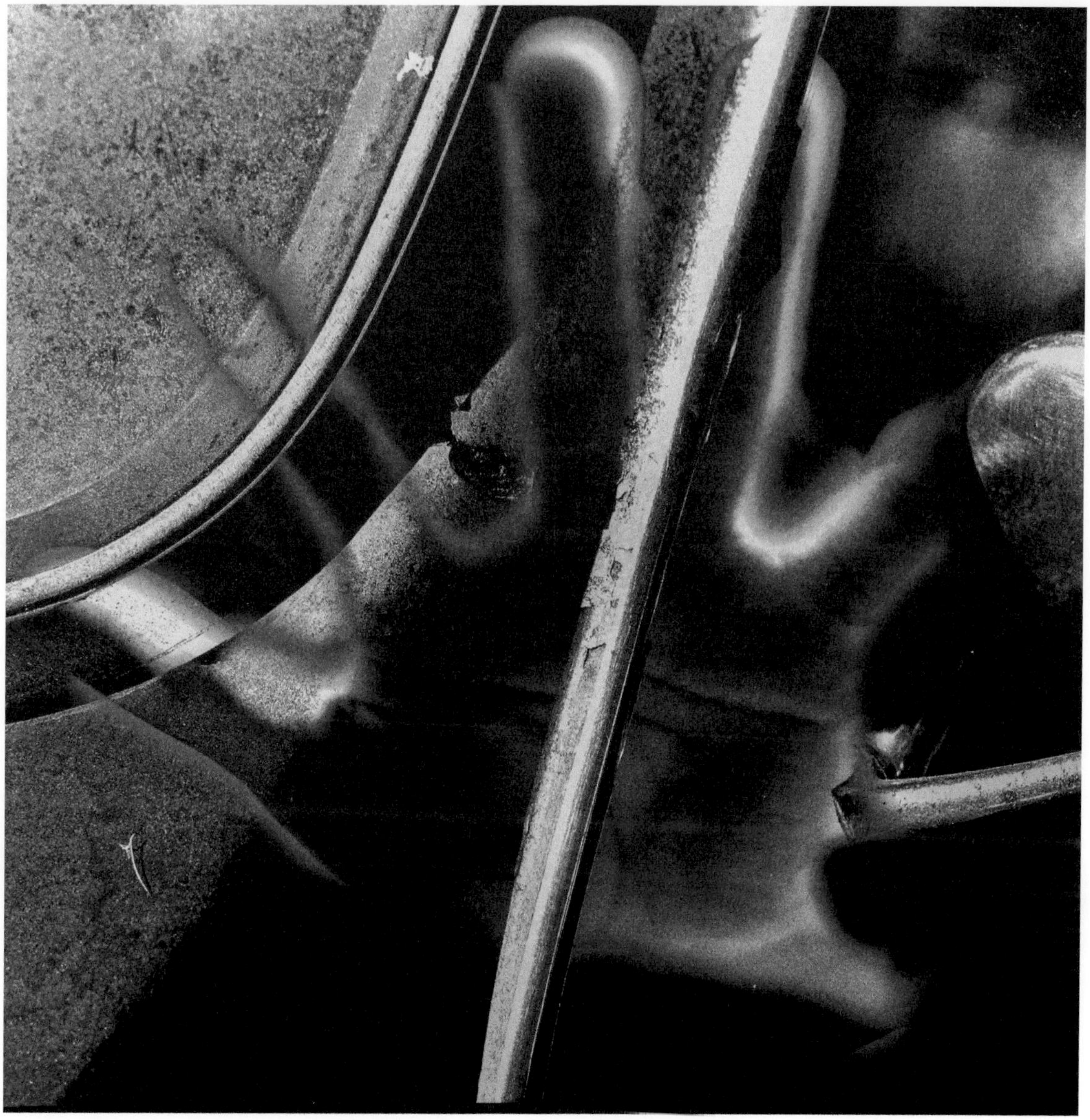

Figure 11.34 Pablo Kolodny, *DON'T*, 1989. Platinum print on Arches Platine 300 gsm paper, m from PiezoDN negative, 25 x 25 cm. Courtesy Pablo Kolodny. Brush-coated.
Fe:Pt:Pd = 1:1:0, hydrated at ca. 75% RH / 24°C.

Gilles Lorin

I use early photographic processes, and this forms my visual signature. My work has been described as a form of pictorialism, transported into the twenty-first century. I neither document nor travel around the world in search of the best snapshot. My photographs are not meant to be contemporary documents. Instead, I prefer to arrange, compose, manipulate light, and leave nothing to chance before I capture a sketch with the camera. The real work follows in the darkroom where I experiment with chemicals, papers, and techniques until I arrive at a desired way to depict my subject. The visual effect of this approach sometimes resembles that of a drawing or a painting. Most of my works are platinum-palladium prints and cyanotypes exposed on handmade papers such as Japanese Gampi or Arches. Some variants are backed with gold leaf. Altogether, this leads to a confrontation with being and non-being; I seek a transient or a supernatural atmosphere for most of my series. It results in a very personal view of the fragile beauty of our earth, revealed through peaceful, aesthetic photographic work.

Advice about making printing-out Pt/Pd prints:

I find that the printing-out process has many more advantages over the development Pt/Pd process. The light sensitive ammonium ferric oxalate is much easier to prepare, and more stable over time than the traditional ferric oxalate. This also leads to better accuracy and control of the prints. Using digital negatives, I find that I do not need to linearize as often as I used to with the development Pt/Pd process. There is slightly less granularity in the highlights. There are no black dots, which randomly plague development Pt/Pd platinum/palladium prints. I can get additional color control through humidity. The process is not compatible with some papers such as Japanese gampi—while it is still possible to print on these papers with printing-out, the development Pt/Pd process leads to much better results. Regardless of the type of process, start from the beginning and learn the tested formulas until you master them. From there, explore and make it your own. Only change one parameter at a time, so you understand what is happening. Photographic processes offer a rich language for artistic expression, with endless possibilities.

Gilles Lorin is an artist-alchemist who creates timeless images, carefully elaborated with historic photographic techniques. His platinum-palladium and cyanotype prints, on precious papers and occasional gold highlights, depict landscapes and still lifes with deep and subtle shades that evoke the human condition and the beauty of terrestrial things. Lorin grew up in the French Alps before moving to the United States at the age of fifteen. Educated as an art historian and archaeologist both in France and the United States, he worked for several years as an expert in Asian art before becoming a full-time photographer and printmaker.

Figure 11.35 Gilles Lorin, *Ying & Yang*, 2018. Platinum-palladium print on Arches Platine 300 gsm paper, from PiezoDN negative on Pictorico TPS-100 of full frame digital original, Leica Monochrom, 76 x 56 cm. Courtesy Gilles Lorin.
Fe:Pt:Pd = 4:1:3, hydrated ca. 70% RH at room temperature.

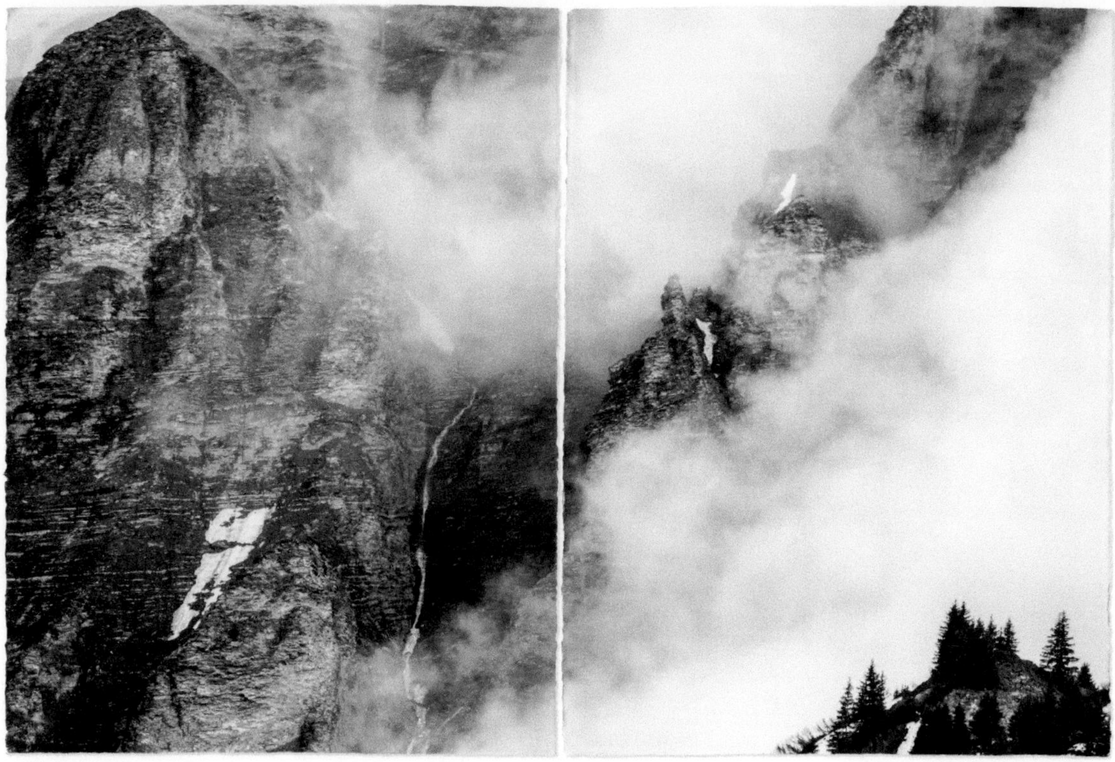

Figure 11.36 Gilles Lorin, *Land of the Gods*, 2018. Palladium print on Arches Platine 300 gsm paper, from PiezoDN negative on Pictorico TPS-100 of full frame digital original, Leica Monochrom, 76 x 112 cm (diptych). Courtesy Gilles Lorin.
Fe:Pt:Pd = 1:0:1, hydrated ca. 70% RH at room temperature.

Paula Luttringer

I am a visual artist creating images and collecting testimonies of political violence and of the ways women survivors of political violence are marked by it. My work reveals an era in history that is specific to Argentina but has features in common with political violence wherever it erupts in the world. Survivors of political violence have some problems in common. We have blank spaces in our memories, and sometimes we get unexpected flashes of memory. The flashes are not words but images, mental pictures. They don't come when we want them to come. We can't erase the mental pictures that arrive, and we can't change them. If we try to not see them, they insist. The flashes arrive when we are relaxed, having good moments in our lives. They appear when we are happy, when we are working on other things, when we are not expecting them. These flashes are images that exist in our minds and cannot be shared with anyone. Photography allows me to create images that suggest what is in my mind. The photographs are not illustrations of what is in my mind, but they are reminders, suggestions. Unlike flashes of memory, photographic images can be shared with others. When I create photographic images, I am finding a way to keep the past which was not recorded, the past which was hidden, the past which intrudes into the present through these flashes of memory. The images that arrive in my mind are not the same as the images that arrive in another woman's mind, but sometimes my photographs have reminded other survivors of their own flashes of memory. And so through photography, I am a keeper of memory, and photography helps me to cope with my gaps in memory, my memories, and those unexpected flashes of memory.

It's difficult to say if it is the detail in both the shadows and highlights, or the matte rendering of the image, that captivates about platinum-palladium prints. Thinking deeper, it is the immersive sensation that I have when I look at a print, as if I could travel deeply within the image. I love the fact that our labor and heart are still so intrinsically necessary for the image to appear, in a lab, using a technique that mocks today's rushing needs for time and productivity. These prints seem to sneer at the passage of time, and—in a world where everything seems to be destined to disappear faster and faster—they mirror our human destinies: they will still be around when we, ourselves, will have quit the ship long ago.

Paula Luttringer is an Argentine woman who was a political prisoner in one of more than 520 Secret Detention Centers (SDCs) in Argentina during Argentina's Dirty War between 1976 and 1983. In 1977 she took exile after having been kidnapped and held for five months in an SDC close to Buenos Aires. Upon returning to Argentina in 1995, she turned to photography as a means of expression. The subject matter of her photographic work concerns the meeting point between her country's recent history and her own. She considers photography and film as a way of engaging in the process of re-creating a life after disappearance. Luttringer received a John Simon Guggenheim Memorial Foundation Fellowship in 2001. In the year 2000, her project *El Lamento de los Muros* was distinguished with the Artistic Award granted by the National Art Fonds (FNA) in Argentina. Globally recognized and honored with awards, her work is included in the collections of the Museum of Modern Art (MAMBA) and the National Museum of Fine Arts (MNBA), in Buenos Aires, Argentina, the Museum of Fine Arts (MFAH) in Houston, TX, the George Eastman House Photography Collection in Rochester, NY, the Portland Art Museum in Portland, OR, USA, and the BnF, Paris France, as well as various private collections.

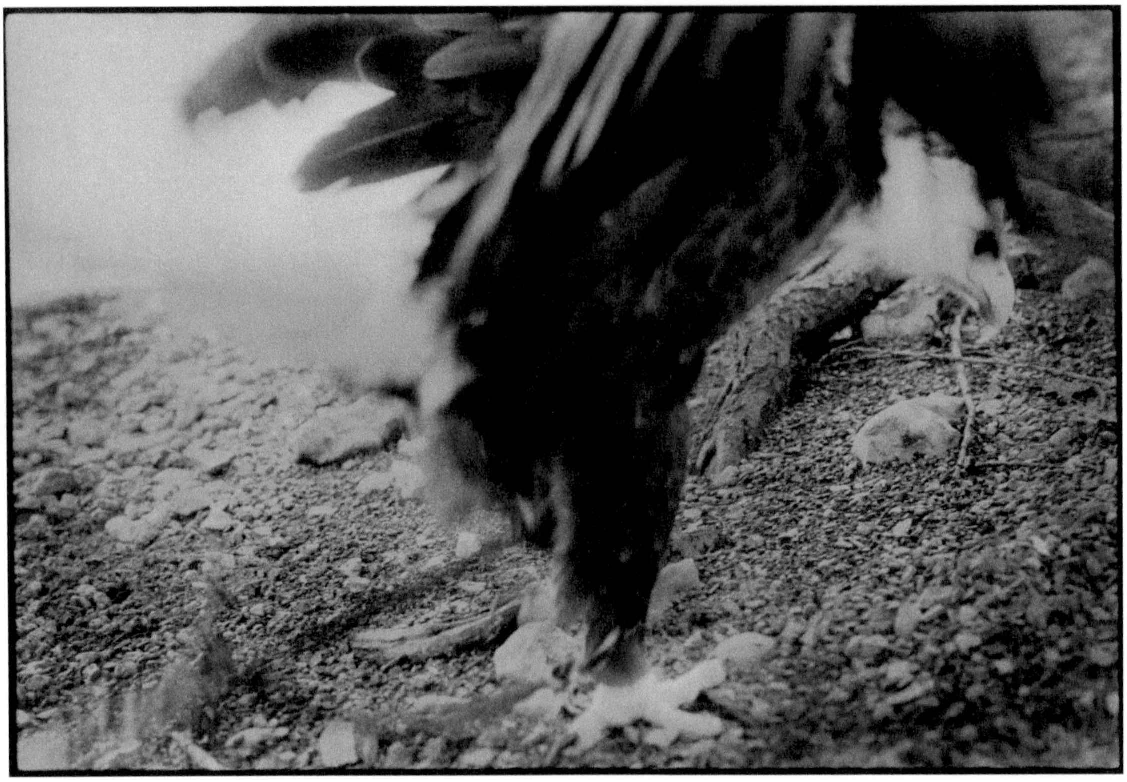

Figure 11.37 Paula Luttringer, from the series *"Bestiario" Aguila,* 2000. Platinum-palladium print on Revere Platinum paper, from PiezoDN negative of 35mm film original, 19.5 x 29 cm. Courtesy name. Fe:Pt:Pd = 2:1:1, hydrated for 90 min. at 76% RH / 20°C.

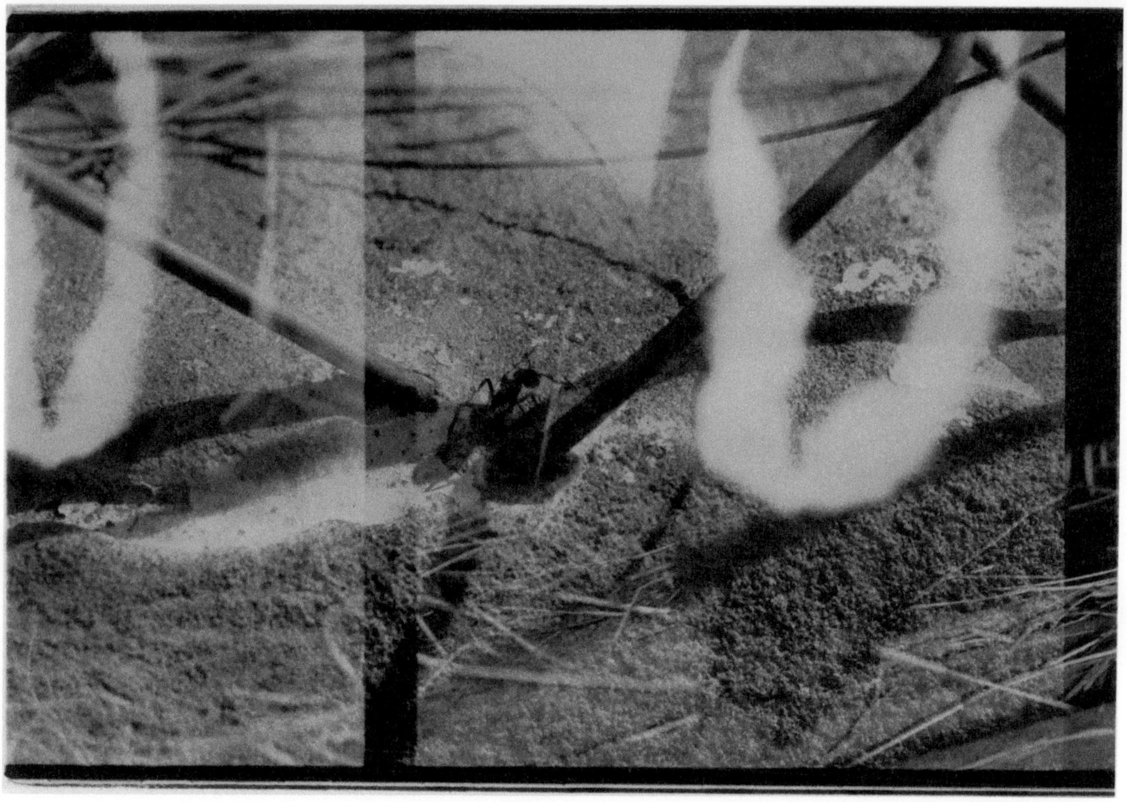

Figure 11.38 Paula Luttringer, from the series *"Bestiario" Hormiga,* 1998. Platinum-palladium print on Revere Platinum paper, from PiezoDN negative of 35mm film original, 19.5 x 29 cm. Courtesy name. Fe:Pt:Pd = 2:1:1, hydrated for 90 min. at 76% RH / 20°C.

Rachel Malde

My photographs describe singular moments when silence and noise collide, when ancient and new grow together, when softness gives way to the edge, when light stabs through darkness. My photographs illuminate these contradictions and offer balance. Reconciliation happens within the frame. I strive for photographs that are beautiful, haunting, delicate, and deep.

I am standing in a grove of ancient olive trees, the sunlight is harsh, unrelenting. The air smells of salt. The earth below me has been worked for centuries, turned in and over, planted and harvested. I am not unaware of the millions of footsteps that have walked these paths before me. Yet, I am there, with my camera. The air has never been exactly as it is now, the light has never shone on the trees exactly as it does now, the combination of salt, water, olive branch, and breath has never been the same. A photograph is made, a token. In that moment, the weight of age and the lightness of the instant meld.

In the most dismal of times, we sense and are perhaps more connected to the darkness around us. At other times, the light has a more powerful pull. In both situations, however, the other is present and palpable. The darker the shadow, the more intense the highlight. Photography has provided the tool I need to see both sides. It has become my voice for these contradictions and the medium between myself and the world I inhabit.

Advice about making printing-out Pt/Pd prints:

Before you start printing, and regularly after you are familiar with the process, make an appointment with the print viewing department of a major museum to look at prints. Look carefully at platinum, palladium, and platinum/palladium prints by historic masters of the process, Frederick H. Evans, Julia Margaret Cameron, Gertrude Käsebier, and many others. Delight in how the photographs are at once gentle and strong, present and elusive.

Rachel Malde is a photographer and educator living in Sewanee, Tennessee, where she teaches visual art to middle and high school students at St. Andrew's-Sewanee School. For many years, Malde has traveled to Kingston, Jamaica to photograph and assist in schools there. She is currently working on a series of photographs about memory that was inspired by Langston Hughes' poem, "Dreams."

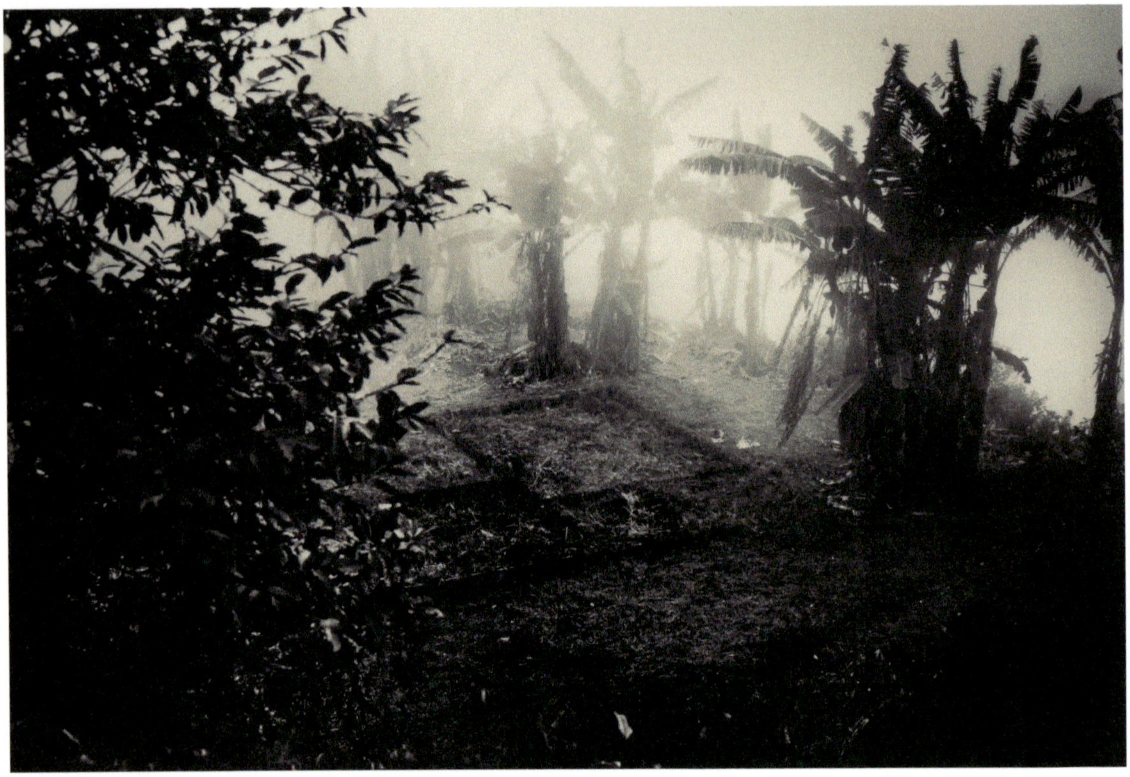

Figure 11.39 Rachel Malde, *Foundation. Blue Mountains*, 2014. Platinum-palladium print on Revere Platinum paper, from PiezoDN negative of full frame digital original, 19.5 x 29.5 cm. Courtesy Rachel Malde. Fe:Pt:Pd = 2:1:1, hydrated for 90 min. at 76% RH / 20°C.

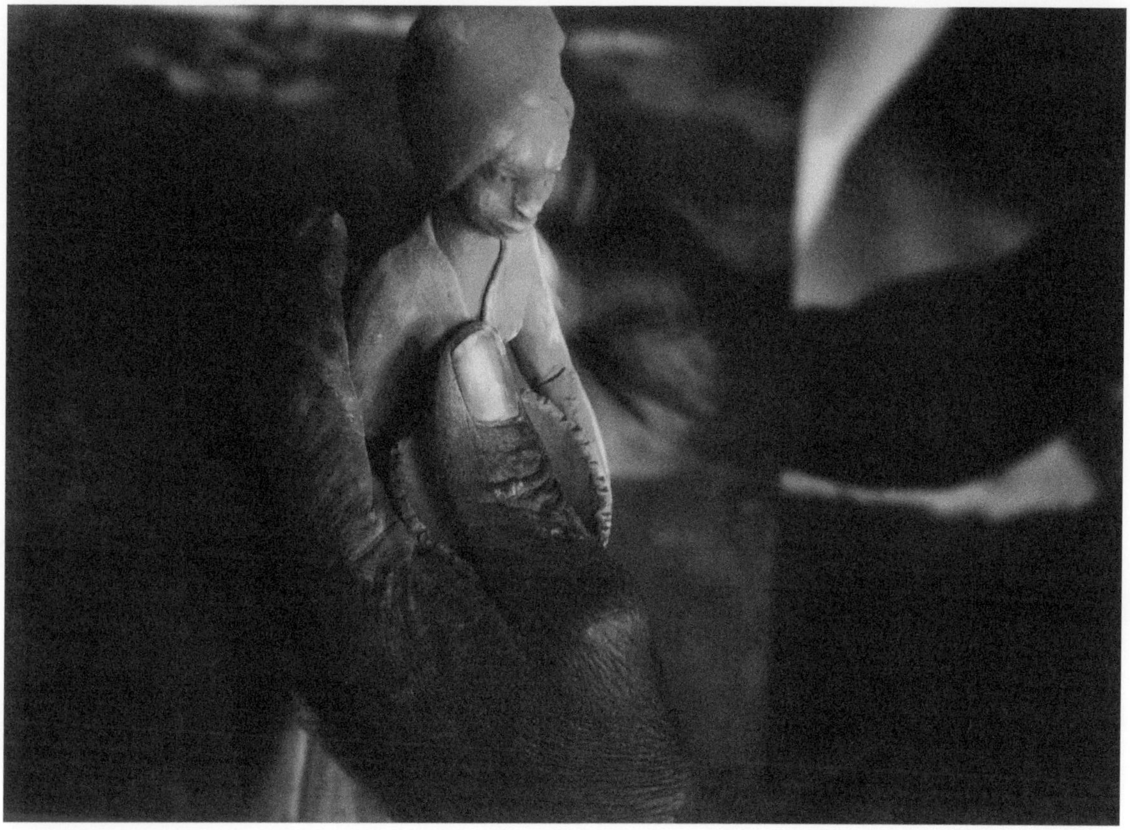

Figure 11.40 Rachel Malde, *Nanny of the Maroons, Trenchtown*, 2016. Platinum-palladium print on Revere Platinum paper, from PiezoDN negative of full frame digital original, 19.5 x 29.5 cm. Courtesy Rachel Malde. Fe:Pt:Pd = 2:1:1, hydrated for 90 min. at 76% RH / 20°C.

Caroline Minchew

My work focuses on the intersection of landscape and personal story, and the idea of psychogeography: how spaces may be understood from how one feels as opposed to what one knows. I work with a large format 8 x 10 inch camera and print using a variety of alternative and historic photographic processes. I heavily rely on hunch and try to not put too much thought into the technical aspects of shooting with an 8 x 10. I really enjoy the process of getting lost in it. In some ways it's a complete pause from reality and lets me translate what I see into what I want to present to the world. And then there is that peaceful state when you look at the negatives. Printing with platinum-palladium continually captivates me. Each print still feels like a gift as it comes out. I love its texture, and how easily I can get lost staring into the print. I also love how viewing other platinum prints encourages me to create more. When you combine 8 x 10 shooting with platinum printing, the process becomes a cyclical movement of choices and intuition that result in an object to get lost in.

Advice about making printing-out Pt/Pd prints:

This is a physical process, in making the sensitizer and coating the paper, and the ability and satisfaction of a clean coated piece of paper. My background is certainly not in chemistry, but I do find joy in following the "recipe" for creating sensitizers, both new and from historic journals. I love the challenge of figuring out the nuances of a print by deciphering it visually. Why is it grainy? Is the hue too brown? Being willing to methodically solve problems and understand how to fine-tune the process is important. I also enjoy that it can be carried out without the necessity of a darkroom. I have even printed in a backyard, washing prints with a garden hose. Reading manuals and instructions is daunting— the process is not a casual one. And it can be costly. But then you make a beautiful print, and it all becomes worth it. Sort out what works for your own process and understanding. You can read how other photographers print, and may be convinced that this must be the only way to produce a beautiful print, but that is not necessarily true. Lean into what works for you, and what you can handle. I don't mean skip necessary steps or compulsively create a sensitizer that won't work chemically. For example, I have recently scrapped the (tank) hydration process from my workflow. I instead rely on the humidity of where I am printing, and now embrace that the location can have a profound impact on the final print. I also would advise you to improvise your setup to the best of your ability. Printing-out Pt/Pd works without a darkroom, and that should be embraced!

Caroline Minchew is a photographer and artist currently working from Virginia. She received her BA in Fine Arts from The University of the South, Sewanee, in 2014. Caroline has exhibited in galleries throughout the United States and has works housed in the permanent collections of the National Museum of American History and the National Gallery of Art's Photograph Study Collection in Washington, DC. Caroline is a contributing author in *Platinum and Palladium Photographs: Technical History, Connoisseurship, and Preservation* (Constance McCabe, ed. Washington: American Institute for Conservation, 2017).

Figure 11.41 Caroline Minchew, *Ephemeral Pool, Sewanee, Tennessee,* 2019. Platinum-palladium print on Arches Platine 145 gsm paper, from original 8 x 10 inch negative developed in Pyro PMK, 20 x 25.5 cm. Courtesy Caroline Minchew. Fe:Pt:Pd = 3:2:1, hydrated in varying room conditions.

Figure 11.42 Caroline Minchew, *Forest, Sewanee, Tennessee*, 2019. Platinum-palladium print on Arches Platine 145 gsm paper, from original 8 x 10 inch negative developed in Pyro PMK, 20 x 25.5 cm. Courtesy Caroline Minchew. Fe:Pt:Pd = 3:2:1, hydrated in varying room conditions.

Elizabeth Motlow

I worked passionately with my photography for about 35 years. I did not seek fame and recognition, but knowledge of myself. One day, at the end of a beautiful artistic residency in New Mexico, I made my last 8 x 10-inch photograph, a portrait of myself disappearing. I made one printing-out platinum-palladium print from the negative, and I knew that photographing my life was finished, or rather it was completed.

Now, I raise bison, sheep, and vegetables. The dilemma we find ourselves in currently, in 2020, has been a very long time coming, although many people are surprised by it. For the past 25 years or so, I have been getting ready for it. I learned to do this, to really see what was happening in the world around me through my photography. What I made pictures of, was what I was thinking about. But what I saw, is what I still see; an unfolding of time and reality through my personal experience.

Advice about making printing-out Pt/Pd prints:

In this career driven world, a good question to ask oneself is why you would want to learn a photography process that was almost archaic one hundred years ago. Just to answer requires commitment, in that you have to get out there and find real platinum prints to look at. It won't be easy, but not impossible. What you will find might amaze you. That is, you won't be able to tell if they were made over a hundred years ago or today. Everything you know as a student of art and art history will be called into question by this simple truth. No other process gives you that connection with a subject, or that lasting connection with yourself as a photographer. A well-made platinum print is close to eternal.

Elizabeth Motlow is a farmer and environmentalist, living in the Appalachian foothills not far from her birthplace in south-eastern Tennessee. She was one of the earliest adopters of the printing-out process as formulated by Mike Ware and Pradip Malde in the mid-1980s. She received an MFA in 1981 from the Rochester Institute of Technology.

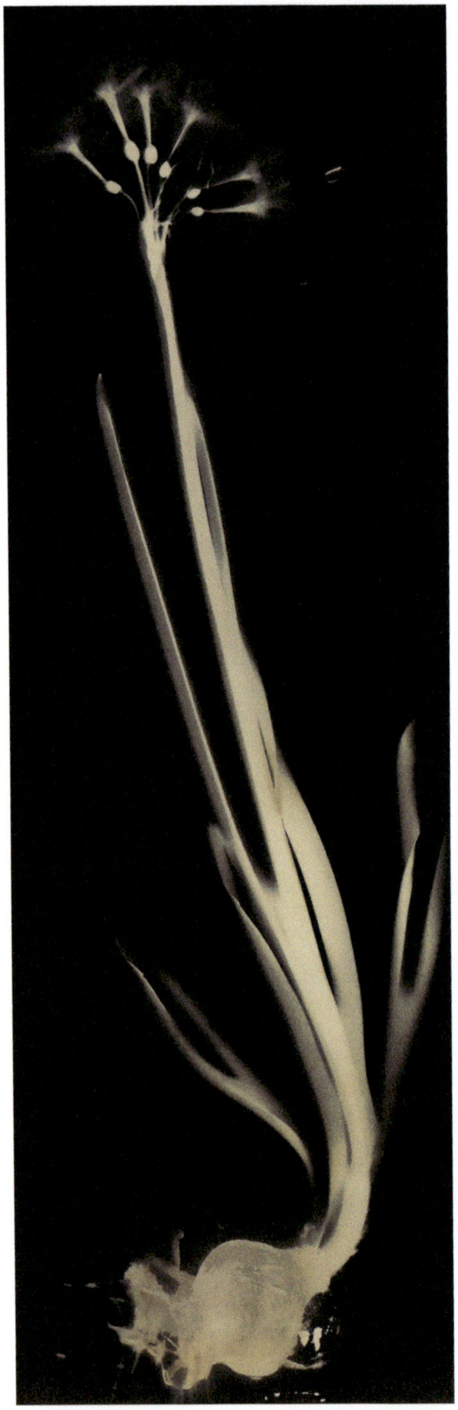

Figure 11.43 Elizabeth Motlow, *Paperwhites*, ca. 2000. Platinum-palladium print on Fabriano 5 paper, photogram, 44.5 x 14.5 cm. Courtesy Pradip Malde. Fe:Pt:Pd = 2:1:1.

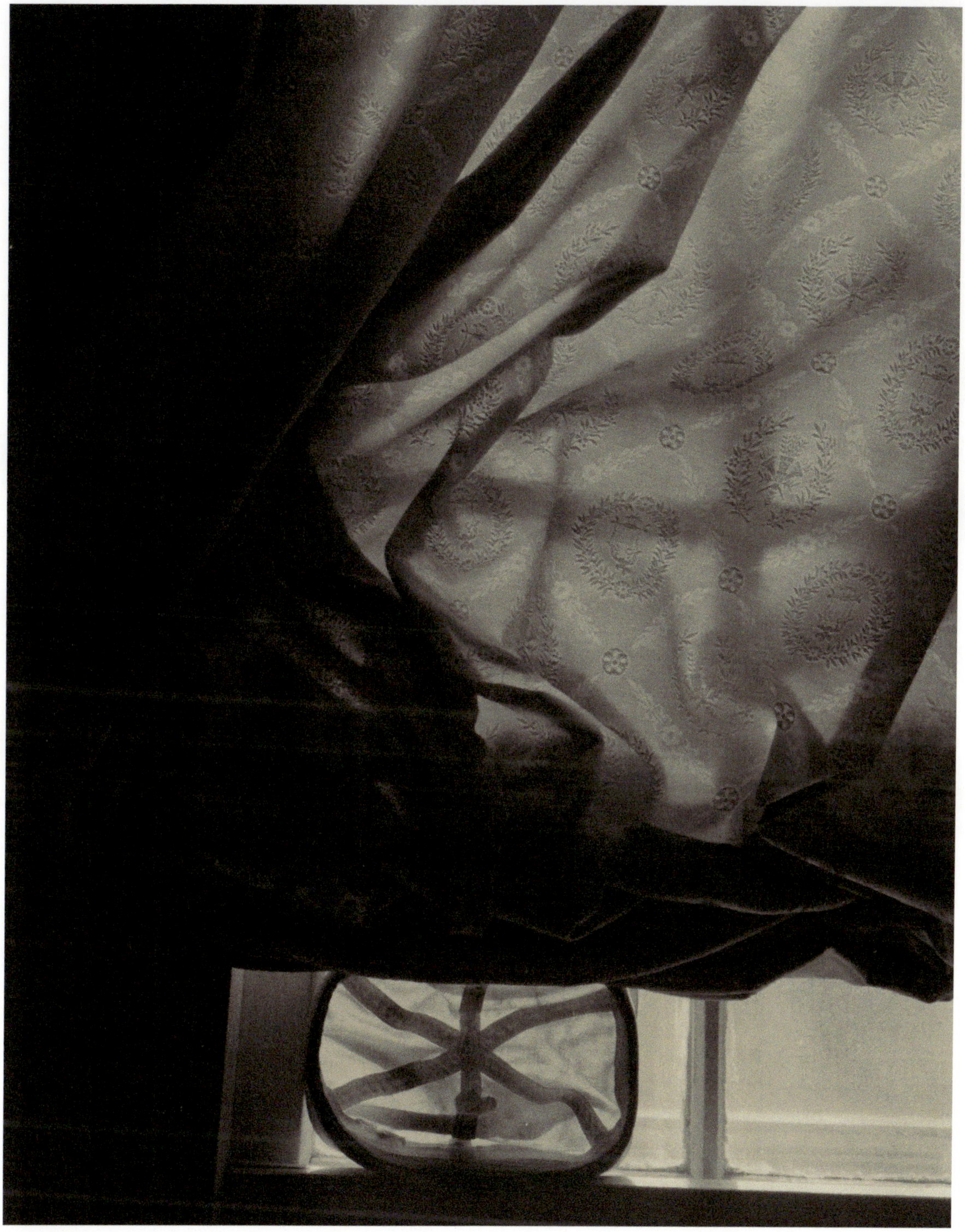

Figure 11.44 Elizabeth Motlow, *Robert Motlow, Crushed by the Weight of his Family*, 1992. Platinum-palladium print on Fabriano 5 paper, from original 8 x 10 inch negative, 24.5 x 19 cm. Courtesy Pradip Malde. Fe:Pt:Pd = 2:1:1, hydrated for 20 min. at 100% RH / 20°C.

Shaun O'Dell

My work is focused on transformational events—geological, historical, and metaphysical—that have significantly impacted the human mind. I examine how our awareness of constant dramatic change affects the ways in which we relate to the world and each other. I am currently researching the origin and decline of our engagement with mythology and its influence on art and consciousness. I am specifically interested in mimesis and its psychological function in our earliest and transformative efforts to produce representational images as a mechanism to examine reality and generate empathy.

Printmaking has been a part of my practice and influenced my painting and drawing process for several decades, as well as sculpture, photography, and video. Remediation has been a way for me to create new ideas for painting. I was first exposed to the Pt-Pd process in 2018 when I collaborated with photographer Angela Berry on a print project produced for the Vulcan Mountain Foundation Residency that combined collage and a range of alternative photographic processes. I have since pursued the use of the medium, because of its unique formal quality that combines characteristics of photography with intaglio printmaking. It has generated a new method for me to work between analog and digital techniques of drawing, painting, and printmaking to ultimately produce an image using UV exposure and a purely photographic process. I like how the print is made with light, but creates an impressive range of black tones that sit with depth on the paper. The process is involved and requires precision chemically.

Shaun O'Dell received a BA from the New College of California, San Francisco, in 2002 and an MFA from Stanford University in 2004. His work explores the intertwining realities of the human and natural orders. His work has been exhibited widely in the US and internationally. O'Dell is represented in New York by the Susan Inglett Gallery, in Houston by the Inman Gallery, and in San Francisco by Gallery 16 and on Long Island by Halsey/McKay. He has won numerous awards and honors including the Tournesol Award (2009, Headlands Center for the Arts), Diebenkorn Teaching Fellowship (2006, SF Art Institute), Artadia Award (2005, San Francisco) and The Fleishhacker Foundation Award in 2002 His work is included in a number of permanent collections, including the Museum of Modern Art, New York, the Whitney Museum of American Art, the Bronx Museum of Arts, the de Young Museum, the San Francisco Museum of Modern Art, Berkeley Art Museum, and the DESTE Foundation of Contemporary Art. O'Dell lives and works in San Francisco.

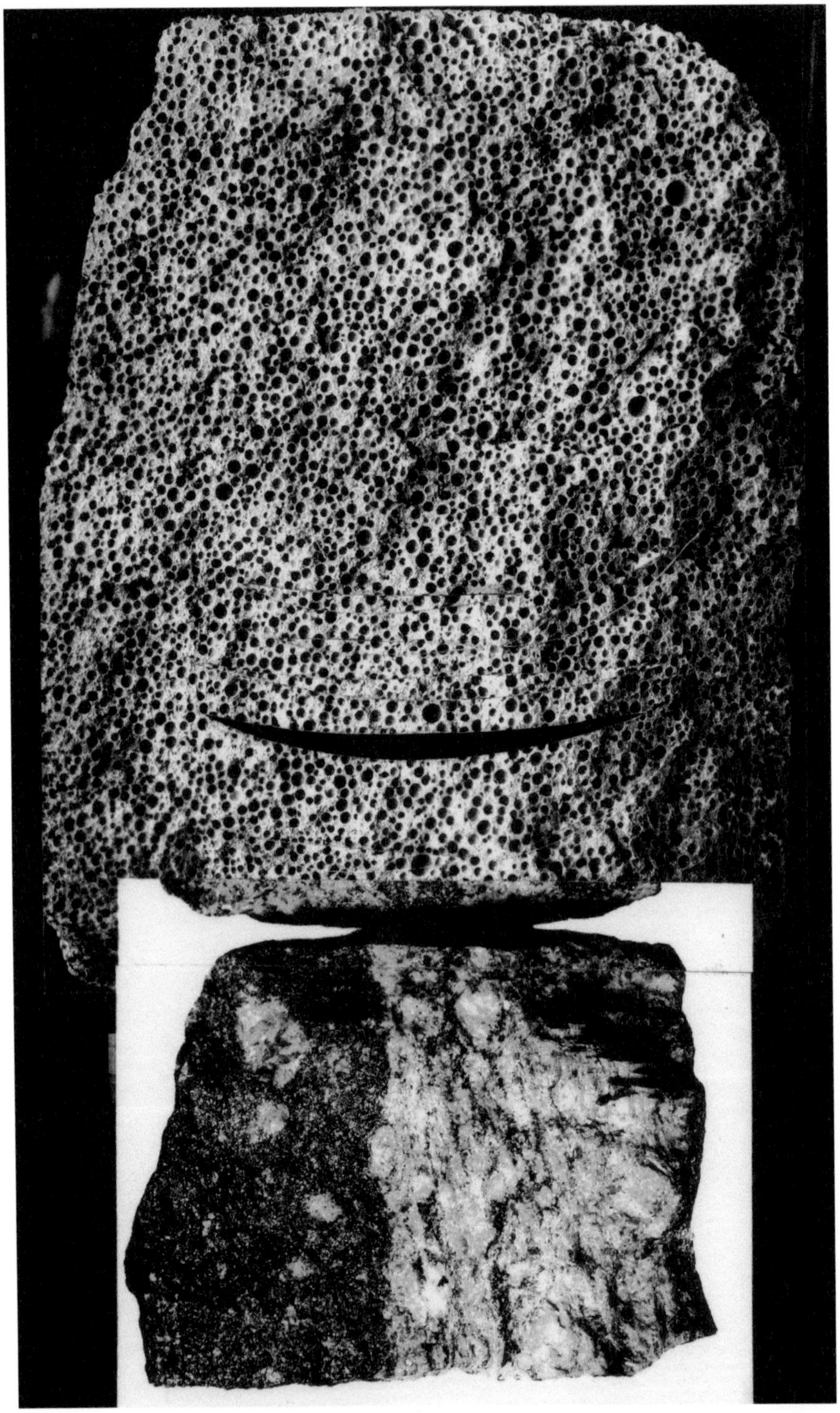

Figure 11.45 Shaun O'Dell, *Untitled*, 2020. Platinum-palladium print on Revere Platinum paper, from PiezoDN negative of digital original, 20.5 x 29 cm. Courtesy Shaun O'Dell. Fe:Pt:Pd = 2:1:1, hydrated for 90 min. at 76% RH / 20°C.

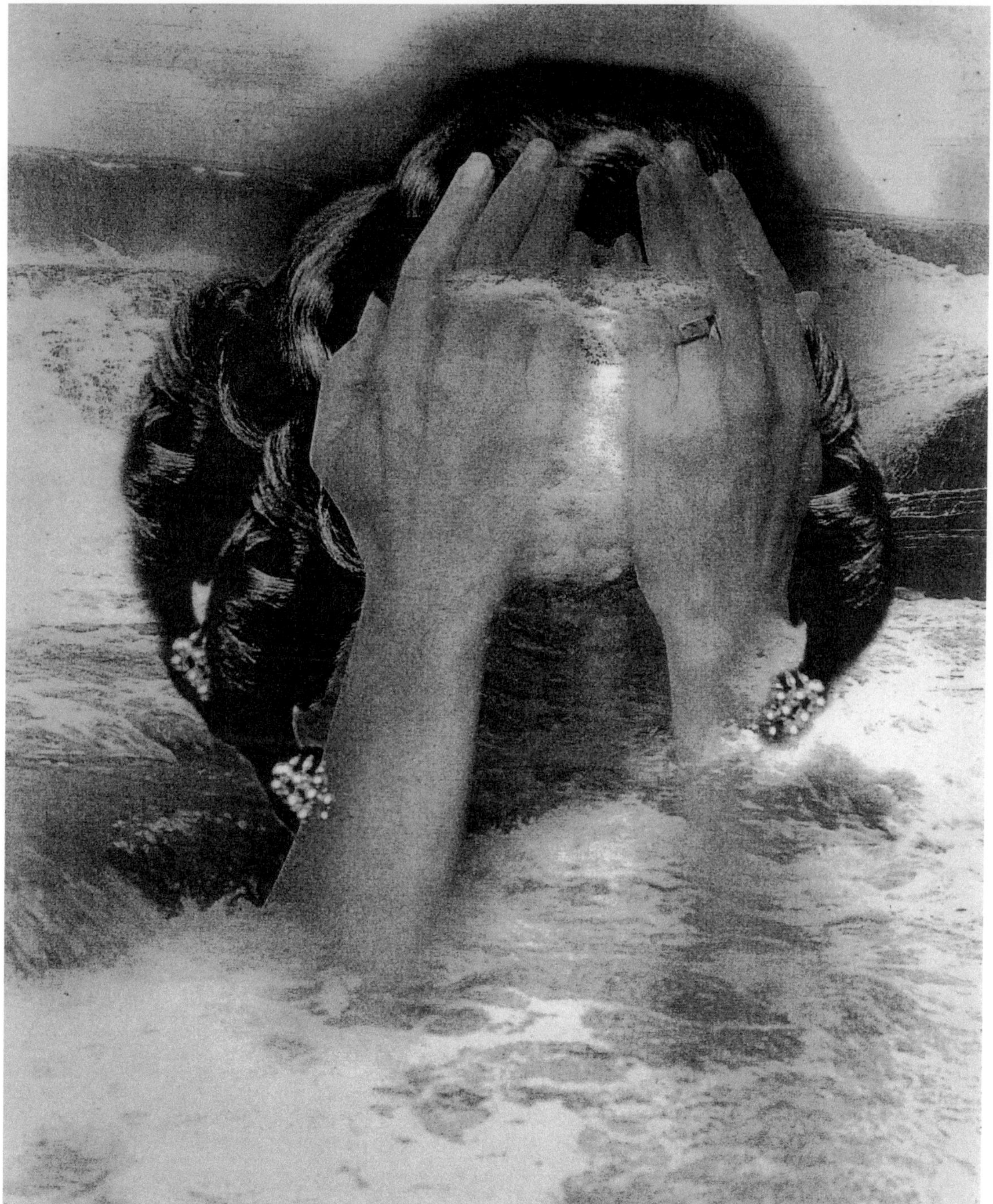

Figure 11.46 Shaun O'Dell, *Margaret Waterfall*, 2020. Platinum-palladium print on Revere Platinum paper, from PiezoDN negative of digital original, 20.5 x 29 cm. Courtesy Shaun O'Dell. Fe:Pt:Pd = 2:1:1, hydrated for 90 min. at 76% RH / 20°C.

Rondal Partridge

I know sometimes I do things people won't like, but I do it anyway. Most people make photographs for other people, and they bypass their own sensibility. I photograph for myself.

I never think of myself as an artist. I think of myself as making a point.

Rondal became interested in the platinum process in the 1980s, while he was printing Imogen Cunningham's negatives for the Trust, and coincidentally, when his own work in photography was becoming more centered around his still-life work in the studio. His studio consisted of a corner of the living room of his house, with a large south-facing window giving him access to natural light. His negatives ranged in size from 2¼ to 11 x 14 inches, depending upon his interests, inspiration, and availability of cameras and film. For Rondal, platinum printing was not a goal unto itself, but a means to a print that carried all the tonality, warmth, and tactile qualities he was after in his images.

Son of the renowned photographer Imogen Cunningham, Ron Partridge (1917–2015) began helping his mother in the darkroom at the age of five, standing on an apple box to reach the sink. By 17 he was a skilled photographer and became Dorothea Lange's apprentice, driving her up and down the back-roads of California as she created her well-known images of migrant laborers during the Great Depression. In 1937 and 1938 he worked with Ansel Adams in Yosemite National Park; carrying Ansel's large format camera up the mountains by day, and printing in his darkroom in Yosemite Valley by night.

Intimately associated with the great California photographers of his time, Partridge absorbed all the techniques his teachers could give him, yet he wore this lineage lightly, dedicating himself to following his own unique path. For over 75 years he was a professional photographer, using his camera quietly for his own reflections, but also as his voice and sometimes his megaphone. His subjects can hardly be categorized, he was so prolific: pursuing telling moments of the changing California landscape, making personal family portraits, creating strong architectural images at the height of mid-century design—all while experimenting with both some of the oldest and also some of the latest techniques in the photographic world.

These printing-out platinum-palladium prints were made by Pradip Malde for the Ron Partridge Archive.

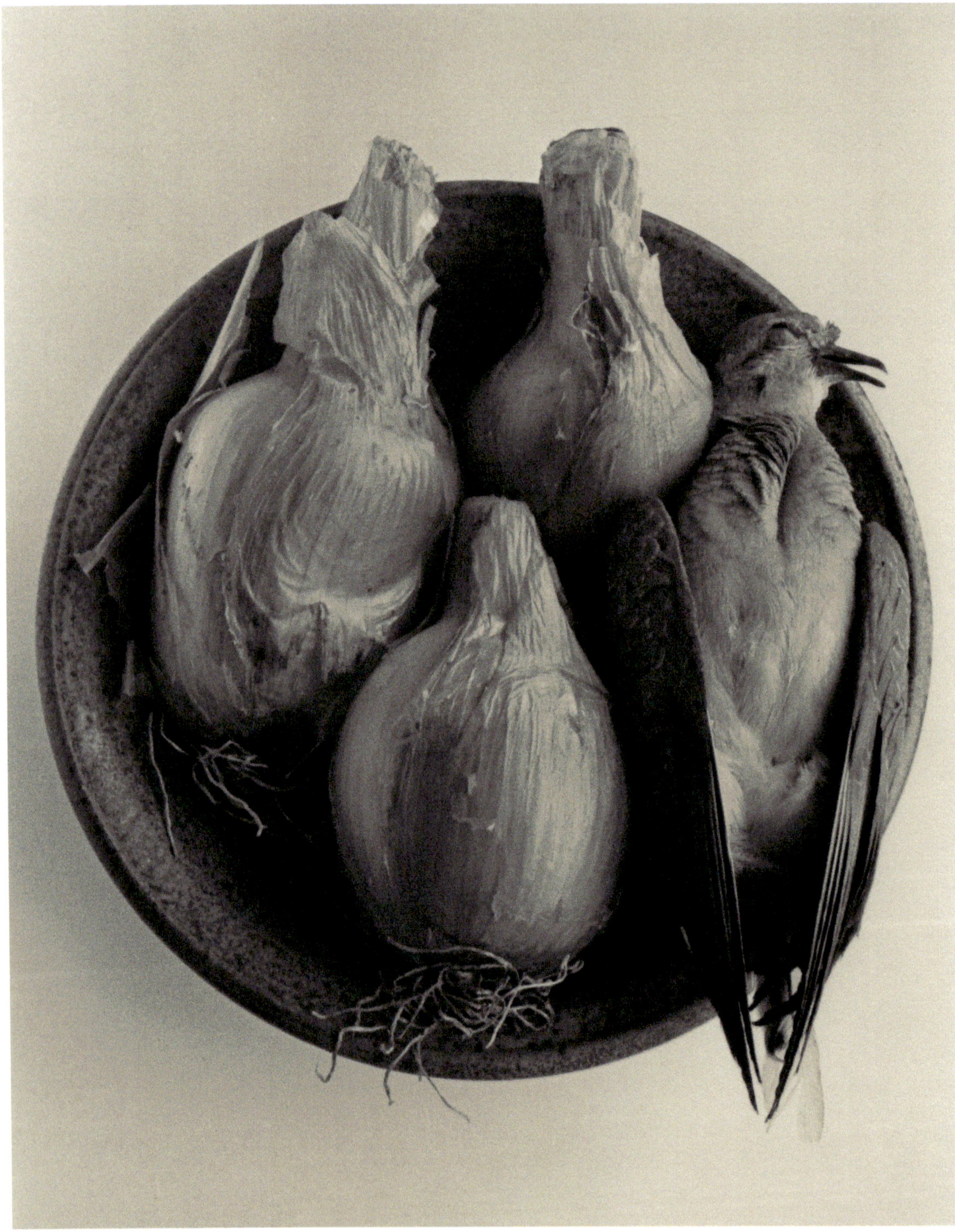

Figure 11.47 Rondal Partridge, *Bird and Onions*, 1989. Platinum-palladium print on Arches Platine 145 gsm paper, from PiezoDN negative of 35mm original, 25.2 x 217.8 cm. Courtesy Rondal Partridge Archive. Fe:Pt:Pd = 2:1:1, hydrated for 90 min. at 76% RH / 20°C.

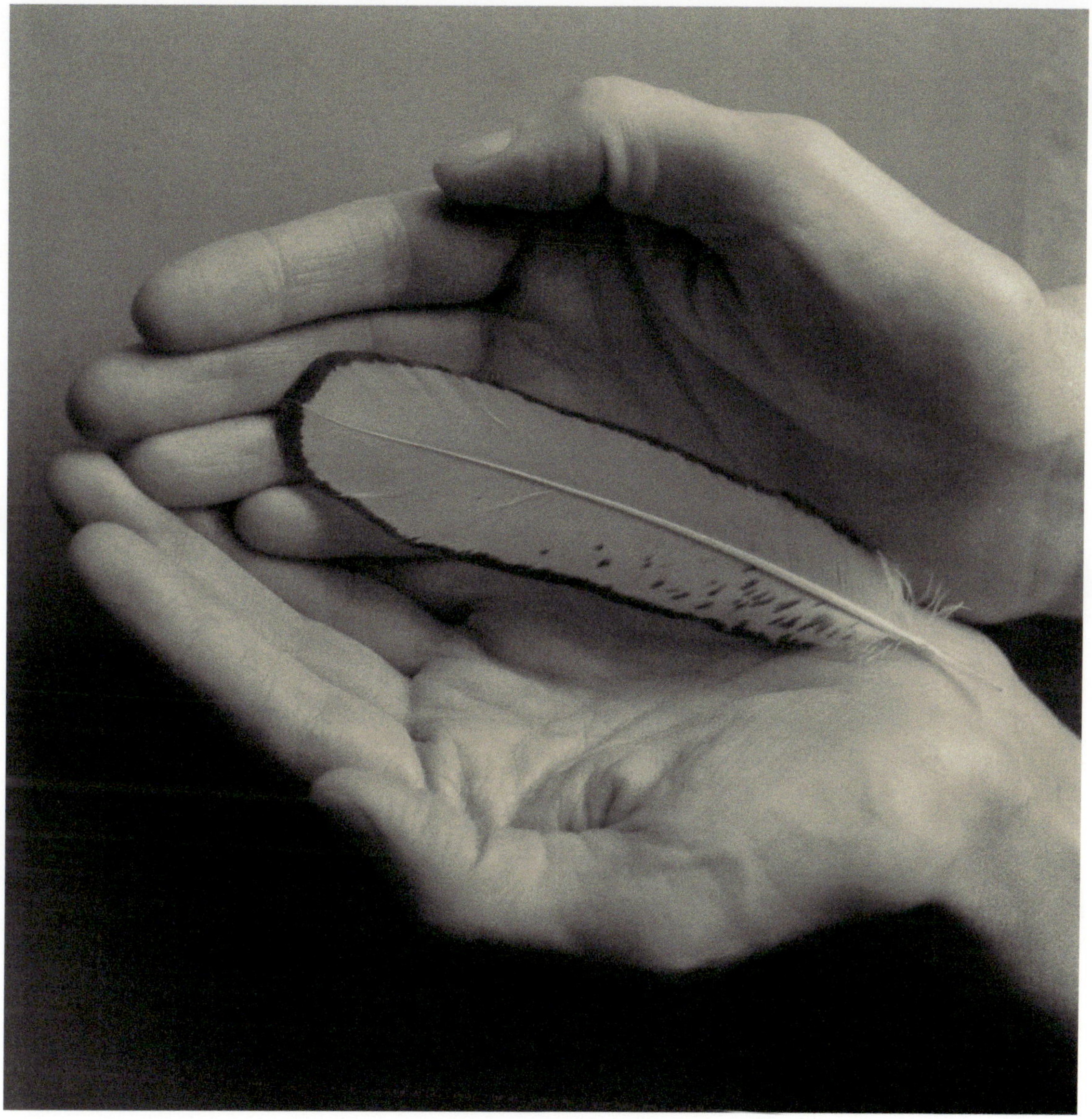

Figure 11.48 Rondal Partridge, *Betsy's Hands and Feather*, 1989. Platinum-palladium print on Arches Platine 145 gsm paper, from PiezoDN negative of 6 x 6 cm. original, H x W cm. Courtesy Rondal Partridge Archive. Fe:Pt:Pd = 2:1:1, hydrated for 90 min. at 76% RH / 20°C.

Greg Petropoulos

Photography is a pursuit that continually intrigues me and draws me in to discover new approaches and new processes. Film photography remains my favorite starting point, but with ever growing choices, I find that I am always a student experimenting with both old and new technologies. I enjoy being an explorer within a field that is constantly changing, but with a rich and varied history to learn from.

Advice about making printing-out Pt/Pd prints:

There are two advantages to the printing-out process that I appreciate. The first is the level of control over tone (brown to black), contrast, and other important factors. By controlling paper hydration, I can create, and recreate, the image I intend. The second, which is by far the most important, is the lovely tonal gradation that the printing-out process provides. While many photographers, and viewers, prefer high contrast images, the beauty of subtle changes in the shadows, highlights and mid-tones is often indescribable. To be able to render these subtleties is a gift, almost akin to magic. I made the most progress with the printing-out process when I committed to a daily routine for several weeks. I would make two to four exposures with my 4 x 5 camera during the late afternoons. Negatives were developed and dried during the following morning. After lunch I would print whichever negatives I felt were worth printing, but always making at least two prints. Then I would go out and make more exposures for the following day. This schedule allowed me to understand and adapt to the relationship between the negative and the print in short, immediate cycles. As a result, my exposures became better suited for the printing-out process, and the prints helped me to identify which pictorial elements made the best Pt-Pd images.

Greg Petropoulos is an amateur photographer living in Sewanee, Tennessee. He enjoys working in all film formats and appreciates the possibilities offered by the ever-evolving digital workspace. He currently serves as Sculpture Studio Technician at the University of the South.

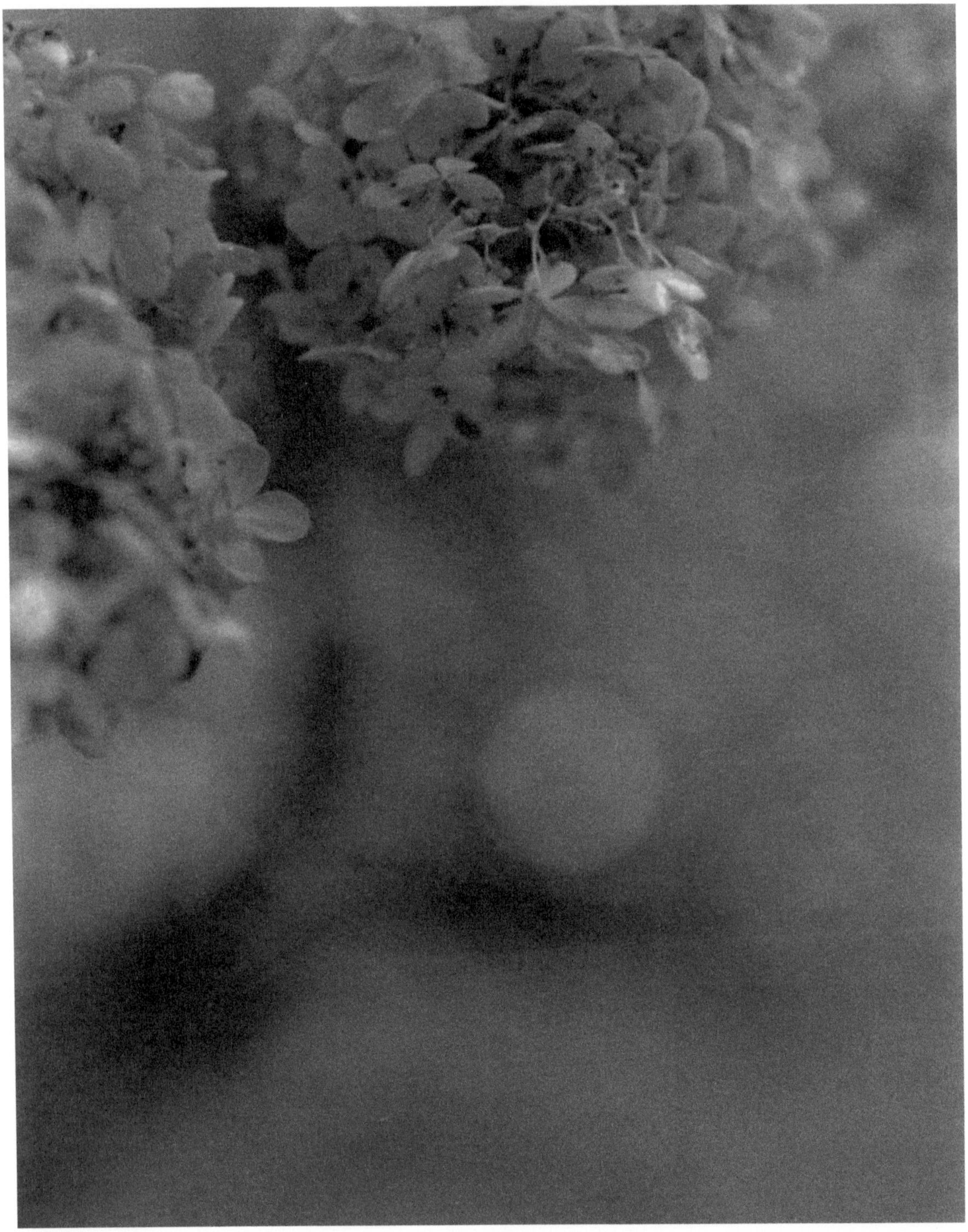

Figure 11.49 Greg Petropoulos, *Hydrangea 2*, 2018. Platinum-palladium print on Revere Platinum paper, from PiezoDN negative of 35mm original, 13 x 10 cm. Courtesy Greg Petropoulos. Fe:Pt:Pd = 2:1:1, hydrated for 30 min. at 100% RH / 21C.

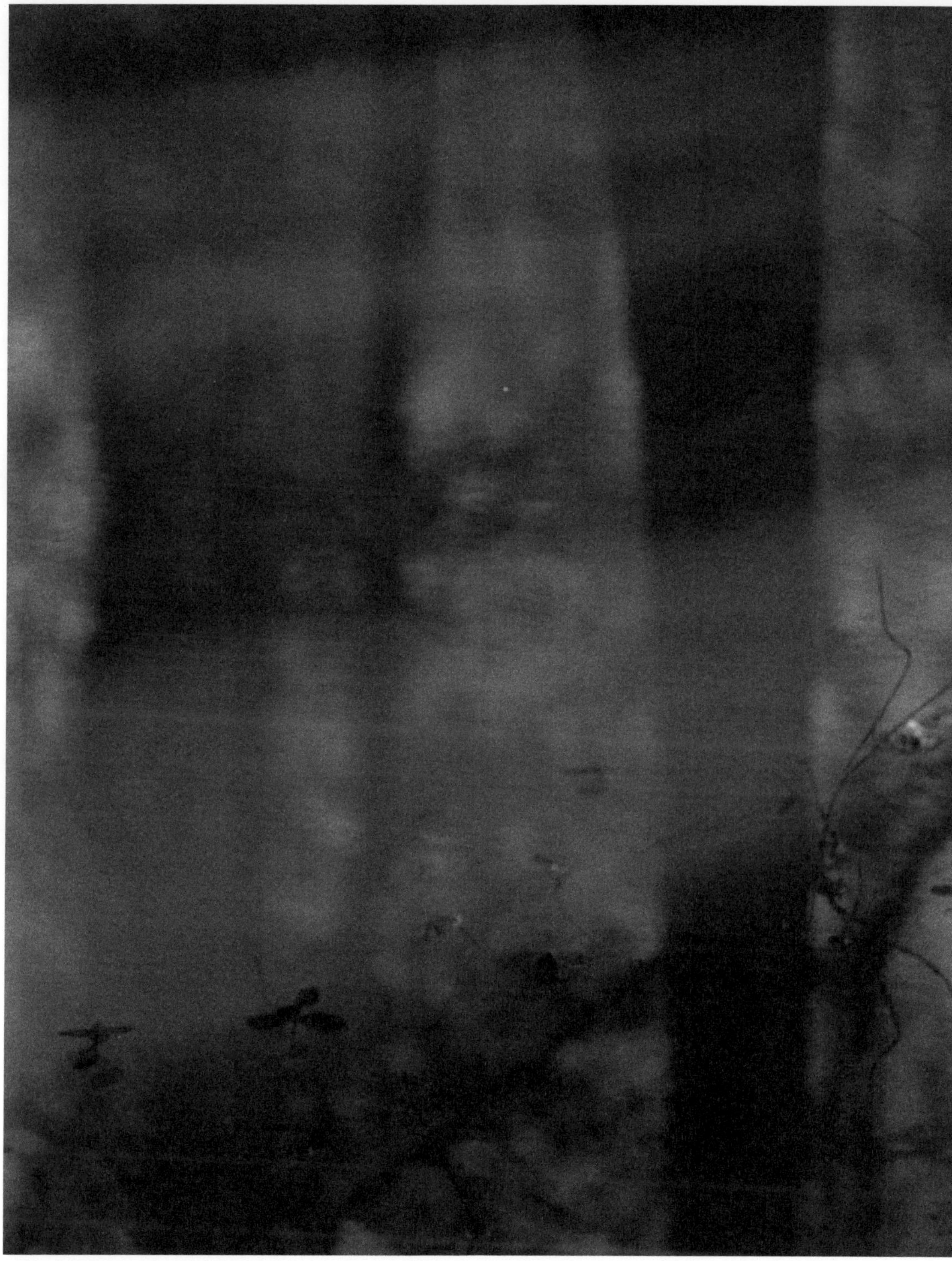

Figure 11.50 Greg Petropoulos, *Reflection in Flood Waters*, 2018. Platinum-palladium print on Revere Platinum paper, from PiezoDN negative of 35mm original, H x W cm. Courtesy Greg Petropoulos. Fe:Pt:Pd = 2:1:1, hydrated for 30 min. at 100% RH / 21°C.

Stavros Pippos

My interest in photography has evolved over a lifetime, a passion kindled when I was a 12-year-old boy with a camera cycling the dusty back tracks near my country home. My desire to follow a creative life ultimately led me to television where I gained a position as a property boy with a local television station. I found television a perfect fit, and it wasn't long before I began to work my way up through the ranks. Eventually, many years later, I became Managing Director of the company for which I had initially been employed to sweep studio floors. I can now see just how relevant my time as a studio cameraman and later as a director/producer was to my post-television life as a photographer. Following my retirement from television, the South Australian wilderness beckoned. Expeditions into remote locations, crossing deserts with camels and navigating wild unregulated rivers and creeks all exposed me to pristine environments never-before photographed.

Advice about making printing-out Pt/Pd prints:

The printing-out process lets me monitor exposure times, and allows me to make incremental adjustments until I get to the desired outcome. With care, even dodging and burning is possible. Because I live in the driest state of the driest continent on Earth with minimal humidity I prefer to print in our winter months when the relative humidity levels are generally higher and more controllable. I have devised a routine which enables me to maintain ambient humidity levels throughout the printing session within a 60–65% RH range as follows:

The evening before printing I mark up each sheet of paper and hang in order of coating sequence above the darkroom sink.

5:15am next morning. Mix the iron, palladium, and platinum solutions and leave for one hour. Run hot water into sink to raise RH to 80% to enable smooth coating. This is not necessary for all papers.

6:15am. Start coating and release water from sink. RH begins to drop, thereafter stabilize RH with air conditioner to settle in a range of 60–65% RH. Hang coated paper in the dark for at least one hour, check to confirm stable RH throughout.

8:30am approximately, commence printing. Ambient conditions of 24° C and RH of 60–65% controlled if necessary with air conditioner for the duration of the printing session.

Stavros Pippos is a fine art photographer based in Adelaide, Australia. He has published six photographic books featuring landscapes of his home state, South Australia. His most recent book comprised mainly monochrome images printed by him in his darkroom. These included platinum-palladium, gelatin-silver, and silver-chloride prints along with panoramic color photographs. All of his work is created using analogue cameras ranging from medium-format to 8 x 10 inch. Pippos now intends to concentrate exclusively on his platinum-palladium work with an emphasis on still life and landscape subjects.

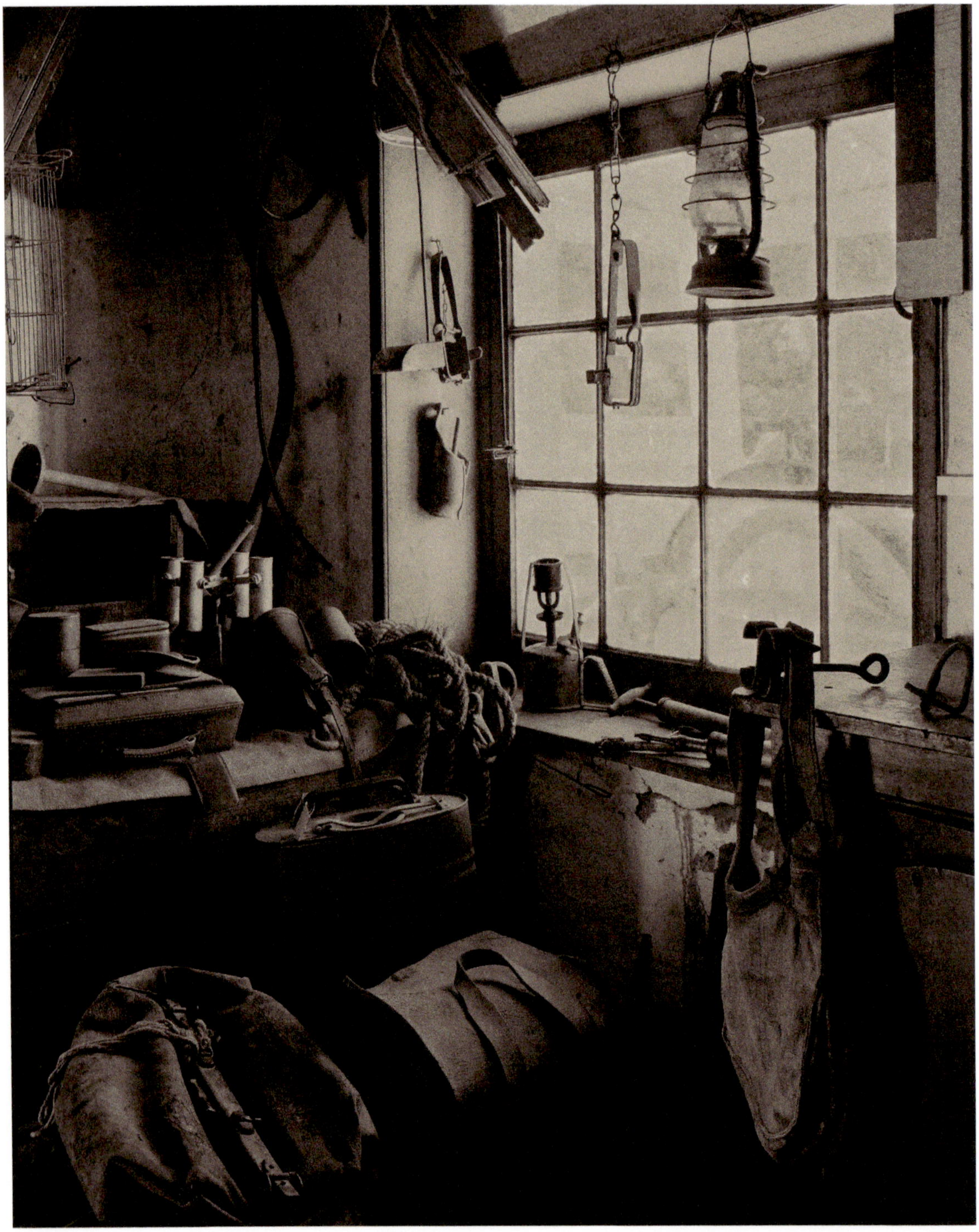

Figure 11.51 Stavros Pippos, *Old Store, Bungaree Ranch, Clare, South Australia*, 2013. Palladium-platinum print on Arches Platine paper, from original 8 x 10 inch negative developed in ABC Pyro Plus, 25.5 x 20.5 cm. Courtesy Stavros Pippos. Fe:Pt:Pd = 4:1:3.

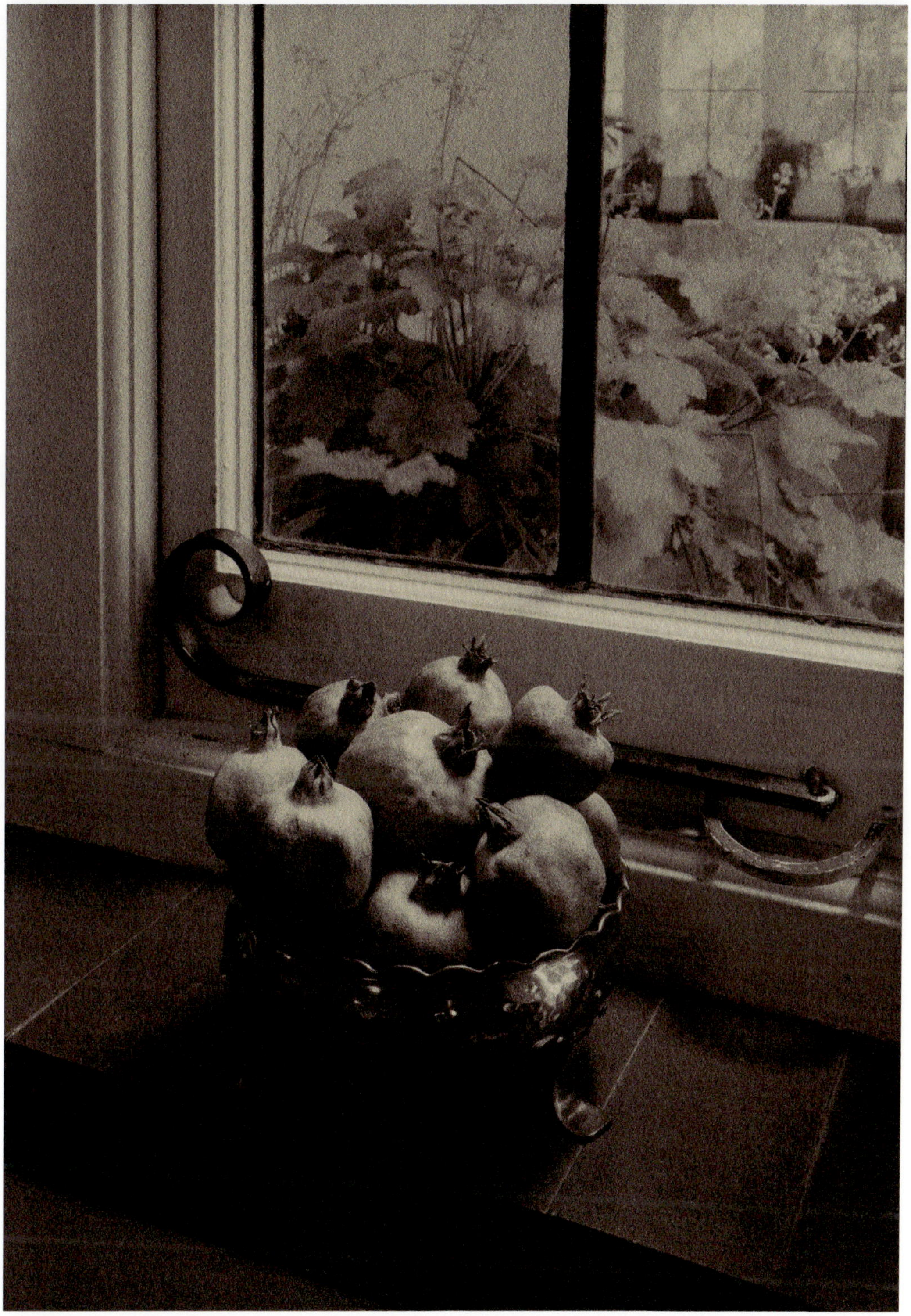

Figure 11.52 Stavros Pippos, *Still Life with Dried Pomegranates, Hahndorf, South Australia*, 2016. Palladium-platinum print on Arches Platine paper, from original 8 x 10 inch negative developed in ABC Pyro Plus, 25.5 x 20.5 cm. Courtesy Stavros Pippos. Fe:Pt:Pd = 3:1:2.

Robert Poole

I am fortunate in now having the time and resources to immerse myself in photography. My scientific and laboratory background beckons me, not to digital technologies, but to the traditional hands-on routes of making photographs and prints. To me, the photograph is an image, whether it is on film, paper, or a digital file. It can be shared digitally or online and it's still a photograph. But the print … that's something quite different. I can lift it up or hang it on a wall, let it reflect or absorb light, direct or glancing, dull or bright. I can purchase or sell one or make a gift of one. It is tactile, the paper having a distinctive color and texture; like our skins, it lives. A print should be an accessible treasure and the platinotype is the pinnacle of printing. Such a fine and expressive print is all I strive for.

Advice about making printing-out Pt/Pd prints:

To me, the major advantage of the printing-out process is exactly that—the "printout." The ability to visualize an image, emerging in real time after discrete periods of ultraviolet exposure is invaluable. It allows much more precise judgment of density, greater reproducibility and contributes to economy with these expensive reagents. It allows new combinations of paper, sensitizer, and other conditions to be evaluated without numerous reagent-consuming trials. This process can be used with platinum and/or palladium on a wide variety of papers, including delicate Japanese Gampi. I have full confidence in the rigorous chemical basis of the reactions and processes, without recourse to additives, and ill-founded ideas. As a scientist, I feel this is how I wish to work. I value the ability to control print color and contrast by simple means—hydration. The methods are well-documented. View original platinotypes and good reproductions and be inspired and driven to produce beautiful images yourself. Look at prints by Fredrick Evans, Laura Gilpin, Irving Penn, Alfred Stieglitz, Paul Strand, George Tice and many others. Visit galleries and museums where these prints can be seen. Attend a hands-on course but be aware that printout methods are not often those taught. Start small: all my early efforts were in 4 x 5 format and this size (and smaller) can yield exquisite gems. Above all, I would like to record my heartfelt thanks to Mike Ware. I am embarrassed to admit that decades were to pass between my first workshop with Mike and taking up the glass spreader again. How I regret those decades.

Robert Poole is an (almost-) retired University Professor, working in the UK. His passion is photography, particularly the "alternative" printout processes using platinum, palladium, or gold but also monochrome silver-gelatin printing. Since attending a workshop with Mike Ware on the former methods long ago, in 1990, Robert shamefully put these practices aside to concentrate on color and later monochrome printing. He now invests considerable time in developing his art and skills in printout methods in precious metals, working from pyro-developed negatives from 4 x 5 and 5 x 7 cameras and darkroom enlargements on film. He has attended workshops with other monochrome masters including John Blakemore, Fay Godwin, Alan Ross, and John Sexton. His prints are in collections in Australia, Canada, New Zealand, the United States of America, and the United Kingdom.

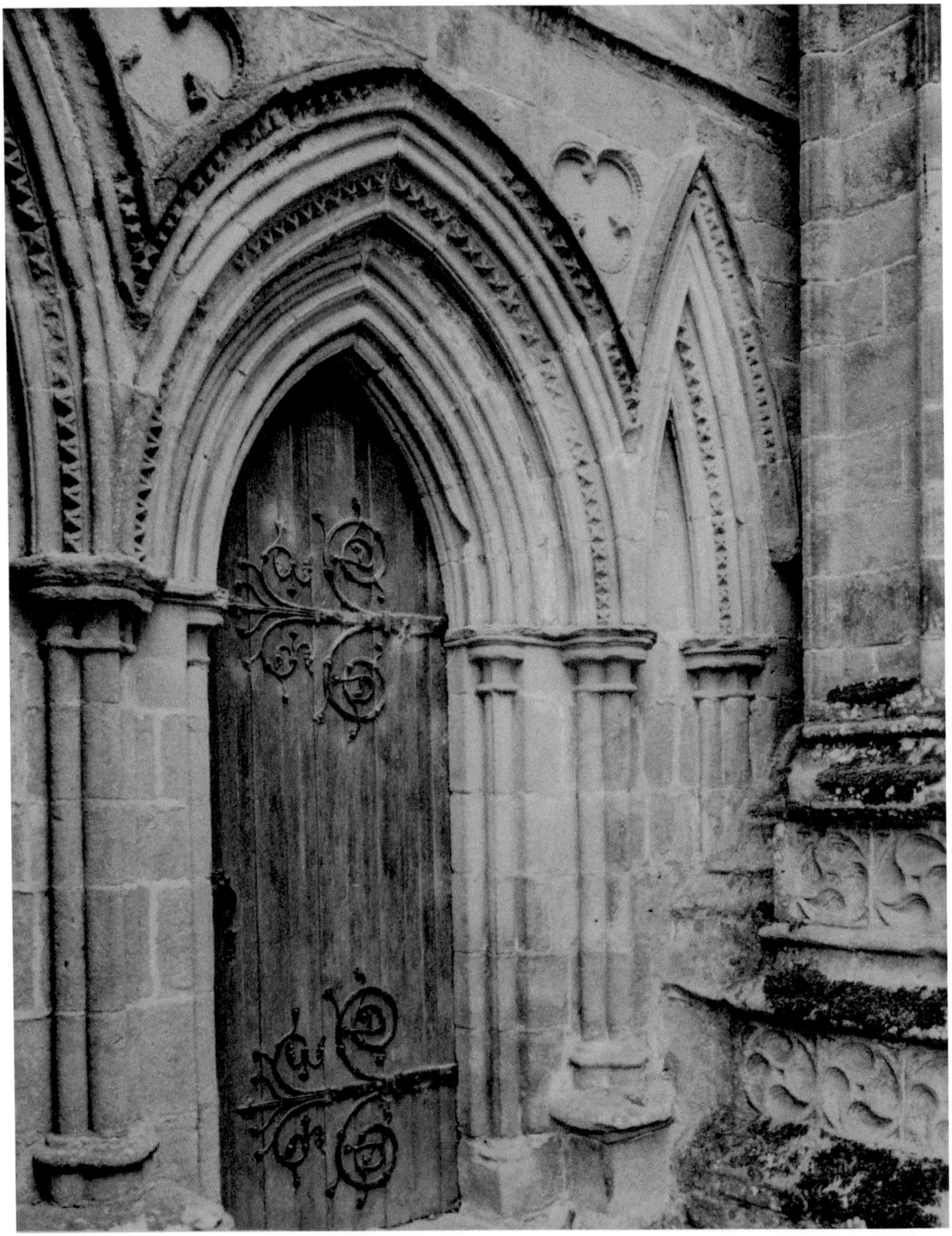

Figure 11.53 Robert Poole, *Church of St Mary and St Cuthbert, Bolton Abbey, North Yorkshire, UK*, 2017. Platinum-palladium print on Ruscombe Mill "Herschel" paper, from original 4 x 5 inch negative developed in PMK Pyro (N+2), 12 x 9 cm. Courtesy Robert Poole. Fe:Pt:Pd = 1:0:1, hydrated for 90 min. at 80% RH / 20°C.

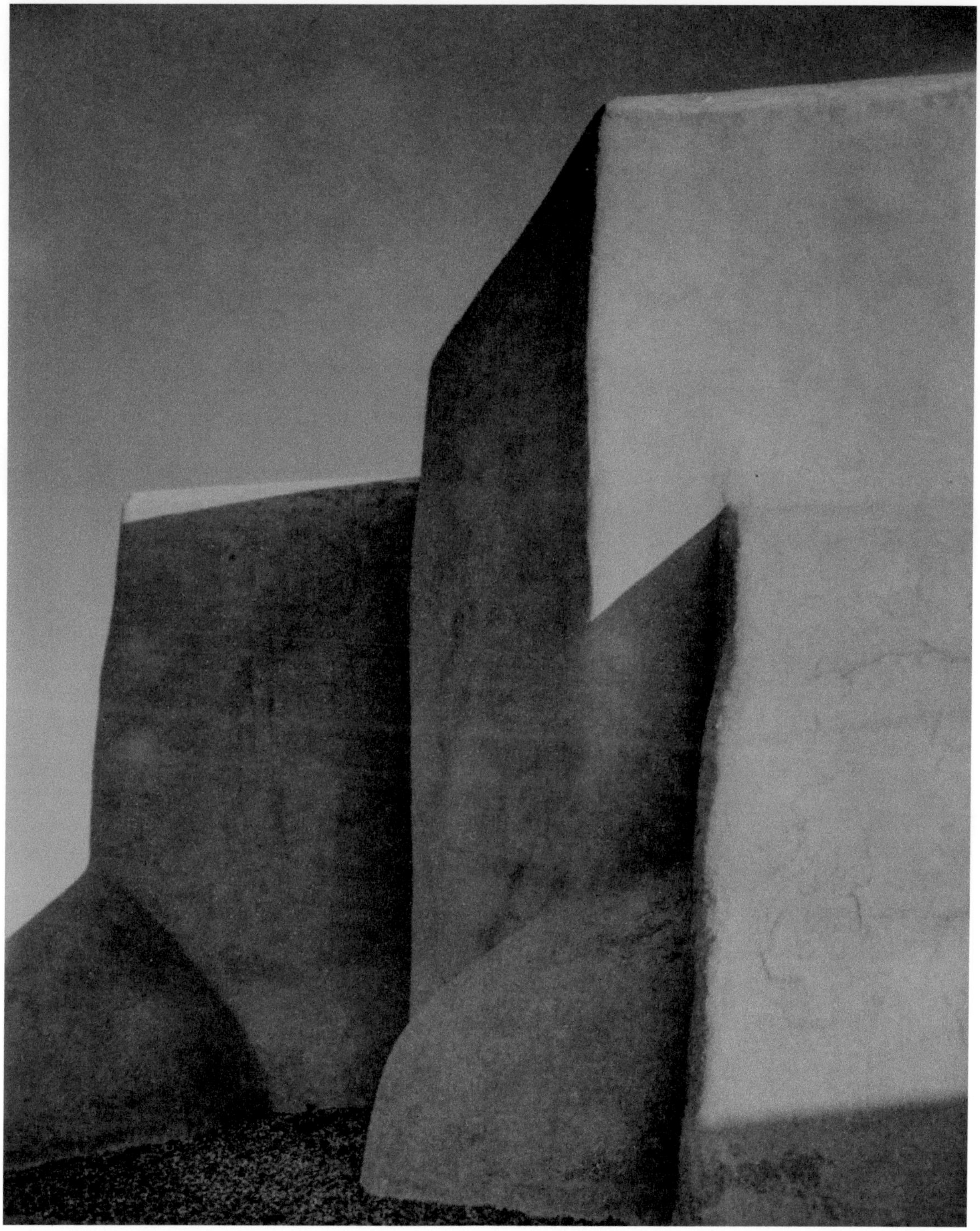

Figure 11.54 Robert Poole, *San Francisco de Asis Church, near Taos, New Mexico, USA*, 2018. Platinum-palladium print on Hahnehmüle Platinum Rag paper, from copy 4 x 5 inch negative developed in PMK Pyro (N+2) of original transparency, 12 x 9 cm. Courtesy Robert Poole. Fe:Pt:Pd = 4:3:1, hydrated for 30 min. at 55% RH / 20°C.

Allan Razo

I learned how to platinum print because I was interested in the step-by-step process and the discipline it takes to create each print. The research, the many experiments, and the years of study were all enjoyable to me since I wanted to develop photographs in a style that is unique and pure. I use large format cameras and studio lighting in my work, and this style of photography is, for me, as much about the process as it is about the final output. I've taken many portraits but I'm also interested in still life, particularly bottles with interesting shapes. I love to pre-visualize, play around with my subject, and take my time composing each element before I begin. I want to create something I believe is worthy of outlasting me.

Advice about making printing-out Pt/Pd prints:

It is easy to control density, contrast, and exposure through humidity levels, and it's very flexible when it comes to combining platinum and palladium prints. A con is that it's difficult to achieve a pure platinum print. Relative humidity in your work space is a big factor. A relative humidity of 50–70% RH is perfect for the platinum/palladium printing-out process. In addition, if you want an almost pure platinum print, a drop of palladium will make it work perfectly. Since it's a contact print, the negative must yield an almost perfect density to make a beautiful print.

Allan Z. Razo studied at the New York Institute of Photography and started out as an advertising photographer before he started platinum printing. He has since exhibited his work in Manila and in Vienna and has conducted workshops and lectures on large format photography and platinum printing.

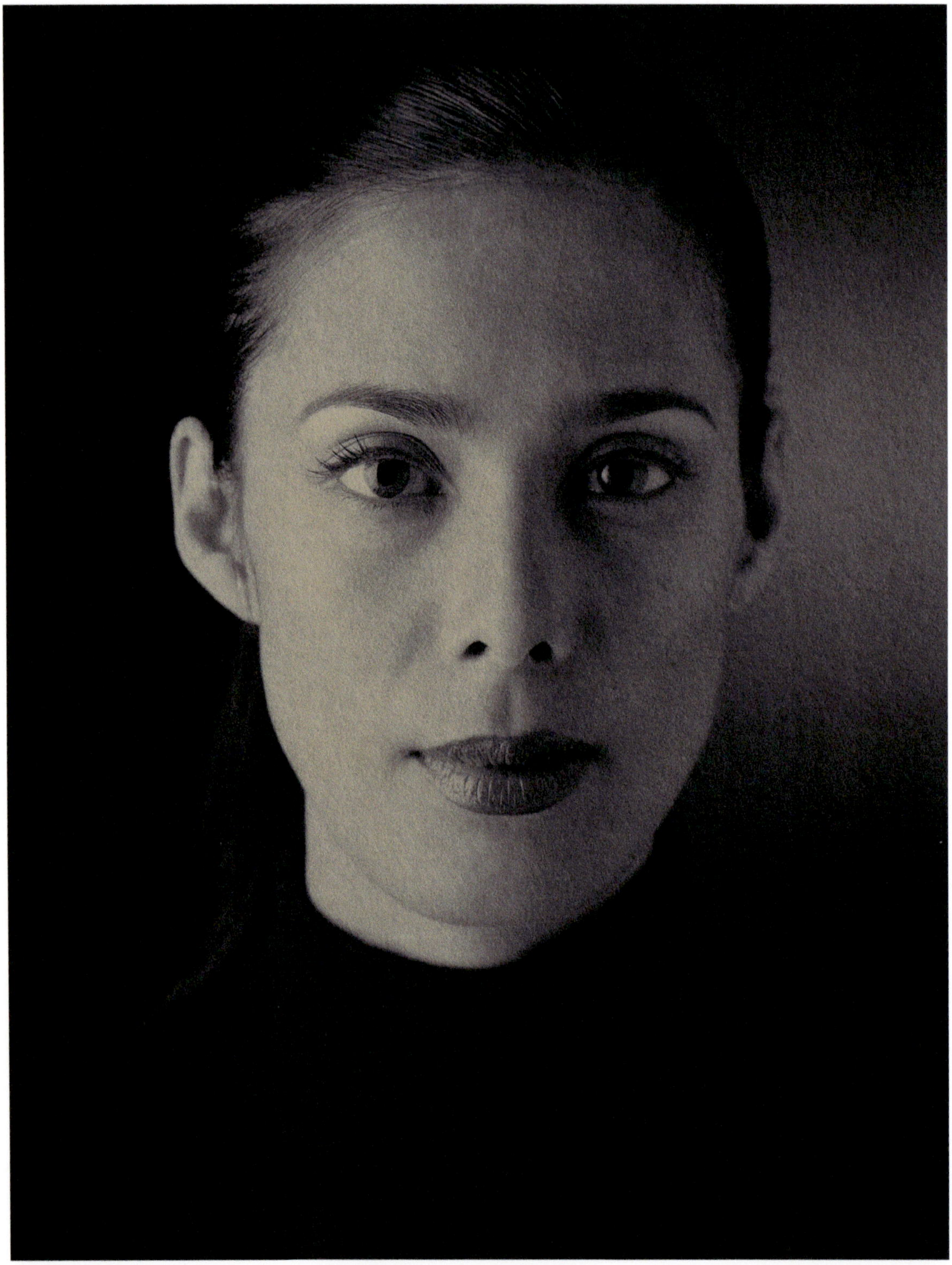

Figure 11.55 Allan Razo, *Angel*, 2006. Platinum-palladium print on Twinrocker paper, from original 8 x 10 inch negative developed in Kodak D76, 24.5 x 19.5 cm. Courtesy Allan Razo. Fe:Pt:Pd = 1:1:0.005*. (*Razo says: "2ml platinum+1 drop palladium".)

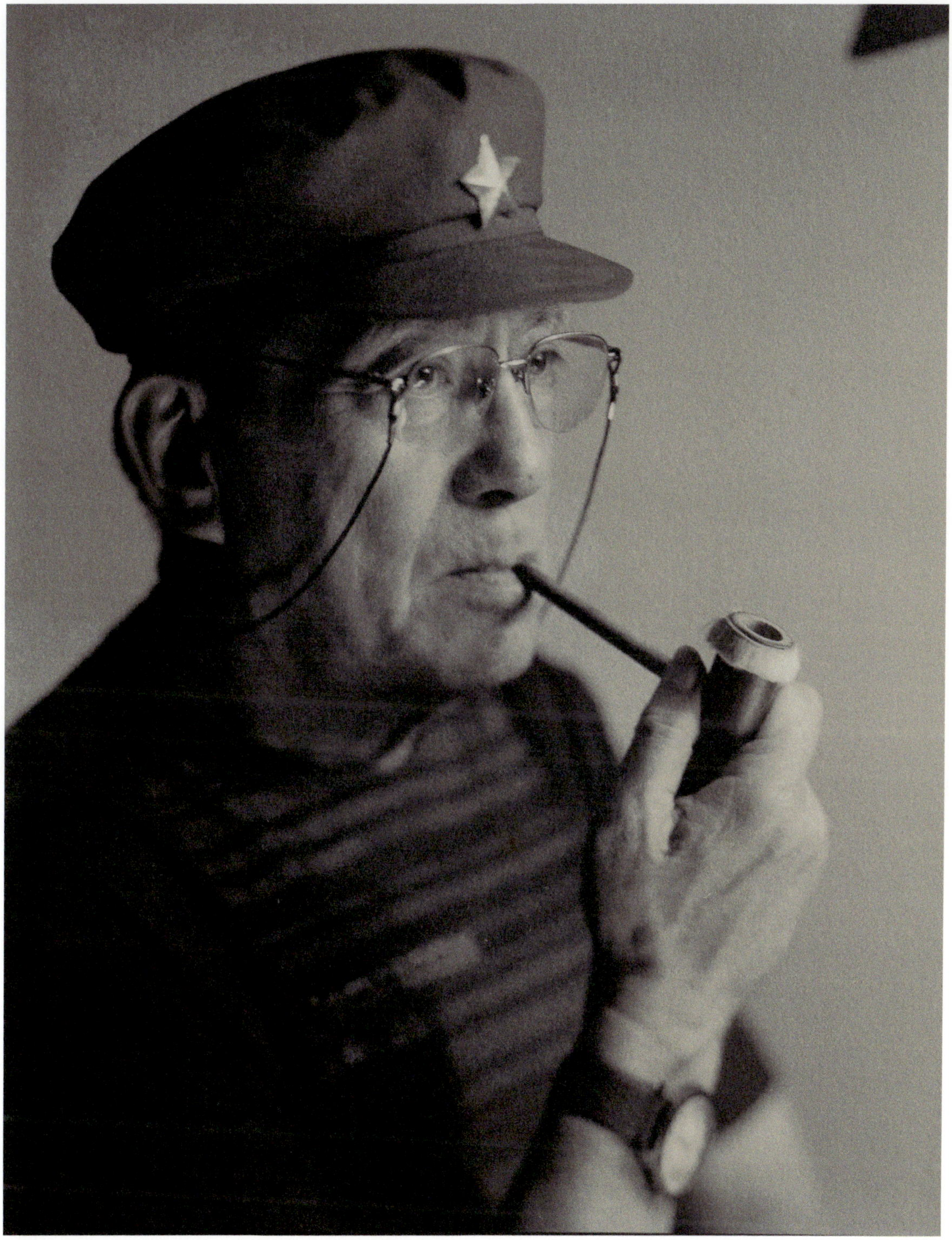

Figure 11.56 Allan Razo, *Direk*, 2007. Platinum-palladium print on Arches Platine paper, from original 8 x 10 inch negative developed in Kodak D76, 24.5 x 19.5 cm. Courtesy Allan Razo. Fe:Pt:Pd = 2:1:1.

Whitt Sellers

In 2016 I fulfilled a dream of mine to go to Burning Man. I spent months researching and planning the trip, and still had no idea what I was in for. Those ten days in the desert changed me forever. On a personal level, I gained new understandings of architecture, light, and space; became more in tune with my body, more confident expressing myself, and had my eyes opened to a new way of life. My main form of expression at the time was photography, and these new understandings let me create work that wielded light and form with greater control. They also changed how I communicated through my work; it became important to me that people participate with my work, which led to more immersive, participatory experiences that coincided with a growing appreciation for humanity's connection to technology.

Advice about making printing-out Pt/Pd prints:

The relative scarcity and cost associated with the process are challenging, so one really has to be committed to making this kind of image. Much of making a good print is in delicate, deliberate steps that require a lot of attention, patience, and skill that is developed over time. Iterating through ideas over series of prints can take weeks instead of hours. It forces people to slow down, be mindful of the choices they are making, understand their materials, and ultimately respect and enjoy the process as much as the end product. Take it slow and be mindful of each step, label everything, take copious notes, make sure the work room is at the proper humidity and temperature, be aware of the chemicals and safe practice, and have fun!

Whitt Sellers is an extended reality developer living in New York City. He is interested in exploring new forms of creative expression and leveraging technology to adapt to changing climates. He primarily works with mobile devices and Virtual Reality headsets using Unity3D and a suite of 2D and 3D design programs.

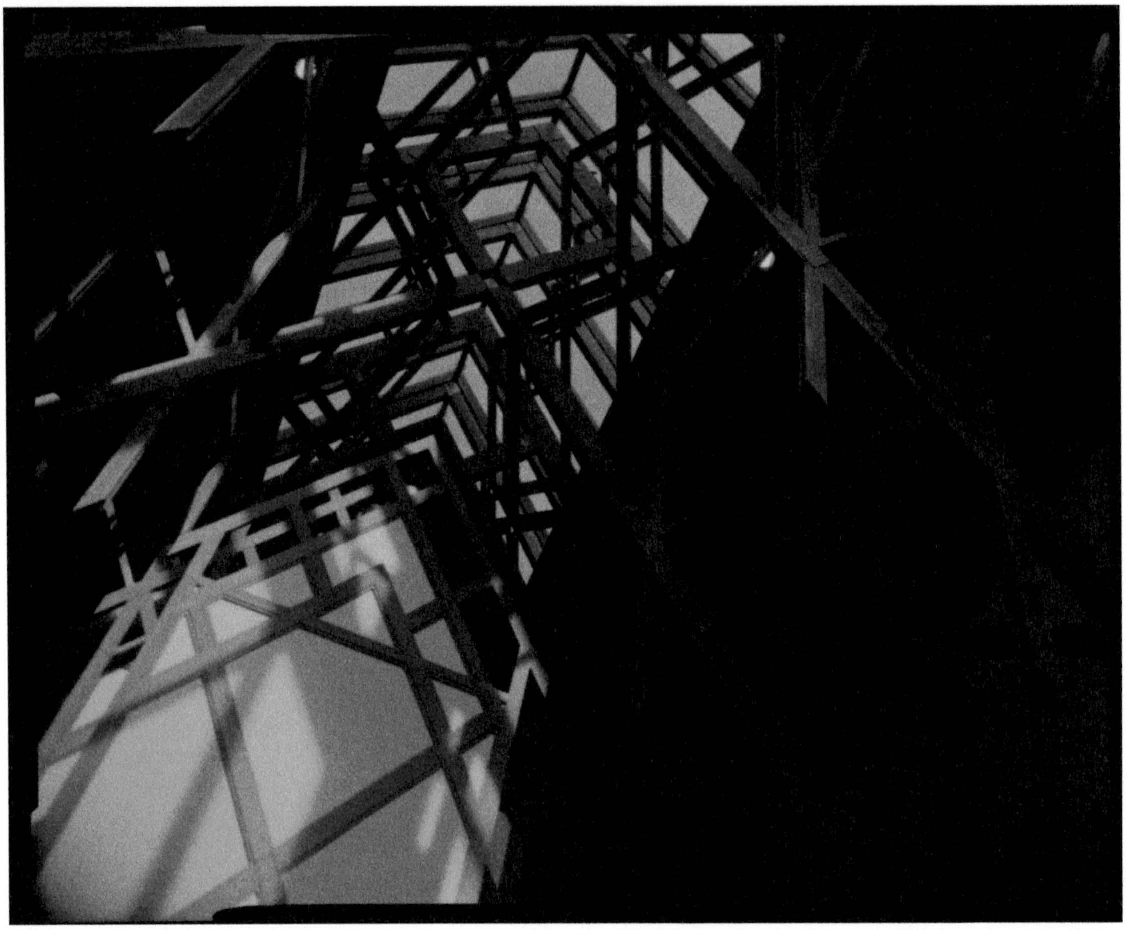

Figure 11.57 Whitt Sellers, *Rafters*, 2015. Platinum-palladium print on Weston Parchment paper, from original 4 x 5 inch negative developed in PMK Pyro, 10 x 13 cm. Courtesy Whitt Sellers. Fe:Pt:Pd = 2:1:1, hydrated for 20 min. at 100% RH / 20°C.

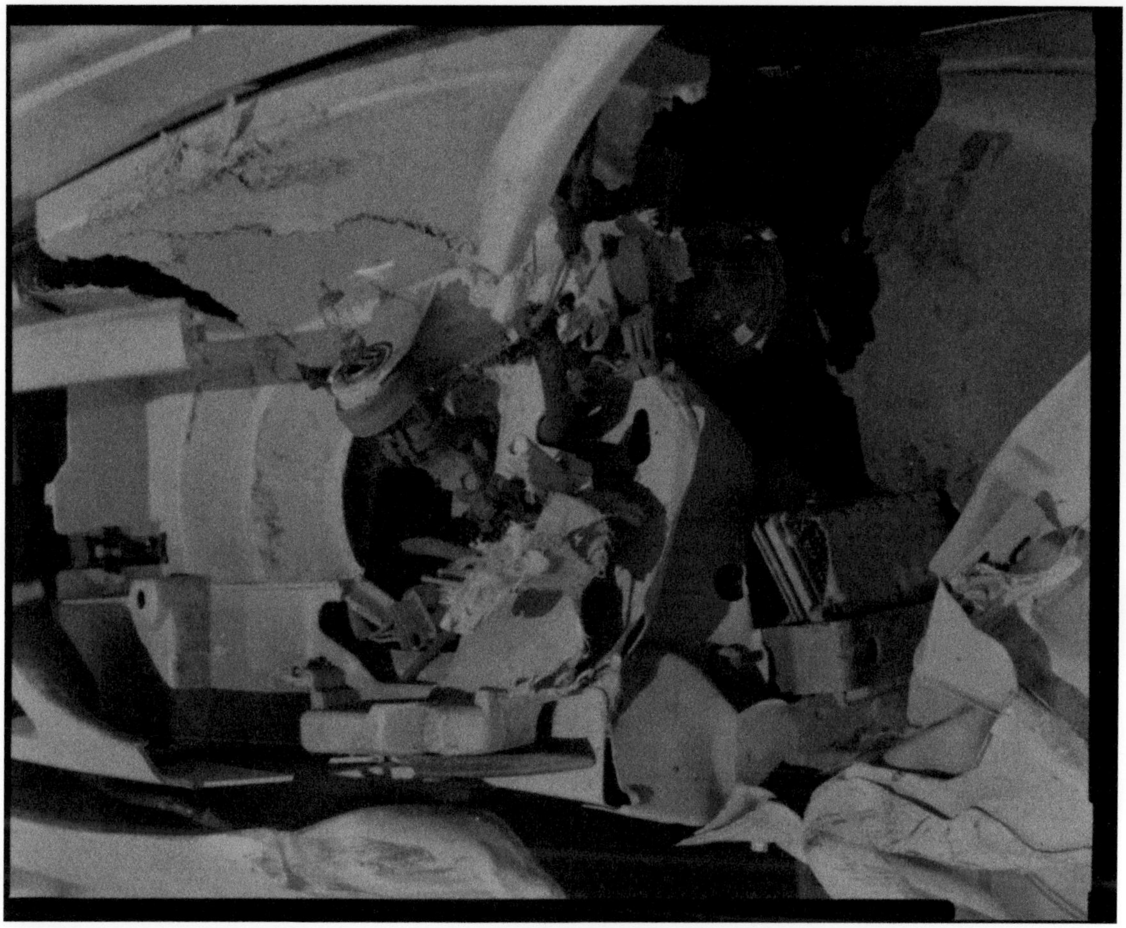

Figure 11.58 Whitt Sellers, *Headlight*, 2015. Platinum-palladium print on Weston Parchment paper, from original 4 x 5 inch negative developed in PMK Pyro, 10 x 13 cm. Courtesy Whitt Sellers. Fe:Pt:Pd = 2:1:1, hydrated for 20 min. at 100% RH / 20°C.

Mark Severson

These are photographs from western New York, where I have lived for the last 12 years. I find my best photographs come from places I've been to many times, and as a result these are mostly from areas close to where I live. I enjoy spending time outdoors, walking and hiking in relatively natural areas as well as in the city, and I enjoy photographing in these areas. Probably due to my background as a chemist, I enjoy the hands-on chemistry of alternative processes, and I think that the platinum/palladium process suits these images well.

Advice about making printing-out Pt/Pd prints:

The control of image color by varying humidity gives more consistent results than use of different developers—as one has to do with the traditional development Pt/Pd platinum/palladium formulation. However, sometimes, proper control of the paper humidity is difficult, especially in the dry winters where I live. Pay careful attention to humidification of paper; ideally, work in a space with well-controlled humidity.

Mark Severson is Professor of Chemistry and Dean Emeritus at Buffalo State College, an institution of the State University of New York. He has been a photographer since 1980, concentrating mostly on alternative process methods for the last ten years, including kallitype, Van dyke, cyanotype, and most intensively platinum/palladium printing.

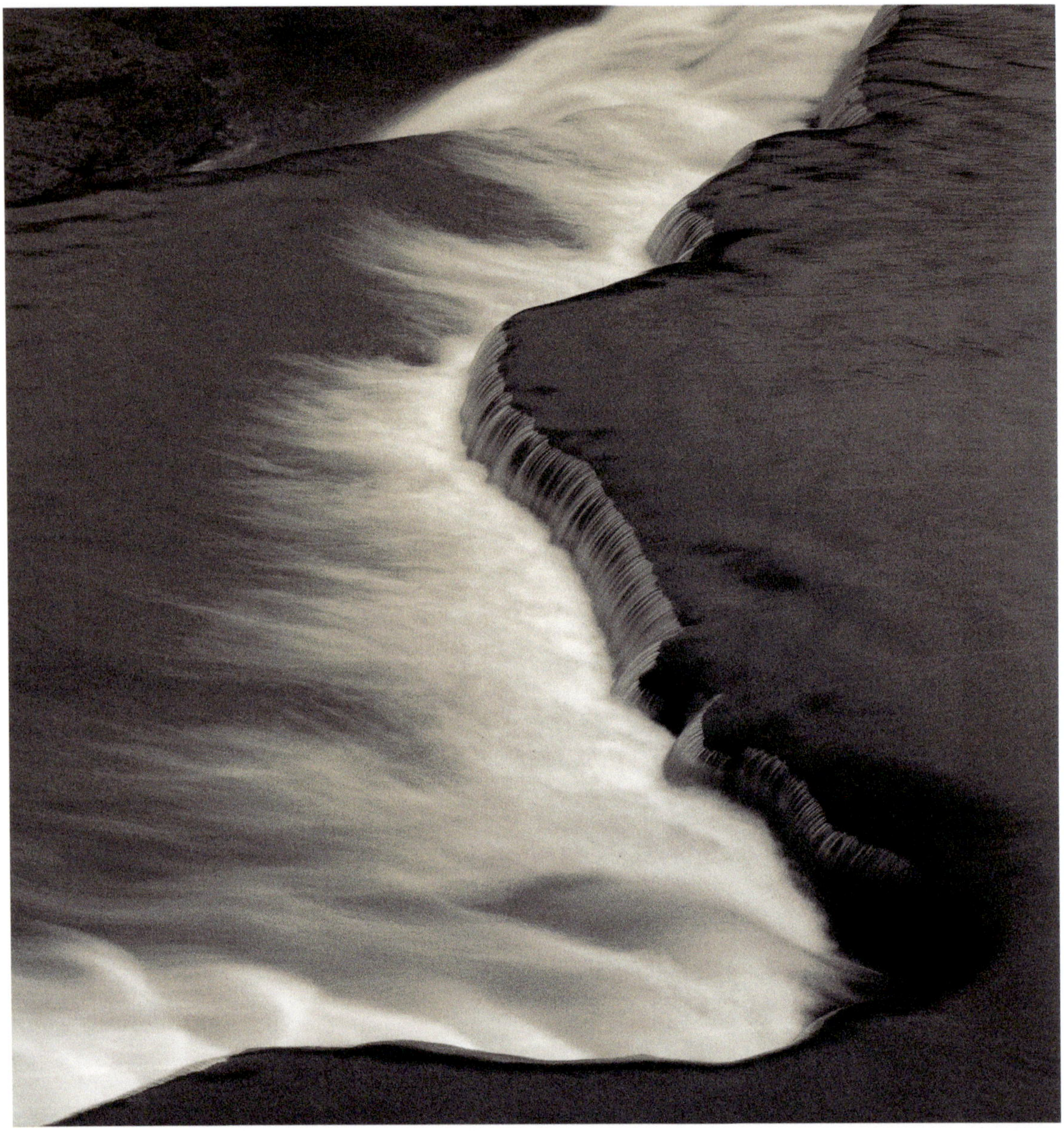

Figure 11.59 Mark Severson, *Above the Middle Falls, Genessee River, Letchworth State Park, New York*, 2012. Platinum-palladium print on Arches Platine paper, from QuadTone RIP negative of 6 x 6 cm. original, 22.5 x 22.5 cm. Courtesy Mark Severson. Fe:Pt:Pd = 4:3:1, hydrated for 60 min. at 57% RH / 18°C.

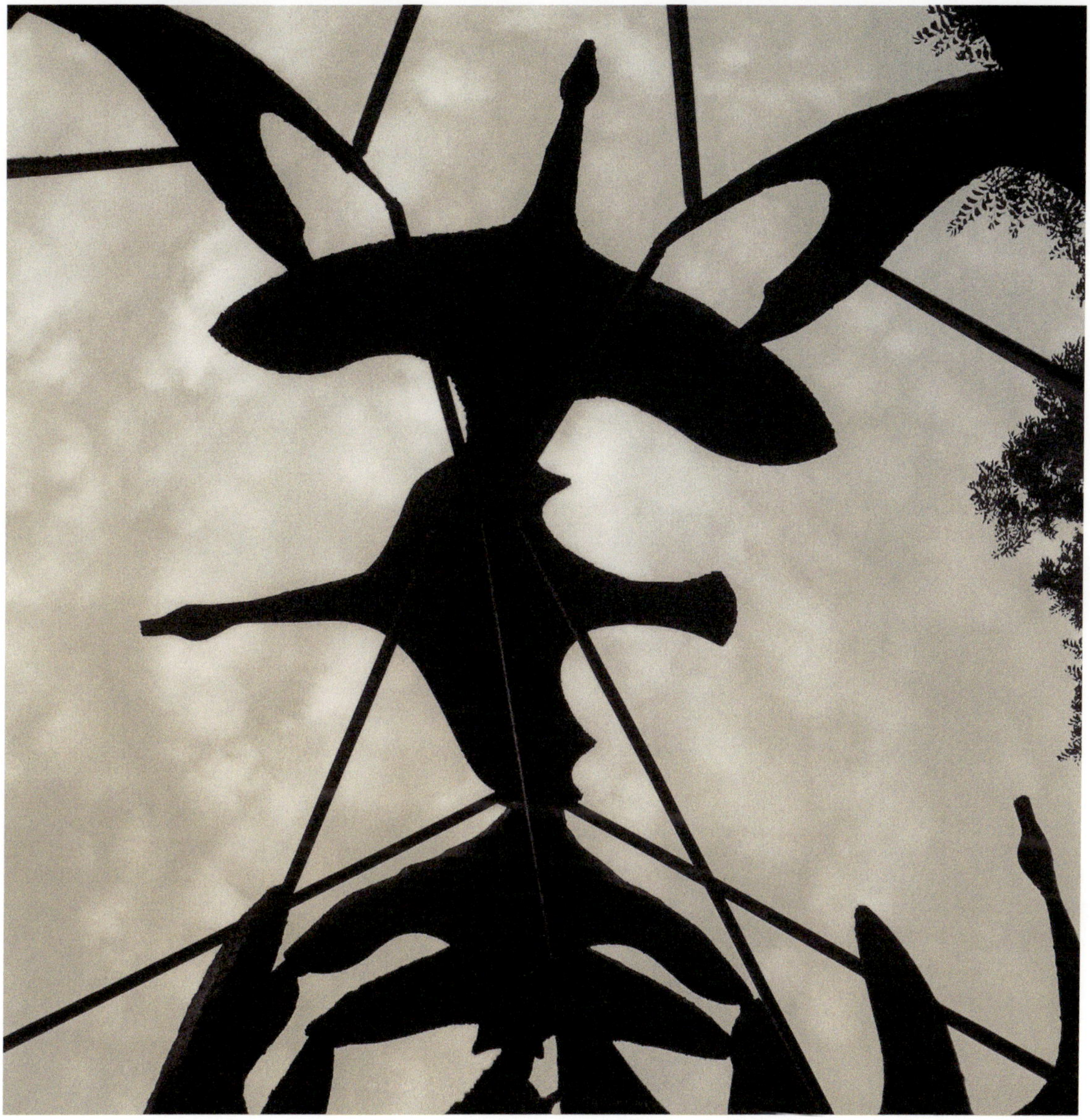

Figure 11.60 Mark Severson, *Statue, Bidwell Parkway, Buffalo, New York*, 2019. Platinum-palladium print on Arches Platine paper, from QuadTone RIP negative of 8 x 10 inch original, 24 x 22 cm. Courtesy Mark Severson. Fe:Pt:Pd = 4:3:1, hydrated for 50 min. at 80% RH / 18°C.

Sara Silks

I combine darkroom printing processes, large and small format photography, and digital tools, with a visual arts background and interests, which include advanced degrees in Art History, drawing, painting, and printmaking, to make personal statements about fragility, vulnerability, and determination. Each body of work is also driven and informed by my recognition that time and memory are precious commodities, and by my desire and need to express this.

Advice about making printing-out Pt/Pd prints:

Platinum/palladium printing is my go-to process, and I have found the tactile application of the metal salts on the rich surface of the paper, along with careful exposure and thoughtful processing, to be rewarding. The surface quality and tonal range of platinum/palladium cannot be attained with silver gelatin or digital printing. The paper also allows me to add gum layers or gum pigment to the surface to add to the expressive qualities of the print. Inspection of the printing-out image allows me to make artistic judgments about exposure and tone— this is more difficult and time consuming with traditional development Pt/Pd platinum/palladium printing. I also appreciate that the clearing process is environmentally friendly and less toxic to practitioners.

Sara Silks is a fine art photographer known internationally for her work in a variety of alternative and historical processes. Silks exhibits her work both nationally and internationally in museums and galleries, and had her first invitational solo show in New York at the Soho Photo Gallery, October 2017.

Silks has been selected as a finalist in Photolucida's Critical Mass for the past three years, and was the international winner in two categories of the Julia Margaret Cameron Awards, 2017. Her work and interviews have been featured in *SHOTS Magazine*, *Lenscratch*, *L'Oeil de la Photographie*, and more. Her work is currently in major private collections across the country and internationally.

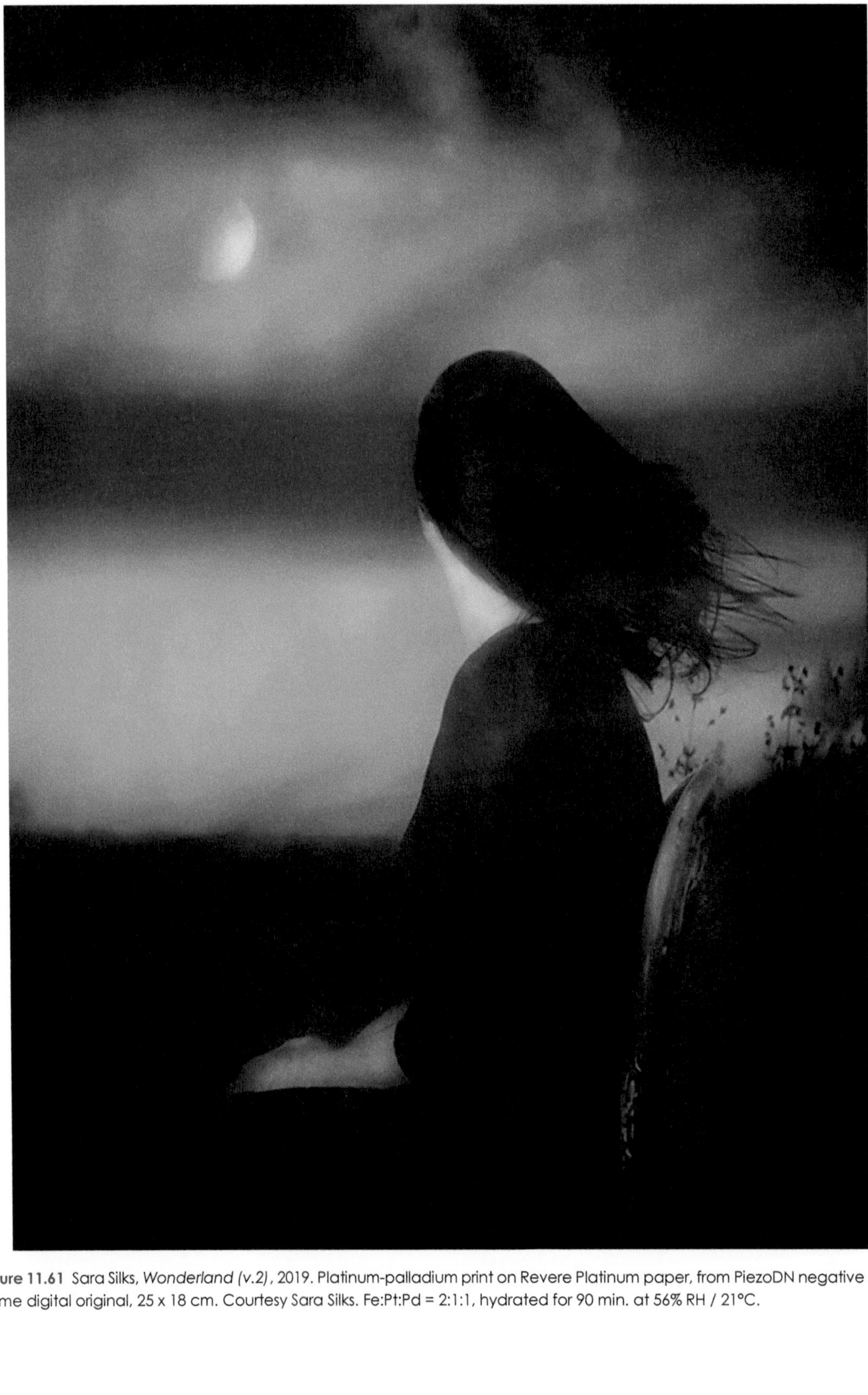

Figure 11.61 Sara Silks, *Wonderland (v.2)*, 2019. Platinum-palladium print on Revere Platinum paper, from PiezoDN negative of full frame digital original, 25 x 18 cm. Courtesy Sara Silks. Fe:Pt:Pd = 2:1:1, hydrated for 90 min. at 56% RH / 21°C.

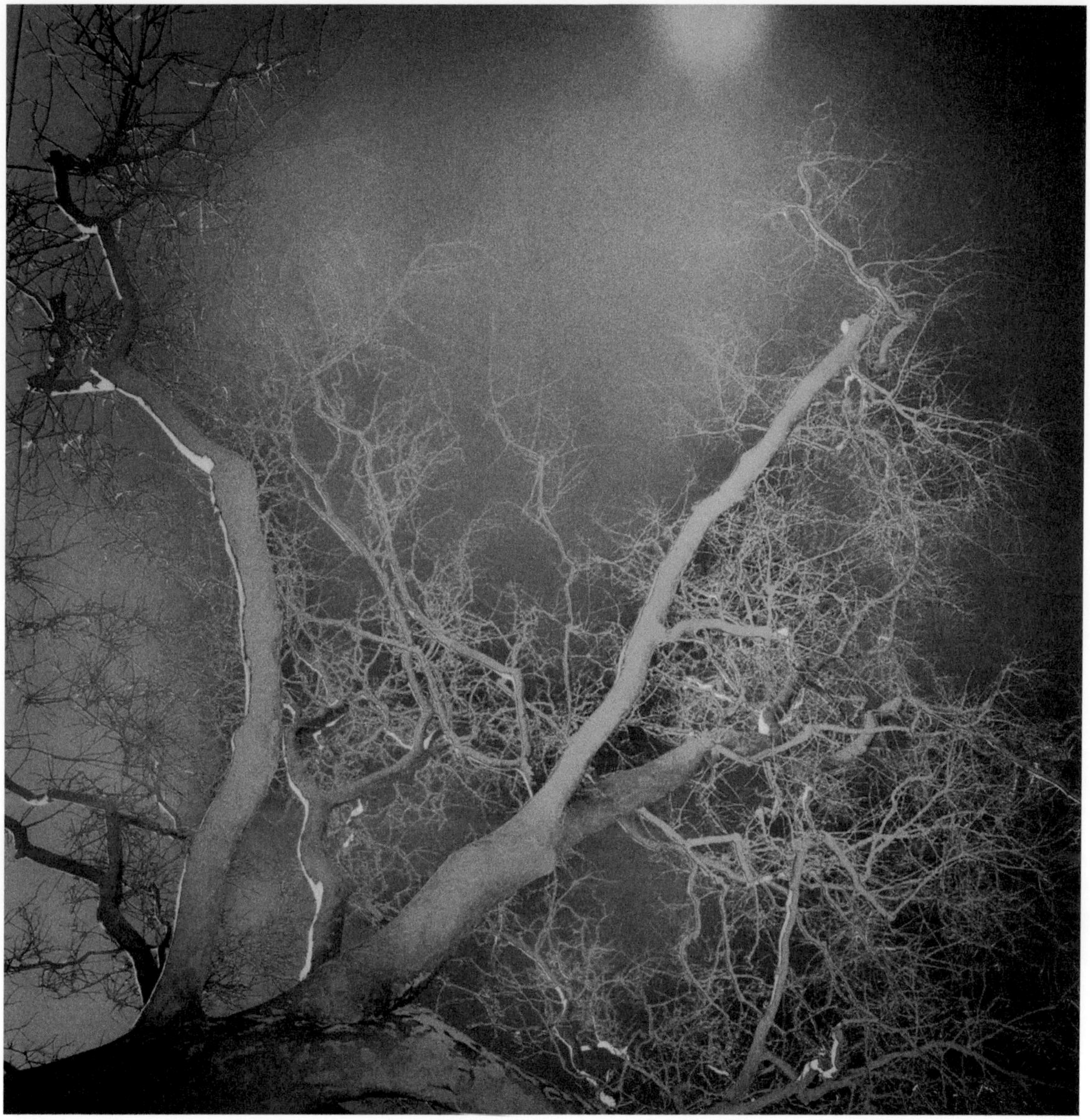

Figure 11.62 Sara Silks, *Family Tree*, 2019. Platinum-palladium print on Revere Platinum paper, from PiezoDN negative of full frame digital original, 20.3 x 16.5 cm. Courtesy Sara Silks. Fe:Pt:Pd = 2:1:1, hydrated for 90 min. at 56% RH / 21°C.

Larry Smukler

In the unquiet stillness of my photographs, I explore the ache of longing, the tenderness of intimacy, the foreboding of loss, the inevitable measure of time's passage. Past, present, and future fuse into an immediate whole.

I am drawn to the light and the shadows—a burst of sun through a doorway, a beckoning window in a night-time storm, a woman fading in the fog. The light does not illuminate the rich black background; rather, it is the subject itself, allowing me to create an emotional space poised on the edge of bursting. Ordinary moments become extraordinary as they are transformed into photographs where time and memory pulse in an endless now.

Advice about making printing-out Pt/Pd prints:

I love the beautiful and unique tonality of the prints; the stability of the chemistry; and the benign environmental effect. The need for rigid control of factors such as temperature and humidity to achieve consistency will be challenging, but dive in—you won't want to come out.

Larry Smukler is a New Hampshire based artist who received his MFA from Maine Media College in May of 2019. His part of the two-person exhibition, *Immediate Whole*, was recognized for evoking feelings of loss, longing, and the passing of time. His photographs transform ordinary moments into expressions of mystery.

Smukler was introduced to photography by his father. He has made photographs ever since. Now retired from the judiciary—a profession that required him to be detached, thoughtful, and measured—Smukler embraces the opportunity for self-expression.

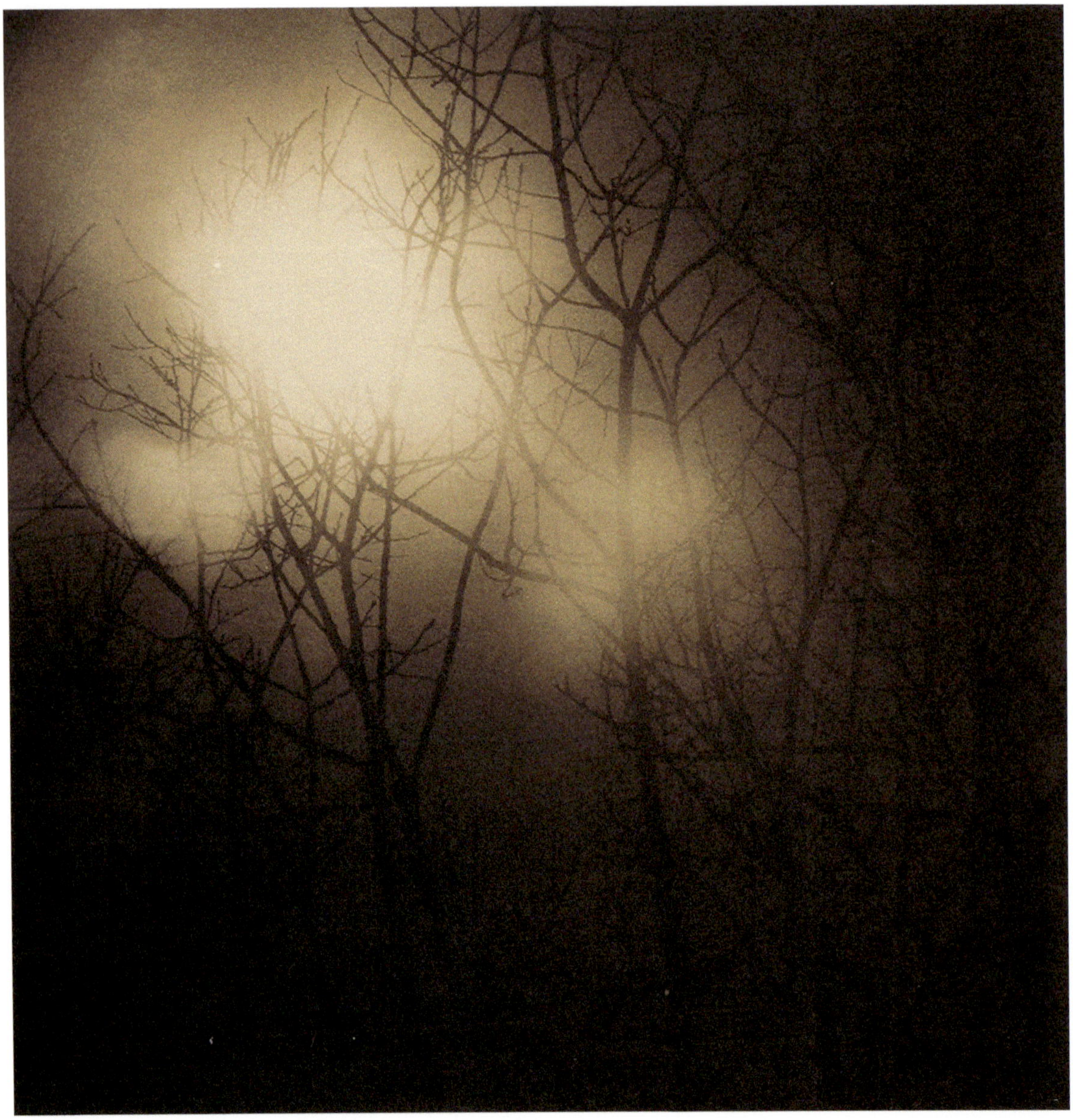

Figure 11.63 Larry Smukler, *Fog*, 2018. Platinum-palladium print on Revere Platinum paper, from PiezoDN negative of full frame digital original, 20.3 x 20.3 cm. Courtesy Larry Smukler. Fe3:Pt:Pd = 2:1:1, hydrated for 60 min. at 34% RH / 20°C.

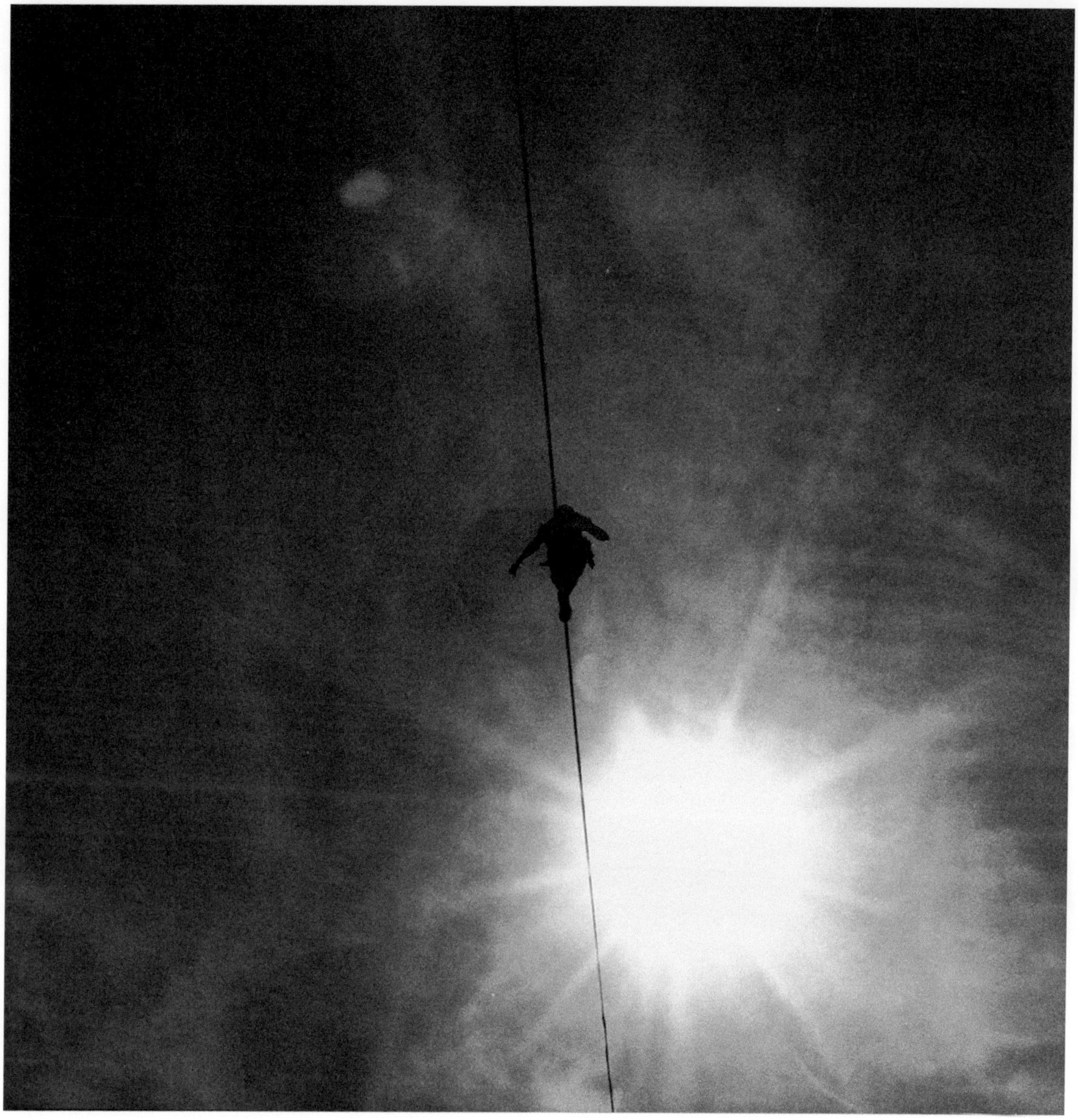

Figure 11.64 Larry Smukler, *Tightrope*, 2018. Platinum-palladium print on Revere Platinum paper, from PiezoDN negative of full frame digital original, 20.3 x 20.3 cm. Courtesy Larry Smukler. Fe1:Fe3:Pt:Pd = 1:1:1:1, hydrated for 20 min. at 100% RH / 20°C.

Mark Stracke

My introduction to the richness and expressiveness of New York City happened in 1971 in a barn in Newfane, Vermont—in the studio of the painter Gandy Brodie. Gandy's paintings, with their luxurious, tactile surfaces, indistinct color fields, and subjects that appeared and disappeared as he worked on them, showed me how history and experience can be written on a seemingly static surface. Since then, I've pointed my cameras at walls, clouds and trees, building fronts, standing stones, and views out of New York City windows. A sense of the present moment, coupled with both the visible and invisible past, beckon and challenge me to capture something of what witnessing and being a part of the river of time means. After having experimented with platinum/palladium prints in the early 90's, when one could buy pre-coated paper, and abandoning it when that option disappeared, I have returned to these materials in recent years. The tactile nature of the paper and the beautiful tonality of the precious metals have the capacity to mimic the play of light on buildings in the city, and the sense of time being written out on a surface.

Advice about making printing-out Pt/Pd prints:

The printing-out process, also known as the Ammonium Method, produces lovely tone that is easily adjustable by varying the water content in the paper and/or the formula for the sensitizer; no more hot potassium oxalate for warm toned images (as one uses in some versions of the development Pt/Pd, potassium-based process)! It offers continuous tone areas, such as skies, that are free of black spots. Sensitizer components (including the ammonium ferric oxalate) can be mixed together in advance and stored for months without fear of fogged prints or other hard to track down anomalies. There is no significant contrast control in the printing stage—negatives must fit the process. However, contrast can easily be adjusted by making digital negatives. Systematize your workflow—follow it without variation until you have your printing well under control. Take detailed, accurate notes as you are printing and have them available in the darkroom for guidance and review. When you experiment, change only one variable at a time. Use the salt hydration method; it's the best. If you can afford it, buy an ultraviolet light meter for determining exposure of your prints. Light sources vary for all sorts of reasons and knowing how much ultraviolet struck your printing frame is very helpful. It's true that one can determine print exposure by examining the image in the printing frame. However, I've found it hard to get an accurate sense of exposure fluctuations when looking at the uncleared yellowish sensitizer. Finally, clean your coating equipment scrupulously before, during, and after each printing session.

Mark Stracke has lived and worked in New York City since 1972. He teaches photography at the Ethical Culture Fieldston School in the Bronx and lives near the school. He has worked with film cameras from 35mm to 8 x 10 and currently works with digital images, both scanned film and digital capture, which are used to make digital negatives for platinum/palladium printing. While based in New York, Mr. Stracke has also traveled to New England, Scotland, Arizona, New Mexico, Florida, and California on shooting excursions.

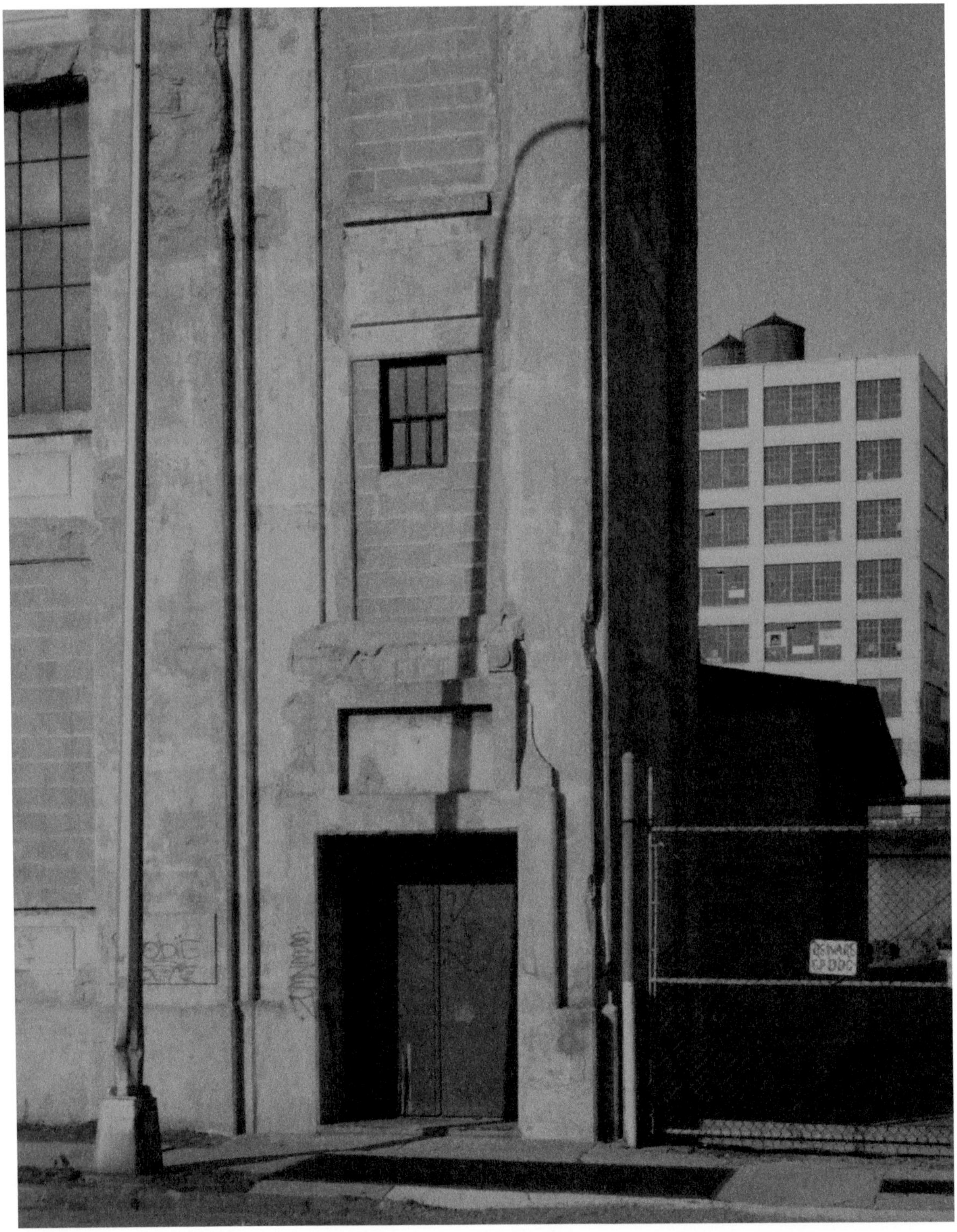

Figure 11.65 Mark Stracke, *Bronx, NY*, Negative 1991, Print 2018. Platinum-palladium print on Revere Platinum paper, from original 4 x 5 inch negative developed in Kodak D-23, 12.5 x 10 cm. Courtesy Mark Stracke. Fe1:Pt:Pd = 2:1:1, hydrated for 10 min. at 100% RH / 20°C.

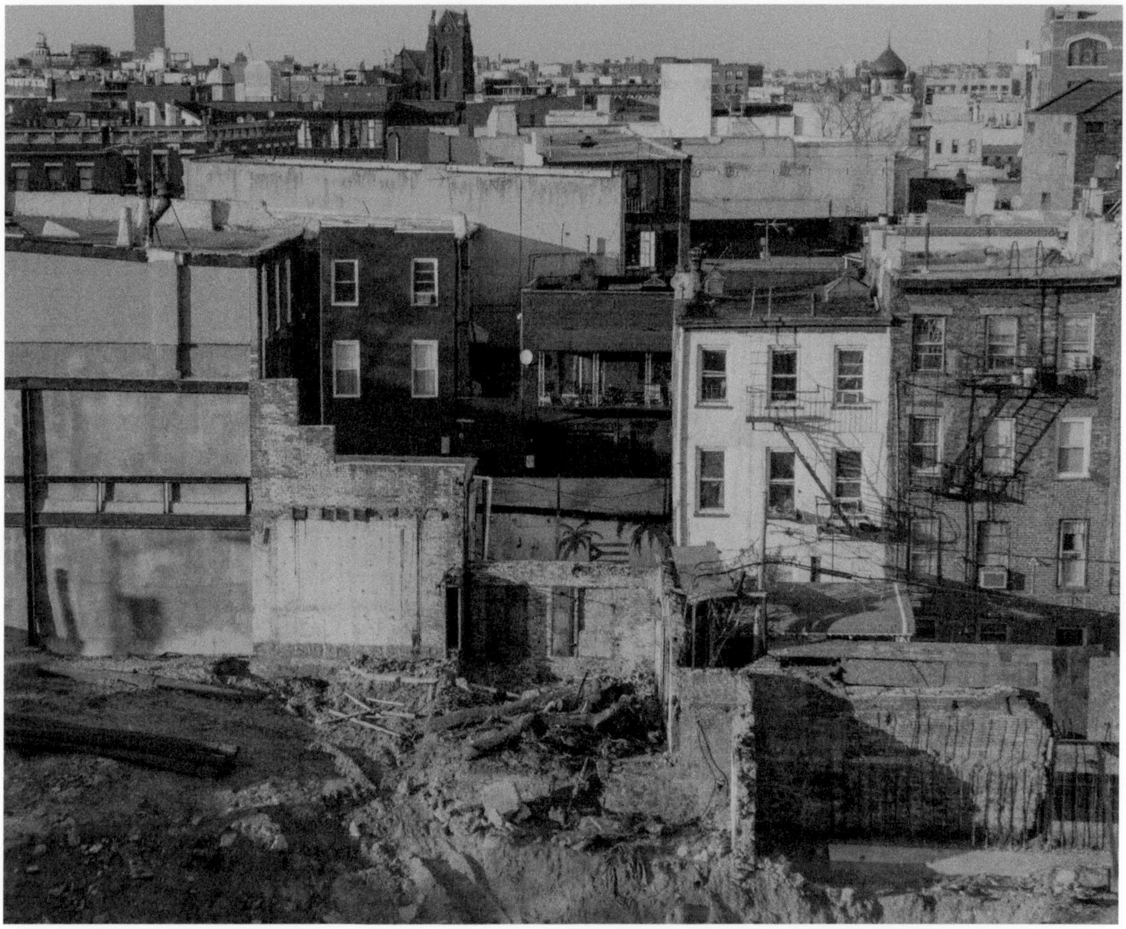

Figure 11.66 Mark Stracke, *Brooklyn,* Digital capture 2008, Print 2018. Platinum-palladium print on Revere Platinum paper, from PiezoDN negative of full frame digital original, 15.2 x 21.3 cm. Courtesy Mark Stracke. Fe1:Pt:Pd = 2:1:1, hydrated for 10 min. at 100% RH / 20°C.

Hunter Swenson

I am a place photographer. I am particularly drawn to places in which I find quietness and awe. I generate these images using small to large format 8 x 10 cameras and then make platinum-palladium prints from the images.

Advice about making printing-out Pt/Pd prints:

The printing-out process produces prints with great subtlety, and is able to render delicate tonal shifts and minute shadow details, all the while keeping the total image sharp. The difficulty, however, is that the execution of this subtlety requires a lot of time—from editing the digital file/processing the film all the way to the final wash. I would suggest looking at as many prints as you can and being sure to take your time making prints.

Hunter Swenson is a Colorado native and a graduate of The University of the South, Sewanee, Tennessee (2018) majoring in Philosophy, and Art (with Honors). Influenced by continental and phenomenological philosophy, he works primarily with photography and sculpture to examine the intersections of quietness, clouds, daydreams, the land, and spaces. Swenson has been awarded Sewanee's Shepard Award for Excellence in Photography, and has worked in rural Tennessee, Colorado, Haiti, Jamaica, and Tanzania.

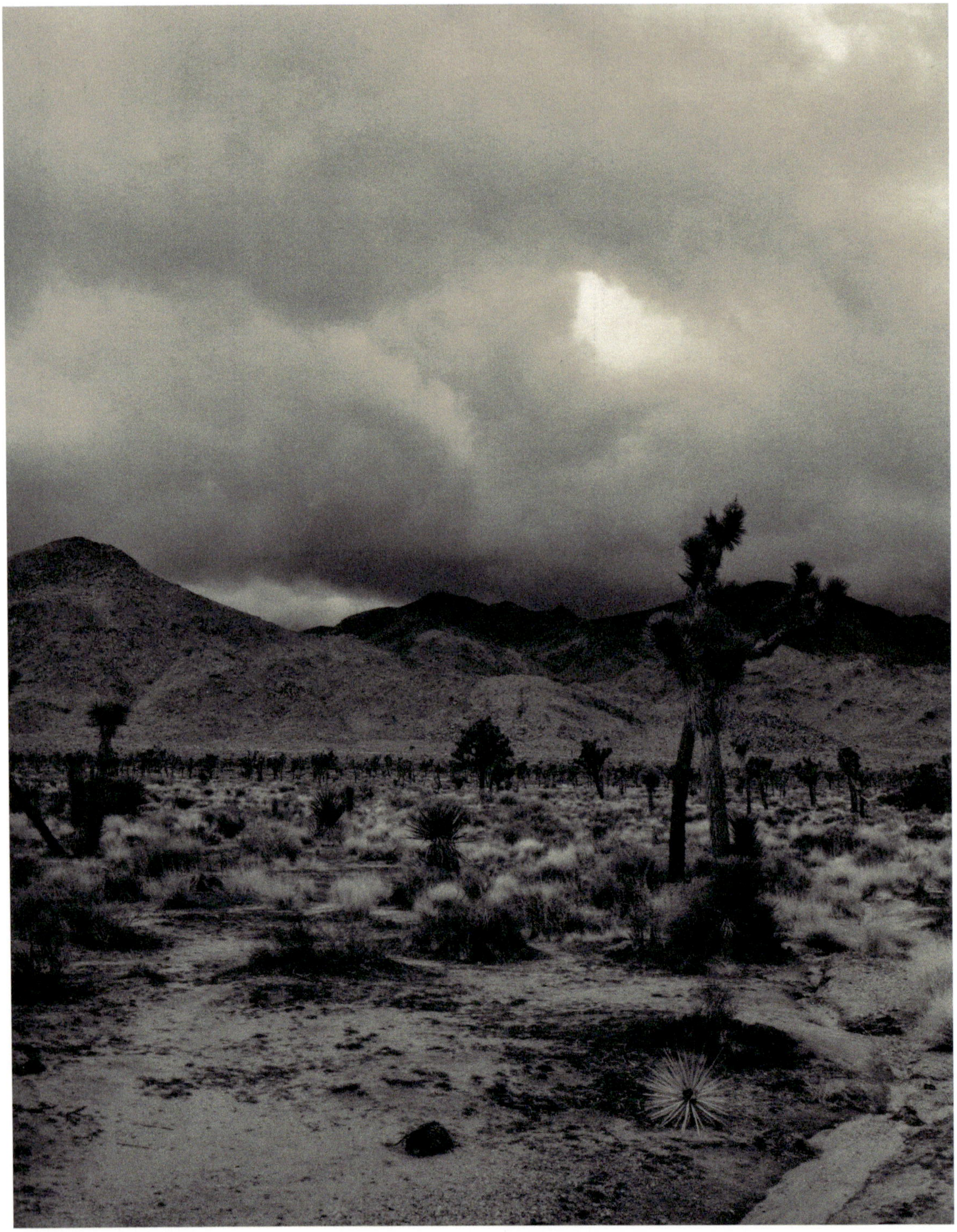

Figure 11.67 Hunter Swenson, *Joshua Tree, California 3*, 2019. Platinum-palladium print on Revere Platinum paper, from PiezoDN negative of full frame digital original, 25.4 x 20.3 cm. Courtesy Hunter Swenson. Fe1:Pt:Pd = 2:1:1, hydrated for 90 min. at 76% RH / 20°C.

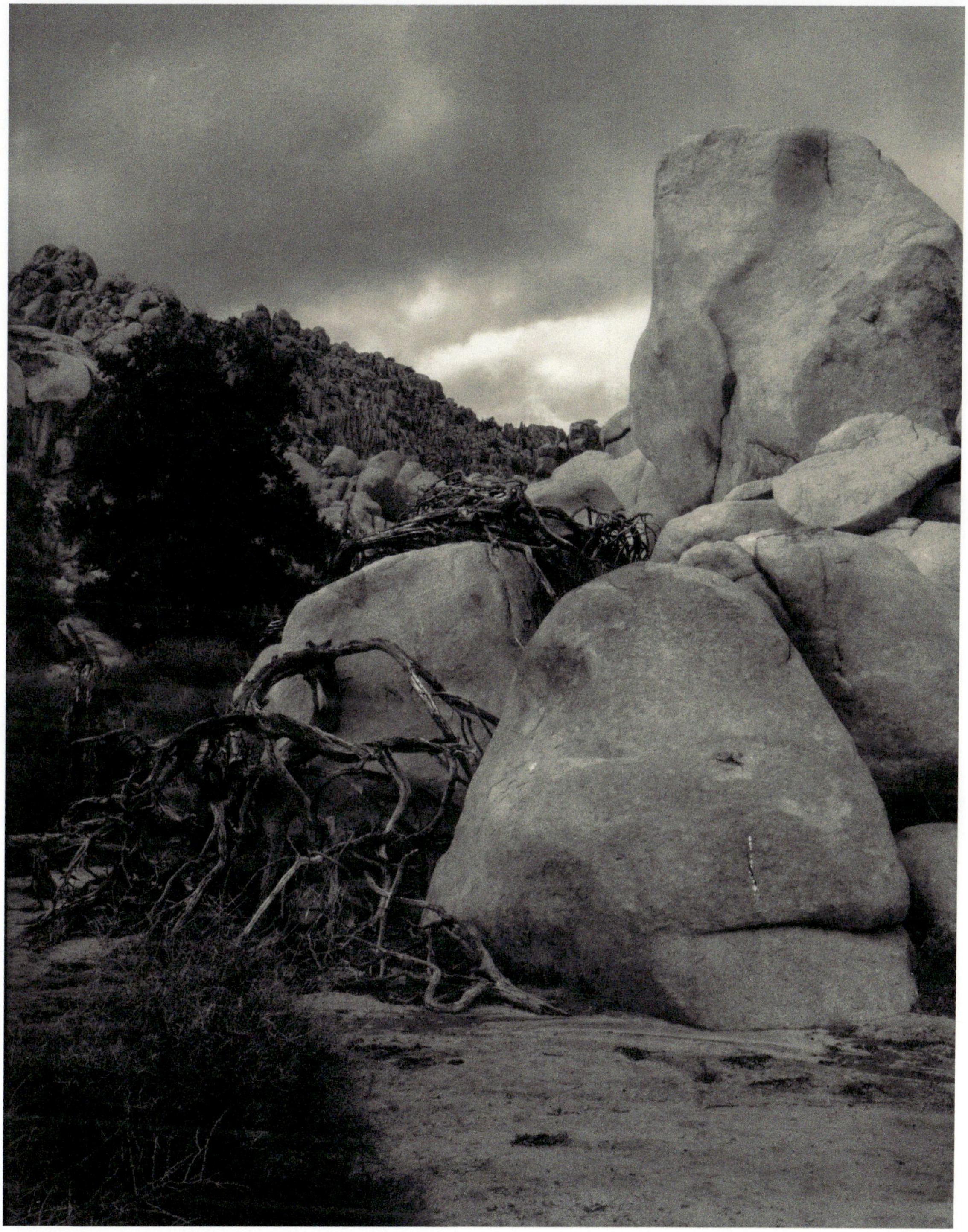

Figure 11.68 Hunter Swenson, *Joshua Tree, California 1*, 2019. Platinum-palladium print on Revere Platinum paper, from PiezoDN negative of full frame digital original, 25.4 x 20.3 cm. Courtesy Hunter Swenson. Fe1:Pt:Pd = 2:1:1, hydrated for 90 min. at 76% RH / 20°C.

Basil Tahan

Inspired by the great artists of landscape photography, I enjoy nothing more than hiking in the beautiful outdoors and making images of what presents itself. Last winter, I traveled to the Japanese alps and made a series of 40 images as an homage to Michael Kenna. As I am now residing in a big city, I look for landscape in the city. I have presented here an avalanche of motor parts shot in Chinatown, Bangkok.

In Thailand, the humidity is quite high. We don't have a problem with dry paper. The day we made those prints, the humidity was around 70%.

Most "exotic" chemistry is imported from the US. In this case, Pradip gave the chemistry to Jill Enfield who came for a visit.

Basil Tahan is an American photographer and educator, currently working as a photography teacher in Bangkok, Thailand. He enjoys exploring the history of photography through alt processes. His current project is taking him around Thailand with a portable darkroom he built in the back of his truck and several large format cameras. He is creating wet plates of the local people, landscapes, and culture.

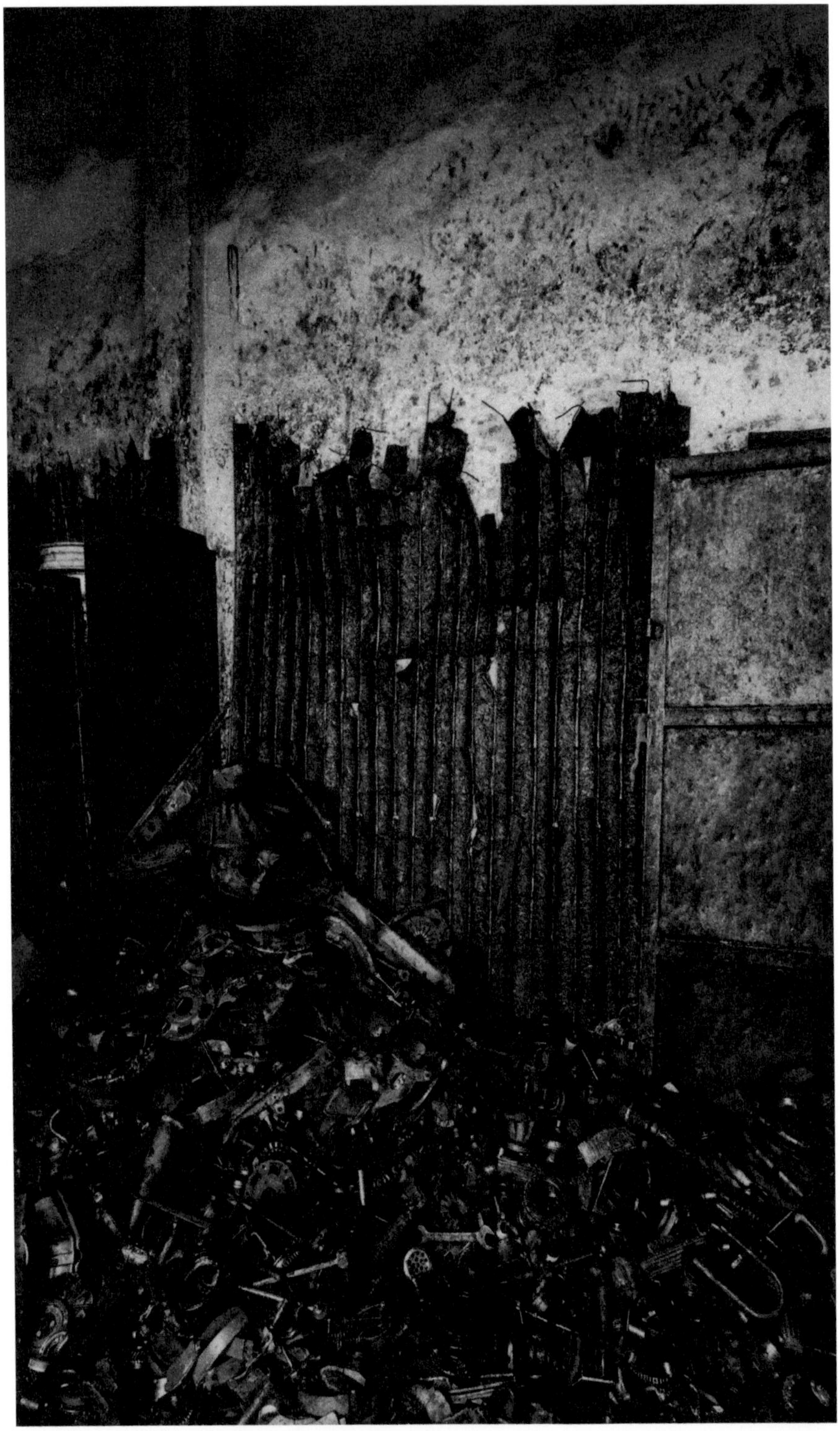

Figure 11.69 Basil Tahan, *Avalanche*, 2019. 24 x 14 cm. Platinum print.

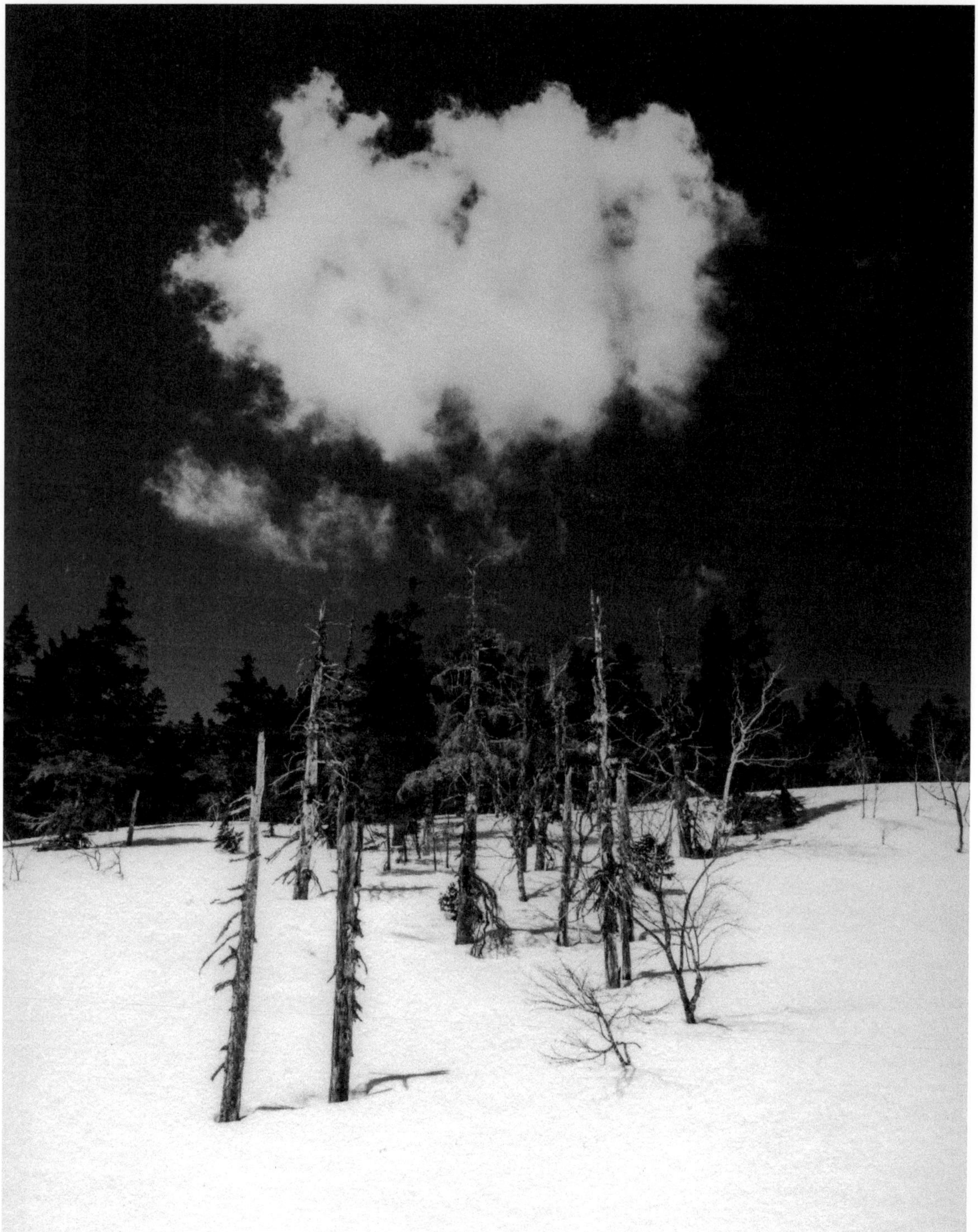

Figure 11.70 Basil Tahan, *Above Tsugaike*, 2019. 24 x 16 cm. Platinum print.

Roger Vail

I honestly do not think there is any contradiction between craft and discovery. For me one ultimately flows out of the other. "Ars sine scientia nihil est." The poet and close friend, Dennis Schmitz writes:

"In the world of these (Vail's) astonishing photographs, all the events are simultaneous because one plane of attention is slowed to be penetrated by the next. So, sometimes the imagery is like the sets of transparent overlays in a medical text in which the bone structure appears under the overlay of the circulatory system, which itself is under the transparent musculature. The effect—appropriately—dizzies us, though we're in visual stasis. If you argue this way, even the titles themselves, which should orient us to the literal, are ironic warnings, sly Kafkean humor."

Advice about making printing-out Pt/Pd prints:

Be very, very patient.

Roger Vail earned his BFA and MFA degrees from the Art Institute of Chicago. He taught photography at California State University, Sacramento from 1970, and is now retired. In 1970, Vail began photographing carnivals and their thrill rides with his 8 x 10 inch view camera. His pictures were made in the evening hours with long exposures times, resulting in extended moments which depict the trajectory of the ride with a sense of wonderment that is both tangible and otherworldly. One such image graced the cover of *Life* magazine, introducing a picture essay with an accompanying text by Garrison Keilor titled, *A Magical Spin on a Summer Night* (2006). Vail's carnival rides are described and transformed through the act of photography. He allows the viewer to experience the flux of the ride in a single scene, rendering both the atmosphere of the night and the energy of his subject, against the recognizable background of the state fair.

Vail's photographs are in the collections of numerous institutions, including: Art Institute of Chicago, Museum of Contemporary Photography, Nelson-Atkins Museum of Art, Museum of Modern Art, NY, San Francisco Museum of Modern Art, Princeton University Art Museum.

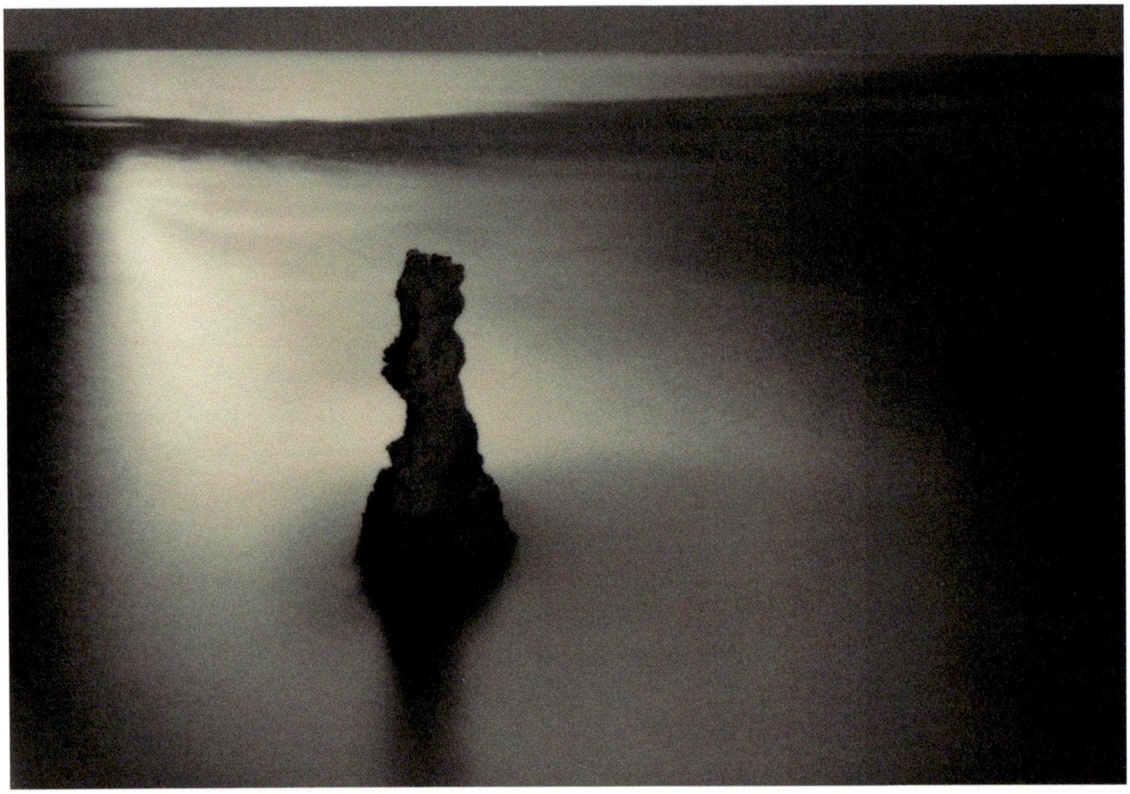

Figure 11.71 Roger Vail, *Santa Cruz, Moonlight*, 1981-96. Platinum-palladium print on Wyndstone Vellum paper, from original 5 x 7 inch negative developed in PMK Pyro, 13.5 x 20 cm. Courtesy Roger Vail. Fe1:Pt:Pd = 2:1:1, hydrated for 20 min. at 100% RH / 20°C.

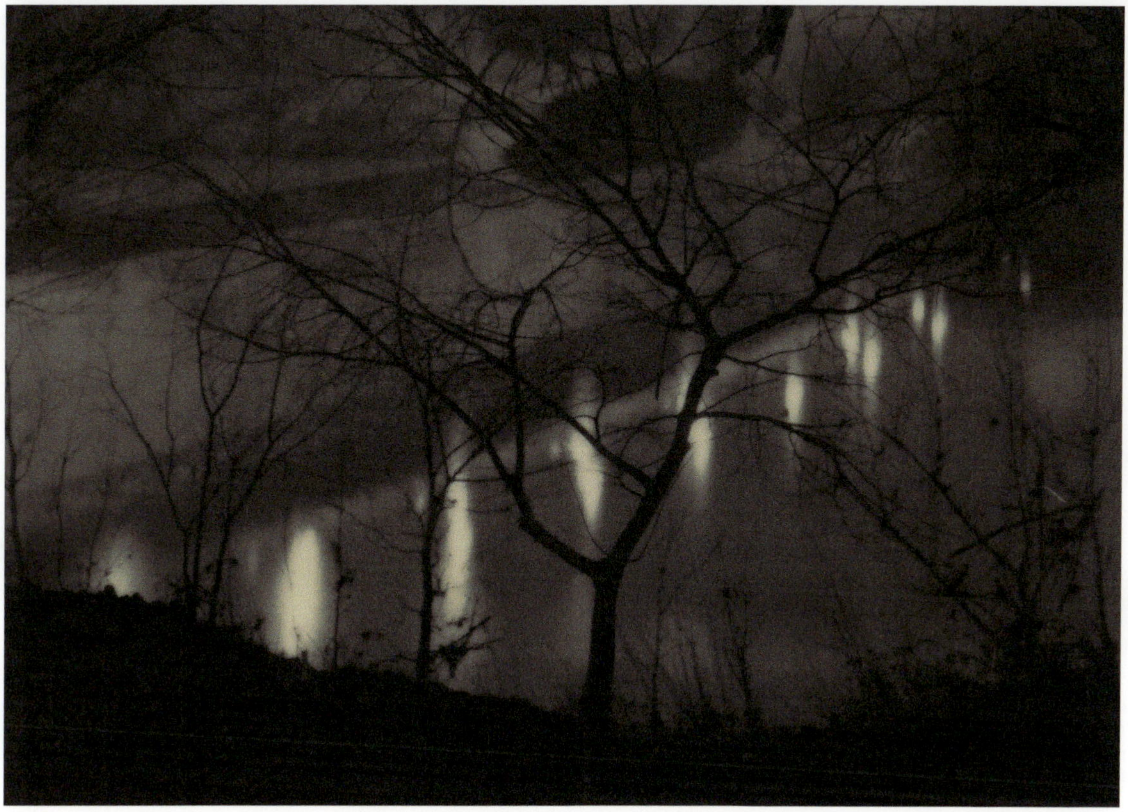

Figure 11.72 Roger Vail, *2.1.97*, 1997. Platinum-palladium print on Wyndstone Vellum paper, from original 5 x 7 inch negative developed in PMK Pyro, 11.5 x 16.5 cm. Courtesy Pradip Malde. Fe1:Pt:Pd = 2:1:1, hydrated for 20 min. at 100% RH / 20°C.

William Watson

After a number of years exploring East Africa and the Caribbean, experimenting with the digital medium in documentary photography, I journeyed ever closer to working with large format film and platinum/palladium processes. This transition paralleled personal journeys, inward and away from the patriarchal empires of global development and aid. I now make work for the sake of the work, where the camera is the eye into a liminal space which I would not be able to access without the intention and care required by manual, tactile processes. My work can hardly be called mine as it is only an outpouring of relationship: between myself and the natural world, between the inner world and the outer world—an interrogation of perception and experience of difference and sameness. Though I have explored the Western USA in my work, my attention has narrowed, focusing largely on clusters of trees and of roots out my front door in Sewanee, Tennessee.

Advice about making printing-out Pt/Pd prints:

I love the slowness and singularity of attention with Pt-Pd printing. Sometimes it bothers me that it might take days to make a single image or a set of images, but this kind of slowing down redefines productivity and reframes what we think of as valuable. I am going to make one print now, and likely no one will ever see it. And that is more than enough. The process itself, then, is the way itself. Production is less about the numbers of prints or images and more about the feeling and the experience and the inquisitiveness one brings with them to the work. Slow down. Stop and pause. There are interesting waves of hurry up and wait in these processes. Preparation for days and hours culminates in the intense focus and flow of coating the paper. Many elements of the process are forgiving enough to get started, and only in the practice of doing will you learn what parts to iterate and experiment with. "There is no try, there is only do" said the little old green dude with pointy ears. If you ever see these words, all that's left to do is get after it.

William Watson is a photographer and organizational designer based in Sewanee, Tennessee. His interests include resilient human and organizational systems, individual and organizational change in complex environments, and the liminal spaces between the human and the natural, if one can discover any distinction between the two these days. His photographic work is largely medium and large format, using the platinum-palladium printing-out process. His work resides in the collections of both the University of the South and the Smithsonian Museum of American History. He lives in Sewanee with his wife, Keri, who is a professor at the University of the South, and their two children, Amelie and Calvin.

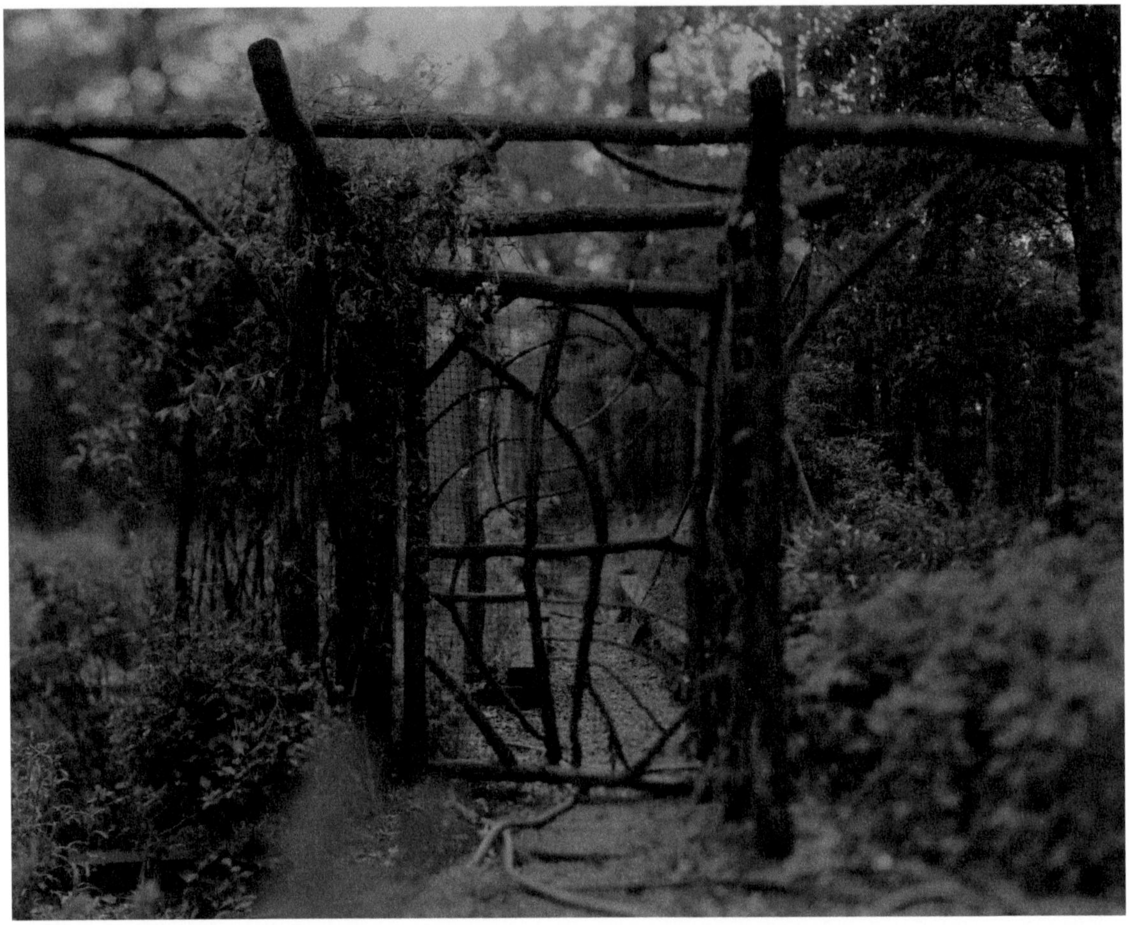

Figure 11.73 William Watson, *How is the Passing?*, April, 2012. Platinum-palladium print on Revere Platinum paper, from PiezoDN negative of 8 x 10 inch original processed in PMK Pyro, 19 x 24 cm. Courtesy William Watson. Fe:Pt:Pd = 2:1:1, hydrated for 90 min. at 56% RH / 20°C.

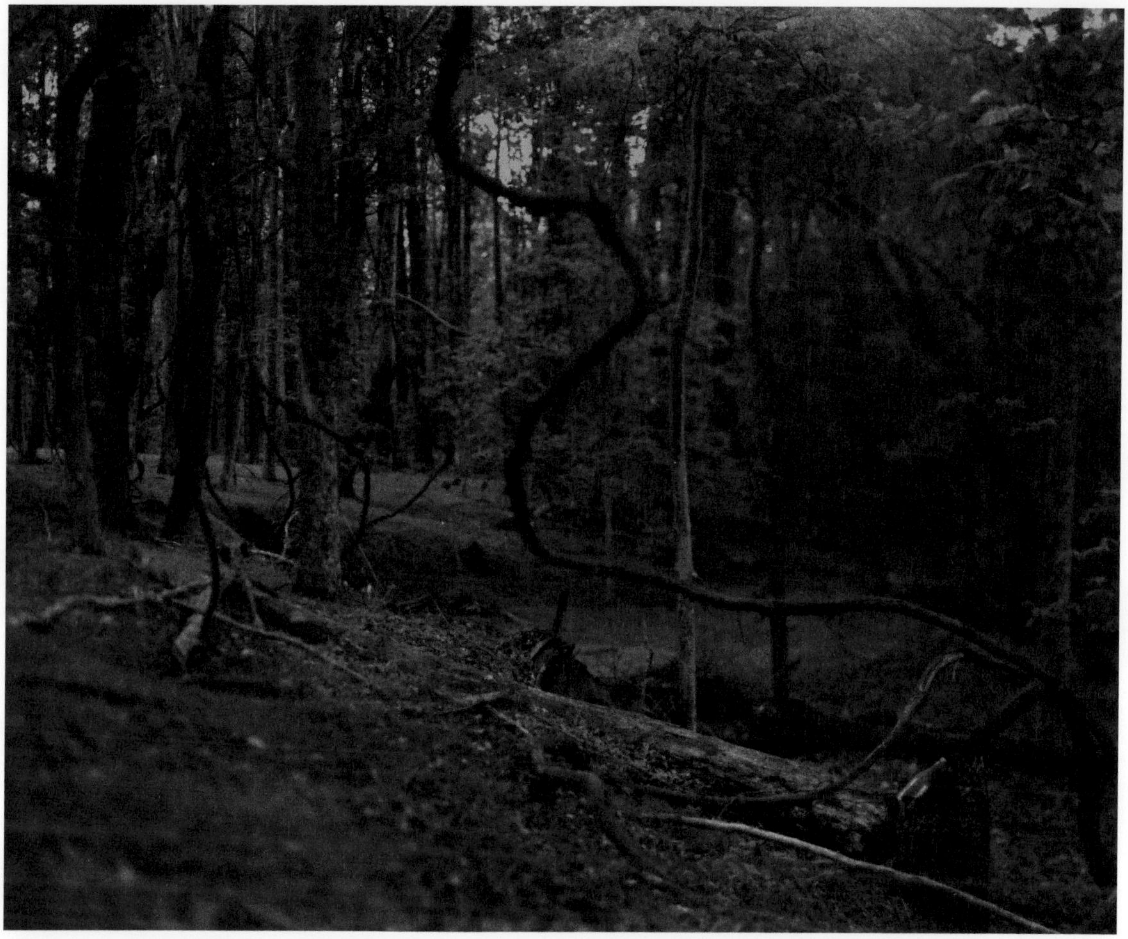

Figure 11.74 William Watson, *Now What I Will Part With*, April, 2013. Platinum-palladium print on Revere Platinum paper, from PiezoDN negative of 8 x 10 inch original processed in PMK Pyro, 19 x 24 cm. Courtesy William Watson. Fe:Pt:Pd = 2:1:1, hydrated for 90 min. at 56% RH / 20°C.

Elisabeth Weissensteiner

My fascination with the space between perception and the perceived has led me into all sorts of experiments with image, material, and technique. Over the years, my art practice has come to contain various photo-based techniques like encaustic photography, polyester coated photography, photography for the outdoors, and recently, palladium printing, alongside of a variety of sculpture practices. Indeed, my interest in photography originated in the digital world. Working with purely immaterial images which only exist in the beholder's mind deepened my interest in the space between the image and the object: an image must always "materialize," that is, it needs a body before it can be observed, thus rendering the boundary between two and three dimensions fluid. Navigating this boundary over and over again has become constitutive for my art making.

Advice about making printing-out Pt/Pd prints:

The printing-out process involves me directly in the creative process. Generally, the creative process is a dialogue between the artist and the material, in which the material has as much to say as the artist. A process that permits "to get one's hands dirty" makes this dialogue balanced and exciting. These kinds of processes need patience and frustration-tolerance.

Elisabeth Weissensteiner is an Austrian artist who divides her time between Melbourne, Australia, and Vienna, Austria. Her path into art started by studying German language and art history at the University of Vienna. She also trained as a graphic designer. Her long-standing practice in the visual arts comprises solo and group shows, public acquisitions, public art commissions, art-in-residence invitations and art-science projects. Her work is held in public collections and has been funded by the Austrian government, the Deutsche Forschungsgemeinschaft and others. In 2013, she also received the "Prämie" of the "Outstanding Artist Award for Interdisciplinary Practice" in Austria and in 2010 a photo-based work was Finalist "Best Contribution to Urban Art," City of Port Phillip, Australia.

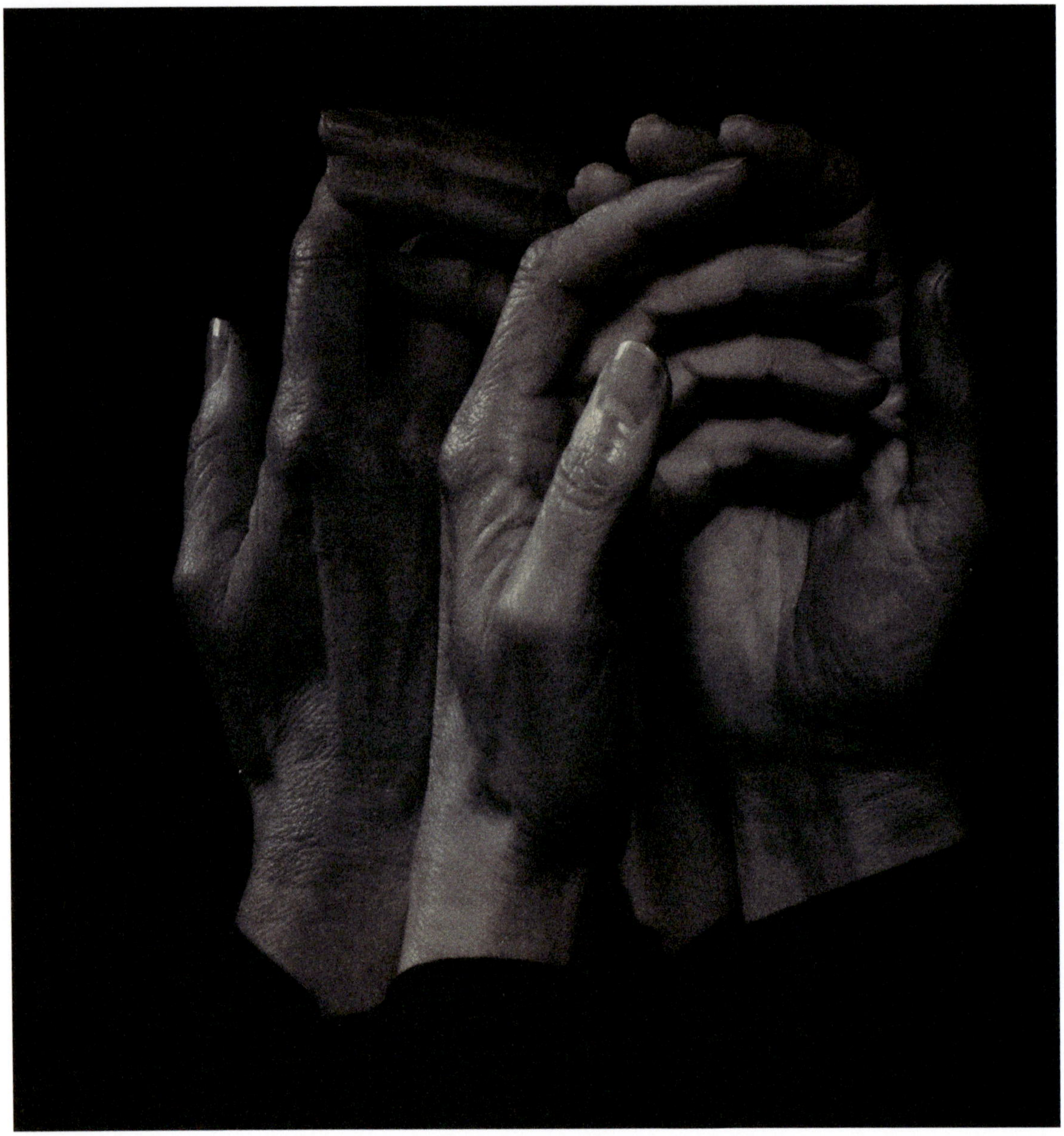

Figure 11.75 Elisabeth Weissensteiner, *Hands*, 2019. Palladium print on Bergger COT 320 paper, from digital negative of full frame digital original, 18.5 x 18 cm. Courtesy Elisabeth Weissensteiner. Fe:Pt:Pd = 1:0:1.

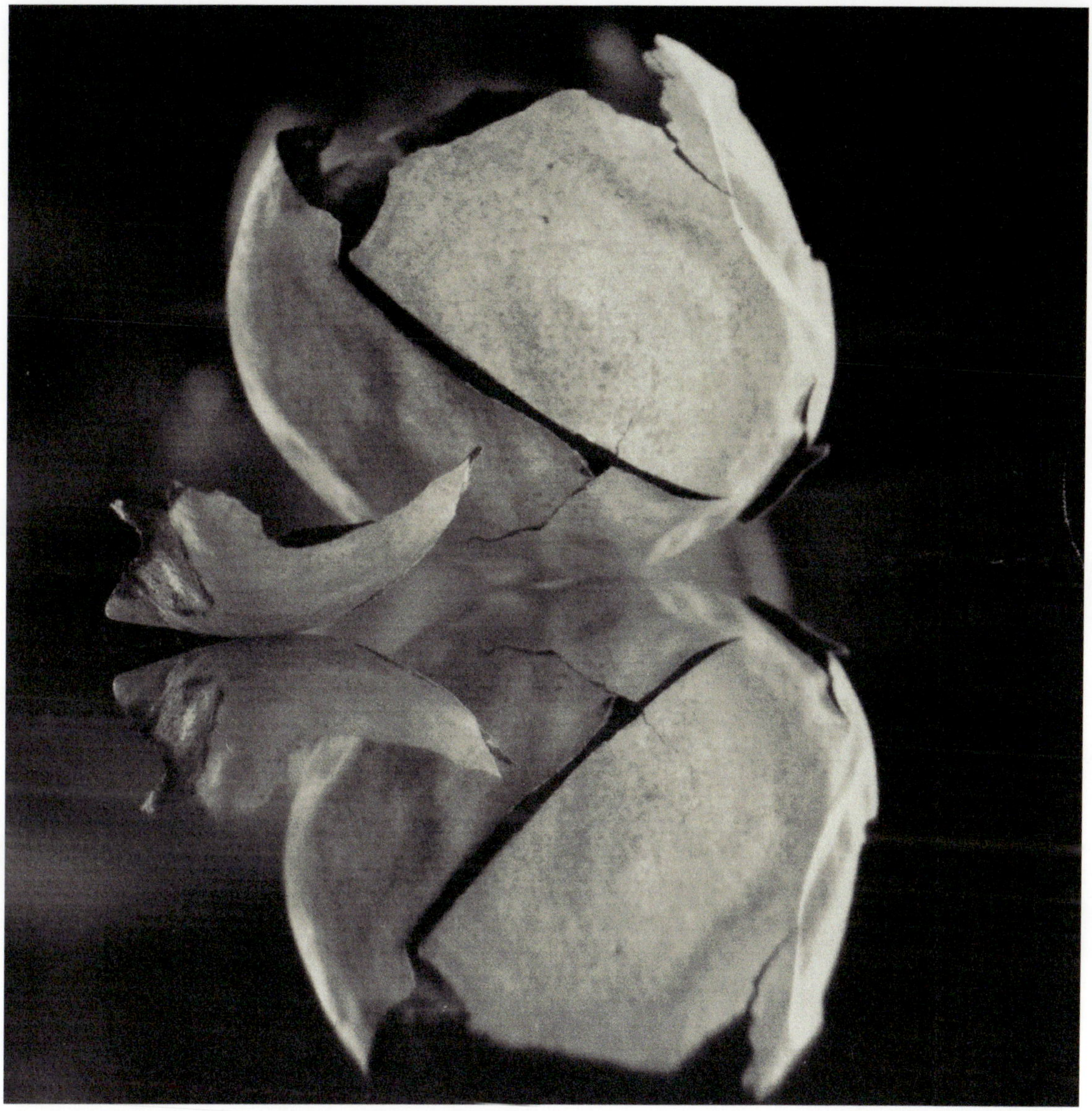

Figure 11.76 Elisabeth Weissensteiner, *Avocado Pit*, 2019. Palladium print on Bergger COT 320 paper, from digital negative of full frame digital original, 14.5 x 18 cm. Courtesy Elisabeth Weissensteiner. Fe:Pt:Pd = 1:0:1. (Striations are a part of the image.)

Kirk Williams

Light meditations and camera interventions. My work is part meditation and part adventure; looking for the edges of the unknowable in our existence. Capturing those ephemeral, glittering moments when light becomes something other than a reflection in physical space; slowing time, extending the exposure and layering time and light.

Advice about making printing-out Pt/Pd prints:

With this process you can make prints either with a really well equipped or a minimalist darkroom. If you follow the instructions and set up a good workflow, both precisely and mindfully, you will quickly get amazing prints that have an intangible quality to them. You do have to slow down to make these prints; it takes time to hydrate paper to the exact humidity, but this added time also gives you a chance to pause and catch your breath and maybe read a book or play the piano or guitar. There is a limited choice of papers, but this has helped me to focus on the images and create a synergy between image and paper.

Kirk Williams, a native of West Texas, has been photographing and traveling extensively for more than 30 years. He holds a BFA from the Rochester Institute of Technology and an MFA from the School of the Art Institute of Chicago. Kirk studies the classical guitar and shares his photo experiences by teaching. He has extensive knowledge of alternative photographic printing processes. Currently, Kirk is working on multiple long-term projects.

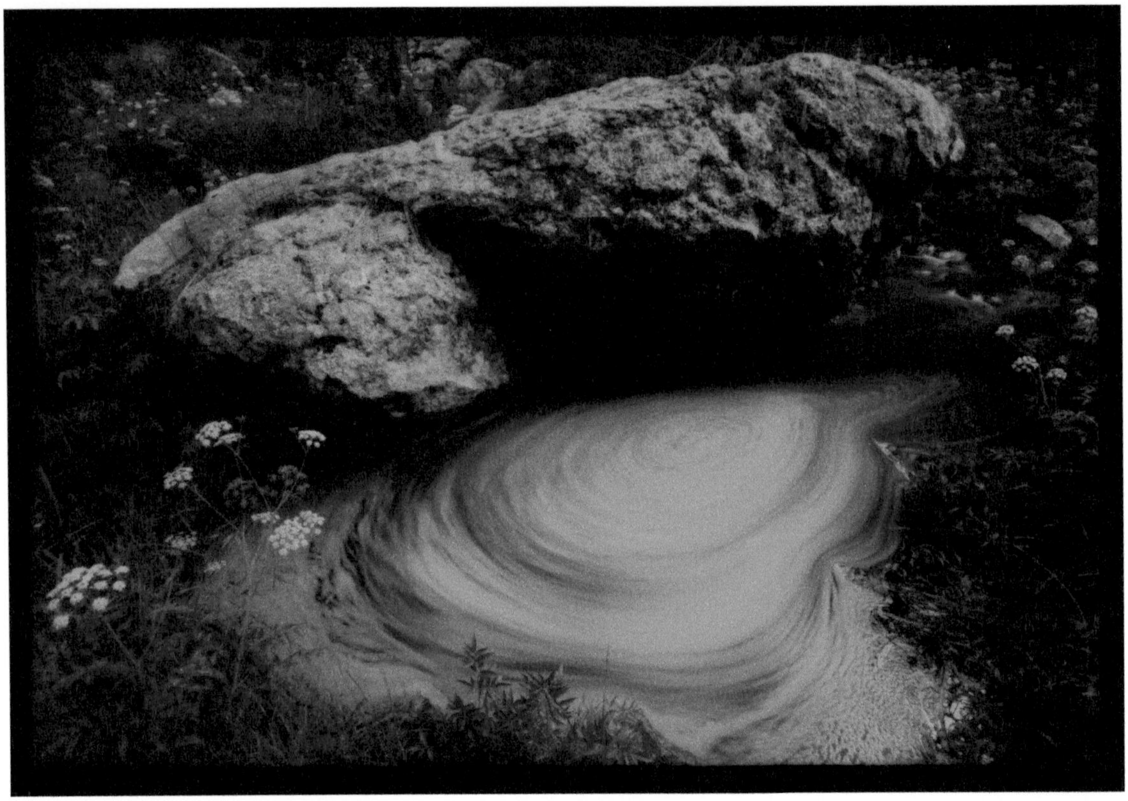

Figure 11.77 Kirk Williams, *Event Horizon, Northern New Mexico*, 2020. Platinum-palladium print on Reich CT 48# paper, from PiezoDN negative of full frame digital original, 17.8 x 22.8 cm. Courtesy Kirk Williams. Fe:Pt:Pd = 2:1:1.

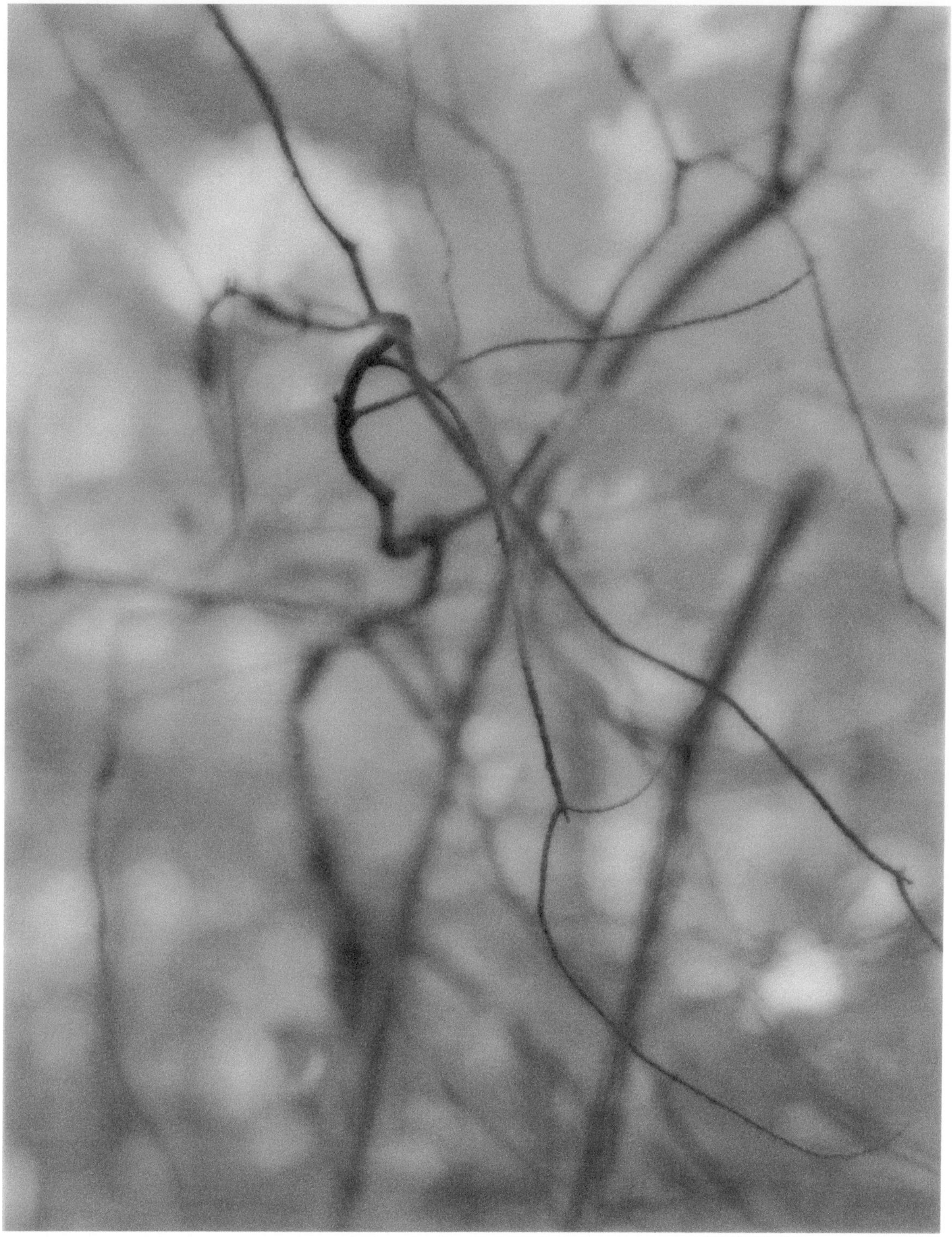

Figure 11.78 Kirk Williams, *Zen Twig, Redux,* from the series *Ephemerati,* made in a forest preserve outside Chicago, 2000. Platinum-palladium print on Wyndstone Vellum paper, from original 4 x 5 inch negative developed in Kodak HC110, 12.1 x 9.6 cm. Courtesy Pradip Malde. Fe2:Pt:Pd = 2:1:1, hydrated for 20 min. at 100% RH / 20°C.

Lucy Wimmer

When I was a sophomore in college, I decided to visit a small rural church for a photo assignment. I was nervous with the prospect of becoming an observer to a community that I had no part in, but instead, I was welcomed in with open arms. I became a part of the church's family, attending services, lunches, meeting the children of the friends I made at this church. This experience has forever changed my perspective on community, and opened my heart to a deep empathy that I pour into every image I make.

My work is driven largely by a love of community, whether that be communities that I have built, communities that have welcomed me in, or communities that I do not belong to. Community is vital, beautiful and full of comfort, and I seek to express the tender, often unseen, moments that make up this aspect of all of our lives. I use photography as an avenue to explore relationships, communities, and the human yearning for comfort. My work reminds us that we are never truly alone.

Advice about making printing-out Pt/Pd prints:

There are many variables to think about and deal with, and it can be a tedious process. However, it is so rewarding (and magical) to put in the time and effort and produce a print that is unlike any other type of print. I love how physical the process is, and how the physical print changes the meaning of the image itself—once you have created a platinum-palladium print, it is impossible to then untangle the image from the print itself. Be patient! Know that you are probably not going to make a perfect print on your first try, but if you stay calm and patient and persevere, you will *make perfect prints. Also, keep detailed notes. This may seem tedious, but it will help you make that perfect print, and it will become invaluable in the future as you continue to make prints, and as you submit work to be shown.*

Lucy Wimmer is a photographer from Austin, Texas who graduated magna cum laude from The University of the South in 2020 with a Bachelor's degree in Art History and a minor in Art. She works with medium and large format (8 x 10) film, digital cameras, and prints in silver-gelatin and printing-out platinum-palladium. She is passionate about visual literacy and access to art education, and has worked in Haiti with the Zanmi Foto visual literacy program, and started a photography program with a summer camp for immigrant and refugee children in Nashville. Her work has been published in the *Mountain Goat Literary Journal*, the Sewanee Purple newspaper. She is the 2020 recipient of Sewanee's Shepard Award for Excellence in Photography.

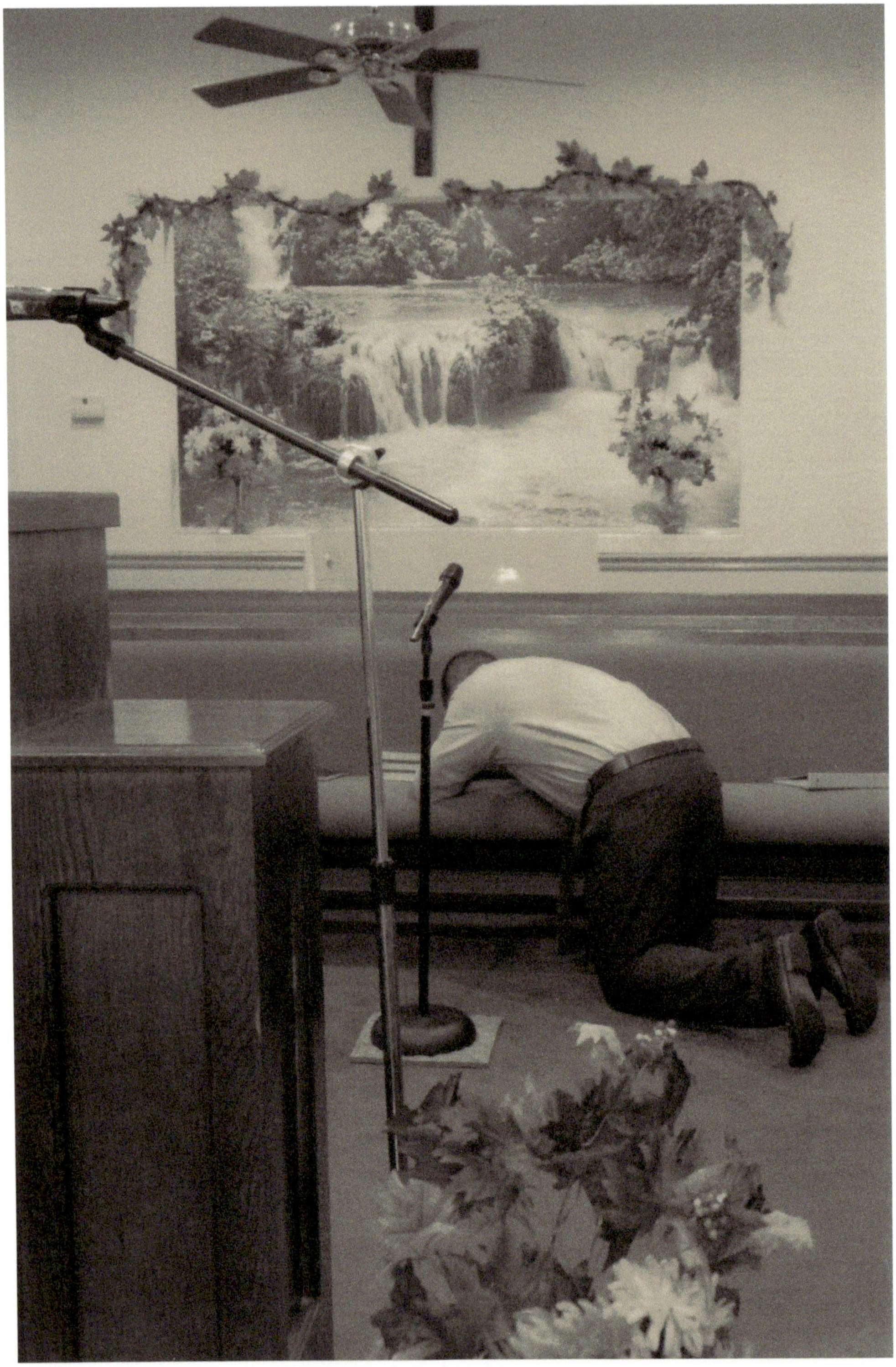

Figure 11.79 Lucy Wimmer, *Brother Freddie*, 2018. Platinum-palladium print on Arches Platine paper, from PiezoDN negative of full frame digital original, 24.7 x 19.6 cm. Courtesy Lucy Wimmer. Fe:Pt:Pd = 2:1:1, hydrated for 20 min. at 100% RH / 20°C.

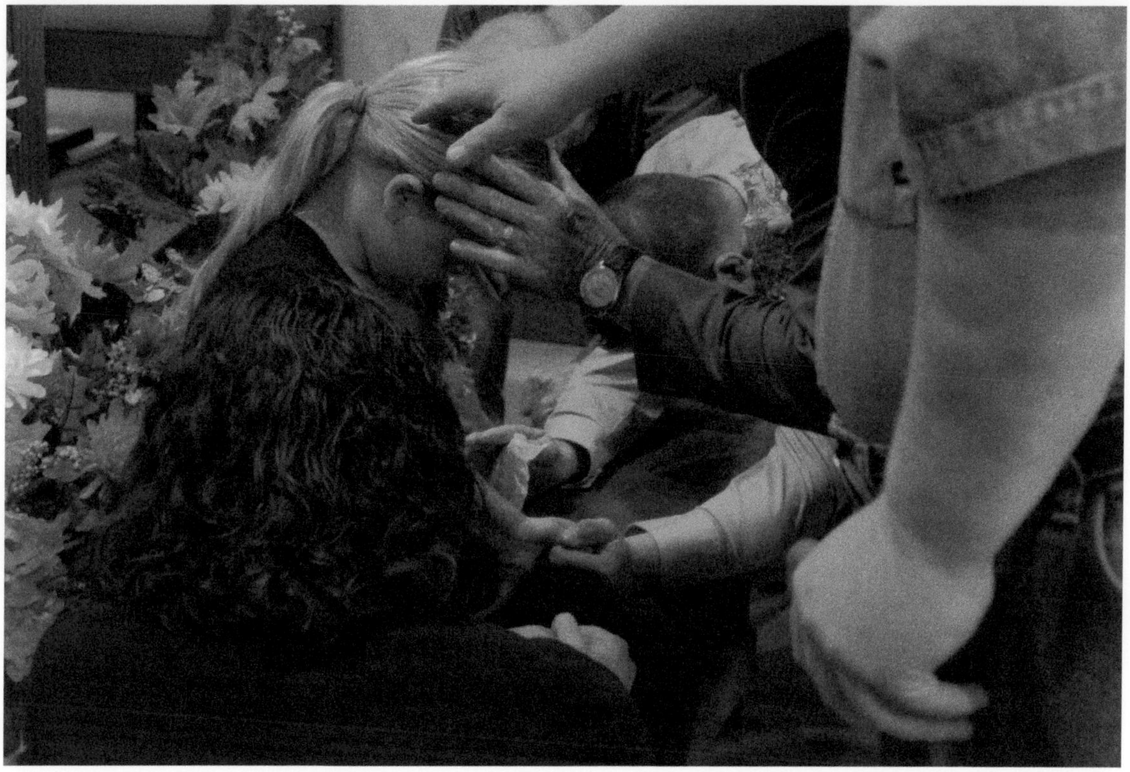

Figure 11.80 Lucy Wimmer, *Revival*, 2018. Platinum-palladium print on Arches Platine paper, from PiezoDN negative of full frame digital original, 24.7 x 19.6 cm. Courtesy Lucy Wimmer. Fe:Pt:Pd = 2:1:1, hydrated for 20 min. at 100% RH / 20°C.

Ellie Young

I am interested in exploring the inherent abstraction of the real world. These works simultaneously capture the bold, graphic qualities of my subject and the nuances of texture and tonality. The images appear both as powerful, gestural forms from a distance, and as rich topographies with minute details and subtlety up close. The simple still life works introduce a tension between the real and the abstract.

Advice about making printing-out Pt/Pd prints:

When exposing, the printing-out effect easily helps you decide about the correct exposure. It is easy to create rich blacks, and the hydration methods provide control of the print's color. There is a remarkable tonal range with a gentleness that most other processes do not appear to have. The choice of paper can make a massive difference to the final outcome—so test papers thoroughly.

Ellie Young is the founder of Gold Street Studios in Victoria, Australia. Since its establishment in 1999, Gold Street Studios has become the center for hand-made photographic print processes in Australia and New Zealand. Young is the author of *The Salt Print Manual*, 2011 and exhibits widely in Australia, Europe, England, and China. Teaching in Beijing, Nanjing, and other cities in China has allowed these crafts to be shared with the photographic community in China.

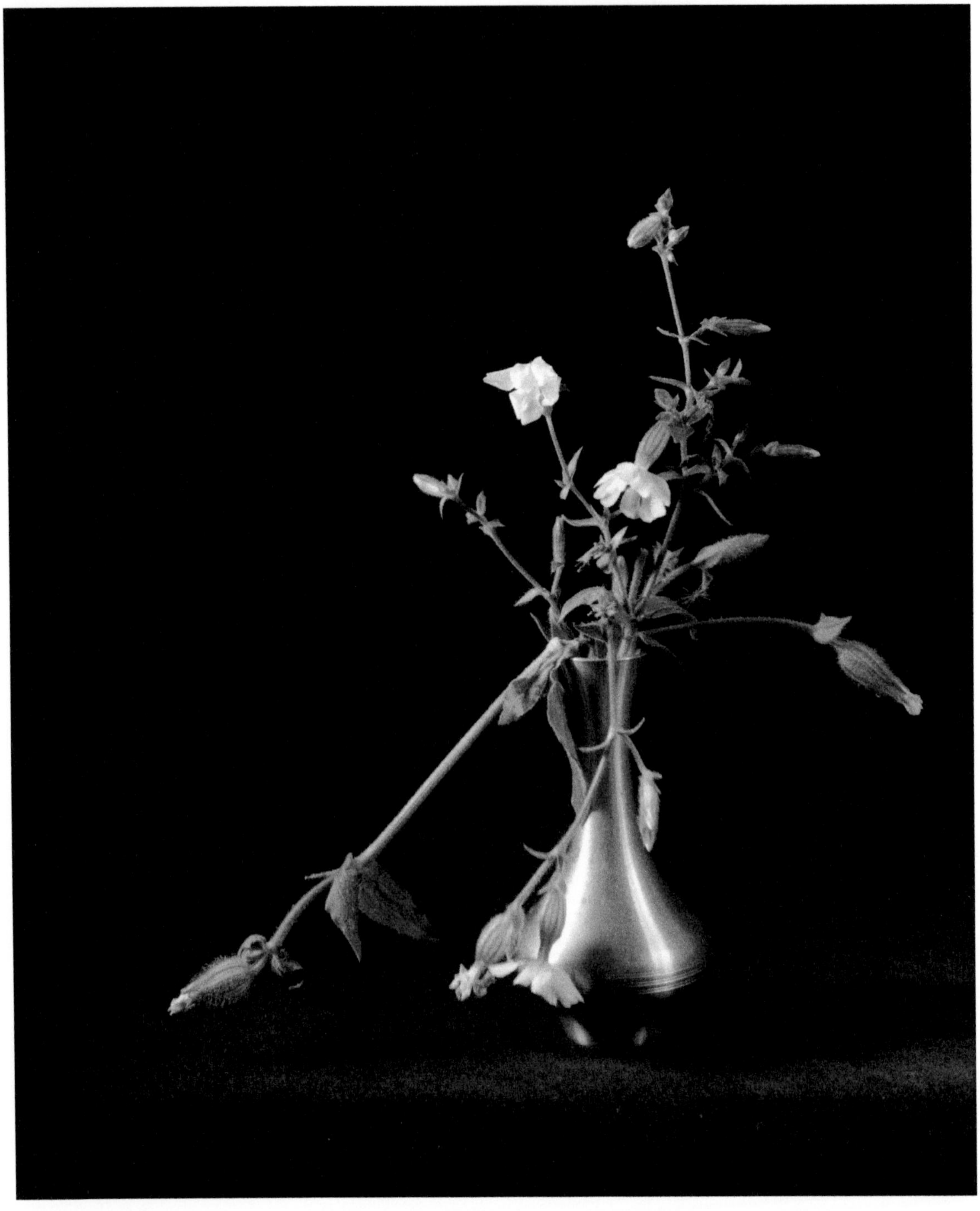

Figure 11.81 Ellie Young, *Chickweed*, 2017. Palladium-platinum print on Bergger COT 320 paper, from original 8 x 10 inch negative developed in Pyro PMK, 25 x 21 cm. Courtesy Ellie Young. Fe:Pt:Pd = 4:1:3, hydrated for 40 min. at 100% RH / 20°C.

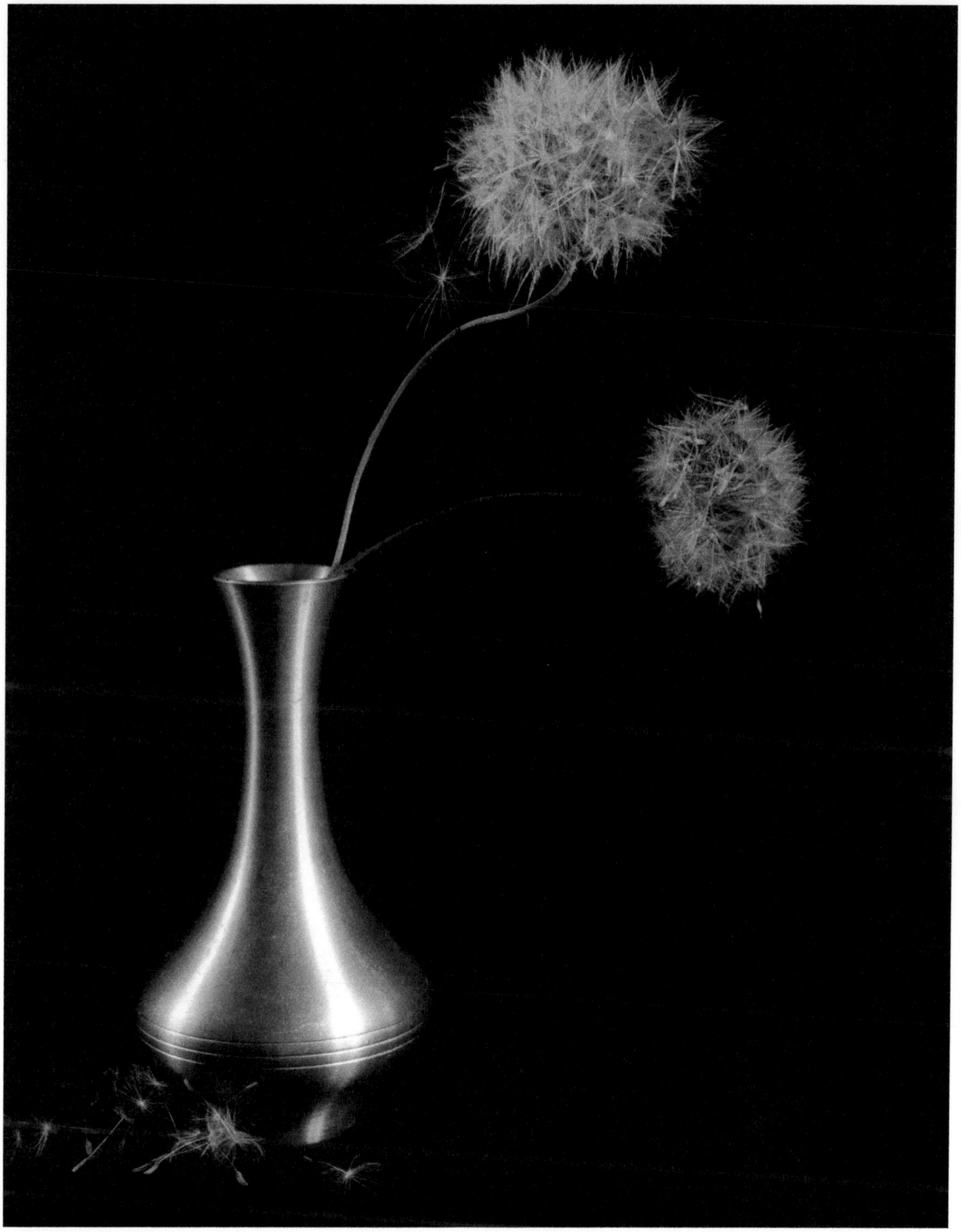

Figure 11.82 Ellie Young, *Dandelion Seed Head*, 2017. Palladium-platinum print on Bergger COT 320 paper, from original 8 x 10 inch negative developed in Pyro PMK, 25 x 21 cm. Courtesy Ellie Young. Fe:Pt:Pd = 4:1:3, hydrated for 40 min. at 100% RH / 20°C.

Notes

1 "Hope is a contraband passed from hand to hand, and story to story. ...I'd rather reject the terms optimistic and pessimistic. They suggest a calculation of how things are going to evolve, and if it's going to evolve in the way you want, you're optimistic. That has very little to do with despair and hope. Hope is not a form of guarantee, it's a form of energy, and very frequently that energy is strongest in circumstances that are very dark."—John Berger, *Bento's Sketchbook: How Does the Impulse to Draw Something Begin?* New York: Pantheon Books, 2011. p. 87.

2 Peter Henry Emerson and Thomas Frederick Goodall, *Life and Landscape on the Norfolk Broads*. Dundee: Valentine, 1885–86. https://www.metmuseum.org/art/collection/search/290484, last accessed May 9, 2020.

3 Interview with Roger Vail, "Roger Vail: In the Still of the Night," *Black & White Magazine*, April 7, 2016, pp 82–90.

4 https://multitimer.net — perfect for workflows where more than one item has to be timed at different stages of the process.

5 https://www.mikeware.co.uk/mikeware/New_Cyanotype_Process.html

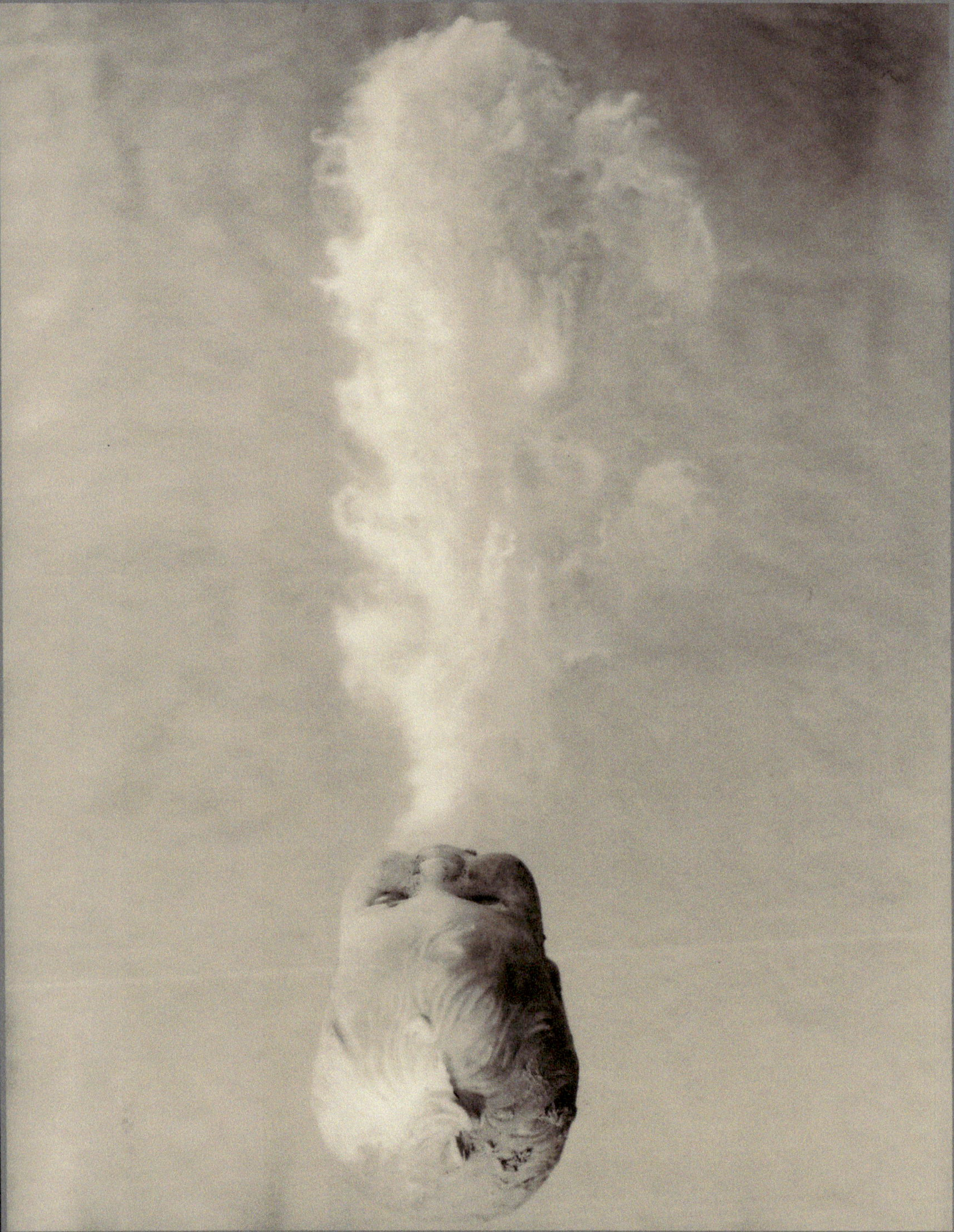

Figure A.1 Pradip Malde, *Expiration II.* 2003. Platinum-palladium print on vellum from 11 x 14 negative. 35 x 27 cm. Courtesy Pradip Malde. Fe3:Pt:Pd = 2:1:1, 20 min. hydration in 100% RH / 20°C

Appendix I
Troubleshooting

Anticipating Problems

The instructions in this book are well tested, and should enable the making of good prints. Yet glitches and aberrations can occur, and learning how to solve such problems is part of the printer's craft. Keeping notes, being consistent and methodical, working with step tablets or targets, and changing only one thing at a time, will all help to resolve problems. These practices also deepen our understanding of the process, which in turn reduces the likelihood of encountering those gremlins that Nancy Rexroth warns us about. Be aware of your assumptions. If suddenly your prints look much warmer, and nothing was different from the last time you made prints, then what are you assuming? Did the weather shift, and trigger a different set of environmental conditions in your work or storage space? Did you do something as innocuous as move the paper stack last week—from a low shelf to a work surface that happens to be next to a radiator?

It will become apparent that most problems are rooted in coating techniques or—that greatest of imponderables—the paper. Be extra-vigilant when working with a new batch, or type, of paper. It will help to have at least two types of papers on hand, and to be able to make acceptable prints from each. Then, if one paper begins to present problems, a logical first step would be to put the other type through the same workflow in order to resolve whether or not the problem is paper related. For these reasons, it is always a good idea to replenish paper supplies well before a working batch runs out. The following suggestions and solutions anticipate problems that may be encountered within the parameters described in this book. The discipline of documentation and sound scientific methodologies becomes essential if you are to experiment and deviate beyond this.

Preparing Chemicals

Ammonium Ferric Oxalate Crystals are not Bright Green

Crystals of pure ammonium ferric oxalate should appear bright green (see Figure 4.8). If they are a dull green, or even brownish, the chemical may be old, exposed to ultraviolet, contaminated, or the wrong formula. Do not use the batch, and replace with a fresh supply from a reliable source.

Ammonium Tetrachloroplatinate Solution has an Orange Scum

After preparing a solution from the crystalline solid, there may be a fine, light brown to orange scum across the surface. It's probably a small amount of platinum diamine dichloride, an almost inevitable water-insoluble decomposition product. This is normal, but the scum should be removed as much as possible. Any one of these methods, or a combination, may be used:

- Filter the solution with a Whatman #1 filter. Use a 125 mm diameter filter with a filter funnel of around 64 mm diameter. Fold the disk-shaped filter in half, and then once more into a quarter-circle, then open one side of the fold to form a cone. Rinse the filter funnel with distilled water, then settle the cone into the funnel, and place it into the mouth of a clean vessel or the final storage bottle. Gently pour the ammonium tetrachloroplatinate(II) solution into the funnel.

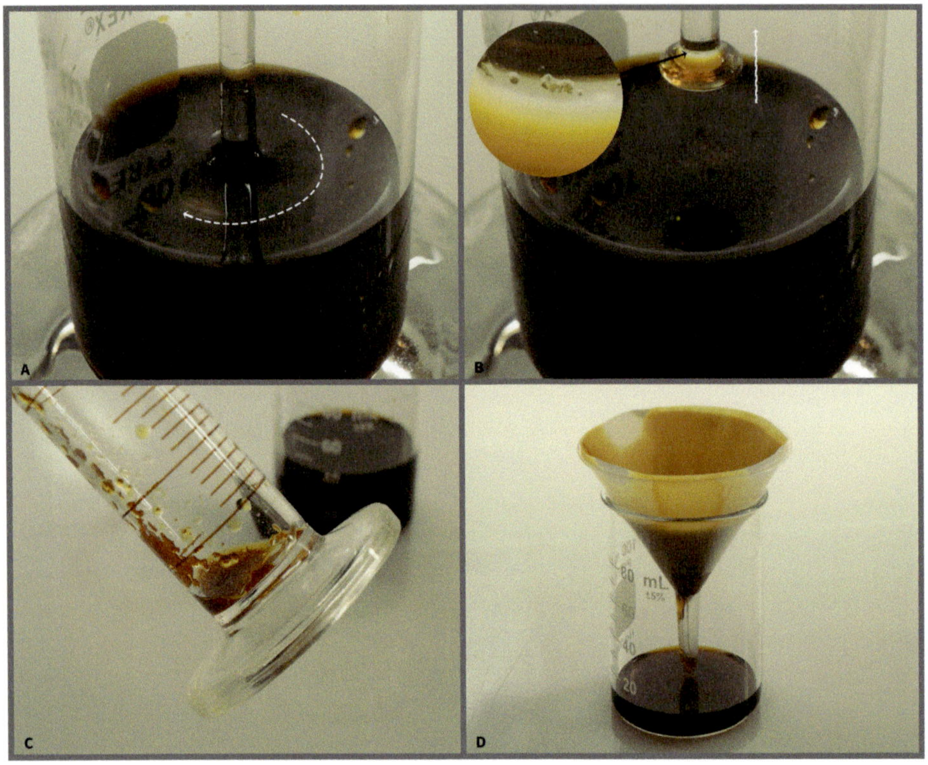

Figure A.2 Three methods for removal of scum in ammonium tetrachloroplatinate(II) or other solutions. A. and B. stirring and drawing up.

C. decanting.
D. Filtration.

Wait for another five minutes after the last of the solution has passed through the filter to minimize waste. Discard the filter.

- Using a clean, dry glass rod, gently stir the solution. Hold the rod vertically in the center of the vortex, and slide it down into the solution. Wait for the scum to coalesce around the rod, then gently draw it up and out of the solution. Surface tension will cause much of the scum to be pulled up with the rod. Wipe the rod with a clean piece of paper towel or blotting paper, and repeat until most, if not all, of the scum has been removed.
- Gently pour the solution into a tall 50 or 100 ml glass measuring cylinder. Let the scum come to the surface, then very gently decant the solution—catching the scum on the bottom and sides—into another clean glass cylinder of similar size. Repeat until most of the scum is removed.

Processing Chemicals are not Dissolving

- Check that solutions are being prepared at the recommended working temperatures. It is best to warm solutions slowly, bringing them up to the right temperature through the use of water baths or a carefully regulated laboratory-grade hot plate.
- If temperatures are as recommended, then check that the right chemicals are being used. Compare chemical formulae on labels with those in the recipe.

- Always add ingredients in the same order as described in a recipe, and work with implements and containers that are clean and freshly rinsed with distilled or purified water.

Coating the Paper

Sensitizer is Running out in Less than Four Passes

See Chapter 6: Testing Papers for Pt/Pd
For optimal print quality, paper should have the capacity to absorb 20–30 milliliters of sensitizer solution per 1 square meter of surface area (20 – 30 ml / m^2). In practical terms, this translates to 1.2–1.5 ml for an 8″ x 10″ / 20 cm x 25 cm area. Volumes for brush-coating are harder to quantify, but may generally be double the amounts just described.

If the sensitizer is running out in less than four passes with a coating rod, or before an even distribution is achieved by brush, then consider the following options:

- If the solution is running out between two and four passes, increase the sensitizer volume by up to 30% of the starting volume. Coating with more than 1.7 ml for an 8″ x 10″ may be expensive and other paper options should be considered.
- Ensure that the paper is being stored at 40 – 60% RH at room temperature. Acclimate each sheet for 10 minutes in a hydration environment of 50% RH (or greater) before coating. We refer to this as a "pre-coat hydration." See Chapter 6: Humidifying.
- Use a 15% solution of glycerin in water, adding an amount equal to ca. 1/30th of the sensitizer volume. e.g. combine 1.5 ml of sensitizer with one drop (about 0.05 ml) of 15% glycerin. Glycerin increases liquid viscosity, and will slow down the sensitizer's penetration into the paper. Be aware that excessive glycerin in the sensitizer can prevent it from drying, and lead to damaged negatives.
- If the solution is running out in less than two passes, the paper is far too absorbent. Consider using another type of paper.

Sensitizer is Crystallizing on the Paper

Small, discrete crystals may form on the paper during or soon after coating for several reasons. See Figure 6.5.

The sensitizer solution has become supersaturated, and the friction between the coating tool (glass rod or brush) and paper surface is "seeding" small crystals. These often appear as sparkling bands following the coating direction brushing pattern. Coating with a rod may feel "gritty." Supersaturation may be caused by a temperature drop, which usually lowers solubility, arising from the solution or paper being chilled in a cold room or kept in a fridge.

Supersaturation can also arise if a puddle of excess sensitizer solution is allowed to stand and evaporate, especially in a warm environment, thus concentrating the solution beyond its solubility limit.

Some remedies, and a few more possible causes are as follows:

- Check the chemistry for contamination: e.g. ensure no potassium (K+) salts are present with an ammonium ferric oxalate sensitizer.
- Check that the quantities used for making up the solutions were correct, and not exceeded.
- Acclimatize the sensitizer solution and paper sheets to a standard room temperature of 20–22°C/68–72°F before coating.
- Ensure that the paper is not too dry by giving it a pre-coat hydration.
- Use the correct volume of sensitizer, avoid excess, and see that all is absorbed into the paper on coating, or that any inadvertent excess is blotted off promptly.
- Apply the sensitizer using a syringe, followed by rod coating, rather than pouring it out randomly on the paper and brushing.
- Avoid puddling of the sensitizer, and leaving puddles to evaporate, concentrate, and crystallize. Use a strip of blotting paper to remove excess solution from the edge.
- If crystals are forming in a very specific line and the coating rod generates a scratchy feeling,

examine the rod carefully—it may be scratched. Replace the rod.
- Dilute the sensitizer in 5% increments until the problem is resolved—but only as a last resort, because this may lead to loss of Dmax.

Sensitizer is Puddling on One Side at the Start of Coating

- Adjust the level of the coating surface. When using a glass rod, it is especially important that the surface is level in the perpendicular of the coating direction. i.e. if the coating direction is away and towards your body, then the left-right direction needs to be very level.

Sensitizer is not Spreading out Evenly

- Check your coating technique: you may be applying insufficient pressure and lifting the coating rod up at times, or not creating an even "wash" with a brush.
- The coating rod may be warped. Look along its length from one end, against the light, to see any irregularities. Try flipping it over and coating with the other side. If the problem persists, replace the rod.
- Adjust the level of the coating surface using an accurate and appropriately sized instrument.

Sensitized Paper Looks Blotchy

Also see Figure 6.5. The printed image shows fluctuations in density in adjacent areas of the paper which should otherwise be smooth and uniform. i.e. it shows fibrous, "woolly" or "cloudy" patches that are not in the original image. Some call this "splotching," but we prefer "blotching." It may be more conspicuous in homogenous areas of very fine tonal separation in the high values. The problem manifests with certain papers and paper batches. It has been seen on some batches of Arches Platine and Revere Platinum, but never with Buxton, Herschel, Wyndstone Vellum, or Reich CT papers. Blotching must be due to some local non-uniform property of the paper sheet, which lessens the image's optical density in the paler fibrous regions. Note that aqueous sensitizer solutions—particularly if filtered—are perfectly homogeneous and cannot be the cause of this non-uniformity; but once the image pigment is formed, it can disclose or amplify the paper variations visually owing to its greater optical density.

Consider one or a combination of the following to resolve the problem:

- Pre-humidify the paper sheet over water at room temperature for 10 minutes
- Use 15% glycerin, adding an amount equal to ca. 1/30th of the sensitizer volume. e.g. combine 1.5 ml of sensitizer with one drop of 15% glycerin, to help slow down and even out the penetration of sensitizer in paper fibers
- Use 15% Tween, adding an amount equal to ca. 1/30th of the sensitizer volume. e.g. combine 1.5 ml of sensitizer with one drop of 15% Tween, to help penetrate the fibers uniformly
- Try the reverse side of the paper
- Dilute the sensitizer with distilled water, in increments of about 5% of the original concentration, until the problem is resolved. For example, to 1.5 ml of sensitizer, add 0.075 ml of distilled water—practically, it is difficult to measure such a small volume without a 100–1000 μl adjustable micropipette, and a proximate volume may need to be measured out in drops: either one (0.05 ml) or two (0.10 ml) drops as a starting point. Make a test print, and if blotches are still apparent, increase the dilution by another 5% or so. As the sensitizer dilutes, a corresponding increase of base exposure will be necessary. The sensitizer may be diluted by as much as 25%, at which point exposure may need to be twice that of the original base exposure. It may be difficult to achieve strong Dmax by any further dilution[1]
- Select a better paper

Sensitizer Soaks All the Way through to the Back of the Paper

The sensitizer passes all the way through the paper, and cannot be evenly distributed across the desired image area, while certain tissue-like papers with

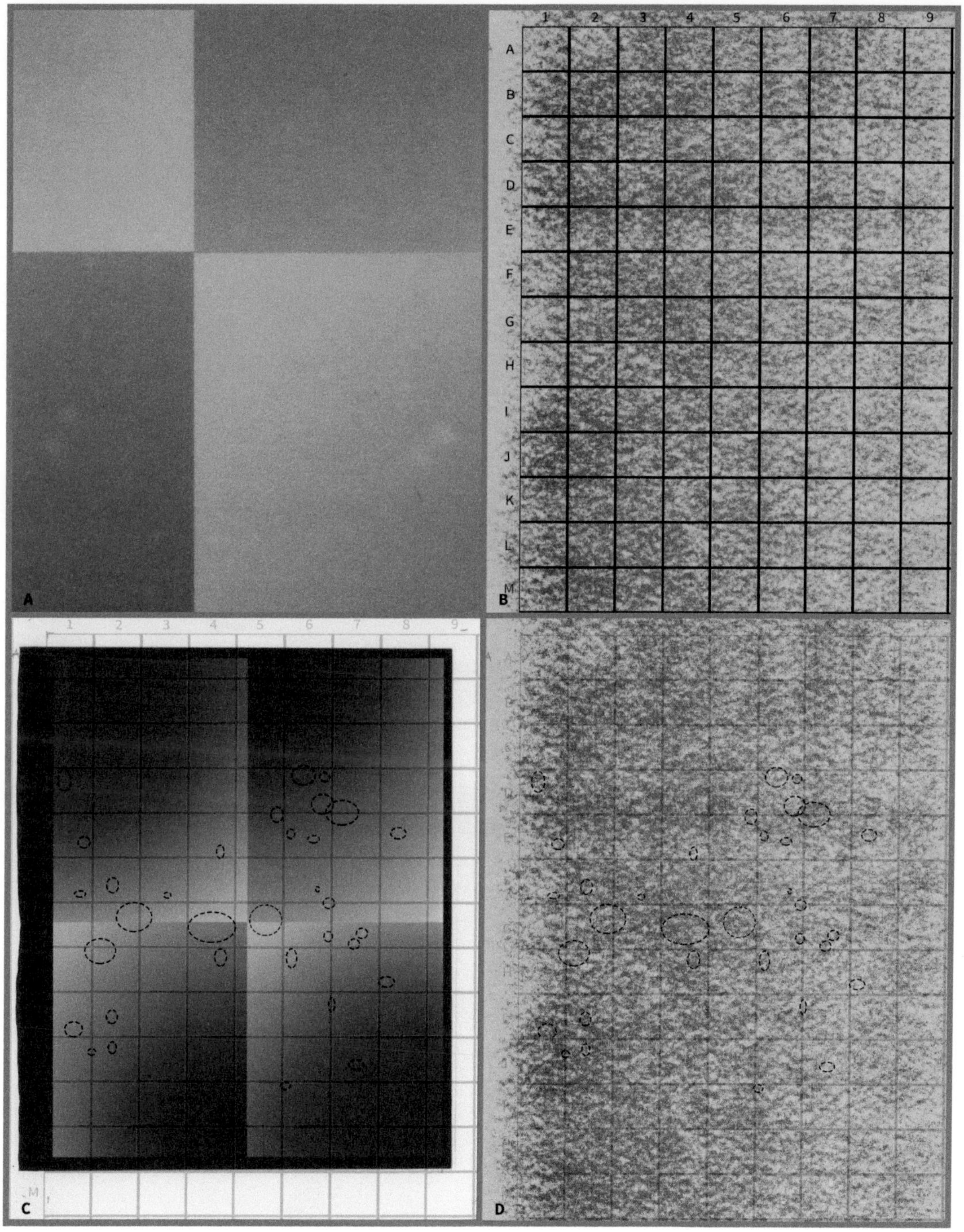

Figure A.3 Blotches not in the original image can appear as soft-edged and cloudy (A) and tend to be related to unevenness in paper structure. B, C, ad D establish this relationship clearly.

B. a sheet of Arches Platine was registered with a grid and scanned as a transparency.

C. the sheet was coated and printed, then reflection-scanned again. Blotches are circled.

D. Using the grid as a register, the blotches are observed to manifest in the densest parts of the paper. See Figure 6.5

manageable wet-strength, such as Japanese-style gampi, kozo, and mitsumata papers are highly absorptive but may be successfully coated with special methods.

- Select another paper
- If working with tissue or thin Japanese papers:
  - tape the edges down on a glass sheet
  - adjust the sensitizer volume to minimize wastage
- For brush-coating, measure a volume of sensitizer into a shallow, wide-mouthed container; coat the desired area by dipping and brushing in a rapid, parallel set of strokes, then in a perpendicular direction to the first set of strokes, and so on until evenly coated
- for glass rod-coating, apportion the sensitizer into three or more small syringes; place the rod along one edge of the area, distribute the contents of one syringe in front of the rod, and immediately begin coating; as soon as solution begins to run out, use the next syringe, and so on, passing up and down until the area is evenly coated
- Areas of slight unevenness may be ignored as they will either smooth out during the drying process, or be hidden by the surface properties of the dried print

Exposing and Processing Prints

Paper Feels Damp before Exposure

The sensitized paper has not been allowed to dry, or has been over-hydrated.

- If it must be exposed, protect the negative by placing a thin sheet of optically clear acetate between it and the paper. Exposures may need to be increased, depending on the nature of the acetate. The thickness of acetate can have a detrimental effect on sharpness—see Chapter 4, Archival Polyester Film
- Allow the freshly coated paper to dry out. Leave the coated sheet in the dark (covering it with a portfolio box lid will suffice) in room conditions for 10 minutes. Or, if room conditions are above 60% RH, allow the sheet to dry in a controlled, gentle air flow of 40°C / 100°F for 5 minutes (while the exact temperature and time may be adjusted ±10%, consistency is very important). Then, proceed with hydration
- Reduce the amount of hydration. See Chapter 4: *Hydration chambers* and Chapter 8: *Hydration Reminder*
- If glycerin was added, check the strength and proportion of glycerin being used in the sensitizer. See Chapter 6: Humectants and surfactants

Negative Feels like it Stuck to Paper after Exposure

Upon separating the negative from the exposed print, you feel or hear slight sticky-crackliness. Let your heart sink: the negative is irreversibly damaged, with small, dense spots attached to the contact surface. See the above for possible causes and preventions.

Printing Frame Glass Feels Hot

Glass, especially when backed by dark material, will absorb and retain heat. Excessive heat will change the hydration dynamics of sensitized paper during exposure, and can lead to inconsistent results. The following steps may be taken to reduce heating:

- If exposing in sunlight, consider pointing the frame to the North (and the opposite when in the Southern Hemisphere). This will increase exposures but under clear skies, there will be sufficient ultraviolet to allow for full exposures.
- If exposing with a bank of four or more fluorescent lights, increase the distance between frame and tubes by as much as twice the original distance. Note that the commonly understood relationship between a point source of light and its reduction of intensity over distance (the "inverse-square law") does not apply to "light beds," such as these. Intensity, with a diffuse source in this setup (contact printing), may

drop by just a few percent instead. Carry out exposure tests.

- Increase cooling airflow across all electronic components.
- The backing material in a contact frame may be changed from black to white. While this may reduce heat within the frame's environment, there is a caveat: translucent papers display a slight loss of contrast and edge definition when printed against white. This is the result of ultraviolet passing through and bouncing back into the paper during exposure. Exposure times will also change.

Black Deposits Migrate into Highlight Areas

These appear, in the early moments of the first clearing bath, as a smear or bleeding of the photoproduct (platinum/palladium particles) from very dense areas into adjacent highlights, such as the black border around an image bleeding into a sky area.

- Use a fresh, accurately prepared EDTA disodium solution for the first clearing bath.
- Apply immediate, robust and thorough agitation during the first minute of processing in EDTA disodium solution.
- Mask the negative with ultraviolet-blocking material such as Rubylith®.

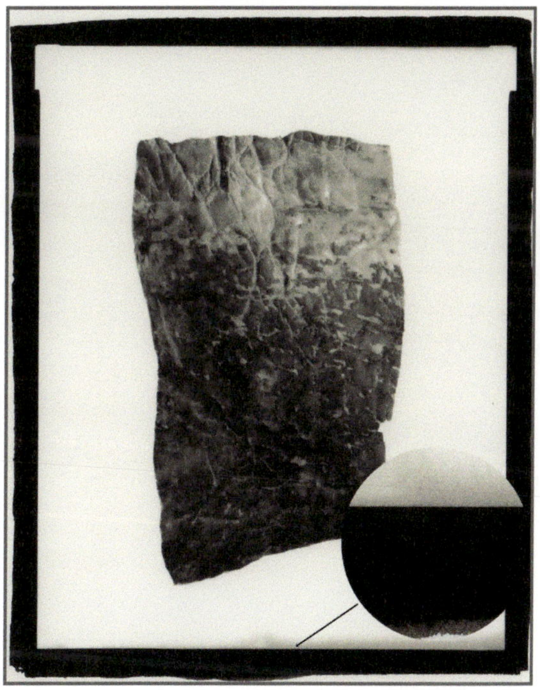

Figure A.4 Black deposits migrate into highlight areas during the early moments of the first clearing bath, as a smear or bleeding of the photoproduct (platinum/palladium particles) from very dense areas into adjacent highlights, such as the black border around an image bleeding into a sky area.

Thin, Curved Line(s) across Image after Processing

These tidemarks appear when a print is not rapidly and completely immersed in the first clearing solution.

- If working with paper heavier than 200 gsm, slide the sheet smoothly but rapidly into the solution, while applying a gentle rocking motion to the tray. Make sure the print is completely immersed with this first pass.
- Papers that are fragile, or lighter than 200 gsm, are best placed in an empty tray, and the clearing solution swiftly but smoothly poured across the entire print surface. Apply immediate agitation.

Image Appearance

Highlights Appear Dull

- Review the methods for making a negative. See Chapter 7.
- Check the work space for possible sources of ultraviolet fogging. See Chapter 3, Lighting Conditions.

Highlights Appear Bleached out

- Ensure that negative densities are not blocked up. Review the methods for making a negative in Chapter 7.
- Check the base exposure times of the ultraviolet source, as described in Chapter 7, Base exposure time.

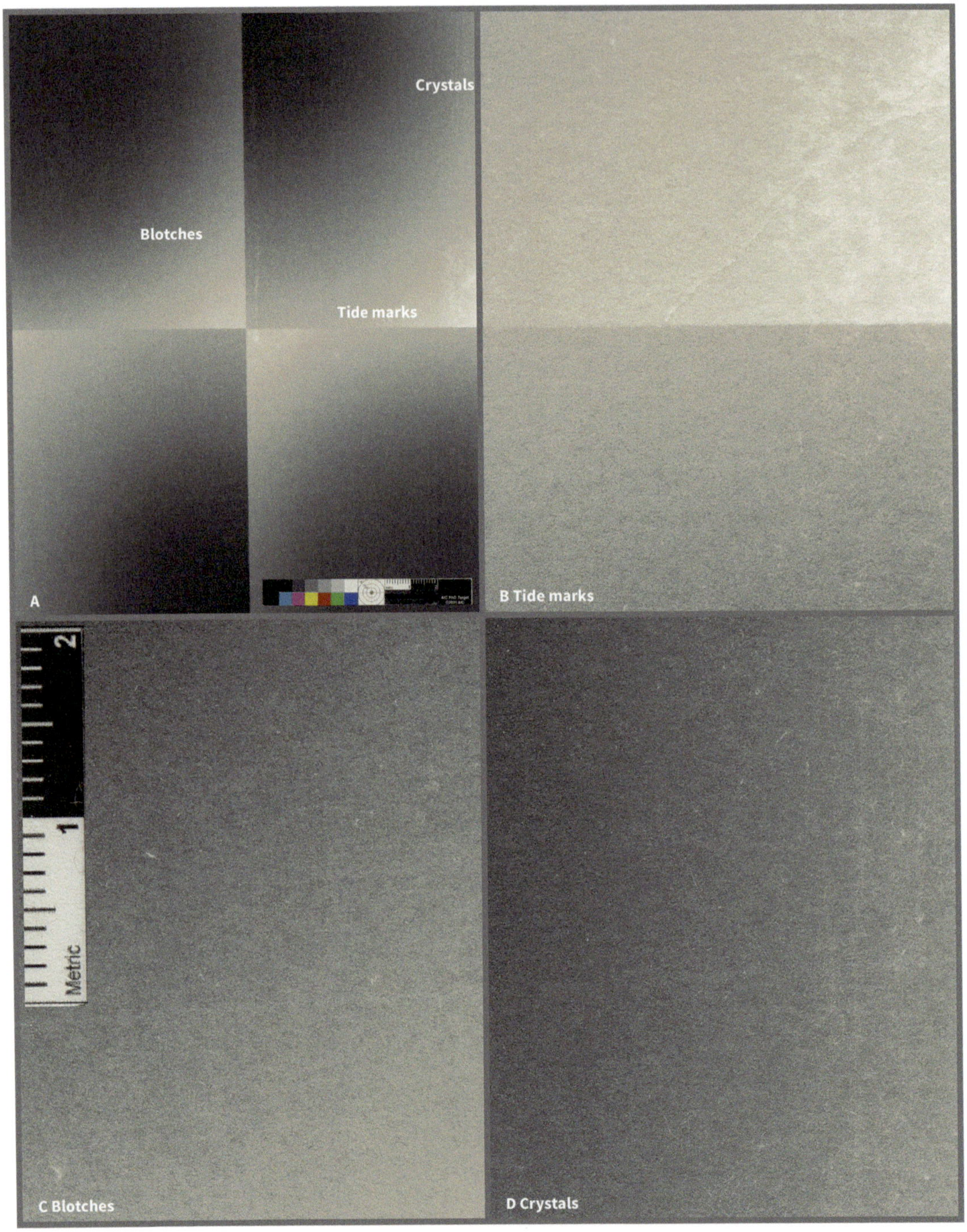

Figure A.5 The most commonly encountered faults. A: A single print displays blotches, crystals and tidemarks. B: Tidemarks appear when a print is not rapidly and completely immersed in the first clearing solution. C: Blotches look like cloudy areas of lighter tone that are not in the photograph, and are generally the result of varying distributions of paper mass. D: Crystals look like reflective speckles prior to exposure, and show as small white spots in the finished print. They can also permanently damage a negative.

- If highlight details are visible after exposure, but seem to bleach out during the first clearing bath, check that you have the correct solution.
- If this is a paper being used for the first time, or a new batch, then see Chapter 6, Testing Papers for Pt/Pd.

Shadows are Reversing Values

- The print may be being overexposed. See Chapter 7, Base exposure time.

Blacks are not Dense

- The print may be being underexposed. See Chapter 7, Base exposure time.
- The coated paper may be receiving a different hydration from previous prints. Sensitivity changes with hydration levels, and there should be a corresponding change in the base exposure time. See Chapter 7, Base Exposure Time and Chapter 8, Table 8.1.
- The sensitizer may be too dilute. See Sensitized Paper Looks Blotchy, above.
- The negative may be fogged. See Chapter 7.
- The sensitized paper may have been fogged. See Chapter 3, Lighting Conditions.

Printed Image is Blotchy

- See Sensitized paper Looks Blotchy, above.

Printed Image is Grainy

- Check to see if this corresponds to the negative. If so, review Chapter 7, and if working from digitally enlarged negatives, reconfigure the scanning methods, or reduce the degree of enlargement, or employ more aggressive noise and grain reduction.
- Grain is sometimes mistaken for what is actually the result of crystal formation during the coating stage. When the image is printed and processed numerous small white, or light, specks can become evident anywhere in the coated area: they are essentially tiny photograms—often angular with sharp edges—of small discrete crystals, formed on the surface of the paper during or soon after its coating. See Sensitizer is Crystallizing on the Paper, above.
- Check the paper storage conditions. It may have been too dry when coated. Ensure that the paper is being stored at 40–60% RH at room temperature. Acclimate each sheet for 10 minutes in a hydration environment of 50% RH (or greater) before coating. We refer to this as a "pre-coat hydration." See Chapter 6: Humidifying.
- The sensitized sheet may have been hydrated at less 30% RH at room temperatures. Increase hydration to at least 40% RH.
- Check the coating technique. Too much pressure, a scratched coating rod, or excessive brushing can cause grain-like noise in the highlight areas.
- Test against another paper (or batch) that is known to work.

Image is Streaky

- Check to see if this corresponds to the negative. If so, review Chapter 7, and if working from digital files or scanned negatives, review the scanning methods and output procedures.
- If the streaking corresponds to the coating direction, either by rod or brush, then see the solutions to crystalizing, puddling, and spreading problems above.

Image Appears to Have Patches of Differing Color

- The sensitized paper is not receiving uniform hydration. Check hydration methods and times. Ensure that the hydration chamber or environment is not being disturbed by strong airflow or localized heating/cooling.
- Ensure that there is a thin sheet of impervious material across the hinge of the contact printing frame.
- Check the temperature. The contact frame or platen glass may be getting very hot during exposure. See Printing frame glass feels hot, above.

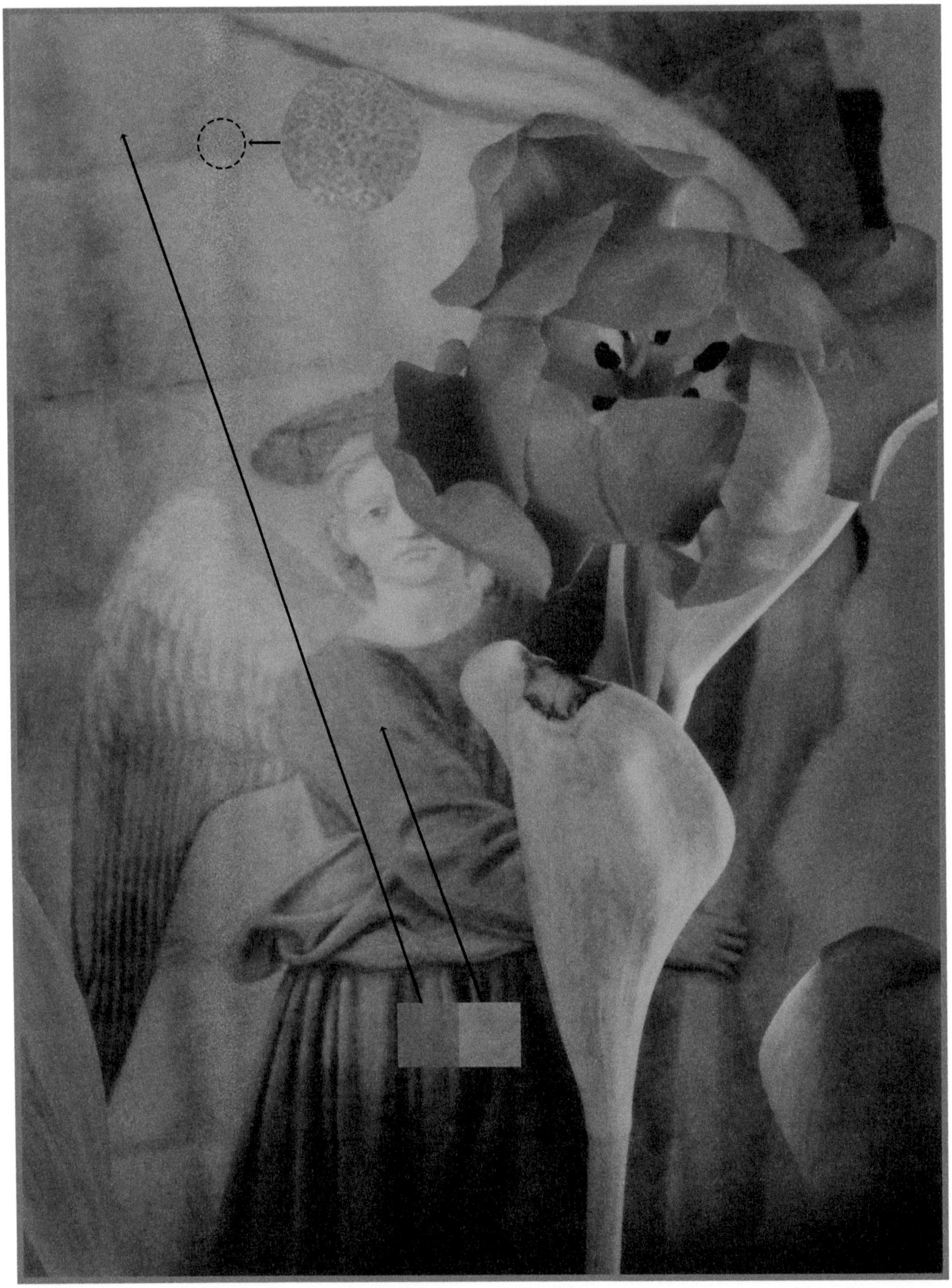

Figure A.6 Print with areas of varying color due to sensitized paper not receiving uniform hydration. The encircled magnification also indicates 'leoparding' from a faulty digital negative.

Image Appears out of Focus in Places

- If working with translucent papers such as Reich CT or Wyndstone Vellum, reduce any edge cockling by trimming off uncoated areas.
- Check that the contact-printing frame is pressing negative to paper. A good indicator of this is when the negative edges look slightly indented into the paper. Use a thin felt pad or soft impermeable material like Neoprene as backing—but see the next point.
- All the siderotype processes evolve carbon dioxide gas during exposure, as a product of their primary photochemical reactions. In the almost-closed environment of a glass-negative-paper-backing sandwich, carbon dioxide may coalesce into bubbles of significant size, with sufficient pressure to separate the negative and paper. This can lead to areas that appear less than sharp. The phenomenon is particularly significant with thinner papers, translucent papers, and large prints. Large areas of Dmax will evolve more carbon dioxide than lighter-tone zones—a good reason to mask off print edges with Rubylith® or similar material. Another solution, and perhaps one to be employed as a standard practice, is to place a sheet of any porous material, such as a sheet of 4-ply museum mat board, or a thin conservation-grade felt blanket (the sort used by papermakers), between the sensitized paper and the frame's back. A third solution, if resources permit, is to use a vacuum frame.

Printed Image has Odd Marks, often in the Same Place, but are not on the Negative

- Thoroughly clean the printing frame glass (wear gloves!) with a residue-free, hand dishwashing soap. Rinse the sheet, wipe dry with paper towel, then carefully wipe both sides with a lens-cleaning solution, or an alcohol-based film cleaner. Check the sheet against glancing light for any smears, residual adhesive or grease.

Print Finishing and Care

Print is Staining after Some Time

- This usually stems from inadequate processing. With care, the print can be run through the clearing bath cycle described in Chapter 8, Printing Workflow, steps 18–25. This may not remove the stains but at the very least will chelate some of the iron complexes out of the paper and increase the archival properties of the paper. More aggressive measures should only be undertaken by a qualified paper conservator.[2]

Print is Cockled or Buckled

- Trim off as much of the cockled edge as possible. Press the sheet between archival museum board in a 40–60°C / 100–140°F dry mounting press. Alternatively, lightly mist front and back of the print with distilled water, then place it between clean sheets of blotting paper, and press under weighted boards for an hour or two. Replace the blotting paper with fresh dry sheets, and repeat until the print is dry.

Notes

1 My thanks to Gilles Lorin for sharing his findings with regard to Arches Platine and blotching. He notes that by adding 12.5 ml of water to 50 ml of sensitizer, and a full 1-stop increase of exposure, he achieves a Dmax of 1.48. (an email exchange with the authors on May 3, 2020.)

2 Matthew L. Clarke and Dana Hemmenway, "Investigating Chelating Agents for the Treatment of Platinum Prints," in *Platinum and Palladium Photographs: Technical History, Connoisseurship, and Preservation*, edited by Constance McCabe, pp. 258–259 and passim. Washington D.C.: Photographic Materials Group of the American Institute for Conservation of Historic and Artistic Works, 2017.

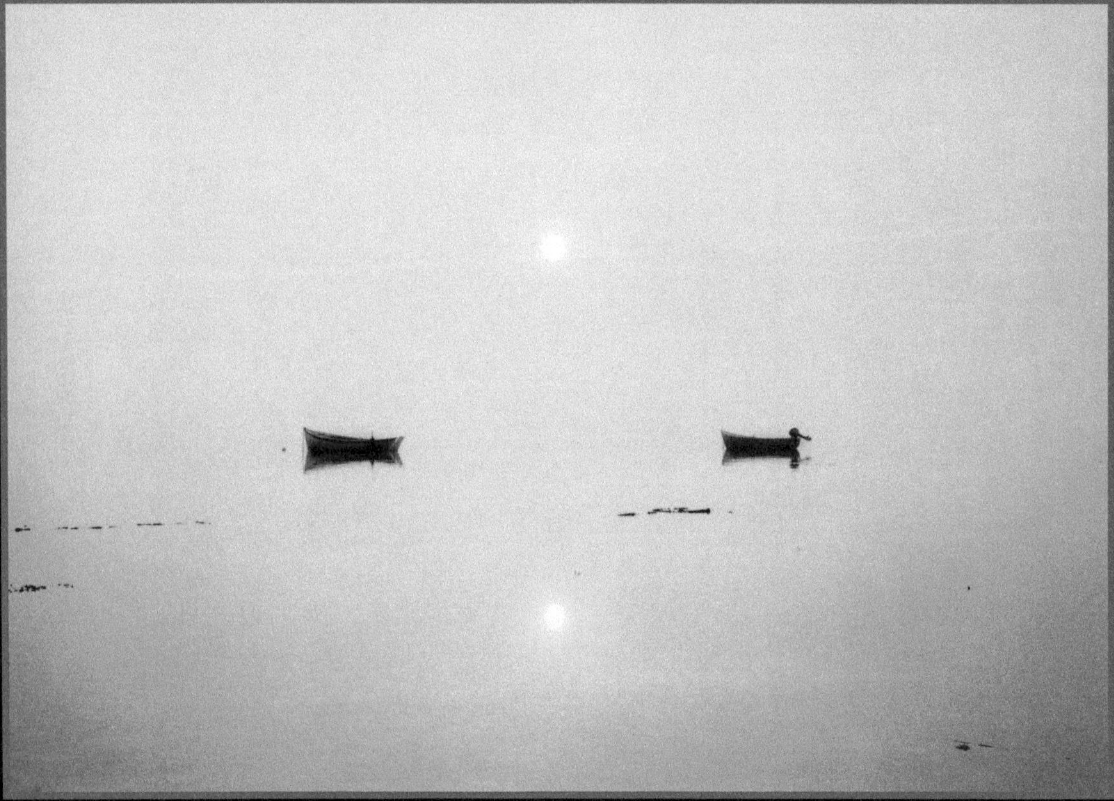

Figure A.7 Mike Ware, *Widewall Bay, Herston, Orkney*, 1984. Platinum-palladium print on Revere Platinum from enlarged digital PiezoDN negative from 6 x 9 cm. Original. 20.2 x 28.8 cm. Fe3:Pt:Pd = 2:1:1, hydration for 90 min at 76% RH / 20°C. Courtesy Mike Ware

Appendix II

Workflow Summaries

Chemicals: Best Practice and Emergencies

Chapters 3, 5

Emergency

Always let someone know when you are working with chemicals, and especially when preparing solutions from hazardous concentrates.

Post a sheet of paper with the following information in a prominent place in your work area:

- Emergency contact name and phone
- Primary care physician
- Nearest medical emergency room
- Telephone of Poison Control Center / Helpline
- Locate a folder of all MSDS near this sheet. Keep the contents alphabetically organized, and up to date with your stock of chemicals

Labels and Storage

Chapter 5

All chemicals should be labeled with:

1. Chemical name
2. Formula if convenient, especially for dry compounds
3. Dilution and preparation date if in solution
4. Supplier and batch number if pre-prepared
5. Hazard Warning if significant

Protection

Chapter 5

- Work in a ventilated space, with air flowing away from your face
- Wear goggles, gloves, apron when preparing solutions
- Wear a surgical face mask if working with crystalline or powdered compounds
- Wear gloves or use print tongs when processing
- Restrain loose clothing and hair
- Protect eyes from exposure to ultraviolet

Solution Preparation

Chapter 5

1. Read the formula.
2. Calculate weights and volumes if scaling up or down.
3. Write out the list of ingredients in a notebook, with exact weights and volumes, in the order of usage.
4. Prepare the work area.
5. Assemble all chemicals and equipment.
6. Mentally run through the process.
7. Prepare the solution.
8. Label all containers.
9. Store safely.
10. Clean up.

Fogging Test

Chapter 3

1. Coat paper in dimmer-than-normal conditions, hydrate as normal.
2. Lightly expose the sheet to UV source.
3. Cut the exposed sheet into at least four pieces.
4. Place one portion in a drawer or light-tight box.
5. Arrange others under normal ambient light. Place an opaque disc in the center of each piece.
6. Wait for 10 minutes, and then process one piece of the test, along with the first piece that

was placed in total darkness as normal. Do the same with succeeding sheets at 10-minute intervals up to at least 30 minutes, but ideally an hour.
7. Compare the tests.

Paper Acidifying

Chapter 6

1. Prepare 10% w/v sulfamic acid solution, place in a tray.
2. Immerse the paper for 20 minutes, with gentle but thorough agitation throughout this time.
3. Wash the paper in five complete changes of water over 20 minutes.
4. Dry.
5. Wait at least 24 hours before using.

Paper Testing

Chapter 6

1. Acquire paper samples from a known stock/supply, get batch or production run number.
2. Acclimate the paper.
3. Identify felt or wire side.
4. Sensitize paper.
5. Check for surface crystals and blotching.
6. Hydrate and expose with a standard step tablet, look for good Dmax.
7. Process and dry, unexposed areas should clear to paper-base white.
8. Inspect print for smooth mid- and highlight tones.
9. Optionally, test reverse side with steps 1–8.
10. Acquire paper if it is suitable, from same batch or production run number.

Printing: Base Exposure Test

Chapter 7

1. Sensitize and hydrate a sheet of paper (see Chapter 8, and Printing: Sequence and Data Template below).
2. Lay a clear piece of your standard film (e.g. UP OHP Pictorico or Ilford HP5) across two-thirds of sensitized area, close the printing frame.
3. Place opaque board, ideally 4-ply black mat board, *across* both portions of the sensitized area, then uncover about 20% of the area.
4. Make a series of two-minute or 100-dose exposures by uncovering in 20% increments.
5. Process and dry the print.
6. Base exposure is indicated by the darkest step after which no further change can be discerned between the areas covered and not covered by film.
7. Repeat test, reducing or extending exposures, to fine tune if necessary.

Printing: Sequence and Data Template

Chapters 8, 9

1. Pre-coat hydration: in at least 50% RH, 5–10 minutes. Choose side, wire or felt, and mark up coating area.
2. Prepare sensitizer: measure volumes of iron and platinum/palladium solutions—these may have already been pre-mixed at least one hour prior to coating; add Tween or glycerin if needed.
3. Coat paper: ca. 1.5 ml per 10″ x 8″ area, four to six passes with a coating rod.
4. Air dry: five minutes, in dark, ambient conditions.
5. Hydration: one to two hours in controlled RH, in hydration chamber or in dark ambient area.
6. Load contact printing frame with negative and paper, mask if necessary.
7. Expose: to ultraviolet source until highlights are just visible, inspecting the print as necessary.
8. Post-exposure hydration: optional, humidify the print a hydration chamber.
9. Develop and clear #1: Disodium EDTA 5% for 10 minutes.
10. Rinse in fresh water 30 seconds.
11. Reduce and clear #2: Sodium sulfite 2.5% for 15 minutes.

12. Rinse in water 30 seconds.
13. Final clear #3: Tetrasodium EDTA 5% for 15 minutes.
14. Wash: at least 30 minutes or 3 static baths.
15. Drain and dry.

Processing Sequence for Analogue Negatives with Pyro PMK developer

Chapter 7

Constant agitation except as noted.

1. Water Presoak: one to five minutes. After use, discard or keep for step 3: Stop Bath.
2. Developer, Pyro PMK: See Table 7.2 for time in minutes. After use, keep for step 5: Stain Bath.
3. Stop Bath: One minute. Water or Presoak bath. After use, discard.
4. Fixer, F-24: Five to six minutes. After use, keep or dispose as per manufacturer's recommendations.
5. Stain Bath: Two minutes. Re-immerse in the developer solution. Agitate gently every 30 seconds. After use, do not reuse the developer—discard.
6. Wash: 20–30 minutes in gently running water,same temperature as process solutions ±2°F/ ±1°C.
7. Final Rinse: One minute water. Gentle agitation (use purified or distilled water if tap water is unfiltered).
8. Wetting Agent: One to two minutes Kodak Photo-Flo™ or similar. Very gentle agitation.
9. Dry, clean up.

Production Sequence for Digital Negatives

Chapter 7

Acquire by:

A. Camera

1. Camera Raw, maximum resolution settings.
2. Import into digital asset management program such as Adobe Lightroom Classic.

Or

B. Scanner

1. Clean surfaces.
2. Position original.
3. Preview scan.
4. Adjust Scan settings to at least 24-bit, 300 dpi, target document size to 20 x 25 cm / 8" x 10".
5. No color correction.
6. Select save location.
7. Image format as TIFF, no compression and embed ICC.
8. Scan.
9. Import into digital asset management program such as Adobe Lightroom Classic.

Then

1. Calibrate the monitor.
2. Control the print viewing area.
3. View image in Photoshop, as a positive.
4. Either apply a custom built "Printing Curve" (see Chapter 7, *Creating an adjustment curve*) or use the Proof Preview mode (Preview > Proof Setup > Custom) with a custom-built ICC such as Cone's PiezoDN_PtPd_Default.icc if using their custom inks.
5. Adjust the image, preferably mostly with Image > Adjustments... / Curves...
6. Send the file to print with the appropriate layout and printer driver, such as Adobe Photoshop's built-in Print Settings tool, or Quad Tone RIP, and a matched profile or "quad."

Naming and Numbering Images, Negatives and Prints

Chapter 9

- Establish a numbering system that can be applied consistently to all *original* image types
- Establish a separate numbering system that can be applied to all *derivations* from the original image, such as prints and digital negatives
- Assign a unique number based on these systems to all digital files, analogue negatives, and prints,

Table A.1 Printing Data Sheet with example entries. Records should be kept for every print.

Description	Example
Print Number	52897
Sensitizer type and batch	Fe 1 #200608: Pt #190606: Pd #190606 = 2:1:1
Negative identifier / Title	3001190 / Flowers with Portrait
Ambient conditions	20 C, 64% RH
Paper name, weight, batch	Revere Platinum, 320 gsm, VOR 160425 – 15014781
Pre-coat hydration method and duration	Ambient, ~ 4 hours
Coating date and time	June 13, 2020, 12:15 PM
Coated area	8″x10″
Coating method (only if using more than one method)	*Glass rod*
Sensitizer volume	1.2 ml
Additives and volume	Tween 15%, 0.05 ml, Glycerin 15%, 0.05 ml
Number of passes (if using coating rod)	*5*
Ambient drying time	5 minutes
Hydration method and duration	Calcium nitrate 56%, 90 minutes
Exposure Source (only if using more than one type)	*Fluorescent bank*
Exposure time or dose	1600 units
Post-exposure hydration method and duration	Ambient, 5 minutes
Processing solutions, sequence and time (if varying from the standard 3-step clearing sequence)	*Standard*
Wash temperature and duration (not necessary if this is constant)	*20C, 30 minutes*
Any other observations and notes	Used a hand-drawn acetate mask, #3001190-1
Current location of print	In print box #4

Note: Italicized rows may not be necessary, as noted in parentheses. It is strongly recommended that records be kept for every print. The data set forms a key troubleshooting tool, but also facilitates re-printing and editioning. This particular example describes the exact printing conditions for the final print in Figure 7.4.D.

Table A.2 Printing Data Sheet: blank template

Print Number	
Sensitizer type and batch	
Negative identifier / Title	
Ambient conditions	
Paper name, weight, batch	
Pre-coat hydration method and duration	
Coating date and time	
Coated area	
Coating method (only if using more than one method)	
Sensitizer volume	
Additives and volume	
Number of passes (if using coating rod)	
Ambient drying time	
Hydration method and duration	
Exposure Source (only if using more than one type)	
Exposure time or dose	
Post-exposure hydration method and duration	
Processing solutions, sequence and time (if varying from the standard 3-step clearing sequence)	
Wash temperature and duration (not necessary if this is constant)	
Any other observations and notes	
Current location of print	

coordinating print numbers to any printing data records such as the one illustrated above

- Assign titles only as necessary
- Collect all digital images, and scans of analogue negatives, in one image asset manager (or digital asset manager, 'DAM') such as Lightroom Classic. Keep these organized by date or number. Use the DAM to create sub-categories (collections) by subject or title for example

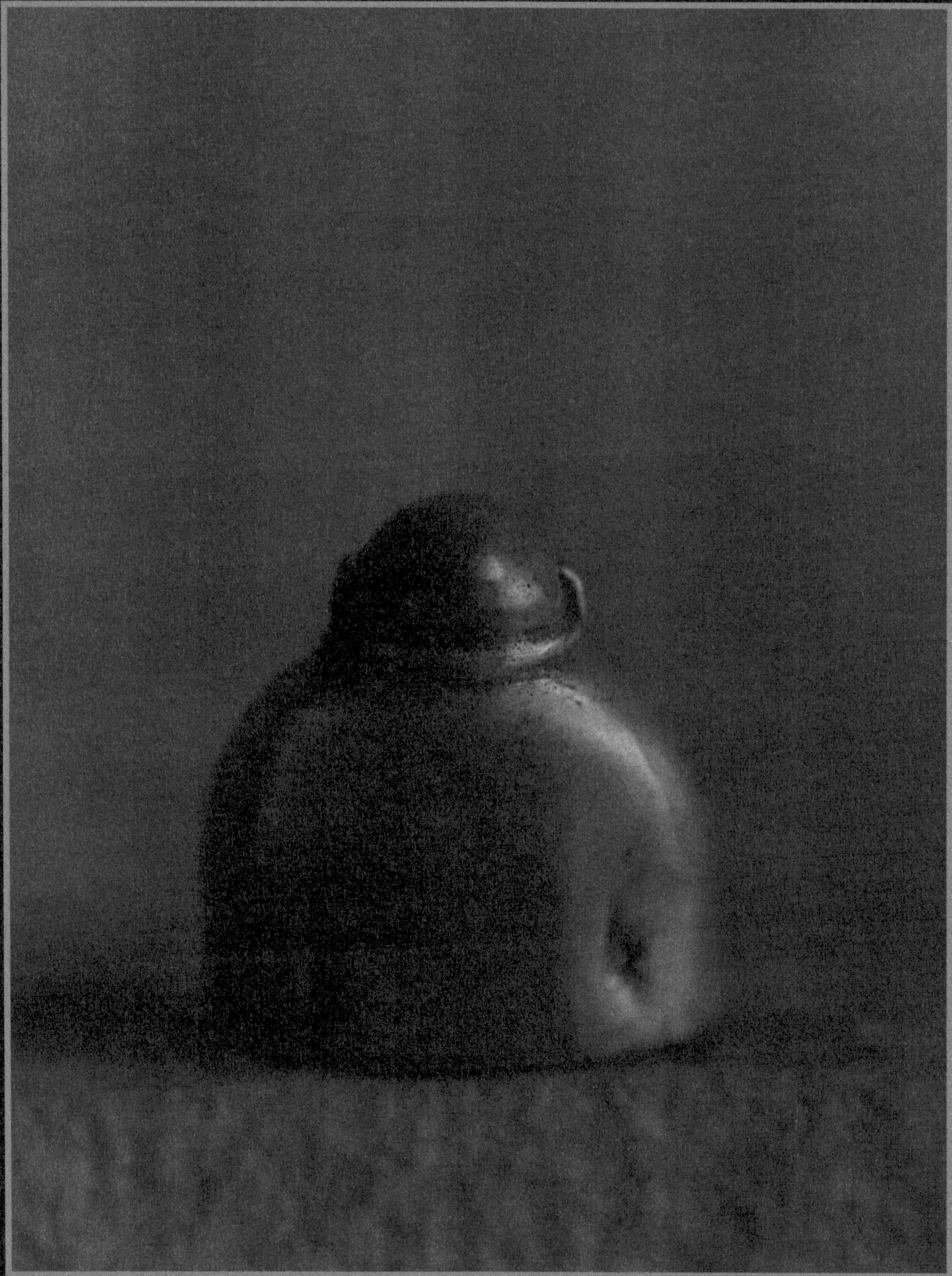

FIGURE A.8 Pradip Malde, *Buddha, Back*. 1999. Platinum-palladium print on Wyndstone vellum from 11 x 14 original negative.

Appendix III

Glossary

Text in **bold face** refers to another Glossary entry

Acid A substance releasing hydrogen **ions** (H^+) in **aqueous** solution. Acidic solutions have a **pH** value less than 7

Actinic Of radiation or light capable of bringing about a chemical reaction; usually confined to the **ultraviolet** and visible light of short wavelengths (blue and green)

Albumen The clear white of hen eggs: a 10% solution of ca. 40 different proteins (macromolecular polypeptides), collectively referred to as albumin, used as a colloidal binder for photographic printing paper of the same name.

Alkali A substance releasing hydroxide **ions** (OH^-) in **aqueous** solution. Some alkalies achieve this by removing hydrogen **ions** from water and binding them, so leaving excess hydroxide **ions**. Alkaline solutions have a **pH** value greater than 7.

Alkyl ketene dimer A synthetic internal sizing agent applied during the production of alkaline or neutral paper. Trade name, Aquapel.

Alum Double salts of aluminium sulfate with alkali metal **cations.** The most common is potassium aluminum sulfate dodecahydrate, $KAl(SO_4)_2{\cdot}12H_2O$. Chiefly used in photography and papermaking for hardening coatings of **gelatin**. This "potash alum," used in fine handmade papers and photographic raw stock, should be distinguished from the lesser quality "papermaker's alum," which is often contaminated with iron.

Aluminosilicate A group of minerals containing aluminum and silicon linked by oxygen, and sometimes metal **cations**; typically, "clays" such as **kaolin**.

Ambient Referring to the properties of the surrounding environment; e.g. its temperature, or **relative humidity**

Amorphous A solid not possessing a regular crystalline structure, e.g. a glassy substance.

Anion A negatively charged molecule or atom, arising through gain of **electron**(s) by the neutral entity. It must co-exist with an equal but oppositely charged atom or molecule; see **cation**. The charge on an anion is always a whole number, written with a superscripted x^{n-}, where x is the element and n- is the number of units of charge of the **electron**.

Aqueous Watery; usually referring to a **solution** of a substance in water.

Argentotype A **siderotype** process invented by Sir John Herschel in 1842 in which the **photoproduct** from exposure of ammonium ferric citrate reduces silver nitrate to **nanoparticle** silver metal, which constitutes the image.

Atom The smallest neutral (uncharged) particle of a chemical element that retains the essential identity of that element, the **atomic number**.

Atomic number A sequential number allocated to each element when they are ordered in the series of ascending **atomic weight** (approximately). More exactly, it is numerically equal to the number of **protons** in the atomic nucleus and, therefore, also equal to the number of orbiting **electrons** in the neutral **atom**.

Atomic weight The averaged mass of the atoms of an element on a scale relative to the most abundant isotope of carbon, having a value of 12.0000. On this scale the oxygen atom has a value of 16.000 and the hydrogen atom 1.008. Now called the **relative atomic mass**.

Baryta Barium sulphate, $BaSO_4$, a dense white pigment used in photographic papers, situated between the emulsion and the paper.

Base A substance that reacts with an **acid** to give a **salt**, plus water. All **alkalies** are bases, but not all bases are significantly alkaline (e.g. some water-insoluble metal oxides).

Buffer A substance or mixture of substances that helps to maintain a fixed, predetermined **pH**. The pH value of its solution is only slightly affected by the addition of moderate amounts of either **acid** or **alkali**.

Buffered paper Certain types of high-quality paper containing a slightly **alkaline** buffer (calcium carbonate = chalk) at ca. pH 9 to neutralize any acid and so enhance their archival qualities.

c Centi-. Prefix meaning one hundredth of.

ca. Abbreviation for **circa** (Latin for "about"). Indicates that a quantity need only be measured to low precision, of around ± 10%.

Calendering The process of creating a smooth and compact surface in a sheet of paper by passing it through a stack of metal rollers, often heated (HP = "hot pressed").

Catalyst A substance that accelerates a chemical reaction but is left unchanged at the conclusion of it. The process is called catalysis.

Cation A positively charged **molecule** or **atom,** arising through removal of **electron**(s) from the neutral entity. It is written with a x^{n+}, where x is the element and $n+$ indicates the number of units of the charge of the **proton** that are present (equal and opposite to the charge of the **electron**). See **anion**.

cc Abbreviation sometimes used for cubic centimeters, more correctly written cm^3. 1 cc = 1 ml.

Cellulose The chief constituent of plant cell walls and therefore of cotton fibers. It can be obtained from materials such as wood pulp, cotton, and flax. Chemically, it is a polysaccharide of formula $(C_6H_{10}O_5)_n$, where n ≈ 15,000.

Celsius The temperature scale also known as Centigrade, which has now replaced the Fahrenheit scale throughout the world, except in the USA. The conversion formulas are:

°C = 5 x (°F – 32)/9
°F = 32 + 9 °C/5

Pure water freezes at 0 °C (32 °F) and boils at 100 °C (212 °F) under normal atmospheric conditions. Normal room temperature is ca. 20 °C (68 °F).

Characteristic curve Also known as the H&D curve (after Hurter and Driffield) or the D/logH curve. The characteristic curve plots the **sensitometric** response of a photographic material, with the horizontal axis as the $\log_{10}$ of the relative **exposure**, H, and the vertical axis as the **optical density,** D.

Chelate A metal compound, or **complex**, in which the metal **cation** is bound by a **ligand** via two or more atoms within the same **molecule**. It derives from the Greek word for "crab's claw." **Oxalate** can act as a bidentate ("two teeth") chelating ligand.

Chine collé A print-making technique in which an image printed on a delicate paper is adhered to a thicker paper, resulting in robustness and finer detail.

Chrysotype A **siderotype** process invented by Sir John Herschel in 1842 in which the **photoproduct** from exposure of ammonium ferric citrate reduces gold chloride to nanoparticle gold metal, which constitutes the image.

Citrate An organic **anion** derived from citric acid, $C_6H_8O_7$, which is usually tribasic, i.e. losing up to three acidic **hydrogen ions (H^+)** in forming **salts**. Exceptionally, it can lose a fourth H^+.

Clearing A term used generally in association with **siderotype** processes to denote the removal from a photographically printed image of all the excess unreacted chemicals, and reaction products, other than the image substance. It is essentially the same as fixing or fixation.

cm^3 Abbreviation for cubic centimeter. An SI unit of volume. Also written as cc. It is now identically the same as a milliliter or ml, since the liter was redefined in 1964 as 1000 cc.

Coating weight A measure of the surface concentration of a substance (e.g. a sensitizer or image pigment) as the mass per unit area of a paper sheet (or other substrate). The standard unit is grams per square meter, g/m^2.

Collodion A solution of cellulose nitrate dissolved in a mixture of diethyl ether and ethanol, and used as a binder layer for photographically sensitized plates and papers.

Colloid A state of solid matter which is aggregated in very small particles, intermediate between molecular dimensions and normal crystals, with a size smaller than the wavelengths of visible light, and now preferably called **nanoparticles.** The term **colloid** tends to be reserved for "gluey" substances, macromolecules such as gum, **gelatin, collodion** or **albumen,** used as binders in photographic materials. A solid or liquid mixture containing particles ranging between 1 and 1000 nanometers in diameter, which remain evenly distributed throughout and do not settle out or precipitate. Common examples are milk, cheese, paint, and whipped cream.

Complex A chemical compound formed from one or more **ligands** binding to a metal **cation.**

Complexing agent Another name for a **ligand.** See **chelate.**

Compound The pure substance resulting from two or more elements entering into chemical combination.

Concentration In general, the amount of one substance dispersed within a perfectly **homogeneous** mixture. Most commonly, it applies to solutions in water or other liquids and can be measured in several different ways: %w/w; %w/v; %v/v; g/l; molarity (mol/l).

Contrast Refers to a photographic image or material developed under specified conditions. It is measured by the slope of the **characteristic curve, D/logH,** or approximately by the **density range** divided by the **exposure scale.**

Corning™ glass Heat-resistant glass, used for cookware and chemical equipment such as flasks and beakers. The equivalent in the UK is **Pyrex™ glass.**

Crystalline A solid that has a regular repeating lattice structure of **atoms, molecules** or **ions.**

Cyanotype A **siderotype** process invented by Sir John Herschel in 1842 in which the **photoproduct** from exposure of ammonium ferric citrate reacts with a ferricyanide to give Prussian blue, ferric ferrocyanide, which constitutes the image.

D/logH curve See **Characteristic curve**

Deckle A removable wooden frame placed on top of a hand papermaking mold, to contain the papermaking pulp within its edges. Pulp trapped beneath the deckle creates a thinned irregular chamfered edge to the sheet, known as a "deckle edge."

Deliquescence Property of a solid having a tendency to absorb water from the atmosphere, and then dissolve in it, to form a solution.

Density (physical property) The property of a substance expressed by its mass per unit volume. The usual units are g/cm^3. See also **optical density.**

Density (optical) Universally given the symbol D, and loosely referred to in photographic contexts simply as density, it is defined as the **logarithm** (to base 10) of the **opacity**:

$$D = \log_{10}(I_o/I_t) \text{ or } 10^D = I_o/I_t$$

where I_o is the intensity of incident light and I_t is the intensity of light transmitted.

Optical densities of superimposed layers are additive, whereas transmittances and reflectances or opacities must be multiplied. Three conditions of measurement must be defined for optical density: the geometry of the light beam may be specular, diffuse, or doubly diffuse; and the wavelength response of the photoreceptor may be spectral, visual (photoptic), printing, or arbitrary.

Density range The difference in **optical density** between the lightest and darkest parts of an image.

Desiccant An inexpensive substance that readily absorbs water from the atmosphere, such as anhydrous calcium chloride or silica gel, used to create a dry chamber e.g. for the storage of sensitized but unexposed papers. Contrast with humectant.

Developer A chemical reagent used to treat an exposed photographic paper in order to bring out the final image in a stable substance. For the development platinum-palladium process it is usually a **chelating agent**, such as oxalate or citrate ions, which solubilizes the iron(II) photoproduct.

Developers for silver images are usually mild reducing agents, such as organic amines, capable of reducing the exposed silver halide grains constituting a latent image, entirely to silver metal—the process of development.

Dmax See **maximum density**

Effervescence The dignified word for "fizzing"—to describe the evolution of a gas, as bubbles, from a liquid.

Electron The fundamental particle of negative charge.

Element A fundamental constituent of matter that cannot be split into simpler constituents by chemical means (i.e. at low energies) consisting of one type of **atom** only, with a characteristic **atomic number**.

Emulsion Small droplets of one liquid suspended and distributed throughout another, immiscible liquid. Compare **suspension.** The word is also used widely (but incorrectly) to describe the suspension of silver halides in gelatin used to make photographic films and paper.

Evolve The process of giving off gas during a reaction

Equimolar Having equal amounts or concentrations on the molar scale of two chemical components, i.e. equal numbers of molecules.

Excess A quantity exceeding that needed to complete a balanced chemical reaction.

Exposure (photographic) Intensity of light multiplied by the duration of the exposure.

Exposure = Illumination x time.

Usually given the symbol "H," the basic units of exposure in photography are lux seconds.

Exposure scale The number of **stops** of exposure needed to transform a photographic material from the minimum **optical density** (usually white) of the paper base to the maximum density (black or other color). It may also be expressed as a $\log_{10}$(Exposure) value, abbreviated to logH, where each numerical step of 0.3 = 1 stop, and the scale is additive.

Ferric/ferrous The names given to the element iron when chemically combined in its two most common **oxidation states** of +3 and +2, respectively: the **cations** Fe^{3+} and Fe^{2+}, now designated by iron(III) and iron(II).

Fluid ounce An ambiguous obsolete unit of liquid volume:

in the UK the British Fluid ounce = 28.412 cm^3
in the USA = 29.573 cm^3.

Formula (chemical) The composition of a pure substance written in chemists' shorthand form, using the symbols for the chemical **elements**, with their relative numbers of **atoms** in the **molecule** attached to each as subscripts.

Formula weight The sum of the **atomic weights** of all the **atoms** making up the **formula** of the substance; also known as **relative molecular mass**. In earlier texts generally referred to as the Molecular Weight.

g Abbreviation for gram(s) the **SI** unit of mass (or, loosely, weight).

gr Abbreviation for **grain**(s).

G Abbreviation for Giga-, one billion times (1,000,000,000 x).

gsm Abbreviation sometimes used for grams per square meter, g/m2. See **weight of paper**.

Gelatin A very complex organic **macromolecular** material (a protein, or polypeptide, consisting of long chains of linked aminoacids) obtained from animal skins and bones (called ossein). Used as a **colloidal** binder in the preparation of photographic emulsions, and as a surface **sizing agent** for some papers. In the refining process, gelatines can be acidic or alkaline and oxidized or not.

Gradation Usually refers to tonal gradation: the visible separation of areas of perceptibly different optical **density** in an image.

Graduate In the USA, a measuring cylinder, usually of glass, calibrated by volume in ml or cc.

In the UK, one who has been awarded a university degree.

Graduations On chemical glassware: the marks to indicate the volumetric calibration.

Grain An obsolete unit of weight seen only in old recipes. Abbreviation: **gr**. Avoid confusing it with **grams**, abbreviation **g.**

1 grain = 0.0648 grams.

Grey Card The standard neutral mid-grey optical **reflectance** of 18% = **density** of 0.72, usually on a card.

Halides The singly charged **anions** —fluoride, chloride, bromide, or iodide—of the group of elements in the periodic table known as the halogens: fluorine, chlorine, bromine, or iodine, in combination with other elements. All metals can form halides.

Hardening The process of rendering **colloids** such as gelatin or gum insoluble in water. Also resembles tanning.

Homogeneous Uniform throughout, with no internal boundaries or surfaces—consisting of a single phase, either solid, liquid, or gas.

HP Abbreviation for Hot Pressed—see **calendering**.

Humectant A substance that tends to attract moisture. See **hygroscopic**.

Humidify To take up water vapor from the atmosphere; to increase the level of moisture in the air, usually in a closed container or space.

Hydrate (noun) A solid containing water of crystallization; a hydrated salt.

(verb) To allow to take up water, often from the vapour. To **humidify**.

Hydration The process of taking up or absorbing water.

Hydrogen ion The ionised hydrogen **atom,** H^+. Since this is a bare **proton**, it will attach itself to the most available **molecule**, e.g. with water forming the hydronium ion H_3O^+, or with ammonia forming the ammonium ion H_4N^+.

Hydrolysis Decomposition of a substance by water or **alkali** (OH^-). In the case of a metal **salt**, to give the metal hydroxide or (sometimes) oxide.

Hydrometer A device for measuring the relative **density** (specific gravity) of liquids by means of the partial immersion of a calibrated float.

Hydrophilic Having an affinity for water.

Hydrophobic Having the property of repelling water.

Hygrometer An instrument for measuring **relative humidity** (RH) in the air.

Hygroscopic Having a tendency to absorb water from the atmosphere. Compare with **deliquescent**.

Hypo An obsolete nineteenth-century name, still used by some photographers, which is short for "hyposulphite of soda"—the substance now called "sodium thiosulphate." It is universally employed as a fixing agent in silver photography because of its ability to dissolve silver halides, a property discovered in 1819 by Sir John Herschel.

Hypo Clearing Agent A proprietary mixture to increase the speed and efficiency of the washing of fixed silver-gelatin photographic materials. With platinum-palladium and gold printing, the agent aids the clearing of residual iron compounds. (Proprietary names include Ilford Washaid™, Permawash™, and Speedwash™.) It consists mostly of sodium sulfite and metabisulfite.

Ion An **atom** or **molecule** carrying an electric charge, due to gain or loss of **electrons**. The charge is written superscripted following the chemical symbol, e.g. Fe^{3+}. See **anion** and **cation**.

IR Infrared radiation; this is electromagnetic radiation having wavelengths ranging from 700 nm to about 1 mm, thus of longer wavelengths and lower energy than visible light. Absorption of infrared radiation is responsible for generating heat.

Joule The **SI** unit of energy, or work, abbreviated to J. 1 calorie = 4.184 Joules

k Abbreviation for kilo-, one thousand times (1,000 x).

Kallitype A **siderotype** process invented by W.W.J. Nicol in 1889 in which the **photoproduct** from exposure of ferric oxalate reduces silver nitrate to give a silver image.

Kaolin The rock composed chiefly of the soft, white mineral kaolinite, known as china clay, which can be fired to make porcelain. Kaolinite is also employed as a filler in papermaking to provide a glossy surface. Chemically it is an **aluminosilicate**, $Al_2Si_2O_5(OH)_4$.

Kelainotype A **siderotype** process invented by Sir John Herschel in 1842 in which the **photoproduct** from exposure of ammonium ferric tartrate reduces mercuric chloride or nitrate to mercury metal, which constitutes the image, but which proves highly impermanent due to its volatility.

Laminar structure Having a layered structure—particularly in the context of photographic prints on paper, which are characterized as having one, two or three layers.

Latent image The initial effect of light on a crystal of silver halide to produce a sensitivity speck consisting of a few atoms of silver, which renders the whole crystal reducible into a grain of silver metal by a suitable **developing agent**. The latent image is invisible to the naked eye.

lbs Abbreviation for pounds weight.

Ligand A **complexing agent**, i.e. any **molecule** or **ion** capable of binding chemically to a metal **cation** to give a molecular compound called a metal complex. E.g. chloride is the ligand in $PtCl_4^{2-}$.

Lignin A component of the cellular structure of wood. It is a complex, polyphenolic macromolecule, susceptible to oxidation and discoloration.

Logarithm The logarithm of a number, X, (to a specified base: 10 is most commonly used) is written as $\log_{10}X$. It is equal to the power to which the base must be raised to give that number:

$10^{(\log_{10}X)} = X$. When numbers are multiplied, their logarithms must be added.

Luer taper connections A standardized system for connecting small-scale medical and laboratory equipment, such as a needle to a syringe, using interlocking male and female parts. There are two sub-systems: the lock style with threaded screw collars, and the slip style with smooth interconnecting collars relying on friction.

Macromolecule A large molecule (of high molecular weight) consisting of many smaller molecular units, which may differ, chemically bonded together. The repetition of units may be irregular. Compare with polymer.

Material Safety Data Sheet See **MSDS**

Maximum density Often called **Dmax**: the highest value of optical density that is present in an image, or that a given photographic material can yield.

m Abbreviation for milli-, one thousandth of.

M Abbreviation for mega-, One million times (1,000,000).

μ Abbreviation for micro- (Greek letter mu), one millionth of.

Molarity The **concentration** of a solution expressed as the number of **moles** of solute dissolved in one liter of the solution.

Molar ratio The relative proportions of two substances expressed in terms of their respective numbers of **moles**, i.e. their relative numbers of **molecules**. It is then obvious why chemicals generally react or combine in simple proportions when expressed as molar ratios.

Mole The chemist's unit to measure the amount of a pure substance. A mole of any substance is equal to the **formula weight** or **relative molecular mass** measured in grams. It is more fundamental than just weight or mass, because one mole of any substance always contains the same number of **molecules** (called Avogadro's number = 6.022×10^{23}).

Molecular formula The chemist's shorthand designation to identify a pure substance, which is done by combining the numbers of **atoms** of the different chemical **elements** within its **molecule**, and sometimes an indication of the way they are bonded together.

Molecular weight Another name for **formula weight**. Now more correctly referred to as **relative molecular mass**.

Molecule The smallest and simplest particle of a chemical substance or compound that can exist and still retain the identity of the whole. Molecules are composed of **atoms** chemically bonded in fixed proportions, and have definite three-dimensional shapes.

MSDS An acronym for **Material Safety Data Sheet,** providing information for the safe handling of chemicals, and their associated hazards. MSDS are now called SDS (Safety Data Sheets) under the Globally Harmonized System.

Nanometer (nm) An **SI** unit of length: one billionth (10^{-9}) of a meter, convenient for measuring the wavelength of visible light, which approximately spans the region of the spectrum between 400 nm (violet) and 780 nm (deep red).

Nanoparticle A particle of a substance having a size intermediate between that of atoms and the wavelength of visible light, i.e. in the region of 1 to 200 nm. See **colloid**.

Neutral Neither **acid** nor **alkaline,** having no electrical charge. Neutrality prevails when the **hydrogen ions** and **hydroxide ions** in a water solution are equal in number, i.e. pure water is neutral, having **pH** 7.

Noble metal Alchemical name for metals which are substantially unchanged by fire; i.e. metallic elements that resist oxidation and corrosion, such as platinum, palladium, gold, and silver.

Opacity The ratio of the intensity of the light falling on an object to that passing through it:

Opacity = Intensity of Incident light /Transmitted light = I_o/I_t

It is the reciprocal of the **Transmittance,** I_t/I_o.

Optical density Universally given the symbol D, and loosely referred to in photographic contexts simply as density, it is defined as the logarithm (to base 10) of the **opacity**.

Oxalate A **salt** containing the oxalate **anion,** $C_2O_4^{2-}$, derived from oxalic acid, $H_2C_2O_4$, which occurs naturally in rhubarb, sorrel, and spinach.

Oxidation The removal of **electrons** from an **atom** or **molecule,** in consequence of which it becomes more positively charged (or less negatively charged); an increase in the **oxidation state** of an atom or **ion**. Contrast with **reduction**.

Oxidation state A formal electric charge that may be assigned to an **atom,** either free or in a **molecule,** which is calculated by the number of **electrons** it has effectively lost or gained, compared with the neutral atom. Shared electrons are always formally allocated to the more electronegative of two bonded atoms. Conventionally, the oxidation state is written within parentheses, in Roman numerals, immediately following the name or symbol of the element, e.g. iron(III), or Fe(III), and pronounced "iron-three."

Palladiotype A **siderotype** process invented by William Willis in 1917 in which the **photoproduct** from exposure of ferric oxalate is made to reduce sodium tetrachloropalladate(II) to give a nanoparticle palladium image.

Parchmentizing A treatment of **cellulose** paper with strong sulfuric acid which renders it more translucent and rather brittle, improving its resistance to water, heat, oil, and abrasion, with a hard surface resembling parchment. It destroys the regular crystalline structure of the fibers and renders the cellulose **amorphous**.

Periodic Table A sequential arrangement of all the 118 known **elements,** in order of increasing **atomic number,** in which elements having similar outer **electron** configurations, and therefore similar chemistry, are tabulated in the same vertical column called a "Group." The number of elements in each row of the Periodic Table is 2, 8, 8, 18, 18, 32, 32; each commencing with an alkali metal and terminating with an inert gas.

pH A measure of the **acidity** or **alkalinity** of a solution. It is equal to the **logarithm** (to base 10) of the reciprocal of the **hydrogen ion molarity**:

$pH = \log_{10} 1/[H^+]$ or $pH = -\log_{10}[H^+]$.

Neutrality in water corresponds to pH 7; acidic solutions have smaller, and alkaline solutions greater pH than 7, with a range usually lying between 0 and 14. Every unit decrease in the pH represents a ten-fold increase in the acidity.

Photochemistry The study of chemical changes brought about by the absorption of light. See **actinic**.

Photogram A photographic image formed on a light-sensitive surface by the shadow cast by an object in direct or close contact with it, i.e. there is no lens to form the image.

Photon The fundamental particle of light and of electromagnetic radiation in general.

Photoproduct A substance produced by a **photochemical** reaction.

Photosensitizer A substance capable of responding to light by undergoing some physical or chemical change.

Pipette A calibrated glass tube used for the controlled delivery of small accurately measured volumes of liquids.

Platinic/platinous The names given to the element platinum when chemically combined in its two most common **oxidation states** of +4 and +2, respectively. Modern terminology uses "platinum(IV)" and "platinum(II)."

Platinotype A **siderotype** process invented by William Willis in 1873 in which the **photoproduct** from exposure of ferric oxalate is made by an oxalate developer to reduce potassium tetrachloroplatinate(II) to give a platinum image.

Polyester A form of inert plastic **polymer** used in transparent sheets for archival sleeves for negatives and prints.

Polymer A **macromolecule** formed by the regular repetition of one (or, in the case of a copolymer, two) molecular unit(s), bonded chemically together many times over.

Precipitate An insoluble solid thrown down in a solution due to a chemical reaction between the soluble components.

Printing-out process A phenomenon in which the photographic image appears during the **exposure** to light, without the use of developing chemicals. Printing-out materials usually require a very heavy exposure, compared with those like silver halides, forming a **latent image** which is developed by chemical reaction after the exposure.

Proton A fundamental positively charged particle, the nucleus of the hydrogen **atom**, i.e. the **hydrogen ion**, H^+. Its charge is equal, but opposite in sign to that of the **electron**.

Pyrex™ glass Heat-resistant glass, used for cookware and chemical equipment such as flasks and beakers. An equivalent in the USA is **Corning™ glass**.

Reagent A reactive chemical, usually dissolved in a **solvent** (solution) at a **concentration** suitable to perform a test or carry out a reaction.

Redox reaction A chemical reaction involving the transfer of electrons from one atom, molecule, or ion, to another. Shortened from "reduction-oxidation."

Reduction (chemistry) The addition of **electrons** to an **atom** or **molecule**, in consequence of which it becomes more negatively charged (or less positively charged). A decrease in the **oxidation state** of an atom or ion. Contrast with **oxidation**.

Reduction (photography) A process of diminishing the **optical density** of an image by dissolving away some of its substance. It is an unfortunate source of confusion that such a process is usually, chemically speaking, an **oxidation** (e.g. of silver metal).

Reflectance The ratio of the intensity of the light reflected diffusely from an object, to that falling on it.

Reflectance = Intensity of diffuse reflected light / Incident light = I_r/I_o

Reflection density The definition is analogous to that of **optical density** in transmittance:

reflection density is the **logarithm** (to base 10) of the reciprocal of the **Reflectance:**

$D = \log_{10}(I_o/I_r)$ or $10^D = I_o/I_r$

where I_r is the intensity of light reflected diffusely (scattered) by the surface.

The term is used to describe tonal values in a print.

Relative Atomic Mass Abbreviated to **RAM**. The mass (loosely, weight) of an atom of an element on a scale relative to the carbon atom as 12.0000. Also called **atomic weight**.

Relative Humidity Abbreviated to **RH**. The ratio, defined as a percentage, of the amount of water vapor present in the atmosphere to the amount that would have been present if the same atmosphere were saturated (i.e. contained all the water-vapor it could hold) at a specific temperature.

Relative Molecular Mass Abbreviated to **RMM**. The sum of the **relative atomic masses** of all the **atoms** making up the **molecular formula** of the substance. Also called "**molecular weight**" or "**formula weight**."

Reversal A **sensitometric** response in which the **optical density** diminishes with increasing exposure. See **solarization**. Also, a method of processing a photographic material in a positive-working sense.

Salt Any substance composed of **cations** and **anions**; an **ionic** compound. "Common salt" is sodium chloride: $Na^{+}Cl^{-}$. Salts result from the reactions of **acids** with **bases**.

Salted paper print A print formed on paper sensitized with silver chloride, formed by first soaking in common salt solution (sodium chloride) ca. 2%, drying, then brushing over with excess silver nitrate solution ca. 12%.

Saturated solution A solution that cannot contain any more **solute** at a specified temperature. Marked by the presence of excess solid in contact with it.

Sensitizer The solution (usually **aqueous**) of light-sensitive chemicals that is coated onto a paper (or other surface) to make a photographic printing material.

Sensitometry The assessment of **photochemical** response in a photographic material by **exposure** to a measured illumination for a given time, which is then compared with the measurement of the **optical density** of the image formed, according to specific exposure and processing conditions.

Siderotype A collective name, originally due to Sir John Herschel, for all the **photochemical** printing methods based on a **photosensitive** iron **salt**.

Sizing agent A glutinous preparation (usually of a **colloid** such as **gelatin** or starch) used to treat paper or fabric to improve water resistance, increase strength and dimensional stability, and influence print quality. Some sizing agents, such as resins and Aquapel™ are **hydrophobic** (water-repellent).

Solar enlarger An optical device using a large condensing lens to gather the sun's light to project the image of a negative via a second lens onto photographic papers of low sensitivity, such as **Platinotype**, for printmaking purposes. A tracking device is needed to follow the sun's apparent motion over the lengthy exposures.

Solarization A photographic phenomenon in which increasing **exposure** diminishes the **optical density** of the image—also called **reversal**.

Solubility The maximum weight of the substance that will dissolve completely in a given volume of **solvent** at a specific temperature. Usually given in units of g/100 cm^3 of solution, or g/100 cm^3 of solvent. N.B. these differ.

The weight of substance needed to make 100 cm^3 of **saturated solution**.

Solute The substance which is dissolved in a **solvent** to make a **solution**.

Solution A homogeneous liquid composed of a solute dissolved in a solvent. True solutions are always clear; i.e. they do not perceptibly scatter light (though they may absorb it, i.e. be highly colored). Contrast with a **suspension**.

Solvent A liquid capable of dissolving other substances (solid, liquid or gaseous) to form **solutions**.

Spatula A chemists' tool for dispensing small amounts of solids; usually a small, shaped strip of inert plastic or nickel-plated metal or stainless steel. Spatulas come in many shapes and sizes.

Stalagmometry The study of surface tension by measuring the weight of drops of liquid.

Stop In photography, one stop more is a doubling, and one stop less a halving of the **exposure**. The term is also used to describe differences in **optical density**: one stop = difference in optical density of 0.3. This corresponds to a doubling or halving of the **transmittance** or **reflectance** because $\log_{10}2 = 0.301$.

Surfactant An agent that greatly reduces the surface tension of a liquid (usually water), enhancing its ability to wet objects or penetrate the **cellulose** fibres of paper. E.g. Tween 20™.

Suspension Small particles of an insoluble solid dispersed in a liquid or a gel.

Supersaturated A solution, at a specified temperature, that contains more **solute** than is permitted by the equilibrium **solubility**. If scratched, jarred, or "seeded," it may rapidly crystallise or **precipitate**.

Système Internationale Abbreviated to **SI**. A rationalized system of units of physical measurement now agreed worldwide and used by scientists to the exclusion of all previous systems. The seven basic physical quantities (from which all others can be derived), their units, and their approved abbreviations are:

Quantity	Unit	Abbreviation
length	meter	m
mass	kilogram	kg
time	second	s
electric current	ampère	A
temperature	kelvin	K
luminous intensity	candela	cd
amount of substance	mole	mol

Tare In weighing out a substance: the weight of the container or receptacle which must be deducted from the total.

Teratogen A substance known to cause congenital (birth) defects, i.e. in the development of the embryo of any animal.

Transmittance The fraction of the intensity of the incident light passing through an object.

Transmittance = Intensity of transmitted light / Intensity of incident light = I_t/I_o.

It is the reciprocal of the **opacity**, I_o/I_t.

Transmittance is sometimes expressed as a percentage: 100 I_t/I_o %

Ultraviolet radiation Abbreviated to **UV.** Electromagnetic radiation with wavelengths from 400 nm to about 50 nm, and thus of shorter wavelength and higher energy than visible light.

Conventional divisions of the ultraviolet spectrum are:

UVA 400-315 nm; **UVB** 315-280 nm; **UVC** 280-200 nm.

The so-called "near ultraviolet" UVA is the most suitable **actinic** radiation for **siderotype** processes.

Unit prefixes The following **SI** abbreviations, used as prefixes, indicate the fractions or multiples of the unit quantity they attach to:

p	pico	one thousand billionth 0.000000000001	10^{-12}
n	nano	one billionth 0.000000001	10^{-9}
µ	micro	one millionth 0.000001	10^{-6}
m	milli	one thousandth 0.001	10^{-3}
c	centi	one hundredth 0.01	10^{-2}
d	deci	one tenth 0.1	10^{-1}
D	deca	ten 10	10^{1}
K	kilo	one thousand 1,000	10^{3}
M	mega	one million 1,000,000	10^{6}
G	giga	one billion 1,000,000,000	10^{9}
T	tera	one trillion 1,000,000,000,000	10^{12}

Watt The **SI** unit of power, abbreviated to W. Power is the rate of doing work, or the rate of transfer of energy. One Watt is equal to one **Joule** per second. 1 W = 1 J/s.

Water of Crystallization Water **molecules** that are incorporated in the crystalline lattice of a solid substance

Wavelength The property of light that determines its hue: the distance between successive crests (maxima) of the waves of electromagnetic radiation. Conveniently expressed in **nanometer** units of length, the following typical wavelengths represent the colors of the spectrum:

red 700, orange 620, yellow 580, green 530, blue 470, violet 420 nm.

The concept extends to all electromagnetic radiation, beyond the limits of visibility (400–780 nm). The shorter the wavelength, the higher the intrinsic energy of the radiation. See **ultraviolet** and **infrared**.

Weight of paper A measure (thickness x density) of a paper sheet. Its units are weight per unit area: grams per square meter, g/m^2 (sometimes written as **gsm**). In the United States it is given as the weight in pounds of 500 sheets, which is an unspecified measure as it depends on size. For imperial-sized sheets (30 x 22 inches) the relationship is:

Weight in gsm = 2.13 x weight in lbs.

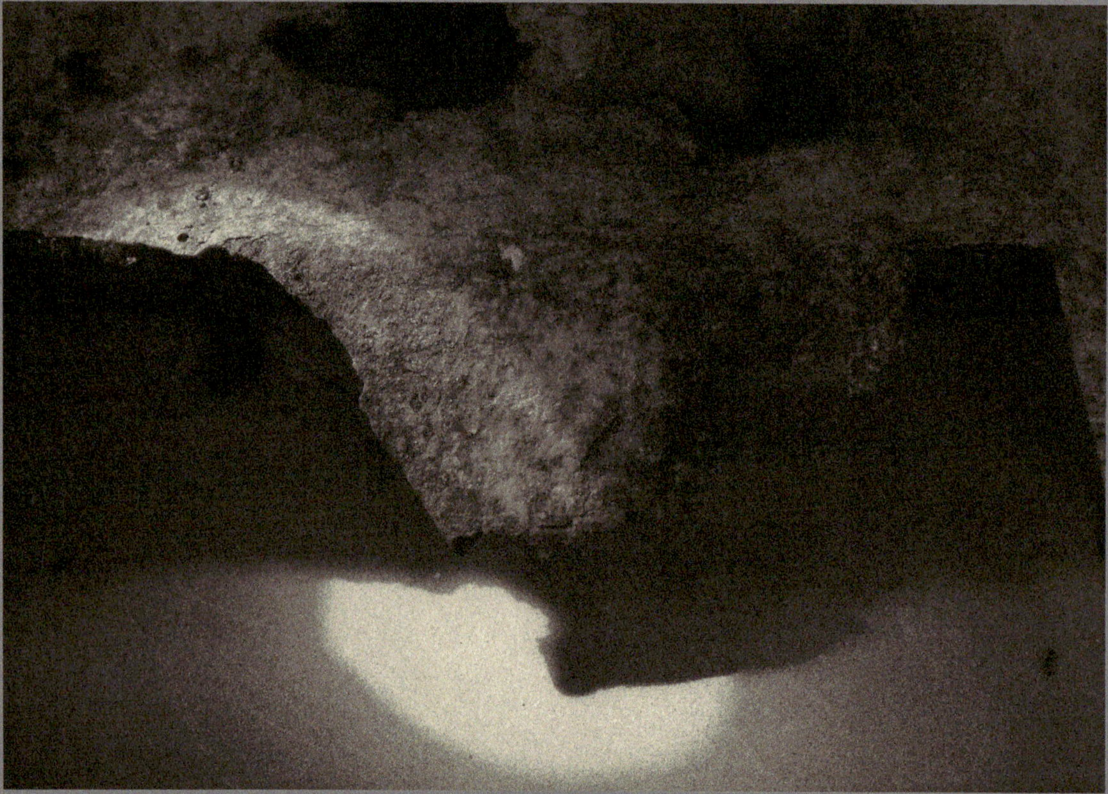

Figure A.9 Mike Ware, *Dies Irae, Orkney*, 1983. Platinum-palladium print on Buxton from enlarged digital negative of 6 x 9 cm original, ca. 30 x 40 cm. Courtesy Pradip Malde. Fe:Pt:Pd = 2:1:1, hydration in ca. 60% RH / 18°C for 140 min. then 80% RH / 18°C for 33 min.

Appendix IV

Equipment, Materials, and Resources

Some chemical suppliers only fulfill institutional orders, but many will let private customers set up an account as it helps establish a track record and facilitates ordering. While almost all suppliers are listed with their primary geographic zone of operation, most will fulfill international orders, and details are provided on their web sites.

See pradipmalde.com/technical-notes for an up-to-date list, including a list of suppliers by specific items.

Name	Region	Web site	Provisions
Adobe	Global	adobe.com	Lightroom and Photoshop digital image management and processing software
Alfa Aesar	Global	alfa.com	Laboratory-grade chemicals, Tween and labware
Alternative Photography Directory	Global	alternativephotography.com	Extensive, loosely curated directory of practitioners, services and suppliers
Amazon	Global	amazon.com	Adjustable micropipettes, hydration chamber components, labware, syringes, and chemicals including EDTA, sodium sulfite and some other chemicals, and labware
Archival Methods	USA	archivalmethods.com	Archival boards, enclosures, frames, sleeves
ArtCraft Chemicals	USA	artcraftchemicals.com	Chemicals
Awagami Factory	Japan	awagami.com	Japanese papers
Aztek	USA	aztek.net	Optically clear mylar film to interleave between negatives and sensitized paper, film scanning supplies
B & H Photo	USA	bhphotovideo.com	Photo and processing equipment and supplies
Bergger	France	Bergger.com	Films, paper
Blanco Negro	Australia	blanconegro.com.au	Sheet film
Bookshop		bookshop.org	Online access to a global collective of local bookstores, new and used
Bostick and Sullivan	USA	bostick-sullivan.com	Chemicals and paper
Capture One	Denmark	captureone.com	Digital image processing software
Cone Editions Press	USA	cone-editions.com	Atelier-quality printing services for cyanotype, inkjet, photogravure, and platinum/palladium; residencies and workshops in all of these processes. See also Inkjet Mall

Name	Region	Web site	Provisions
Disactis	France	disactis.com/store/fr	Chemicals—almost everything except disodium EDTA, paper and supplies
Edwards Engineered Products	USA	eepjon.com	Coating rods, printing frames, and ultraviolet exposure units
Freestyle	USA	freestylephoto.biz	Glass coating rods, printing frames, ultraviolet exposure units, chemicals, and paper
Gaylord Archival	USA	gaylord.com	Archival boards, boxes, sleeves
George Eastman House	USA	eastman.org	Workshops and historic references
Gerstaecker	Gemany	gerstaecker.de	Paper—Arches Platine
Gold Street Studios	Australia	goldstreetstudios.com.au	Alternative process kits, chemicals, paper, photo film, printing frames, printing services and materials, workshops
古典摄影和黑白反转[1]	China	shop34134667.taobao.com /?spm=a230r.7195193.1997079397. 2.ffe6114c6ZUzd9 Phone: +86-186-3385-9860 (Mr. Zhang Yajun)	Chemicals, equipment, paper
富阳新晟机械[2]	China	shop66294779.taobao.com/?spm=a 230r.7195193.1997079397.2.67e250 6dldawcM +86-139-6816-1990 (Mr. Shen Shicheng)	Ultraviolet exposure units
Hiromi Paper	USA	hiromipaper.com	Japanese papers
Ian Leake Studio	Switzerland	ianleake.com	The Light Counter ultraviolet exposure controller and integrator
Inkjet Mall	USA	inkjetmall.com	Brushes, digital negative/positive software, custom inks for inkjet printers, chemicals for printing-out platinum/palladium and other alternative processes, film, paper. ***To date, the only global supplier of ready-to-use printing-out Pt/Pd sensitizer and components**.
Jerry's Artarama	USA	jerrysartarama.com	Brushes and paper
John Purcell Paper	UK	johnpurcell.net	Paper
JoopStoop	France	joopstoop.fr	Inkjet film, paper, step tablets, ultraviolet exposure units
Laboratorium	Switzerland	laboratorium.ch	Printing platinum/palladium
Laboldtech	Italy	laboldtech.com	Chemicals and some precious metals in solution, paper
Labotech2000	Italy	labotech2000.it	Chemicals and some labware
Le Géant des Beaux-Arts	France	geant-beaux-arts.fr	Paper—Arches Platine
Legion Paper	USA	legionpaper.com	Arches Platine, Revere Platinum and other papers
LightMeasure	USA	lightmeasure.com	The lightmeasure ultraviolet exposure meter
Lowes	USA	lowes.com	Glass sheet and hydration chamber components
Maine Media	USA	mainemedia.edu	Study, workshops and Tim Wheelan's Bookstore for photography books and guidance
Mamut	Czech Republic	mamut-photo.com	Some chemicals, papers, printing frames

Name	Region	Web site	Provisions
McMaster-Carr	USA	mcmaster.com	Optically clear mylar film and specialty hardware and project material
Michaels Camera Store	Australia	michaels.com.au	Ilford film, photo equipment, and supplies
Museum Services Corporation	USA	museumservicescorporation.com	Archival boards, batting paper, boards, non-woven polyester fabric, optically clear acetate, tissue paper
Negative Lab Pro	USA	negativelabpro.com	Adobe Lightroom plugin for digital processing of scanned negatives
PGI	Japan	pgi.ac	Brushes, coating rods, supplies
Photographers' Formulary	USA	photoformulary.com	Glass coating rods, printing frames, chemicals and labware
Precision Digital Negatives	USA	precisiondigitalnegatives.com	Digital negative printing software, 31-step table, guides, workshops
Professional Plastics	USA	professionalplastics.com	Acetate, Type-D mylar film, plastic sheets, Rubylith® film
Quad Tone RIP	Global	quadtonerip.com	Digital negative/positive software for inkjet printers
Reich Paper	USA	reichpaper.com	Reich CT Clear and other papers
Ruscombe Paper Mill	France	ruscombepaper.com	Buxton Rag and Herschel Platinotype paper
S. K. Grimes	USA	skgrimes.com	Specialist camera, lens and repair services
ScanScience	USA	wetmounting.com	Wet mount scanning equipment, anti-newton glass
Sigma-Aldrich		sigmaaldrich.com	Chemicals and labware
Silverprint	UK	silverprint.co.uk	Chemicals and papers
Strand Books	USA	strandbooks.com	Books, new and used
Stouffer Industries	USA	Stouffer.net	Step wedges/tablets
Talas	USA	talasonline.com	Bookbinding supplies and papers
Taos Photographics	France	taosphoto.fr	Papers, PiezoDN and Cone inks
Technigraf GmbH	Germany	technigraf.de	Ultraviolet exposure units
Ulano	USA	ulano.com	Rubylith® ultraviolet masking film
Uline	USA	uline.com	Goggles, respirators, storage containers
Washi Arts	USA	washiarts.com	Japanese papers
Wiltronics	Australia	Wiltronics.au.com	Glassware and lab supplies
Wolfgang Moersch	Germany	moersch-photochemie.de	Paper

Notes

1 Gǔdiǎn shèyǐng hé hēibái fǎn zhuǎn (Classical photography and black and white inversion)

2 Fùyáng xīn chéng jīxiè (Fuyang Xinsheng Machinery)

Appendix V

Bibliography and References

Collections and Monographs

Caponigro, Paul. *The Voice of the Print.* Williamsburg, Mass.: Allethaire Press, 1990.

Emerson, Peter H. and T.F. Goodall. *Life and Landscape on the Norfolk Broads.* London: Sampson Low, Marston, Searle and Rivington, 1886.

Emerson, Peter Henry. *Naturalistic Photography for Students of the Art.* London: Dawbarn & Ward, 1899.

Falconer, John. "Maurice Vidal Portman and the Platinotype in India." In *Platinum and Palladium Photographs: Technical History, Connoisseurship, and Preservation.* Washington, DC: Photographic Materials Group of the American Institute for Conservation of Historic and Artistic Works, 2017: 318–331.

Hammond, Anne Kelsey. "Frederick H. Evans: The Interior Vision." *Creative Camera*, 243 (March 1985): 12.

Malde, Pradip. "New Solutions for Platinum Printers." *View Camera* (Sept/Oct 1994): 36–41.

McCabe, Constance, Christopher A. Maines, Mike Ware, and Matthew L. Clarke. "Alfred Stieglitz's Palladium Prints: Treated by Steichen." In *Platinum and Palladium Photographs*: 368–369.

Newhall, Nancy. *P. H. Emerson: The Fight for Photography as a Fine Art.* New York: Aperture, 1975.

Historical

Abney, William de Wiveleslie, and Lionel Clark. *Platinotype: Its Preparation and Manipulation.* London: Sampson Low, Marston & Company, 1895/New York: Scovill & Adams, 1898.

Carranza, Ernest de. "Procédé de tirage des positives par le chlorure de platine," *Bulletin de la Société Française de Photographie* 15 (April 1856): 82–84; ibid. *La Lumière* 6 (8) (23 February 1856): 29; translated in *Journal of the Photographic Society* 3 (40) (21 March 1856): 14.

Clark, Lyonel. *Platinum Toning. Amateur Photographer's Library No. 1.* London: Hazell, Watson, & Viney, Limited, 4th edition, 1895.

Cottington, Ian. "Platinum and Early Photography," *Platinum Metals Review* 28 (4) (1984): 178–188. Idem., *History of Photography* 10 (2) (1986): 131–139.

Herschel, Sir John F.W. "Note on the Art of Photography, or the Application of the Chemical Rays of Light to the Purposes of Pictorial Representation." [Abstract] *Proceedings of the Royal Society* 4 (37) (1839): 131–133. Also briefly reported in *The Athenaeum*, 595 (March 23, 1839): 223.

Hinton, A. Horsley. *Platinotype Printing: A Simple Book on the Process.* London: Hazell, Watson, & Viney, 1897.

Hübl, Arthur Freiherr von, and Giuseppe Pizzighelli. *Platinotype*. Translated by J.F. Iselin. London: Harrison and Sons, 1886.

Hunter, Dard. *Papermaking: The History and Technique of an Ancient Craft.* New York: Alfred A. Knopf, 1947.

Martin, James. "The Theory and Practice of the Photographic Art." In *The Circle of the Sciences*, edited by Marcus Sparling: 7 Practical Chemistry, 142. London: Griffin, Bohn and Co., 1856.

Pizzighelli, Capt. Giuseppe. "The Platinum Print," *Anthony's Photographic Bulletin* 1 (1888): 410–421. See also: idem, *Photographische Correspondenz* 24 (October 1887): 25 (January 1888); idem, *The Photographer's Book*, edited by Holmes and Griswold, 1888.

Pizzighelli, Giuseppe and Arthur Freiherr von Hübl. *Die Platinotypie.* Vienna and Leipzig: 1882.

Schaaf, Larry. "Sir John Herschel's 1839 Royal Society Paper on Photography," *History of Photography* 3 (1) (1979): 47–60.

Schaaf, Larry. *Out of the Shadows: Herschel, Talbot and the Invention of Photography*. New Haven, Conn.: Yale University Press, 1992.

Tennant, John A. "Platinotype Processes," *The Photo-Miniature* 1 (7) (October 1899): 351.

Tennant, John A. "Platinotype Modifications," *The Photo-Miniature* 4 (40) (July 1902): 187.

Wagner, Sarah S. "Manufactured Platinum and Faux Platinum Papers, 1880s–1920s." In *Platinum and Palladium Photographs*: 144–183.

Wessling, Margaret. "Preservation Concerns for Edward Weston's Platinum and Palladium Photographs." In *Platinum and Palladium Photographs*: 274–275.

Willis, William. "A New Process of Photo-Chemical Printing in Metallic Platinum," *The British Journal of Photography* 25 (August 23, 1878): 400.

Willis, William. "A Recent Improvement in the Platinotype Process," *The Journal of the Camera Club* 2 (17) (1888): 47–51; 2 (19) (1888): 99; 2 (20) (1888): 103.

Willis, William. *British Patents: Improvements in Photo-chemical Printing*, no. 2,011, 5 June 1873; *An Improved Process of Photo-chemical Printing*, no. 2,800, July 12, 1878;

Improved Materials and Processes for Photo-chemical Printing, no. 1,117, March 15, 1880; *Improvements relating to Photo-chemical Printing*, 1,681, February 2, 1887;

Improvements relating to Photo-chemical Printing, 16,003, 21 November 21, 1887; *Improvements in or relating to Photographic Printing and Paper therefor*, no. 20,022, September 4, 1913.

[Woodbury, Walter E.]. "Alfred Clements and his Work", *Photographic Times* 27 (4) (October 1895): 216–220.

Wright, Philippa, and John Taylor. "Peter Henry Emerson's Platinum Prints and Photogravures." In *Platinum and Palladium Photographs*: 290–301.

Instructional and Technical

Arentz, Dick. *Platinum & Palladium Printing*. Boston: Focal Press, 1999.

Carver-Kubik, Alice, Saori Kawasumi Lewis, and Andrew Robb, with Erin L. Murphy. "Additives Used in the Platinum Process." In *Platinum and Palladium Photographs*: 92–97.

Clarke, Matthew L. "Characterization, Degradation, and Analysis of Platinum and Palladium Prints." In *Platinum and Palladium Photographs*: 108–114.

Crawford, William. *The Keepers of Light: A History and Working Guide to Early Photographic Processes.* Dobbs Ferry, N.Y.: Morgan & Morgan, 1979.

Enfield, Jill. *Jill Enfield's Guide to Photographic Alternative Processes.* New York: Routledge, 2nd edition, 2020.

Farber, Richard. *Historic Photographic Processes.* New York: Allworth Press, 1998.

Gutierrez, Jennifer J. "Caring for Platinum and Palladium Prints: Storage and Display." In *Platinum and Palladium Photographs*: 264.

Haakman, H.L.T. "Toning paper positives with platinum." *The Photographic News* 1 (28 January 1859): 251; 2 (1859): 204, 263.

Hutchings, Gordon. *The Book of Pyro.* Granite Bay, Calif.: Bitter Dog Press, 1992.

James, Christopher. *The Book of Alternative Photographic Processes.* Boston: Cengage Learning, 3rd edition, 2016.

Leake, Ian. *The Platinum Printing Workshop.* 3rd edition, 2017.

Lundgren, Adrienne. "The History and Use of Glycerine in Platinum Printing." In *Platinum and Palladium Photographs*: 204–217.

Malde, Pradip and Mike Ware. *The Ammonium System: a contemporary method for making platinum and palladium prints, A Workshop Manual*, self-published, Manchester, 1985.

McPhee, Leanne. *Chrysotype: A Contemporary Guide to Photographic Printing in Gold.* New York: Focal / Routledge, 2020

Murphy, Erin L. "A Summary of Early Chemical and Physical Treatment of Platinum Prints." In *Platinum and Palladium Photographs*: 251–257.

Nadeau, Luis. *History and Practice of Platinum Printing.* Fredericton, New Brunswick: Atelier Luis Nadeau, 3rd edition, 1994.

Reeder, Ron and Christina Z. Anderson. *Digital Negatives with QuadToneRIP, Demystifying QTR for Photographers and Printmakers.* New York: Focal / Routledge, 2021.

Rexroth, Nancy. *The Platinotype.* Yellow Springs, Ohio: Violet Press, 1977.

Roberts, J.C. *The Chemistry of Paper.* Cambridge: The Royal Society of Chemistry, 1996.

Tice, George. "Caring for Photographs." In *Life Library of Photography*, 1972, 86–93.

Tice, George. "The Lost Art of Platinum." *Album 11* (December 1970): 12–13, 23.

Ware, Michael J. "An Investigation of Platinum and Palladium Printing," *Journal of Photographic Science* 34 (1986): 165–177.

Ware, Mike. "Platinum Reprinted." *The British Journal of Photography* 133 (41) (October 10, 1986): 1165–1167; ibid. 133 (42) (October 17, 1986): 1190–1194.

Ware, Mike. "Contemporary Platinum-Palladium Printing." In *Ag+ Photographic, Silver Print Manual*, edited by Tom Ang, Eddie Ephraums, and Martin Reed, supplement 2, London: Silver Print manual, 1991.

Ware, Mike. "Photography in Platinum and Palladium: Platinum & Palladium Printing, Second Edition, an essay book review," *Platinum Metals Review* 49 (2005): 190–195.

Ware, Mike, *The Chrysotype Manual: the science and practice of photographic printing in gold.* Brighton: ffotofilm publishing, 2006.

Ware, Mike, *Platinomicon: A Technical History of Photographic Printing in Platinum and Palladium*, self-published, Buxton, 2016.

References and Further Readings

Benjamin, Walter. "The Task of the Translator." In *Selected Writings*, edited by Marcus Bullock and Michael W. Jennings, 258–259. Cambridge, Mass.: Harvard University Press, 1996.

Berger, John. *Bento's Sketchbook: How Does the Impulse to Draw Something Begin?* New York: Pantheon Books, 2011.

Blakemore, John. *John Blakemore's Black and White Photography Workshop.* Newton Abbot: David & Charles, 2005.

Cole, Tatiana. "The Platinum Renaissance: Oral Histories of Platinum-Palladium Printers and Artists." In *Platinum and Palladium Photographs*: 433.

Dawson, George. "The Platinotype Printing Process," *The British Journal of Photography* 27 (August 20, 1880): 399–400.

Feynman, Richard P. *QED: The Strange Theory of Light and Matter.* Princeton: Princeton University Press, 2014.

Gillanders, Robin. *The Photographic Portrait: Techniques, Strategies and Thoughts on Making Portraits with Meaning.* Newton Abbot: David & Charles, 2004.

Hammond, Anne Kelsey. "Aesthetic Aspects of the Photomechanical Print." In *British Photography in the Nineteenth Century*, edited by Mike Weaver, 176, 256. Cambridge: Cambridge University Press, 1989.

Heidegger, Martin. *The Question Concerning Technology and Other Essays.* Translated by William Lovitt, Garland, 1977.

Kurlansky, Mark. *Paper: Paging Through History.* Norton, 2016.

McCabe, Constance. "Preface and Acknowledgements." In *Platinum and Palladium Photographs*: 8.

Scarry, Elaine. *On Beauty and Being Just.* Princeton, N.J.: Princeton University Press, 1999.

Index

Note: References to figures are *italicized*; tables are **bold**; footnotes are indicated by page, letter 'n', and note number.